FOSSIL AND RECENT BIOFILMS

Fossil and Recent Biofilms

A Natural History of Life on Earth

Edited by

Wolfgang Elisabeth Krumbein

ICBM, Carl von Ossietzky University Olgenburg, Germany

David Maxwell Paterson

Gatty Marine Laboratory, University of St. Andrews, Scotland, U.K.

and

Georgii Aleksandrovich Zavarzin

Institute of Microbiology, Russian Academy of Science, Moscow, Russia

KLUWER ACADEMIC PUBLISHERS

DORDRECHT / BOSTON / LONDON

A C.I.P. Catalogue record for this book is available from the Library of Congress.

ISBN 978-90-481-6412-7

Published by Kluwer Academic Publishers,
P.O. Box 17, 3300 AA Dordrecht, The Netherlands.

Sold and distributed in North, Central and South America
by Kluwer Academic Publishers,
101 Philip Drive, Norwell, MA 02061, U.S.A.

In all other countries, sold and distributed
by Kluwer Academic Publishers,
P.O. Box 322, 3300 AH Dordrecht, The Netherlands.

Printed on acid-free paper

Printed in the Netherlands

THIS BOOK IS DEDICATED TO THE MEMORY OF:

Sadi CARNOT **Robert von MAYER** **Wilhelm OSTWALD**

Nicolas Léonard Sadi Carnot (June 1, 1796 – August 24, 1832) first considered the quantitative relationship between heat and work.

Lord Kelvin of Largs regarded Carnot as the founder of thermodynamics.

Julius Robert von Mayer (November 25, 1814 – March, 20,1878) was the first to apply thermodynamics to the inanimate and animate world.

His statement..

"Und so fängt die Krautdecke der Erde das Licht, die flüchtigste aller Kräfte im Fluge ein und verwandelt es in nützliche chemische Energie" (And thus the blanket of herbs catches light, the most evasive of forces, in flight and transforms it into useful chemical energy) already indicates the potential of biofilms to store huge amounts of sun energy in the sedimentary crust of planets. This energy pump proposed by Robert Mayer (or honey pump of J. Beuys) combines physical and planetary thermodynamics with the thermodynamics of life as later elaborated by W. Ostwald.

Wilhelm Ostwald (September 02,1853 – April 04,1932) was Nobel laureate (1909) and is commemorated for his immortal contribution to the understanding of multiple catalytic processes in Nature
(Biofilms have been and are the most efficient catalysts between inanimate and animate Earth)

TABLE OF CONTENTS

Biodata of the Editors xi

Acknowledgements.. xiv

Preface
J. William Costerton and Paul. Stoodley
Microbial biofilms: protective niches in ancient and modern geomicrobiology xv

Chapter 1:
Wolfgang E. Krumbein, Ulrike Brehm, Anna A. Gorbushina, Geroge Levit, Katarzyna A. Palinska
Biofilm, Biodictyon and Biomat - Biolaminites, Oolites, Stromatolites - Geophysiology, Global Mechanism and Parahistology 1

Chapter 2:
Thomas R. Neu, Annett Eitner, M. Luz Paje
Development and architecture of complex environmental biofilms – lotic biofilm systems 29

Chapter 3:
David M. Paterson, Rupert Perkins, Mireille Consalvey, Graham J. C. Underwood
Ecosystem function, cell micro-cycling and the structure of transient biofilms. 47

Chapter 4:
John F. Stolz
Structure of marine biofilms - Flat Laminated Mats and Modern Marine Stromatolites 65

Chapter 5:
Michael Kühl, Tom Fenchel, Jozef Kazmierczak
Growth, structure and calcification potential of an artificial cyanobacterial mat 77

Chapter 6
Pamela Reid, C. D. Dupraz, P. T. Visscher, D. Y. Sumner
Microbial processes forming marine stromatolites - Microbe-Mineral Interactions with a Three-Billion-Year Rock Record 103

Chapter 7:
Lucas J. Stal
Nitrogen cycling in marine cyanobacterial mats....................................... 119

Chapter 8:
Georgii A. Zavarzin
Diversity of Cyano-Bacterial Mats 141

Chapter 9:
Ulrike Brehm, Wolfgang E. Krumbein, Katarzyna A. Palinska
Microbial spheres from microbial mats 151

Chapter 10:
Hartmut Arndt, Kerstin Schmidt-Denter, Brigitte Auer, Markus Weiterer
Protozoans and biofilms.. 161

Chapter 11:
Joachim Scholz, George Levit
Bryozoan morphoprocesses 181

Chapter 12:
Gisela Gerdes
Biofilms and macroorganisms 197

Chapter 13:
Chantal DeRidder, Robin L. Brigmon
"Farming" of microbial mats in the hindgut of echinoids 217

Chapter 14:
Alan W. Decho, Tomohiro Kawaguchi
Extracellular polymers (EPS) and calcification within modern marine stromatolites.. 227

Chapter 15:
David C. Gillan
The Study of a recent Iron-encrusted biofilm in the marine environment. 241

Chapter 16:
Stjepko Golubic, Jürgen Schneider
Microbial endoliths as internal biofilms 249

Chapter 17:
Nora Noffke
Epibenthic cyanobacterial communities in the context of sedimentary processes in siliciclastic depositional systems (present and past)............. 265

Chapter 18:
Derek E. G. Briggs
The role of biofilms in the fossilization of non-biomineralized tissues..... 281

Chapter 19:
Alain Preat, K. Kolo, B. Mamet, Anna A. Gorbushina, David C. Gillan
Fossil and subrecent fungal communities in three calcrete series from the devonian Canadian rocky mountains, Carboniferous of northern France and Cretaceous of central Italy... 291

Chapter 20:
Maud M. Walsh, Frances Westall
Archean biofilms preserved in the Swaziland supergroup, South Africa.. 307

Chapter 21:
Douglas E. Caldwell
Is the feedback between genetic structure and community structure the computational mechanism of biofilm evolution?...................................... 317

Chapter 22:
George Levit, Wolfgang E. Krumbein
Is there an adequate terminology of biofilms and microbial mats? 333

Chapter 23:
Betsey D. Dyer
Hidden in plain sight - A macroscopic approach to biofilms and other visible manifestations of bacteria in the teaching of field microbiology... 343

Chapter 24:
Nickolai V. Shadrin
Is it possible to quantitatively assess the role of algobacterial films in a water body?.. 353

Chapter 25:
Eric P. Verrecchia, Corinne Loisy, Olivier Braissant, Anna A. Gorbushina
The role of fungal biofilm and networks in the terrestrial calcium carbonate cycle 363

Chapter 26:
Karsten Pedersen
Past and present biofilm formation in deep fennoscandian shield groundwater 371

Chapter 27:
Jesus M. Valero, J. Iñaki AlavaAlava
Melanin and chromatic changes in black crusts of sandstone monuments 381

Chapter 28:
Karsten Rodenacker, Bettina Hausner, Anna A. Gorbushina
Quantification and spatial relationship of microorganisms in sub-aquatic and sub-aerial biofilms 387

Chapter 29:
Joseph Seckbach
Life on the edge and beyond 401

Chapter 30:
Yuri M. Svirezhev, Arthur Block, Werner von Bloh
"Active planetary cover" concept and long-term evolution of planetary climate 415

Chapter 31:
Jan Toporski, Andrew Steele, David S. McKay, Frances Westall
Bacterial biofilms in astrobiology: the Importance of life detection 429

Chapter 32:
Frances Westall, Maud M. Walsh, Jan Toporski, Andrew Steele
Fossil biofilms and the search for life on Mars 447

Chapter 33:
A. Yu. Rozanov, E. Zhegallo, G.T. Ushatinskaya
Some aspects of studying of fossil mats and biofilms 467

Index 473

Professor Wolfgang Elisabeth Krumbein is an interdisciplinary scientist (and artist) combining many fields of expertise. He has published as a sedimentologist, microbiologist, molecular ecologist, ecologist, marine scientist, palaeontologist philosopher and conservation scientist in more than 125 different scientific periodicals and more than 15 books. He is the founder of ICBM Oldenburg, a recipient of several honorarium doctorates and international prizes. His main interests, however, were always to promote young scientists in many different fields. He has been involved in more than 75 doctorate promotions worldwide. He feels his major scientific obligation is the promotion of the fusion of Planetary and Life Sciences beneath, above and beyond physical, chemical, and biological borders of planet Earth. He is a leading figure in geomicrobiology and geophysiology as emerging branches of science as well as in Exobiology including the NASA sponsored "Planetary Biology and Microbial Ecology (PBME)" courses. One of his major contributions, however, was understanding the impact of microbes on the conservation and restoration of cultural heritage.

The photo depicts his acceptance speech of the Dr. honoris causa of the University of Gdansk in the Artushof, Gdansk, from where his ancestors traded beer, science, wheat and wood with all Europe and further afield.

Professor David M. Paterson is director of the Sediment Ecology Research group at St Andrews University, Scotland. He has 20 years of experience in interdisciplinary research and has produced 3 books and more than 80 papers concerning the microbiology, sedimentological, physical dynamics,

biodiversity and ecology of aquatic systems. Professor Paterson has developed novel techniques and applied new approaches to determine the impact of biological processes on the physical dynamics of natural systems. Like Professor Krumbein, he is primarily interested in the coupling between the biological and the physical world. His work also covers a range of scales, from the contributions of individual cells to ecosystem processes (low temperature scanning electron microscopy, cellular exudates, carbon fixation), to the role of biodiversity and assemblage structure in ecosystem functioning, and, finally, interpreting system status at a basin scale using spectral remote sensing and ground truthing.

Prof. Georgii Aleksandrovich Zavarzin is a member of the Russian Academy of Science. For two terms Zavarzin was Vice-Chairman of SCOPE when the concept of Global Changes was elaborated. More recently he was appointed by ICSU as a member of the SC for the International Geosphere - Biosphere Program. He served as a Deputy Minister of the Environment in Russia and in many national and international committees. He published more than 150 papers in reviewed Journals and many book chapters and books. He was the first to include budding bacteria, *Metallogenium* and

many other environmentally important microorganisms into the community of microbiologists. He has collaborated for many years with geologists and paleontologists on microorganisms in hot, hypersaline and hyperalkaline environments. Prof. Zavarzin is well known for his work on "global microbiology" which elucidates role of microorganisms in the global biogeochemical processes.

ACKNOWLEDGEMENTS

The editors would never have thought of bringing the authors and topics of this book together had not Anna Besse-Lototskaya of Kluwer Academic Publishers contacted and convinced us that a book on biofilms, microbial mats and stromatolites would be a real modern challenge. Starting from her optimism, one of us (WEK) collected some funds largely from the Hanse Institute for Advanced Study (HWK) in order to bring together scientists working on various aspects of biofilms in an informal workshop. The HWK Delmenhorst is geographically located between the Universities of Bremen and Oldenburg, the Alfred Wegener Institute of Polar and Marine Research (AWI) in Bremerhaven, and the Max Planck Institute for Marine Microbiology in Bremen. Much of the international scientific effort centred on subaquatic and subaerial biofilms is being done in these institutions. Starting from this workshop the book plan was further developed and finalized. This was again supported by a fellowship at HWK for one of us (GAZ). We are especially grateful to Professor G. Roth, and I. Mehser of HWK and all members of the Geomicrobiology group of ICBM, Oldenburg. The efforts of U. Brehm, Th. Dornieden, G. Gerdes, A. A. Gorbushina, G. Levit and K. Palinska have to be especially mentioned. M. Latoschinski helped us in the time consuming work of image processing and editing. Elaboration of the work on the subaquatic biofilms and biofilm and the chances for the editors to meet were supported by grant support from the EU and NERC (to DMP), the DFG and EU (to WEK) and of HWK (to GAZ). To be embedded in the family of science always creates problems in terms of time lost in family contact. Therefore all of us wish to thank our families for patience, understanding and permanent encouragement. All critical readers , who sometimes read several of the chapters, are acknowledged as well as the staff of Kluwer Academic Publishers. Finally and most importantly we wish to thank those colleagues and students, who honoured our work by using the book and reading at least parts of it. We hope that it stimulates new though, imagination and understanding.

PREFACE
MICROBIAL BIOFILMS: PROTECTIVE NICHES IN ANCIENT AND MODERN GEOMICROBIOLOGY

J.W.Costerton and Paul Stoodley
Center for Biofilm Engineering
Montana State University

As this book is published based on discussions of a conference that was held in 2001, it may be useful to provide an update on the most recent revelations about biofilms, so that this excellent exposition of the contribution of microbial biofilms to geological processes may be placed in a modern context. The importance of the contribution of microbial biofilms to global processes is only now being appreciated as it is revealed that all terrestrial surfaces are teeming with microbial life in the form of biofilm communities. These communities live on soil particles, in rock fissures, marine and river sediments and at the very extremes of terrestrial habitats from inside Antarctic ice to the walls of deep sea hydrothermal vents. The contribution of these biofilm communities generally went unrecognized because it was the water that was where microbiologists looked for life, not the surfaces, although, evidence of the early association of microbes with surfaces was in fact present in the fossil record (Rasmussen, 2000; Reysenbach, and Cady, 2001). It is also revealing that biofilm formation is found in prokaryotes from the most deeply rooted branches of the phylogenetic tree in both the *Archaea* and *Bacteria* kingdoms, the *Korarchaeota* and *Aquificales* respectively (Jahnke *et al.* 2001; Reysenbach *et al.* 2000). There are also striking similarities in the biofilm morphology from biofilms grown in widely different environments suggesting adaptive convergence of forms. For example, in high shear flows biofilms tend to form filamentous streamers whether in acid mine drainage (Edwards *et al.* 2000), hydrothermal vents (Reysenbach and Shock 2002), surface hot springs (Reysenbach, and Cady, 2001), or human pathogens in laboratory flow cells (Stoodley *et al.* 2000). Molecular techniques using knock out mutants have also demonstrated that a wide range of surface adhesion proteins and appendages, often in the same organism (one of the best studied being *Pseudomonas aeruginosa*) play a role in initial attachment and subsequent biofilm formation (Hall-Stoodley and Stoodley 2002). However, closer inspection of the papers reporting the knockout of genes "required" for biofilm formation actually show that biofilms still form, but to a reduced degree compared to the wild type. Also models based on *P. aeruginosa* generalize motility as being integral for the biofilm life cycle of 1) attachment, 2) the formation of complex structures and 3) detachment, although the same complex processes also occur in non-motile species such as *Staphylococcus aureus* and *Streptococcus mutans* (Stoodley, P. unpublished result). This redundancy is further evidence of the importance of biofilm formation to many prokaryotes and taken together these data suggest that biofilm formation is an ancient and integral component of prokaryotic life.

Further mysteries concerning biofilms were revealed when it was discovered (Prigent-Combaret *et al.*, 1999; Sauer *et al.*, 2001) that bacterial cells, up regulate large numbers of their genes when they form these sessile communities. Modern proteomic techniques have shown that cells in the biofilm phenotype vary from their planktonic counterparts, in the expression of as many as 70% of the genes in their genomes. Analysis of gene expression in biofilms, using m RNA-based methods (Wagner *et al.*, 2003), tends to confirm this conclusion that biofilm cells are as different from planktonic cells of their own species, as planktonic cells of a given species are from those of a completely different species (Stoodley *et al.*, 2002). The continuing examination of bacterial and fungal biofilms shows that there is no single biofilm phenotype, but that sessile cells vary in gene expression in different locations within biofilms, and that these communities follow a developmental cycle not unlike that of other complex multi-cellular organisms. The bacteria and fungi that formed the biofilm communities that are described in this book grew in phenotypes that bear little or no relationship to planktonic cells in lab cultures, and the best modern models we have for them are the biofilms that we now study in lab reactors and in nature.

Because we are considering microbial biofilms that grew millions of years ago, and have become important parts of the fossil record, it may be useful to ponder the advantages of biofilm over planktonic modes-of-growth in the conditions that existed in the primitive earth. In a volcanic environment, without the "buffering" effects of large amounts of biomass, aquatic environments may have had localized niches that permitted microbial growth while in other areas growth was precluded by extreme temperatures, pressures and/or incompatible chemistry. Microbial life in these permissive/non-permissive streams would favor the formation of stationary sessile communities, because planktonic cells would be swept to their deaths in downstream maelstroms of hostile environments. However, the viscoelastic nature of all of the biofilms we have yet tested (*Desulfovibrio* spp, *P. aeruginosa*, *S. mutans*, *S. aureus*, tap water, pond water, and hot spring biofilms) may also be an adapted and adaptive strategy for survival in flowing environments. Viscoelasticity allows biofilm to absorb some energy elastically but under sustained elevated shear can allow them to flow along surfaces thus remaining attached (Stoodley *et al.* 1999). Even when shear induces the detachment of parts of the biofilm the crude homeostatic environment provided by the remaining attached "mother colony" allows survival and the production of further "seeding" events.

One of the most successful strategies for growth in primitive ecosystems may have been the combination of heterotrophic bacteria with photosynthetic and chemoautotrophic bacteria, in biofilms on sun-lit surfaces. In these metabolically integrated communities, the symbionts would flourish, but they would be subjected to periodic drying and exposure to intense ultraviolet irradiation, when the water splashed and receded in the normal diurnal rhythms. While bacteria in the planktonic phenotype are exquisitely sensitive to drying and UV light, sessile microbes in biofilms are very well protected by their enveloping matrix material, which retains water and blocks UV irradiation (Costerton *et al.*, 1995). It seems, that even multiple species symbiotic spherical films may develop in such biofilm systems (Brehm et al. 2003). These properties of stationary growth in well-protected biofilms have allowed modern organisms to thrive in intermittently moist environments, like tidal zones, and many of the large biofilms that have been described in various geological contexts may have had similar origins. As life on earth developed, in a progressive manner, microbial populations would have been challenged, at some time in their evolution, by both bacteriophage and free-living amoebae. While planktonic bacteria are sensitive to bacteriophage, and are readily taken up and digested by amoebae, sessile bacteria are protected from both of these biological agents and this characteristic of biofilms will have enhanced their predominance in the primitive earth.

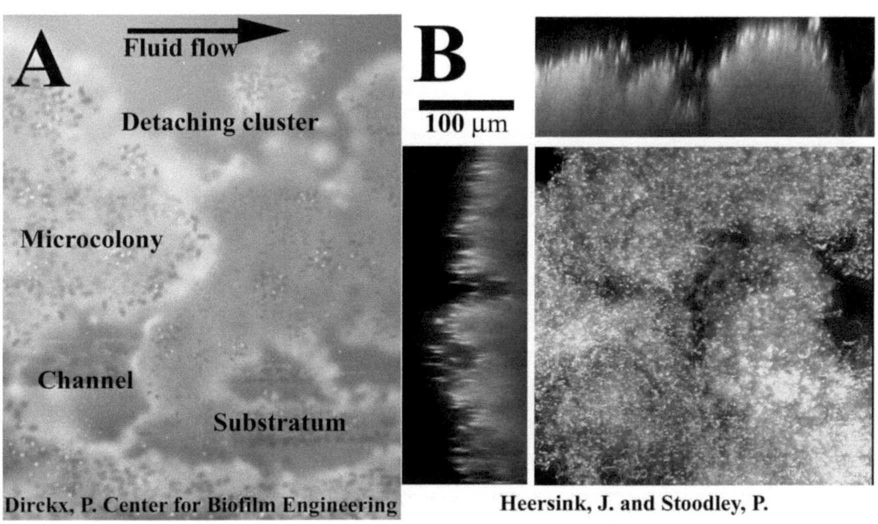

In recent publications (Stoodley *et al.*, 2002) many authors have drawn attention to the fact that biofilms provide a stable platform for genetic interchange, and for metabolic interactions, that are really not possible between planktonic cells as they "wheel and dance" in fluid space. The structure of biofilms, which are composed of matrix-enclosed bacterial micro-colonies separated by open water channels (Figure 1), allows the sessile organisms access to nutrients in the bulk fluid. Within the matrix-enclosed micro-colonies, bacterial cells comprise approximately 15% of the volume, while the matrix occupies the remaining 85% of the space, and cells are apposed to each other in spatial relationships that facilitate genetic exchange and metabolic interaction. We have observed that rates of "horizontal gene transfer" are very high in biofilms (Ghigo, 2001). Shen *et al.* (2003) have suggested that bacteria may even divide up the genomic elements that comprise pathogenicity islands, between planktonic cells of different strains, and re-assemble these elements to make functional genomic structures, when the strains are again apposed in biofilms. Morphological examinations of biofilm communities that carry out complex integrated metabolic activities, like cellulose digestion (Kudo *et al.*, 1987) and methane generation (MacLeod *et al.* 1990), have shown that cells of species with complementary activities are organized in ways that appose co-operative species. Bacteria that remove critical end products of a particular metabolic process often "drive" the overall activity, by being adjacent to primary digestive organisms, and scavenging the end products so efficiently that very high turnover rates can be maintained. In methanogenic biofilms, the methane-producing archaea are directly apposed to concentric rings of heterotrophic organisms that produce the substrates (acetate and hydrogen) for methane production, and many similar metabolic co-operations are reflected by cellular apposition in other sessile communities. If the metabolic activities in a biofilm generate salts that tend to crystallize, like the magnesium ammonium phosphate (struvite) produced by many strains of *Proteus*, the biofilm matrix is gradually filled up and mineralized. This process appears to have produced many of the stromatolites, and other large heavily mineralized structures, that have entered many geological structures as a result of the predominance of biofilms in the primitive earth (Brehm et al., 2003).

In our recent review of biofilms as multi-cellular "organisms" (Stoodley *et al.*, 2002) we considered the probable sequence of the evolution of these complex multi-species communities. Taking the development of plants as an approximate model, we suggested that microbial biofilms may have evolved early, and reached a relatively high level of structural and functional sophistication, before the planktonic phenotype emerged. In the case of plants, the earliest forms occupied the habitats within which they evolved with considerable success, and only developed elaborate means of dissemination much later in their developmental sequence. Flowering plants, with seed dispersal strategies, developed relatively late in evolution, presumably because these characteristics allowed them to colonize favorable distant habitats. Similar variation in dispersal strategies, although more subtle, may be found in prokaryotes. For example, motile species such as *P. aeruginosa* tend to "seed" individual cells, which although unprotected, through chemotaxis may swim to favorable environments while non-motile species such as *S. aureus* tend to detach in clumps, maintaining the protective environment seen in attached biofilms. Spore formation is another dispersal strategy. As we gradually identify the genes that are expressed in the biofilm phenotype, we perceive that biofilms predominate in all natural ecosystems, because they are protected, metabolically integrated, and highly adaptive both morphologically and phenotypically. These sessile communities continuously shed matrix-enclosed "clumps", in which the cells are still in the biofilm phenotype, and these potential propagules drift downstream and may set up new biofilm communities in distant favorable locations. However, modern biofilm communities have much more elaborate dissemination strategies. In any mature biofilm a proportion of the cells assume the planktonic phenotype, digest the matrix material in their immediate location, and leave the community in a regular "programmed detachment" pattern. This programmed detachment is accelerated by certain environmental factors, like stagnation due to low rates of bulk fluid flow, so it appears to be a dissemination strategy. These detaching cells may be individually motile, if the species is capable of flagellar motility or pilus-mediated twitching motility, and the planktonic phenotypes of many species have evolved very elaborate mechanisms of chemotaxis, that help them in their search for a perfect remote habitat. To carry the analogy with plant evolution to its logical conclusion, we suggest that the biofilm phenotype may not only be the predominant form of growth of bacteria in the biosphere, but it may also be the form that evolved first, to exploit scattered favorable niches in the primitive earth. We may look at the large bacterial residues in geological formations as representing very ancient remnants of an earlier and more primitive stage in the evolution of the multi-cellular multi-species biofilm communities that are so successful in the contemporary biosphere.

FIGURES

Figure 1. 3D imaging by confocal microscopy has revealed that biofilms consist of many different types of complex architectures. Panel A) is a schematic highlighting some of the common features such as microcolonies separated by water channels. Schematic provided by Peg Dirckx, Center for Biofilm Engineering. Panel B) Confocal image of a biofilm formed from Streptococcus mutans, an early dental plaque colonizer. The square panel is a plan view and the top and side bars show the structure in cross-section. Image provided by Joanna Heersink and Paul Stoodley, Center for Biofilm Engineering, funding by Philips Oral Health Care.

REFERENCES

Brehm, U, Krumbein, WE, Palinska, KA (2003) Microbial Spheres: a novel cyanobacterial-diatom symbiosis. Naturwissenschaften 90:136-140.

Costerton, J. W., Z. Lewandowski, D.E. Caldwell, D.R. Korber, and H.M. Lappin-Scott. 1995. Microbial biofilms. *Ann. Rev. Micro.* 49: 711-745.

Edwards KJ, Bond PL, Gihring TM, and Banfield JF. 2000. An archaeal iron-oxidizing extreme acidophile important in acid mine drainage. *Science.* 287(5459):1731-1732.

Ghigo, J-M. 2001. Natural conjugative plasmids induce biofilm development. *Nature.* 412(6845): 442-5.

Hall-Stoodley, L. and Stoodley, P. 2002. Development regulation of microbial biofilms. *Curr. Opin. Biotech.* 13: 228-233.

Jahnke, L.L., Eder, W., Huber, R., Hope, J.M., Hinrichs, K.U., Hayes, J.M., Des Marais, daV.J.,

Cady, S.L., and Summons, R.E. 2001. Signature lipids and stable carbon isotope analyses of octopus spring hyperthermophilic communities compared with those of aquificales representatives. *Appl. Envir. Microbiol.* 67: 5179-5189.

Kudo, H., K.-J. Cheng, and J.W. Costerton. 1987. Interactions between *Treponema bryantii* and cellulolytic bacteria in the *in vitro* degradation of straw cellulose. *Can. J. Microbiol.* 33: 244-248.

MacLeod FA, Guiot SR, Costerton JW. 1990. Layered structure of bacterial aggregates produced in an upflow anaerobic sludge bed and filter reactor. Appl Environ Microbiol 56(6):1598-607

Prigent-Combaret C, Vidal O, Dorel C, Lejeune P. 1999. Abiotic surface sensing and biofilm-dependent regulation of gene expression in Escherichia coli. *J Bacteriol* 181(19): 5993-6002

Rasmussen, B. 2000. Filamentous microfossils in a 3,235-million-year-old volcanogenic massive sulphide deposit. *Nature* 405: 676 – 679.

Reysenbach AL, Ehringer M, Hershberger K. 2000. Microbial diversity at 83 degrees C in Calcite Springs, Yellowstone National Park: another environment where the Aquificales and "Korarchaeota" coexist. *Extremophiles* 4(1): 61-7.

Reysenbach AL, Shock E. 2002 Merging genomes with geochemistry in hydrothermal ecosystems. *Science* 296(5570): 1077-82

Reysenbach, A.L., and Cady, S.L. 2001. Microbiology of ancient and modern hydrothermal systems *Trends in Microbiology* 9: 79-86.

Sauer, K., A.K. Camper, G.D. Ehrlich, J.W. Costerton, and D.G. Davies. 2002. *Pseudomonas aeruginosa* displays multiple phenotypes during development as a biofilm. *J. Bacteriol.* 184: 1140-1154.

Shen, K., Wang, X., Post, J.C., and Ehrlich, G.D.: Molecular and Translational Research Approaches for the study of Bacterial Pathogenesis in Otitis Media. In: Rosenfeld, R., and Bluestone, C.D. (eds). Evidence-based Otitis Media, 2nd Edition. Hamilton: B.C. Decker Inc. (In Press) (2003).

Stoodley, P., Hall-Stoodley, L., Boyle, J.D., Jørgensen, F. and Lappin-Scott, H.M. 2000. Environmental and genetic factors influencing biofilm structure. In *Community Structure and Cooperation in Biofilms*. pp. 53-64. (eds. D. Allison, P. Gilbert, H.M. Lappin-Scott and M. Wilson). SGM Symposium Series 59. Cambridge University Press, Cambridge.

Stoodley, P., K. Sauer, D.G. Davies, and J. W. Costerton. 2002. Biofilms as complex differentiated communities. *Annu Rev. Microbiol.* 56: 187 – 209.

Stoodley, P., Lewandowski, Z., Boyle, J.D. and Lappin-Scott, H.M. 1999. The formation of migratory ripples in a mixed species bacterial biofilm growing in turbulent flow. *Environ. Microbiol.* 1: 447-457.

Wagner VE, Bushnell D, Passador L, Brooks AI, Iglewski BH. 2003. Microarray Analysis of Pseudomonas aeruginosa Quorum-Sensing Regulons: Effects of Growth Phase and Environment. *J Bacteriol.* 185(7): 2080-2095.

Chapter 1

BIOFILM, BIODICTYON, BIOMAT MICROBIALITES, OOLITES, STROMATOLITES GEOPHYSIOLOGY, GLOBAL MECHANISM, PARAHISTOLOGY

Wolfgang E. KRUMBEIN, Ulrike BREHM, Gisela GERDES, Anna A. GORBUSHINA, George LEVIT and Katarzyna A. PALINSKA
Geomicrobiology, ICBM, Carl von Ossietzky Universitaet Oldenburg, D-26111 Oldenburg, Germany, wek@uni-oldenburg.de

1. INTRODUCTION

This article is conceived as an introduction to "Fossil and Recent Biofilms". Fossil and recent microbial communities adhering to, overgrowing and enmeshing inorganic substrates are introduced. The biofilm substrates embrace rocks and minerals exposed to the atmosphere or to overlying water bodies of different salinity and include deeply buried sediments and rocks. Biofilm communities active in soil, sediment and rock interstices, cavities, cleavages, and cracks until deep into the Earth crust are dealt with. Definitions and common traits of a biofilm, biodictyon and biomat (microbial mat) are presented and discussed. Fossilised examples as well as ichnofossils (trace fossils) of extant communities are sketched. Some science history reflections are included. New terms such as biodictyon, geophysiology, parahistology, and global mechanisms of microbial communities in past and present planetary ages are defined. The intimate relationship between major rock types such as biolaminated sequences (and ore deposits), stromatolites, oolites, onkolites and also reef type deposits is discussed. The planetary and exobiological dimension and global importance of biofilms, microbial mats and network microbial communities are noted.

2. HISTORY AND EVOLUTION OF TERMS

The first clear and lucid report of field observations and laboratory experiments on biofilms and their rock generating potential stems from Paracelsus (Krumbein, 1993a, 1994; Paracelsus, 1528, 1982). However, Maxim 524 of Titus Livius clearly demonstrates that knowledge existed long before: "Rolling stones gather no moss, i.e. biofilm growth" (which the Rock-Band "Rolling Stones" definitely did). Paracelsus himself wrote:

"For from the mucilago of the water are growing and born all rocks and all pebbles and sands are coagulated into rocks. This is easily eyeable (visualised) for a mucilago sooner or later attaches to any stone, deposited into flowing water. And upon separating such mucilaginous matter from the rock and depositing it into a cucurbite (a kind of an Erlenmeyer) it will coagulate and transform into a rock just like any rock growing in the natural water by self-coagulation and genesis after a long period of time".

Thus Paracelsus did, what happens today in many biological, ecological or geoscience Institutes: Biofilms and microbial mats are observed in field and laboratory experiments. Hereby the modern scientist makes the same observations as our ancestor in science. Namely we observe and bias the formation of biofilms and microbial mats and their eventual transformation into e.g. sedimentary rocks in field and laboratory.

Microscopic evidence of rocks produced by microbial mats was published quite early (Brückmann, 1721; Hooke, 1665). The latter depicted microbial mats and ooids embedded in them. They were, however, less certain than Paracelsus about the origin of these structures. One of the most interesting transfers from medical science and histology into microbial (and global) ecology seems to be the term glycocalyx derived from the phenomenon in the abdomen of many animals including the human body which was later identified as slimy extracellular substances (EPS) of eukaryotic cells in intimate relationship to cell clusters or biofilms of archaea and prokarya. Such glycocalyx-like slime films were later recognized as biofilms (Characklis and Marshall, 1990; Costerton et al., 1987; Harris, 1972; Marshall, 1976, 1984; ZoBell, 1943). ZoBell (1943) was the first author to ascribe the evolution of biofilms to hydrophobic and hydrophilic characteristics of the outer envelopes of bacteria getting attached to surfaces this way. The extra-cellular material into which the bacteria and also algae and protozoa are embedded embraces extracellular compounds mostly consisting of complex polymers of many different carbohydrates to which phospholipids, proteins (exo-enzymes), and highly polymeric partially phenolic substances (lignin, melanin, carotene, and compounds chemically similar to Maillard reaction products) are added. These extra-cellular, but metabolically, cytological and morphologically important substances represent outer layers of cells beyond membrane and cell wall, the classical envelopes of the cellular machinery of life. They were first named extracellular polysaccharides, later extra-cellular polymeric substances (EPS). Paracelsus called them slime, mucilago, or just "Schnoz" not unlike slime droppings from the nose from which he derived the name *Nostoc* for a biofilm producing cyanobacterium (Potts, 1997).

Paracelsus and others thought it were organic meteorites (chondrites) for hundreds of years. Dried and detached biofilms may be transported by the winds and fall down hundreds of kilometres inland. These were initially interpreted as extraterrestrial bodies or "meteor paper". Chondrites, or paper meteorites were found, collected, analysed and described for about 250 years before Ehrenberg (1839) identified them as wind-blown microbial mats or carpets. The interesting point is, that these gelatinous substances were very rapidly swelling and reviving after heavy rainfalls and thus confused with meteorites. However, the most important and still not clearly settled phenomenon in relation to biofilms and biofilm producing microbes is their capacity to produce large amounts of EPS and excrete them so to say as garbage from the cell. These extracellular products in many cases of biofilms and microbial mats make up more than 99 % of the total metabolic products (Cooksey, 1992; Decho, 1990). The idea of slime as a living agent was very productive on the one hand. On the other hand the multiple lines of thought around evolution, spontaneous generation, bacteria as "chemical enzymes" and diseases generated from dirt and mud lead to many an erroneous observation, hypothesis and scientific debate. The most fascinating and heated debate lasted only 7 years and dealt with *Bathybius Haeckelii* (Haeckel, 1877; Huxley, 1868). The "tragic rise and descent of *Bathybius*" was described to some detail by Krumbein (1984). Mucilaginous networks of "living material" were described by Huxley from samples of the Challenger expedition. Huxley named this protoplasmic substance or organism in relation to the monograph of Haeckel on Monera in his name. The original English publications contained the German word "Ur-Schleim" best translated as "primordial living soup or super-organism". It was, however, quite quickly found, that it was an artefact and Haeckel (1877) himself had to admit that a world embracing organism in the sense of Hutton's Super-Organism Earth (Lovelock, 1969) was not existing. Biofilms and microbial mats, however, today take this position as a worldwide ubiquitous power driving mainly the geophysiology of Earth (Krumbein and Schellnhuber, 1990, 1992).

Marshall (1976, 1984) and later Costerton et al. (1987) introduced the term biofilm into biotechnology and microbial ecology. Doemel and Brock (1974, 1977), Krumbein, 1978, and Krumbein et al., 1977a, b) slowly transformed the misleading term "algal mat" into bacterial mat and microbial mat and biofilms as the driving forces behind the formation of microbial sediments and sedimentary rocks (Riding and Awramik, 2000). This was necessary, when blue-green algae turned out to be cyanobacteria (Krumbein, 1979; Krumbein et al., 1979b; Rippka et al., 1979). A merger process started between the geological literature on oolites and stromatolites and the microbiological literature on biofilms and microbial mats (Krumbein, 1983a). This process is still in progress (Brehm et al., 2003). One result of cross-boundary research was a change from the study of individual microbial cells floating separately in water (plankton) studied in liquid and isolated agar culture to a sophisticated systemic approach in the study of surface attached biofilms and mats. This was initiated by Ludwig and Theobald (1852), critically explored by Weed (1889), ZoBell (1943) and introduced into laboratory scale experiments by Caldwell and Hirsch (1973) using chemical gradient agar surface film studies as a new experimental approach. Ludwig and Theobald (1852) in their classical underestimated and thus unnoticed paper added another facet to the fascinating question of the behaviour of benthic surface microbial communities, which only now starts to be elucidated by several research groups. Within many mats and biofilms a change of organisation is observed from a gradient oriented laminated community (biofilm, microbial mat, potential stromatolite) into separate sub-systems within a generally two-dimensional biofilm or multiply laminated microbial mat (Ludwig and Theobald, 1852; Krumbein and Cohen, 1974). This change of organisation creates three-dimensional spherical communities regulated mainly by the community and the signals sent out by the individual members within it (Brehm et al., 2003). Hereby spherical units are formed within a laminated system. These microbial symbiotic spheres, when solidifying into rock generate ooids and oolites within or outside of the context of a microbial mat (petrified as stromatolite or biolaminite). It is the great achievement of early workers (Brückmann, 1721; Ludwig and Theobald, 1852; Kalkowsky, 1908) to have clearly recognised the intimate relationships between laminated and spherulitic growth and the final petrified rock products. Much earlier the mathematician, physicist and microscopist James Hooke (1665) depicted such microbial organogenic structures clearly featuring biofilm structures in and around ooids (Kettering stone plate). Figures 1-3 demonstrate this quite clearly.

Figure 1: Ooids, fungal mycelium within ooids and fungal remains with attached calcium carbonate crystals around ooids indicating the autochthonous genesis of spherulites within microbial mats (micrograph by J. Hooke, 1665).

Figure 2: Ooids generated by fungal activity within a microbial mat. The ooids and the associated microbial mat exhibit the same fungal mycelia. Jurassic Minette iron ore (photomicrograph, Krumbein, 1984).

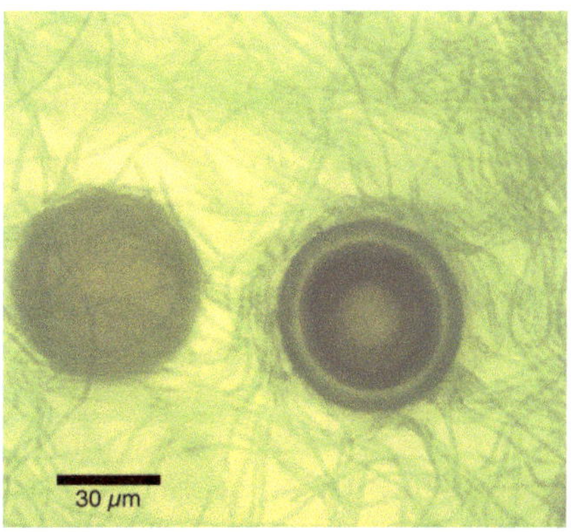

Figure 3: Ooids generated by a biodictyon (a microbial symbiosis between bacteria, cyanobacteria and diatoms from the North Sea coastal Wadden (photomicrograph U. Brehm, 2003).

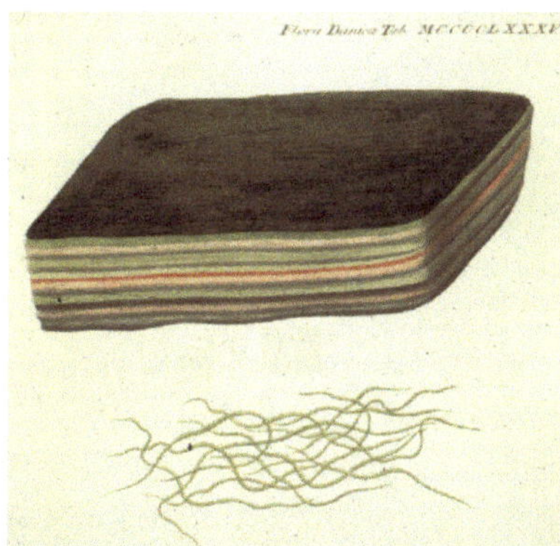

Figure 4: The first coloured presentation of a laminated microbial mat with individual oxygenic phototroph, anoxygenic phototroph and chemotroph layers (hand coloured micrograph, Flora Danica, 1813).

Biofilms and microbial mats can be regarded as interactive microbial communities thriving in the vertical chemical gradients between sediment and water column, rock and atmosphere, bed-rock and soil etc., creating laminated crusts or rocks (microstromatolite, rock rinds, patina, rock varnish, stromatolite, laminated ore deposits bioliths, biolaminites, erosive and accretionary karst, etc. (Gorbushina et al., 1996; Krumbein, 1983a, Krumbein et al., 1994, Stal and Caumette, 1994; Walter, 1976). The microbial sphere in contrast creates spherical networks and globules and ultimately geological products such as oolites, pisolites, onkolites, and spherulites (Brehm, 2001; Brehm et al., 2003; Kalkowsky, 1908; Krumbein et al., 1977a, b; Krumbein, 1994; Ludwig and Theobald, 1852). The laminated two-dimensional system obviously is an orientated growth of microbes within a vertical chemical gradient modified by the community. In some cases these microbial communities are densely interwoven. In other cases a distinct lamination of different communities occurs. The best studied laminations are known from the Intertidal or Littoral of all coastal water bodies. The first coloured presentation was published in the famous Flora Danica (1813). Figure 4 gives an impressive view of the "Farbstreifen-Sandwatt" (colour striped wadden sediment) with the topmost diatom layer often sealing the surface in a parchment-like way, followed by oxygenic cyanobacteria, anoxygenic phototrophs and then by sulafte reducing and methanogenic bacteria (see also Figures 9-13).

The spherical three-dimensional system obviously is an oriented growth and structuring of space through chemical signals or communication systems between micro-organisms imposing their signals and rules to the general chemical and physical environment. The term was introduced by us in the last years (Brehm, 2001, Brehm et al., 2003; Gorbushina, 2001). The structuring of macro-organism symbiotic communities as described in many communications for bioherms and reefs also reflects a more complex relationship between microbial biofilms and mats and macro-organisms. These also yield three-dimensional geological deposits, often erroneously interpreted as the product of imaginary genera and species. With a growing understanding of the role of biofilms within and on organisms (glycocalyx, karies, termite and cattle digestion system, ulcer, arterial clogging, mykorrhiza, skin flora, hair of the green Polar Bear) and especially in all terrestrial ecosystems above and below sediment or rock surfaces it seems appropriate to state, that biofilms, mats and networks are the most ubiquitous phenomenon and expression of life on Earth. Extraterrestrial biofilms in the frame of planetary biology are also close to our perception, as some contributions in this book seem to indicate. Therefore it is appropriate to move from microscopic to macroscopic in the global biogeochemical connotation of biofilms and microbial mats.

3. BIOFILM

Sub-aquatic biofilm (Characklis and Wilderer, 1989; Costerton et al. 1987; Flemming, 1991; Gerdes et al., 1987; Marshall, 1976)

Three different types of biofilms may be defined, (1) sub-aquatic biofilm, (2) sub-aerial biofilm and (3) biodictyon or microbial network. The sub-aquatic biofilm can be characterised as a layer of more than 99% biologically solidified water (Krumbein, 1993a, 1994; Neu, 1992, 1994), within which a structured community of microbial cells of different genera and species modifies the chemical gradient between a solid surface(s) and the adjacent or overlying liquid and moving water in order to exert metabolic activities maintaining life processes. Solid surfaces or assemblages of larger or smaller-sized rock and mineral particles are covered by a mono-to multi-layered biofilm of microbes. The film develops, may mature and show decay or even detachment at certain moments. The internal structure is complex and has been compared to the structure of larger cities with buildings, shops, streets, canalisation and other specialised units. Neu (1994) claims that biofilms and microbial mats are the same. Other authors insist in making a strict difference between a subaquatic biofilm, a subaerial biofilm and subaquatic or subaerial microbial mats (Krumbein, 1994a, b; Gorbushina and Krumbein, 2000a, b). In that case the biofilm sensu strictu should be regarded more or less as a single layer of organisms within their common matrix (EPS). Mature films develop several layers of organisms. In water processing biofilms even chimney-like structures made of bacterial cells forming several "brick" layers around an aeration channel were observed.

Sub-aerial biofilm (Gorbushina and Krumbein, 1999, 2000a,b)

The sub-aerial biofilm in contrast to the sub-aquatic one is best described as a thin coating of 99% organic matter metabolising in the presence of minimal amounts of active water and surviving for extended periods of time without any supply of liquid water. Rock and mineral particle surfaces exposed to the direct Earth atmosphere or to the gas mixture within a soil, rock, or dune sand system are covered by micro-organisms maintaining an optimum of living cells or a maximum of organic matter in the presence or availability of a minimum of liquid or active water. This community maintains life processes at random intervals of water availability. Practically all surfaces exposed to the atmosphere carry a visible or invisible cover of biofilm communities. These embrace lichen communities, algal films, free living fungi, and bacteria assemblages. All biofilms exposed to a gaseous atmosphere have been identified as capable of forming organic and mineral deposits which have been described already by Darwin (1839) and Humboldt as rock varnish or cataract crusts (see Krumbein and Jens, 1981). These subaerial biofilms, when maturing or vanishing leave structures behind which partially consist of organic polymers such as melanins, carotenoids and chlorophyll derived substances and partially of minerals formed through the activities of the biofilm community. These phenomena are usually described as rock varnish, crust, lichen stromatolite, microstromatolite and by the embracing term patina or biopatina (Dornieden et al., 1999; Gorbushina et al., 2001, Krumbein and Jens, 1981). The organisms involved are to be regarded in many cases as the most astonishing survival experts under extreme and extremely changing environmental conditions. In contrast to the classical extremophile restricted either to very low or extremely high temperature, light, salinity these organisms can withstand and survive also extreme changes of all conditions including extremely long periods of lack of any parameter and compound necessary for metabolic activity. Such communities were recently labled as poikilophilic and poikilotroph biofilms (Gorbushina et al., 1999, 2001; Gorbushina and Krumbein, 1999, 2000a, b). Obviously classical intertidal algal or microbial mats running through extremely changing conditions will also exhibit a large number of poikilophilic microorganisms (Gorbushina et al., 1999).

Biodictyn- microbial network (Brehm, 2001)

A biodictyon is a three-dimensional more or less concentric network of filamentous and unicellular micro-organisms which typically is embedded in soil, sediment or rock. The term was derived from the Greek "bios" for life and "diction" or "diktyon" for net. The biodictyon thus is a living network. Biodictyon may also occur within organic tissue, at certain places of roots (mykorrhyza), within decaying wood (fungal infections) or deep within the deep sedimentary or rock biosphere, where the environmental conditions do not allow for the development of a two-dimensional biofilm or multilayered microbial mat. Within the context of work on intertidal microbial mats, spherical meshworks developing within them were found, which often generate ooids or calcispheres. Sub-aerial biofilms often exhibit typical three-dimensional meshworks and create bio-erosional patterns in the form of biopits (Krumbein and Jens, 1981, Gorbushina and Krumbein 2000a, b). Another typical case of biodictyon formation is reported from many ore deposits. Usually organic matter and other energy sources intrude into pre-existing rock deposits creating cracks and crevices and a system of pores for the percolation of water, nutrients and germs. Within the pore system networks of microbes develop and may ultimately be involved in the creation of rich ore deposits. Such evolution of microbial networks was described recently for a uranium mine in the Black Forest (Germany), where pitch-blende and other uranium minerals were obviously deposited by microbial and especially fungal activity (Hofmann, 1989).

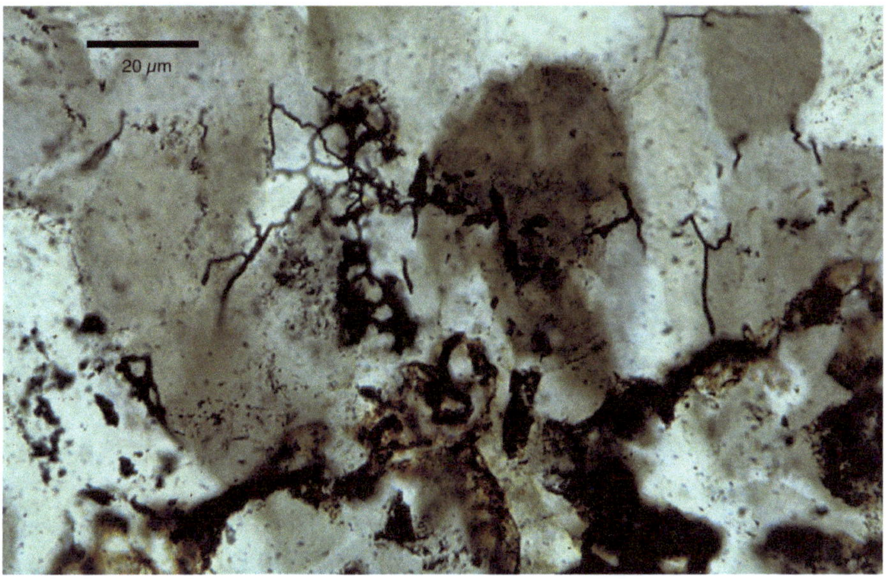

Figure 5: Biodictyon from the uranium deposit Krunkelbach/Menzenschwand, Black Forest (photomicrograph W. E. Krumbein, 1988).

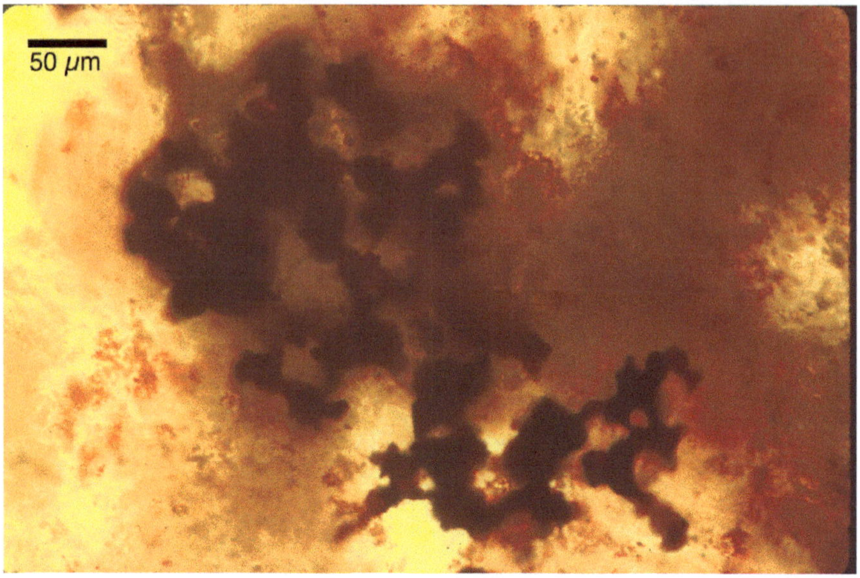

Figure 6: (Bio?)dictyon from the Hammersley banded iron formation (BIF) Precambrian, Australia (photomicrograph Krumbein, 1983).

Descending or circulating phreatic water may have brought organic or inorganic energy sources and micro-organisms to this deeply buried contact zone. The microbes settled at the sites rich in reduced elements. This way a rich microflora developed and infected the rock with a dense network of microbial growth. Figure 5 gives an example of a fungal biodictyon growing about 500m deep below the surface at the contact between the Carboniferous Bärhalde granite and hydrothermal waters ascending in the Tertiary. The microbial per-mineralised network of this uranium deposit is strikingly similar to network structures in the pre-phototrophic world of the Precambrian Hammersley Basin in Australia (Figure 6). Tremendous amounts of iron oxides were deposited in (bio?-) laminated deposits called Banded Iron Formation (BIF). These may have also been deposited by network forming early chemotroph microbial networks. The diameter of the original microbial structures usually is expanded 5 to 10 times by heavy incrustations with iron minerals. This and other rather network or meshwork structured communities we would like to separate from the general picture of biofilms and microbial mats. Such entangled networks were also found by Barghoorn and Tyler (1965) in the Precambrian Gunflint deposits in Canada. Also here it remains doubtful whether phototrophic or chemotrophic biofilms, networks and microbial mats prevailed (see Figures 7 and 8).

4. MICROBIAL MAT, MICROBIAL CARPET, ALGAL MAT, FARBSTREIFEN-SANDWATT, METEOR PAPER

(Black, 1933; Cloud, 1962; Doemel and Brock, 1974; Ehrenberg, 1839; Hofmann Bang, 1826; Krumbein, 1966, 1972, 1983, 1994a, b, 1996; Ludwig and Theobald, 1852; Schulz, 1936).

The origin of all these terms is still not clearly analysed. Without doubt a multilayered microbial community thriving at the sediment water interface and generating rock like deposits was termed an algal carpet (Algenteppich) first by Ludwig and Theobald (1852). Algal mat as a term got very popular, when several teams started to study the potential of rock formation of blue-green algae (later re-named cyanobacteria) in tropical Seas in relation to the modern analogues of Precambrian stromatolites (Black, 1933; Cloud, 1942; Darwin, 1839; Kalkowsky, 1908). Many different definitions and morphological, physiological and geological descriptions of both terms have been proposed and slowly emerged into their individual life. Early work includes also different attempts of definitions and understanding the natural history of these important communities (Black, 1933, Doemel and Brock 1974, Krumbein, 1983a; Walter, 1976). In this context of the evolution of terms it is noteworthy to mention that the referees for the Journal "Science" accepted an article by Doemel and Brock (1974) in which the terms "algal mat", "microbial mat" and "bacterial mat" were used without any discrimination so to say as synonyms for the Yellowstone Hot Spring microbial mat systems. Film, however, does not occur in this publication. Biofilms thus really have to be seen separate from the multidimensional intimately interwoven microbial mat system of recent or potential stromatolites and biolaminites (Gerdes and Krumbein, 1987). A general definition of microbial mat does not exist. Definitions overlap with the term and description of stromatolites (Kalkowsky, 1908; Krumbein, 1972, 1983a).

Without mentioning the term microbial mat Charles Darwin (1839), Alexander von Humboldt (1793, 1807), Christian Gottfried Ehrenberg (1839) and others described these interwoven mats of organisms as fascinating phenomena in intertidal areas, on rock surfaces, and in caves and mines underground. It is also very noteworthy that the transfer of the cyanophyceae from plants or algae to bacteria was made long before on the basis of microscopy of such algal or microbial mats. Further also the geological potential was noted very early in the history of our science. Ferdinand Cohn (1867) deduced a relationship between groups of bacteria, (Phycochromaceae or Cyanophyceae), red algae, and lichen on the basis of pigments, type of cell division, movement, and mode of reproduction. The pigments chlorophyll, phycocyanin, and phycoerythrin were detected by simple chromatographic and spectroscopic methods. On the basis of Darwin's evolutionary theory as well as his own observations, Cohn proposed that the bacterial phylum Phycochromaceae or the cyanobacteria were early inhabitants of the Earth because of their ability to adapt to extreme habitats, their simple way of reproduction, and the fossil records. He also stated the great similarity between colourless (non-photosynthetic) filamentous and coccoid bacteria, such as *Beggiatoa* , *Thiopedia*, *Thiovolum* and their phototroph counterparts (e.g. *Oscillatoria, Synechocystis*). He studied many biofilm and microbial mat communities including microbial mats and meshworks (photosynthetic and chemosynthetic) in water works, waste water, rivers and soils. Figures 7 and 8 give some insight into the hidden world of a microbial mat as viewed by SEM. In such pictures as well as in thin sections of fossil microbial mats from Precambrian to recent it is always difficult to decide whether the individual organisms are phototrophs or chemotrophs.

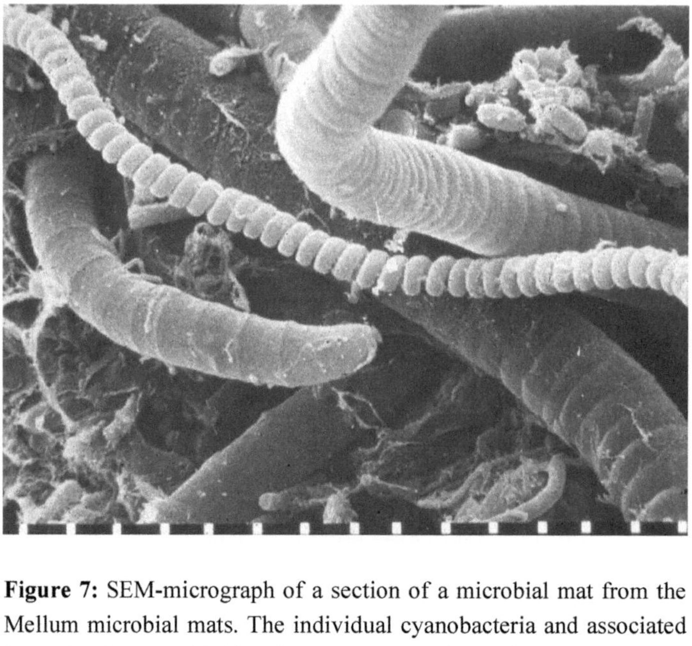

Figure 7: SEM-micrograph of a section of a microbial mat from the Mellum microbial mats. The individual cyanobacteria and associated bacteria form a highly diverse community (photomicro-graph, Krumbein, 1985; bar distance = 3 μm).

Figure 8: Coccoid heterotroph (?) bacteria from Solar Lake microbial mats, Sinai. The extracellular highly structured material (EPS) leaves doubt about the real size of the microorganism. The organism is not in culture but certainly constitutes an important part of the structuring of the microbial mats (photomicrograph Krumbein, 1975).

Microbial mats are intimately interwoven microbial communities including laminated, concentric and network like growth patterns, which by their upward directed growth, physical and chemical gradients, barriers and sticky EPS products trap and embed mineral grains, produce new minerals and, ultimately, laminated and spherulitic sedimentary rocks and structures (Krumbein, 1983a; Noffke et al, 2001). Many microbial mats (especially in the intertidal zones) enmesh siliceous and carbonate sand and mud in multiple layers. Often they compete with and deeply influence the chemistry of evoporitic salt swamp deposits (Gerdes et al., 1987, Friedman and Krumbein, 1985). As an example a general and four individual diagrams and sketches (Figures 9-13) shall demonstrate the multiple functions of sediment structuring microbial mat communities of a siliciclastic environment in the North Sea Wadden deposits. These multicoloured and multipurpose microbial mats were described very early by Hofmann Bang (1813, see Figure 4) under the title "architects of rocks and islands" and by Schulz (1936) as "Farbstreifen-Sandwatt" (multicoloured laminated wadden sediments). Thus already in the 19th Century the old idea of Paracelsus was revived, namely that micro-organisms build rocks and islands. Also Kalkowsky (1908) suggested, that the origin of his stromatolites and oolites was related to bacterial activity or any other kind of micro-organism. Multilayered and often very complex chimney or dome shaped microbial mats have been found in many lakes, in shallow Seas, near Deep Sea hydrothermal vents and in connection with methanogenesis. Some of them transform into laminated rocks, others decay and release their mineral products and entrapped particles into the water body for new sedimentation. Several interesting Symposia have focussed on microbial mats (Cohen and Rosenberg, 1989; Krumbein, 1978, 1979; Krumbein et al., 1994; Stal and Caumette, 1994). The information is tremendous and new facets are coming up daily.

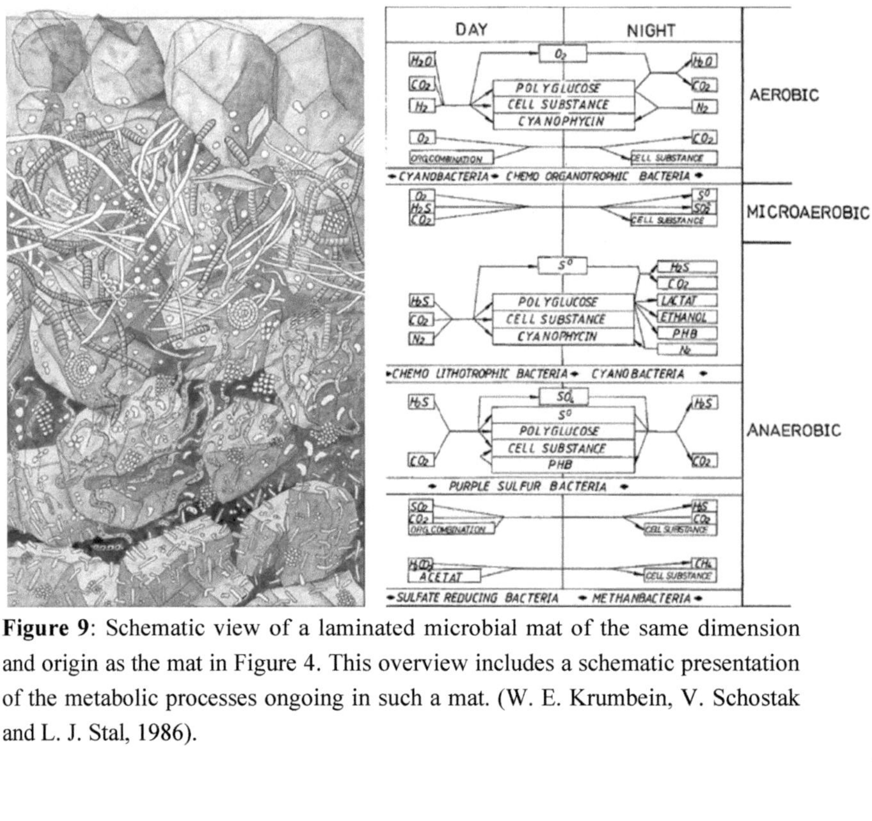

Figure 9: Schematic view of a laminated microbial mat of the same dimension and origin as the mat in Figure 4. This overview includes a schematic presentation of the metabolic processes ongoing in such a mat. (W. E. Krumbein, V. Schostak and L. J. Stal, 1986).

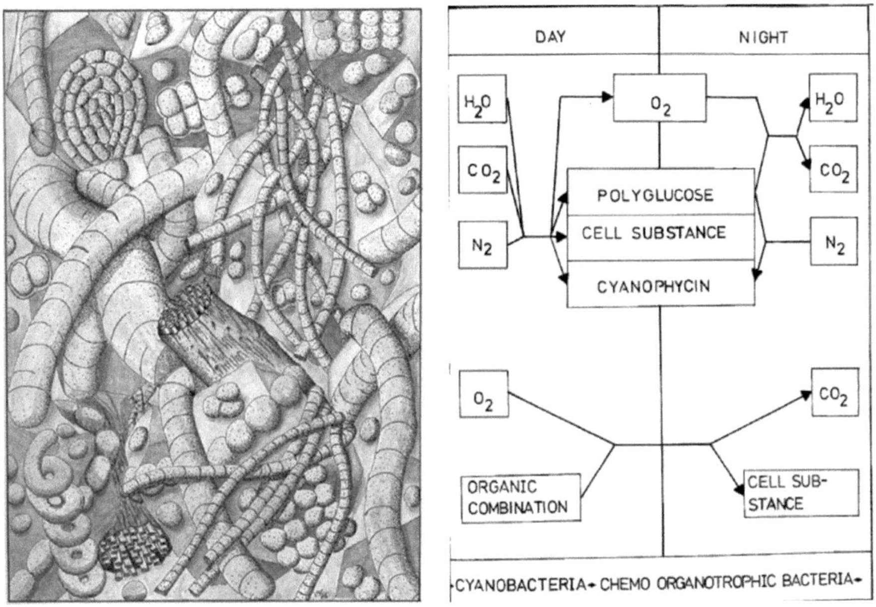

Figure 10: Blow-up of the topmost section of the mat in Figure 9

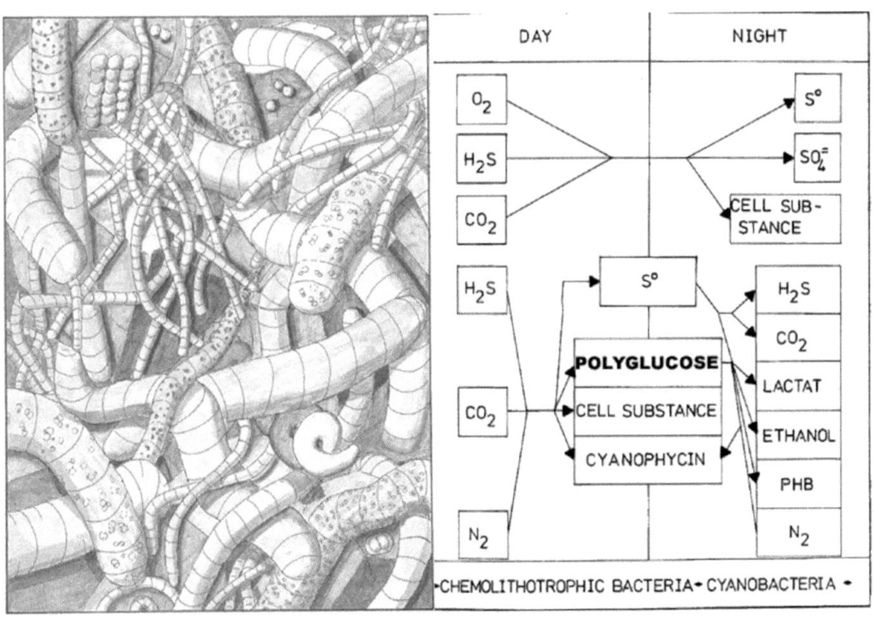

Figure 11: Second layer of the mat in Figure 9.

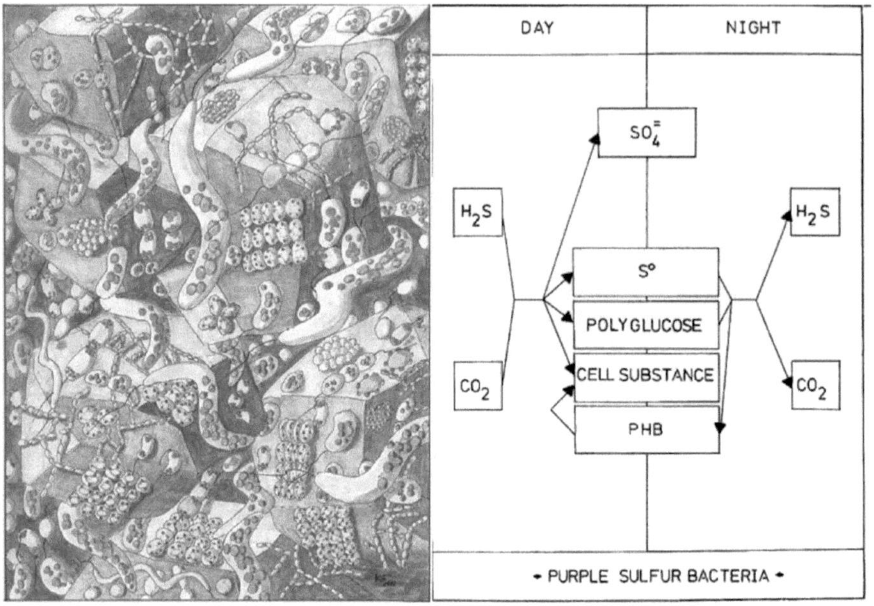

Figure 12: Third layer of the mat in Figure 9.

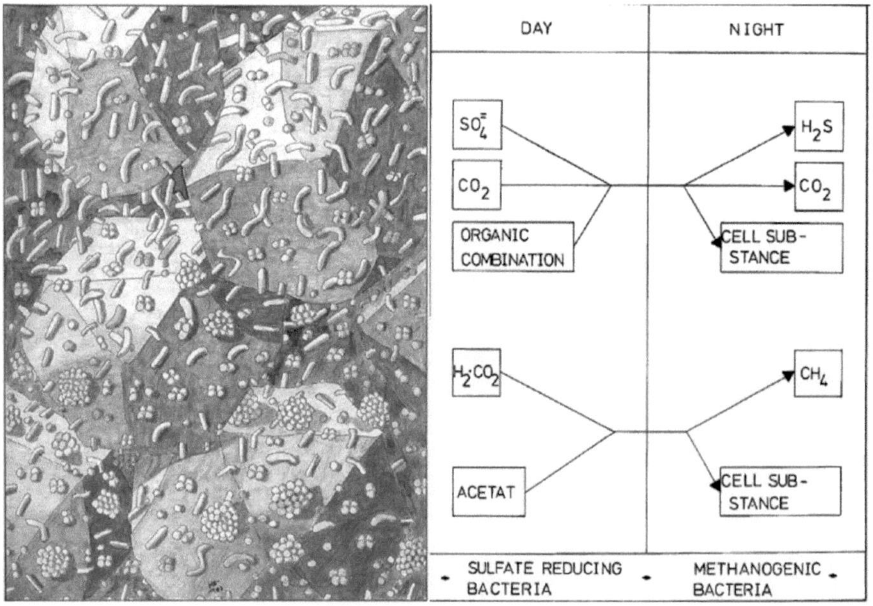

Figure 13: Lowermost layer of the mat in Figure 9.

5. MICROBIALITES (STROMATOLITES, OOLITES, BANDED IRON FORMATION, BIF, LAMINATED ORE DEPOSITS

(Brückmann, 1721, Cloud, 1942; Gerdes and Krumbein, 1987; Kalkowsky, 1908; Krumbein et al. 1994; Riding and Awramik, 2000; Walter, 1976;)

All rocks and rock types, which are visibly generated under the influence of biofilms or microbial mats are microbialites. Work on these structures has evolved over the years and generated many different terms (Gerdes and Krumbein, 1987; Krumbein, 1983a, 1987; Krumbein et al., 1994; Riding and Awramik, 2000; Walter, 1976). Practically all sedimentary rocks and ore deposits are at least microbially modified if not microbially generated as upward oriented growth structures (build-ups) in the sense of bioherms and reefs instead of a concept of physical sedimentation deposited by gravity from a liquid medium. Motivated by the complex structures of microbial mats and networks in shallow and deep Sea environments some authors were also tempted to compare the structural complexity of microbial mats with early attempts of tissue development and tissue-like structures. In discussions with German and Russian paleontologists some of the authors (Brehm, 2001; Gerdes and Krumbein, 1976, Wachendörfer, 1991) expressed the view, that microbial spheres and networks as well as gas and water semi-permeable biofilms and microbial mats by inclosing and trapping not only particles but also anions, cations and molecules with their ballon like and air matrass like morphologies are first steps towards metazoa and other tissue forming organisms. This concept was applied to the Ediacara flora or fauna by Gould (1989) and others. Here a connecting link is opened between histology of individuals, histology of communities and an embracing global or terrestrial parahistology in the sense of Wachendörfer (1991).

Stromatolites are finely laminated sedimentary structures initiated and produced by microbiota by upward directed growth, trapping of and in situ production of mineral particles which include and produce various different forms distinctly of a biogenic origin (Krumbein, 1983a; Noffke et al., 2001; Wachendörfer et al., 1994;). Distinctly bedded, widely extensive, blanket-like build-ups are biostromes. Nodular, biscuit-like, dome-shaped or columnar stromatolites are also referred to as bioherms (Krumbein, 1983a, Krumbein 1984). Benthic films and interwoven structures of cyanobacteria, diatoms and a multitude of other bacteria and eukaryotes (Protozoa) usually represent the modern analogues of these laminated rocks (stromatolites) which are the first extensive biosedimentary rock structures known to exist since the Archean (3.2 – 3.7 GA before present). More and more clearly recognisable forms of filamentous and coccoid or rod shaped cells and cell clusters occur in sedimentary rocks from 3.2 –1.8 GA before present until today. Many important ore deposits such as the iron sulfide and oxide ores, uranium, copper, silver and gold deposits can be mined only because microbial mats enriched the metals within the sediment during their growth period.

6. GEOPHYSIOLOGY, GLOBAL MECHANISM, PARAHISTOLOGY

What is geophysiology? In one of his first books the author of the Gaia hypothesis (Lovelock, 1979, 1989) phrased the idea that if Hutton (1788) talked of Earth as a super-organism and physiology as the method to study Earth stated that physiology was a sub-discipline of medicine and that humankind should be the doctor to treat the problems of the patient Earth (a sick planet). This caused a dispute on the meaning of words in a historical sense. One of the authors of this introductory chapter was biassed into science theory and philosophy in order to analyse the meaning of what Lovelock stated. It turned out, that in a Cartesian sense also a clock is an organism. Worse, physiology at the time of Hutton was defined very different from modern interpretation. We quote from a dictionary of the late 18th Century: "Physiology or Physicks or Natural philosophy is the science of the phenomena and processes of natural bodies". This means in a nutshell that physicks and physiology were synonyms and that natural bodies could be "living" or "dead organisms". Derived from this historical excursion we later defined: "If physiology is the science to study the phenomena and processes of natural bodies, then geophysiology can be regarded as the science to study the phenomena and processes of Earth as a living natural body. Hence, biofilms, microbial mats and networks (Biodictyon) will be the major driving elements of the organisation and maintenance of the optimal conditions for the maintenance of Life and Earth through space and time. This is also the view of modern scientists, who revived the thoughts of Aristotle, Albertus Magnus, Giordano Bruno, Paracelsus and others have to be mentioned in this context. (Krumbein, 1983b, 1996).

How then does histology (to say nothing at first about parahistology) get into the picture? Since biofilms and microbial mats can be compared to tissue in structure, cohesion and distribution of tasks of individual cells histological techniques can also be applied. Since histology is the science of plant or animal tissue and Earth cannot be regarded as a real living organism the study of these structures will have to be regarded as parahistology (not to be confused with parapsychology). Actually geophysiology or physiology of the Earth could also be defined as paraphysiology. The latter, however, is nothing else but global or environmental biogeochemistry (Krumbein, 1996).

Recent mass and turnover balances of biological and geochemical cycles make it clear, that the number and mass of micro-organisms as well as their impact on atmosphere, hydrosphere and geosphere can be identified as 99% or more of the total living matter on Earth. Most of the micro-organisms, however, are organized in biofilm or microbial mat communities. This way it can be easily derived, that despite the considerable impact of the human technical system on Earth and the relation of Earth to the Solar system, biofilms still are the major factor regulating the survival of the delicate equilibrium between external influences (Solar irradiation, morphogenesis of the Earth's outer envelope, geochemical cycling of elements in and out Earth and the Earth's Crust) and the home-made response of living (microbial biofilm) systems interacting with the driving forces of the physical (astrophysical) evolution of the solar system within the Milky Way. It seems, that biofilms, microbial mats, and the microbial network (biodyction) deep in the Crust are the major factors maintaining life on this peculiar planetary body, named "Geos" or "Earth". So much about geophysiology. What is then "global mechanism"? Global mechanism or global dynamics is nothing more or less then the geophysiological and global biogeochemical impact of biofilm and microbial mat communities on the major driving forces of the dynamics of Earth as related to the external factors and stresses imposed on it via astrophysical laws and circumstances. Thus if solar irradiation increases or decreases, the living system (and 99.9% of it are represented by biofilms/microbial mats) will have to respond to it with only one means, i.e. maintain life on the planetary body. This planet was by reasons unknown inoculated or infected with a thermodynamic natural system called "Life" by some people and "Dissymmetry" in the sense of Pasteur and Curie by others (Krumbein, 1996; Krumbein and Schellnhuber, 1990, 1992; Levit et al., 1999). What we should never forget or omit from our vain thoughts is the proportion of human cells to microbial cells, of human geochemical activities as compared to the enormous capacity of global microbial (biofilm/microbial mat) activities. In this frame global biogeochemistry, global carbon and other element cycles are without doubt powered by biofilms and microbial mats and networks rather than by human chemical or physical, "pollution". This way we can feel relatively free to say, that quantity and quality of microbial activities (still) outcompete any activity of macro-organisms including the human ever growing population of a restricted part of the surface of this planet. Planetary biology in the future will teach us to properly understand the biogeochemical difference between humankind and the microbial system.

7. EXAMPLES ON A LOCAL AND GLOBAL SCALE

It is very difficult to find solid surfaces on planet Earth within the range of the highest mountains to the deepest bottoms of the Oceans and within recent sediments and ancient rocks where liquid water is at least sporadically available. The temperature range is between –40°C and + 120°C, especially when the liquid water is under pressure. Biofilm communities, i.e. closely related clusters of micro-organisms attached to a surface and often embedded in EPS thrive not only on rocks, on and in sedimentary granular systems but also in deep rock cracks and interfaces between geological units. The vertical range of these systems spans from 10 km above sea level to more than 20 km below the surface of water the Ocean, and the sediment/water interface beneath. In short: Microbial mats and their products may occur on this planet as a phenomenon of planetary biology at any place offering biologically available water and temperatures below 120°C. This comprises the whole geomorphologic range from approximately 10km above NN to 10km below NN and the crustal range from rock or sediment surface to 20-30km below rock and sedimentary surface level. This includes biofilms and microbial mats or networks exposed to Earth's atmosphere, hydrosphere and upper geosphere (deep biosphere). Figures 9-13 document this potential of biogeomorphogenetic potential as well as the biogeochemical consequences of the individual morphotypes. Unfortunately the elegant and genial work of the founder of biomorphogenesis as a science (D'Arcy Thompson, 1917) despite the similarly elegant memorizing essays by Ruth D'Arcy Thompson (1958) and E. G. Hutchinson (1948) is somehow lost from extent perception of morphotype analysis. The work of Beklemishev (Levit and Krumbein, 1999) and Viles (1984) have repeatedly attracted attention to the tremendous power of the arrangement of individual micro-organisms and macro-organisms in time and space. Recently it was also demonstrated, that the erosional and accretionary features of karst are intimately associated to microbial films and networks including the fungal associations in the hanging drops of stalactites in formation (Viles, 1984, Wang et al., 1993). Biogeomorphogenesis has a mechanical/physical and a biogeochemical or geophysiological connotation and keeps the chemistry and morphology of Earth far from any thermodynamic, stoichiometric and astrophysical equilibrium or constant. The work of biofilms and microbial mats more than the work of any other community including humankind creates, maintains and shapes the morphology and geochemistry of at least the Earth Crust, including ocean and atmosphere dynamics (water movement, climate) Some glimpses of these ideas have already been coined by I. Kant in a very concise manner (Krumbein, 1993b).

For us it suffices to state: microbial biofilms, microbial mats and microbial networks not only are at the basis of any structure and function of macro-organisms (as witnessed by biofilms microbial mats and networks within and around e. g. trees, termites and humans) but also of the relief of this planet. In this last statement it is tentatively proposed that the deviation of the planetary surface of Earth from the ideal spherical shape as constantly organised by plate tectonics and erosion is at least in part a consequence of the global activity of microbial communities labelled as biofilms, microbial mats or biodictyon.

8. CONCLUSION

Biofilms and especially fully developed mature microbial mats embrace all metabolic pathways ever emerging on Earth. These embrace anoxygenic and oxygenic phototrophy, anaerobic and aerobic chemotrophy, organic and inorganic respiration and fermentation (disproportioning), autotrophy and heterotrophy regarding all nutrients. No element of the forces driving life on Earth is not represented within them since more than 3 billion of years. They are thus ultra-conservative and non-evolutionary. Biofilms and Microbial Mats are evolutionary almost totally stable. They are not a part of Darwinian evolution. Biofilms and/or microbial mats create sediments, sedimentary rocks, island and ore deposits. They destroy rocks, monuments and materials. As a "patina biofilm" they protect rocks and monuments from decay. Fossilised microbial mats are stromatolites, oolites, onkolites and many other rocks and ore or petroleum and gas deposits called microbialites. They serve as energy and matter reservoirs, filters, traps and gradient maintaining systems throughout history of life on Earth. They are even regarded as the fueling or powering systems for global tectonics and climate change or stabilisaton (Anderson, 1984, Krumbein and Schellnhuber, 1990). The biofilm and microbial mat systems (sub-aquatic and sub-aerial) are the true driving forces for the establishment, maintenance and future of life on Earth. They even today in the era of eukaryotes (including humankind) outnumber in importance, mass and turnover rates all other living organisms. More than 99% of the living matter (standing crop) on this planet at least as well as more than 99% of the thermodynamic turnover potential on Earth are represented by biofilms, microbial mats and biodictyons of the deep sub-surface layers of Earth.

ACKNOWLEDGEMENTS

The work on sub-aquatic, sub-aerial and (in part) sub-terrestrial biofilms was generously supported by several grants of DFG and the EU. We are indebted to V. Schostak and L. J. Stal for their participation in creating Figures 5-9. All authors have been involved in the planning and organisation of the Microbial Mat Meetings MMMI and MMMII. All of them wish to express their gratitude to the past and present collaboration with many others working at ICBM on biofilms (G. Gerdes, E. Holtkamp, N. Noffke, H. Riege, L. J. Stal, V. Wachendörfer and others as witnessed by many PhD theses and joint publications with the authors). The co-editors D. M. Paterson And G. A. Zavarzin are acknowledged especially deeply. They would have been excellent partners in the formulation of this introductory chapter. However, limits of time and separation in space made this impossible.

REFERENCES

Barghoorn, E.. S., Tyler, S. A., 1965. Microorganisms from the Gunflint Chert. Science 147, 563-577.

Black, M., 1933. The algal sediments of Andros Island, Bahamas. Phil. Trans, Roy. oc. London. Ser B 222,165-192.

Brehm, U., 2001. Untersuchungen zur mikrobiologisch induzierten Strukturbildung, Rezente und fossile Beispiele im Vergleich. Ph. D. Thesis, Oldenburg, 132p.

Brehm, U., Krumbein, W. E., Palinska, K. A., 2003. Microbial Spheres: a novel cyanobacterial-diatom symbiosis. Naturwissenschaften, 90, 136-140.

Brückmann, F. E., 1721. Specimen Physicum Exhibens Historiam Naturalem, Oolithi seu Ovariorum Piscium & Concharum in Saxa Mutatorum, Helmestadii: Salomonis & Schnorrii.23p.

Caldwell, D. E., Hirsch, P., 1973. Growth of microorganisms in two-dimensional steady-state diffusion gradients. Can. J. Microbiol.,19, 53-65.

Characklis, W. G., Marshall, K. C. (eds.), 1990. Biofilms. Wiley, N. Y., 796 p.

Characklis, W.G. and Wilderer, P. A. (eds.), 1989. Structure and function of biofilms. Wiley, N.Y., 386 p.

Cloud, P. E., 1962. Environment of calcium carbonate deposition west of Andros island, Bahamas. U.S. Geol. Surv. prof. papers 350,1-138.

Cohen, Y. and Rosenberg, E., 1989. Microbial Mats Physiological Ecology of Benthic Microbial Communities. ASM, Washington, 494 pp.

Cohn, F. 1867. Beitrage zur Physiologie der Phycochromaceen and Florideen. M. Schultze's Arch. Mikrosk. Anat. 3, 1-60.

Cooksey, K. E., 1992. Extracellular polymers in biofilms. In *Biofilms - science and technology*. (eds. L. F Melo, T. R. Bott, M. Fletcher, B. Capdeville), p. 137-147, Kluwer Academic Publishers, Dordrecht.

Costerton, J. W., Cheng, K. J., Geesey, G. G., Ladd, T. J., Nickel, J. C., Dasgupta, M., Marrie, T., 1987. Bacterial biofilms in nature and disease. Ann. Rev. Microbiol. 41, 435-464.

D'Arcy Thompson, R., 1958. D'Arcy Wentworth Thompson: the scholar-naturalist, 1860-1948 Oxford University Press, London.

D'Arcy Thompson, W., 1917, 1961. On Growth and Form. Cambridge University Press, Cambridge, 345p.

Darwin, C., 1839. Journal of the researches into the geology and natural history of the various countries visited by H. M. ship Beagle. Vol. 1, N.Y., Appleton and Company. (reprint, 1958)

Decho, A. W., 1990. Microbial exopolymer secretions in ocean environments: Their role(s) in food webs and marine processes. Oceanogr. Mar. Biol. Ann. Rev. 28,73153.

Doemel, and Brock, T. D., 1974. Bacterial stromatolites. Origin of laminations. Science, 181, 1083-1085.

Doemel, and Brock, T. D., 1977. Structure, growth and decomposition of laminated algal mats in alkaline hot springs. Appl. Environ. Microbiol. 34,433-452.

Dornieden, Th., Gorbushina, A. A., Krumbein, W. E., 1999. Patina - physical and chemical interactions of subaerial biofilms with objects of art. p. 120-126 in: Ciferri, O., Mastromei, G. and Tiano, P. (eds.) Of microbes and art. Proceedings of the Int.Conference on microbiology and conservation (IMC). FEMS, CNR, Firenze, 279 p.

Ehrenberg, C. G., 1839. Über das im Jahre 1686 in Curland vom Himmel gefallene Meteorpapier und über dessen Zusammensetzung aus Conferven und Infusorien. Annalen derrr Physik und Chemie 16, 187-188..

Flemming, H.-C., 1991. Biofilms as a particular form of microbial life. In Biofouling and biocorrosion in industrial water systems. (eds. H.-C. Flemming, G. G. Geesey), p. 3-9. Springer Verlag, Heidelberg.

Friedman, G. M., Krumbein, W.. E. (eds.), 1985. Hypersaline Ecosystems. The Gavish Sabkha. Springer, Berlin, 484p.

Gerdes, G., Krumbein, W. E., 1987. Biolaminated Deposits. Springer, Berlin, 183p.

Gerdes, G., Krumbein, W.. E., Reineck, H.-E., 1987. Mellum. Portrait einer Insel. Kramer, Frankfurt, 344p.

Gorbushina, A. A., 2001. Who is living on a bare rock? Priroda (in Russian) 9, 37 - 44.

Gorbushina, A. A., Boettcher, M., Brumsack, H.-J., Krumbein, W.E., Vendrell-Saz, M., 2001. Biogenic Forsterite and Opal as a Product of Biodeterioration and Lichen Stromatolite Formation in Table Mountain Systems (Tepuis) of Venezuela. Geomicrobiology J. 18, 117-132.

Gorbushina, A. A., Krumbein, W. E., 1999. The poikilotrophic micro-organism and its environment - Microbial strategies of establishment, growth and survival. p. 175-185 in Seckbach, J. (ed.): Enigmatic micro-organisms and life in extreme environments. Kluwer, Dordrecht.

Gorbushina, A. A. and Krumbein, W. E., 2000a. Rock Dwelling Fungal Communities: Diversity of Life Styles and Colony Structure. p. 317-334 in Seckbach, J. (ed.) Journey to Diverse Microbial Worlds. Adaptation to Exotic Environments, Kluwer, Dordrecht.

Gorbushina, A. A., Krumbein, W. E., 2000b. Subaerial microbial mats and their effects on soil and rock. p.161-170 in Riding, R. and Awramik, S. (eds.): Microbial Sediments, Springer, Berlin.

Gorbushina, A. A., Krumbein, W. E., Hamann, C. H., Panina, L., Soukharjevski, S., Wollenzien, U., 1993. On the role of black fungi in colour change and biodeterioration of antique marbles. Geomicrobiology J. 11, 205-221.

Gorbushina, A. A. Krumbein, W.E., Palinska, K.A., 1999. Poikilotroph growth patterns in rock inhabiting cyanobacteria. pp 657-664 in: Peschek et al. (eds.) The phototrophic Prokaryotes, Plenum, N.Y., 836p.

Gorbushina, A. A., Krumbein, W.E., Vlasov, D., 1996. Biocarst cycles on monument surfaces. p. 319-332 in Pancella R. (ed.): Preservation and restoration of cultural heritage. Proceedings of the 1995 LPC Congress. EPFL, Lausanne, 773p.

Gould, S. J., 1989. Wonderful Life, Norton New York, 347p.

Haeckel, E., 1877. *Bathybius* und die Moneren. Kosmos, 1, 293-305.

Harris, P.J 1972. Micro-organisms in surface films from soil crumbs. Soil Biology Biochemistry 4, 105-106.

Hofmann, B., 1989. Genese, Alteration und rezentes Fließsystem der Uranlagerstätte Krunkelbach (Menzenschwand, Südschwarzwald). NAGRA Technischer Bericht 88-30, Baden, 195p.

Hofmann Bang, N., 1826. Om Confervernes nytte y naturens husholdning. Konegelige Danske videnskabernes selskab. Naturwedenskabelige og mathematiske afhandlinger, pp. 209-220. (submitted 1817, printed 1826)

Hooke, R., 1665. Micrographia, Martyn and Allese, London, 246 p.

Hutchinson, E. G., 1948. In memoriam, D'Arcy Wentworth Thompson. Amer. Sci. 36, 577.

Huxley, T. H., 1868. On some organisms living at great depths in the North Atlantic Ocean. Quart. J. of Microseopy, VIII, New Ser.,203-212+Plate.IV.

Kalkowsky, E., 1908. Oolith und Stromatolith im norddeutschen Buntsandstein. Z. dtsch.. Geol. Gesellsch. 60, 84-125.

Krumbein, W. E., 1966. Zur Frage der Gesteinsverwitterung. Über geochemische und mikrobiologische Bereiche der exogenen Dynamik. Ph. D. Thesis, Würzburg, 106p.

Krumbein, W. E., 1972. Rôle des microorganismes dans la genèse, la diagenèse et la dégradation des roches en place: Rev. Ecol. Biol. Sol 9, p. 283-319.

Krumbein, W. E. (ed.), 1978. Environmental Biogeochemistry and Geomicrobiology. 3 volumes. Ann Arbor Sci. Publ. Inc. Ann Arbor, Michigan 1050p.

Krumbein, W. E. (ed.), 1979. Cyanobakterien - Bakterien oder Algen. I. Oldenburger Symposium über Cyanobakterien. BIS-Verlag, Oldenburg, 130.

Krumbein, W. E., 1983a. Stromatolites-Challenge of a term in space and time. Precambrian Res. 20, 493-531.

Krumbein, W. E. (ed.), 1983b. Microbial Geochemistry. Blackwell, Oxford, 330p.

Krumbein, W. E., 1987. Die Entdeckung inselbildender Mikroorganismen. pp. 62-75 in: Gerdes, G., Krumbein, W. E., Reineck, H.-E. (eds.). Mellum -Portrait einer Insel., Kramer, Frankfurt, 344p.

Krumbein, W. E., 1993a. Paracelsus und die mucilaginischen Substanzen - 500 Jahre EPS-Forschung. DGM-Mitt. Jg. 1993, 8-14.

Krumbein, W. E., 1993b. Microbial Biogeomorphogenesis – an appraisal of Immanuel Kant – p. 483-488 in Guerrero, R.., Pedros-Alio, C. (eds.): Trends in Microbial Ecology, Spanish Society for Microbiology, Barcelona, 717p.

Krumbein, W. E., 1994. The year of the slime. P. 1-7 in: Krumbein, W. E., Paterson, D.M.., Stal, L. J. (eds.), Biostabilisation of Sediments, BIS-Verlag, Oldenburg, 526p.

Krumbein, W. E., 1996. Geophysiology and Parahistology of the Interactions of Organisms with the Environment. Marine Ecology 17, 1-21.

Krumbein, W. E., Buchholz, H., Franke, P., Giani, D., Giele, C. and Wonneberger, K., 1979a. O2 and H2S coexistence in stromatolites. A model for the origin of mineralogical lamination in stromatolites and banded iron formations: Naturwissenschaft 66, p. 381-389.

Krumbein, W. E., Cohen, Y., 1974. Biogene, klastische und evaporitische Sedimentation in einem mesothermen monomiktischen ufernahen See (Golf von Aqaba): Geol. Rdsch. 63, p. 1035-1065.

Krumbein, W. E., Cohen, Y., 1977. Primary Production, Mat Formation and Lithification: Contribution of Oxygenic and Facultative Anoxygenic Cyanobacteria. In Fossil Algae, p. 37-56, edited by E. Flügel, Springer, Berlin, 375 p.

Krumbein, W. E., Cohen, Y., Shilo, M., 1977. Solar Lake (Sinai) 4. Stromatolitic cyanobacterial mats: Limnol. Oceanogr. 22, p. 635-656.

Krumbein, W. E., Jens, K., 1981. Biogenic rock varnishes of the Negev Desert (Israel) an ecological study of iron and manganese transformation by cyanobacteria and fungi. Oecologia 50, 25-38..

Krumbein, W. E., Rippka, R., Waterbury, J. B., 1979b. Schematische bakteriologische Gliederung der Cyanophyten im Vergleich zur phykologischen. In Cyanobakterien - Bakterien oder Algen? 1, p. 107-130, edited by W. E. Krumbein, Universität, Oldenburg, 130 p.

Krumbein, W. E., Paterson, D.M., Stal, L. J. (eds.), 1994.. Biostabilization of Sediments. BIS-Verlag, Oldenburg, 526p.

Krumbein, W. E. and Schellnhuber, H. J., 1990, Geophysiology of Carbonates as a Function of Bioplanets. In Facets of Modern Biogeochemistry, Chapter 2, p. 5-22, edited by A. V. Ittekott, S. Kempe, W. Michaelis and A. Spitzy, Springer, Berlin.

Krumbein, W. E., Schellnhuber, H.-J., 1992, Geophysiology of mineral deposits - a model for a biological driving force of global changes through Earth history: Terra Nova, 4, No. 3, 351-363

Levit, G., Gorbushina, A. A., Krumbein, W. E., 1999. Geophysiology and Parahistology of benthic microbial mats with special reference to the dissymmetry principle of Pasteur-Curie-Vernadskij. Bulletin de l'Institut Oceanographique, Monaco, special issue 19, 175-196.

Lovelock, J. E., 1979. Gaia - A New Look at Life on Earth. Oxford University Press, Oxford, 157p.

Lovelock, J. E., 1989. The Ages of Gaia. A Biography of our Living Earth. Oxford University Press, Oxford, 252p.

Ludwig, R., Theobald, G., 1852. Über die Mitwirkung der Pflanzen bei der Ablagerung des kohlensauren Kalkes. Ann. Phys. Chem. 87,91-107.

Marshall, K. C., 1976. Interfaces in microbial ecology. Harvard University Press, Cambridge.

Marshall, K. C. (ed.)., 1984. Microbial adhesion and aggregation. Springer, Berlin, 423 p.

Neu, T. R., 1992. Polysaccharide in Biofilmen. In Jahrbuch Biotechnologie Band 4 (eds. p. Präve, M. Schlingmann, K. Esser, R. Thauer, F Wagner), p. 73-101. Carl Hanser Verlag, München.

Noffke, N.,Gerdes, G., Klenke, Th., Krumbein, W. E., 2001. Microbially induced sedimentary structures – a new category within the classification of primary sedimentary structures. Jour. Sedimentary Research, 71, 650-656.

Paracelsus, T., 1982. Werke (ed. by Peuckert). Schwabe and Co., Basel.

Potts, M., 1997. Etymology of the Genus name Nostoc (Cyanobacteria). Int. J. of Systematic Bacteriology, 47, 584.

Poxton, 1. R., 1993. Procaryote envelope diversity. J. Appl. Bacteriol. Symp. Suppl. 74, 1S-11S. In: The Society for Applied Bacteriology Symposium Series No. 22. Microbial cell envelopes: Interactions and biofilms, (eds. L. B. Quesnel, P. Gilbert, P. S. Handley) p. 1S-11S. Blackwell Scientific Publications, Oxford.

Rippka,R., Deruelles, J. B., Waterbury, J. B., Herdman, M., Stanier, R. Y., 1979. Generic assignments, Strain histories and Properties of Pure Cultures of Cyanobacteria. J. General Microbiol. 111,1-61.

Schulz, E., 1936. Das Farbstreifen-Sandwatt und seine Fauna., eine ökologisch-biozönotische Untersuchung an der Nordsee. Kieler Meeresforsch., 1, 359-378.

Stal, L. J., P. Caumette, P., 1994. Microbial mats, structure, development and environmental significance. NATO ASI Series - Vol. Springer Verlag, Berlin.

Stal, L. J., Van Gemerden, H., Krumbein, W. E, 1985. Structure and development of a benthic marine microbial mat. FEMS Microbiol. Ecol. 31, 111-125.

Van Gemerden, H., 1993. Microbial mats: A joint venture. Marine Geol. 113, 3-25.

Van Oss, V. J., 1991. Interaction forces between biological and other polar entities in water: How many different primary forces are there? J. Dispersion Sci. Technol. 12,201-219.

Viles, H. A., 1984. Biokarst. Review and prospect. Prog. Phys. Geogr. 8, 532-542.

Wachendörfer, V., 1991. Parahistologische und sedimentmikrobiologische Unntersuchungen an einem potentiellen silikoklastischen Stromatolithen. Ph. D. Thesis, Oldenburg, 209p.

Walter, M. R. D., Ed., 1976. Stromatolites, Elsevier, Amsterdam, 790 p.

Wang, F. X., Junfa, C. J., Jing, J. L. H. J. W., 1993. Biokarst. Geological Publishing House, Beijing, 140p.

Weed, W. H., 1889. Formation of travertine and siliceous sinter by the vegetation of hot springs. U.S. Geol. Surv. Rep. 9, 613-676.

Wiggins, P. M., 1990. Role of water in some biological processes. Microbiol. Rev. 54, 432-449.

ZoBell, C. E., 1943. The effect of solid surfaces upon bacterial activity. J., Bacteriol. 46,39-56.

Chapter 2

DEVELOPMENT AND ARCHITECTURE OF COMPLEX ENVIRONMENTAL BIOFILMS

- lotic biofilm systems

Thomas R. NEU, Annett EITNER, M. Luz PAJE
Department of Inland Water Research Magdeburg
UFZ Centre for Environmental Research
Germany
neu@gm.ufz.de

1. INTRODUCTION

Slippery pebbles and rocks in creeks and rivers are for many people the first encounter with environmental biofilms. However, most of them are not aware of their significance in lotic systems and even scientists neglected the microbial ecology of streams for a long time (Leff, 1994). Complex environmental biofilms are quite different to pure culture or defined mixed culture biofilms. Nevertheless, much progress has been made in the past using model biofilms in order to explore basic structural features and biofilm processes e. g. (Lawrence, Korber et al., 1991;Möller, Sternberg et al., 1998). However, the complexity of environmental systems adds more than one level of variety in the structure and function of biofilms in nature. The variety in natural systems is based on the presence of a multitude of different bacterial species, the coexistence with a wide range of eucaryotic biofilm organisms, the production of a multitude of different extracellular polymeric substances (EPS) by both, procaryotic and eucaryotic biofilm organisms, as well as the interaction with dissolved, colloidal and particulate compounds derived from the environment.

The investigation of microbial biofilms in lotic systems started in the seventieth but made much progress within the last 20 years. However, the results available are scattered throughout many different scientific disciplines (aquatic chemistry, microbiology, limnology, ecology, geology). On one hand the focus was on chemical analysis or one group of organism e.g. bacteria, algae, macrophytes, invertebrates or fishes. On the other hand the complex interactions and pathways in lotic systems were considered e. g. food webs, detritus processing, primary production, spiraling of elements, nutrients and cells as well as modelling (Calow & Petts, 1992). Nevertheless, to the authors knowledge no recent overview with emphasis on the significance of interfacial microbial processes in lotic systems is available.

2. BIOFILM DEVELOPMENT

2.1 Cell Accumulation

Generally, the development of biofilms in the environment starts at an interface already covered with a conditioning film of organic molecules. By approaching the interface, often referred to as substratum, the bacteria may interact reversibly or irreversibly with the interface. Further the bacteria activate or inactivate specific genes which may change the status of planktonic into sessile cells. Various bacteria will use a variety of adhesion mechanisms which may include specific cell surface molecules (Neu, 1996) or cell surface structures e.g. fimbriae or pili (Austin, Sanders et al., 1998;Suoniemi, Björklöf et al., 1995). In the environment other eucaryotic organisms may interact with an interface as well. Pioneering organisms may include not only bacteria but also algae, fungi and protozoa. Once at the surface the bacteria may metabolize nutrients of various origin e.g. the substratum itself, compounds adsorbed to the substratum or molecules from the bulk liquid phase. Thereby the bacteria start growing into microcolonies and eventually into a biofilm showing a specific architecture. During this process, other molecules and cells will continuously interact with the developing interfacial community. A characteristic component of biofilm systems are extracellular polymeric substances (EPS) of different chemistry which may act as a multifunctional element. As the biofilm builds up into a 3-dimensionally structure, gradients will develop which in turn have an effect on the distribution of specific physiological bacterial groups within the complex arrangement of cells and macromolecules. At any stage of biofilm development single cells or in later stages larger areas of the biofilm may leave the interface thereby allowing the cells to colonize new habitats.

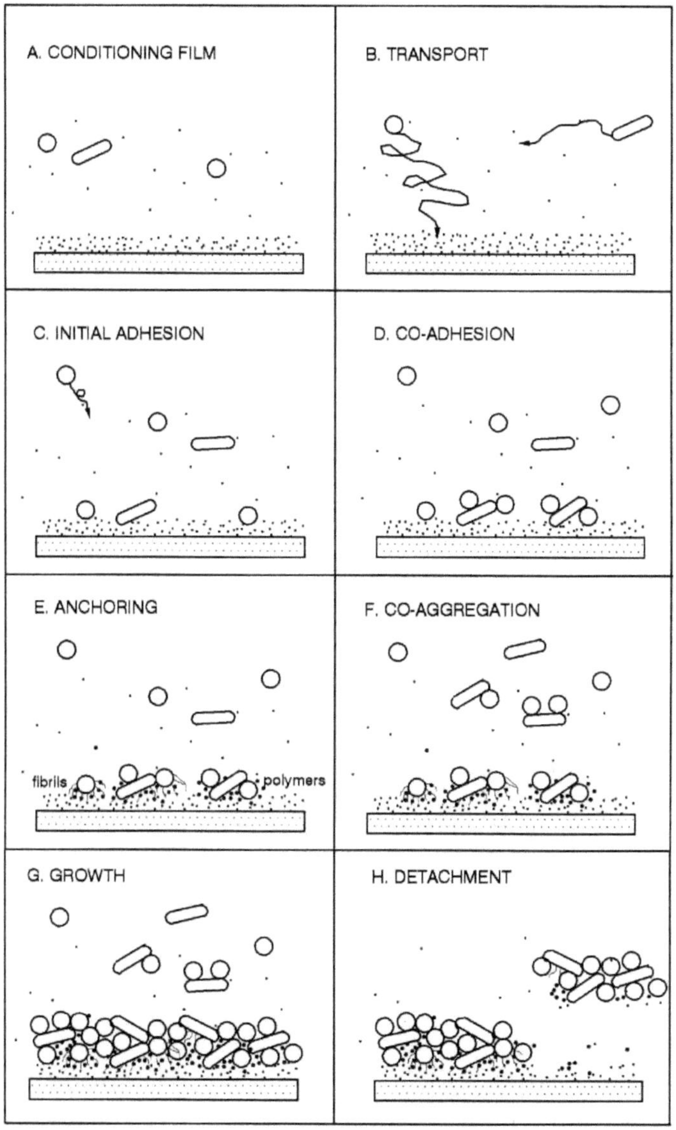

Figure 1 Initial adhesion events and biofilm development. From (Busscher & Mei, 2000).

In addition biofilms are subject to grazing by eucaryotic organisms e. g. protozoa, ostracoda, mayflies, or snails (Lawrence, Neu et al., 1998). The different stages of biofilm development are shown in a drawing by Busscher and van der Mei (Figure 1).

2.2 EPS Accumulation

EPS are produced by different biofilm organisms including bacteria, algae and fungi. Depending on the producers and the specific habitat the EPS may be comprised of numerous types of polysaccharides, proteins, nucleic acids and amphiphilic polymers or a mixture of these (Wingender, Flemming et al., 1999;Neu, 1996). Due to their polymeric chemistry EPS molecules may be involved in many biofilm functions (Wolfaardt, Lawrence et al., 1999) some of which are listed in Table 1.

The characterization of EPS is quite challenging as in environmental biofilm systems a large variety of polymers are building a very complex 3-dimensional matrix (Neu, 1994). So far two major approaches have been used to characterize EPS compounds. Firstly, by applying different treatments to isolate the EPS and separate the polymers from the cellular and other biofilm constituents. Subsequent purification and chemical analysis showed the heterogeneous composition of EPS which is different in every biofilm system (Nielsen & Jahn, 1999). Secondly, fluorescently labeled lectins have been suggested for in situ characterization of glycoconjugates in biofilms (Neu & Lawrence, 1999a;Neu & Lawrence, 1999b). In order to select suitable lectins for staining lectin-specific EPS constituents, all commercially available lectins have been tested on biofilms grown on different substrates (Staudt, Horn et al., 2003). By this approach in combination with laser scanning microscopy, several pieces of information may be collected (Neu, Swerhone et al., 2001): 1) the overall distribution of glycoconjugates within the biofilm and the possible association to certain cell types or microcolonies, 2) due to the specificity of lectins the glycoconjugates may be chemically characterized at the same time, 3) the technique can be combined with fluorescent in situ hybridization (FISH) and may allow to link information on cell identity with glycoconjugate identity (Böckelmann, Manz et al., 2002). The chemistry and function of EPS in biofilms is still an area of intensive research and much more progress has to be made to finally understand the EPS-matrix as an immobilized but dynamic part of biofilms (Sutherland, 2001).

Table 1 Function of EPS in biofilms

Adhesive - irreversible adhesion (attachment) - de-adhesion (detachment)	Structure - polymer network = matrix - cohesive force within biofilms
Protection - predation - desiccation - environmental changes - heavy metals and organic compounds	Recognition - co-aggregation - co-adhesion - microcolony formation - lectin interactions
Sorption - particulate and detritus - nutrients - ionic interactions - hydrophobic interactions	Physiology - regulatory or protective role in enzyme reactions - storage and reserve functions - intercellular transfer of genetic material

3. INVESTIGATION OF BIOFILM ARCHITECTURE

The investigation of interfacial microbial communities is ideally done with culture-independent, integral methods. One of these techniques is traditionally light microscopy which however is only suitable for 2-dimensional imaging. In the meantime light microscopy experienced a renaissance resulting in confocal laser scanning microscopy (CLSM) which is suitable for 3-dimensional imaging of living, fully hydrated biological samples. Very recently, a new technique, 2-photon laser scanning microscopy (2-PLSM), taking advantage of 2-photon excitation is commercially available and further extends the applicability of laser scanning microscopy (Neu, Kuhlicke et al., 2002).

Table 2. Advantages of laser scanning microscopy. Apart from the reflection mode, the advantages of 1-photon excitation do also apply to 2-photon excitation.

1-photon excitation	2-photon excitation
1. examination of fully hydrated, living, microbial communities 2. non-invasive optical sectioning of samples with virtually no light from out-of-focus regions 3. sectioning in horizontal (xy), vertical (xz) and temporal (xt) dimension 4. application of fluorescent (fluorescent mode) and non-fluorescent (reflection mode) probes 5. application of multiple probes and multi-channel recording of digital enhanced signals 6. quantitative analyses of the 3-dimensional organisation of complex microbial communities 7. 3-dimensional presentation and animation of data	1. excitation in the focal region only, excitation volume is extremely small (femtoliter) 2. caged fluorochromes may be excited in extremely localized spots 3. no out-of-focus bleaching and cell damage 4. background fluorescence creates no problems 5. compared to 1-photon excitation, infra red has a higher depth of penetration 6. highly resolved images from deep regions of scattering samples 7. less scattering and less filter problems 8. no pinholes necessary and thus no pinhole throughput loss 9. no UV laser necessary, no UV photo damage possible, no problems with UV optics

Both laser scanning microscopy techniques are ideally suited for the examination of microbial biofilms and allow the simultaneous study of multiple biofilm features in space and time. The advantages of 1-photon and 2-photon laser scanning microscopy are summerized in Table 2. Further details on the application of laser scanning microscopy may be found in recent reviews (Lawrence, Wolfaardt et al., 1998;Lawrence & Neu, 1999;Lawrence, Korber et al., 2002).

4. TYPES OF LOTIC BIOFILM SYSTEMS

4.1 Stationary Films

Biofilms in creeks and rivers may be found on submerged surfaces such as pebbles and rocks, wood, leaves as well as on and within the sediment (Lock, 1993). Early studies showed the significance of sessile bacteria in small mountain streams (Geesey, Mutch et al., 1978). Most biofilms contain algae as a major component having a direct effect on nutrient availability of biofilm bacteria (Haack & McFeters, 1982;Stock & Ward, 1989). The complex coupling between bacterial and algal activity was shown by Kaplan and Bott (Kaplan & Bott, 1989). Other studies focused on the effect of the substratum for colonization of lotic interfaces by bacteria and algae using microscopic or biochemical measures (Baker, 1984;Ferris, Fyfe et al., 1989;Sinsabaugh, Golladay et al., 1991). However, the effect of the substratum for the development and final structure of a mature biofilm is still an issue which is discussed controversially. Apart from the substratum, the hydrodynamic of a lotic system will also have an influence on biofilm biomass and activity (Claret & Fontvieille, 1997). Furthermore, biofilm communities are an important component in river sediments and the hyporheic zone and therefore were investigated with respect to carbon fluxes, activity and EPS (Battin, 2000). Lotic biofilms were also subjected to molecular biological approaches. It was found that during initial development beta-*Proteobacteria* dominated the biofilm. In later stages alpha-*Proteobacteria* and *Cytophaga-Flavobacterium* increased and dominated the mature biofilm population. Beta-*Proteobacteria* constituted the morphologically most diverse group (Manz, Wendt-Potthoff et al., 1999).

4.2 Mobile Aggregates

The significance of aggregates in aquatic systems with emphasis on their general characterization and biology has been treated in detail by several authors (Buffle & van Leeuwen, 1992;Wotton, 1994). Due to the surface to volume ratio in large rivers and streams, biofilms on submerged surfaces play only a very minor role. However, in these lotic systems many microorganisms are associated with mobile interfaces. In analogy to marine snow and lake snow these aggregates have been termed river snow (Neu, 2000). By definition these structures are true biofilms and are important for the self purification of rivers and streams. Furthermore, lotic aggregates are an important component of the so-called spiraling of cells, genes and EPS in lotic systems (Leff, McArthur et al., 1992). In opposition to marine and lake snow, lotic aggregates are exposed to permanent shear forces and therefore they are smaller and more rigid. The literature available on structural features of environmental microbial aggregates has been reviewed in a paper on structural investigation of river snow by laser scanning microscopy in conjunction with fluroescent probes for cellular and polymeric biofilm constituents (Neu, 2000). In addition, the identity of microbial communities associated with lotic aggregates have been characterized using a polyphasic approach. This included traditional culturing techniques as well as molecular biology techniques (Böckelmann, Manz et al., 2000).

5. SIGNIFICANCE OF LOTIC BIOFILM SYSTEMS

Lotic systems are important for many reasons (Leff, 1994). They have a certain potential for self purification which mostly is exhausted in densely populated areas, they are connected via the hyporheic zone with the groundwater and thus have an effect on drinking water quality, they are used for irrigation, they are employed for industrial processes and they carry contaminants. Furthermore, the lotic microbial community is involved in the cycling of elements such as carbon, nitrogen and phosphorus as well as in nutrient re-cycling including dissolved organic matter (DOM) and particulate organic matter (POM) (Lock, 1993).

Rivers and streams are polluted directly or indirectly by various sources such as industrial wastewater, urban waste water treatment plants or intensive agriculture. Both, the cellular and polymeric constituents in biofilm systems play a key role in the adsorption of charged compounds such as heavy metals and adsorption and degradation of toxic organic compounds. Due to their charge properties, some EPS components such as polysaccharides can act like an ion exchange matrix. In addition, the EPS matrix may show hydrophobic locations where organic compounds may be bound and finally become degraded (Winkler, Kopf et al., 1998). Thereby biofilm systems influence the mobility and the transport of contaminants (Admiraal, Tubbing et al., 1995;Lünsdorf, Brümmer et al., 1997). In addition, the presence of contaminants will have an effect on the bacteria in a lotic biofilm community (Brümmer, Fehr et al., 2000;Lemke, Brown et al., 1997;Tay, Hemond et al., 2001). As indicated above, in small creeks and rivers the films on stationary surfaces are important whereas in large rivers and streams microorganisms associated with mobile aggregates are mostly responsible for interaction with contaminants.

Biofilms in lotic systems typically have a significant phototrophic component which is responsible for primary production. Consequently, the primary producers together with the heterotrophic biofilm bacteria represent a food source for higher organisms such as protozoa (Hahn & Höfle, 2001) or invertebrates (Leff & Leff, 2000) and therefore they are very important in the lotic microbial food web (Meyer, 1994). Most grazing studies in lotic systems to date have concentrated on feeding of phototrophic cells e.g. algae or of isolated EPS components and no study compared different grazing strategies with respect to the effect on biofilm structure and biofilm constituents. The nature of selective and differential grazing by a species of gastropod, ostracod and a mayfly has been investigated using natural lotic biofilms. The grazing effect on biofilm architecture, thickness, and community composition was examined and showed a significant reduction of autotrophic and polymeric biofilm constituents whereas the bacterial biomass was enhanced within the grazed region (Lawrence, Scharf et al., 2002).

Biofilm systems, including cells and their EPS may be also involved in mineral formation (Ehrlich, 1999). This has been shown for different habitats e. g. lakes and marine systems (Riding & Awramik, 2000). For lotic systems mineral formation by biofilms has been described in only few studies (Arp, Wedemeyer et al., 2001;Freytet & Plet, 1996;Freytet & Verrecchia, 1998;Merz-Preiß & Riding, 1999;Pitois, Jigorel et al., 2001). It was found that cyanobacteria are responsible for calcite precipitation which was strongly affected by the current velocity at the micro and macro scale (Pitois, Jigorel, & Bertru, 2001). Further data indicate that the carbonate precipitation is driven by CO_2 degassing whereas photosynthetic carbon assimilation is at the macroscale not detectable (Arp, Wedemeyer, & Reitner, 2001).

6. ARCHITECTURE OF LOTIC BIOFILMS

Lotic biofilms usually develop under turbulent flow. Consequently a system suitable for growing this type of biofilm should offer similar hydrodynamic conditions. Rotating annular reactors (RAR) have been used to grow complex lotic biofilms in order to study their architecture and to quantify biofilm parameters (Neu & Lawrence, 1997; Lawrence, Neu, & Swerhone, 1998). In the meantime new annular reactor types are available which are further improved with respect to the characterization of the hydrodynamic conditions at the biofilm surface and adjustment of light conditions (Lawrence, Swerhone et al., 2000).

The development of lotic biofilms in a RAR microcosm with raw river water as inoculum and sole source of nutrients is demonstrated in a series of CLSM images (Fig. 2 a-d). The exposure of polycarbonate slides for few days resulted in the colonisation of the surface by pioneering bacteria of different morphology (Fig. 2 a). Later these bacteria grew into microcolonies, sometimes oriented into the direction of flow (Fig. 2 b). Mostly algae and sometimes fungi are involved in the early colonization process (Fig. 2 b, c). Under the effect of hydrodynamic conditions the microcolonies developed into complex 3-dimensional biofilm structures called ridges. The biofilm comprised of mountains and valleys, which were also oriented into flow direction (Fig. 2 d). Most of the bacterial biomass was found in the outer layer of the ridges which may be due to the nutrient availability in the exposed regions. After about 4 weeks the biofilms are in a sort of plateau phase and do not grow further. Due to nutrient availability and physiological processes in this latter stage sloughing may occur.

A different view of the biofilm is shown in Figure 3 a, b. In these images additional signals have been recorded which included reflective signals caused by mineral particles and the distribution of a fraction of the EPS specific to certain lectins. Firstly, the reflective and autofluorescent properties of the biofilm were employed for imaging the structure (Fig. 3 a). Secondly, the biofilm was stained with a nucleic acid specific stain as well as with a fluor-labeled lectin in order to show the glycoconjugate distribution (Fig. 3 b).

In the microcosm experiments it was also observed that lotic aggregates interacted directly with the reactor surfaces thereby creating a complex inoculum growing into similar biofilm ridges. An aggregate directly collected from the river Elbe is shown in Fig. 3 c, d. Similar to the biofilm images in Figure 3 a, b, the aggregate is presented in the reflection mode including the autofluorescent signal (Fig. 3 c) and after staining with a nucleic acid specific fluorochrome and a lectin specific for glycoconjugates (Fig. 3 d).

In the natural habitat the development of the biofilms followed initially the same regime. However, compared to the RAR biofilms growth in situ was faster. In later stages the effect of grazing by higher organisms became a structure-forming factor. Biofilms which have been grown initially into a 500 µm thick structure were reduced to films of about 40 µm only. The grazed biofilms were dominated by cyanobacteria and it seemed as if they were somehow protected against further grazing activity possibly by the production of toxins (Dowell Kearns & Hunter, 2000;Heyduck-Söller & Fischer, 2001;Kaebernick & Neilan, 2001). Sometimes a cleared grazing pattern became visible on the substratum, thereby new areas were created for subsequent colonisation.

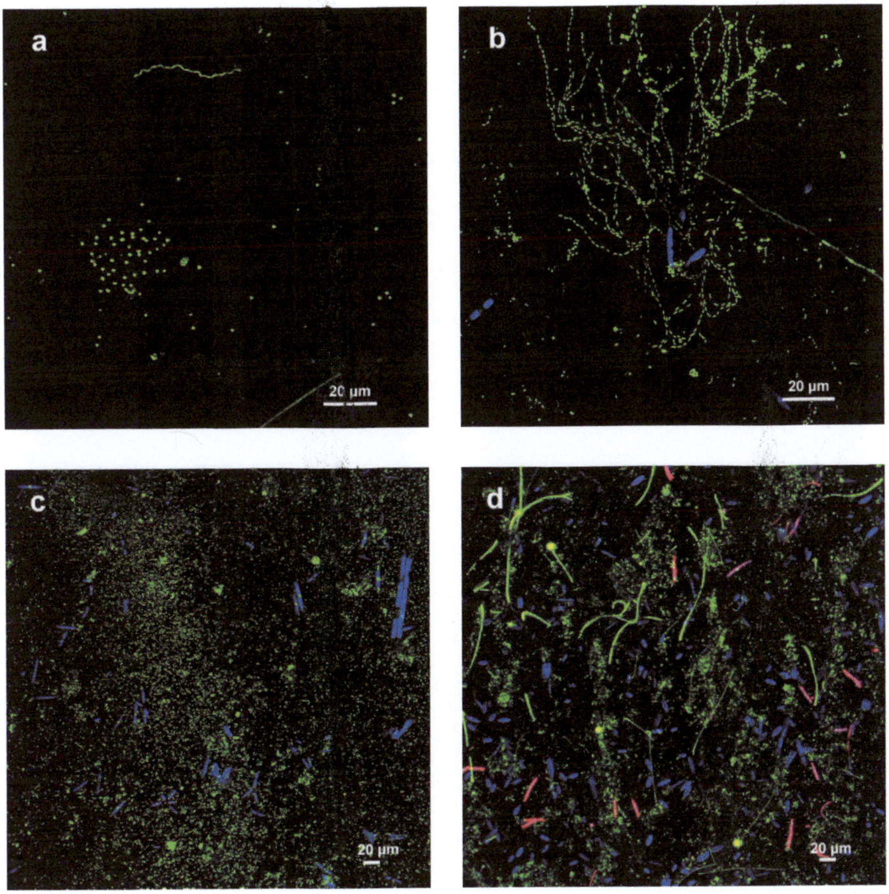

Figure 2 **a-d)** Development of lotic biofilms grown with raw river water in the course of 5 weeks. Confocal laser scanning micrographs showing maximum intensity projections of nucleic acid stained bacteria (green), autofluorescence of algae (blue) and autofluorescence of cyanobacteria (pink).

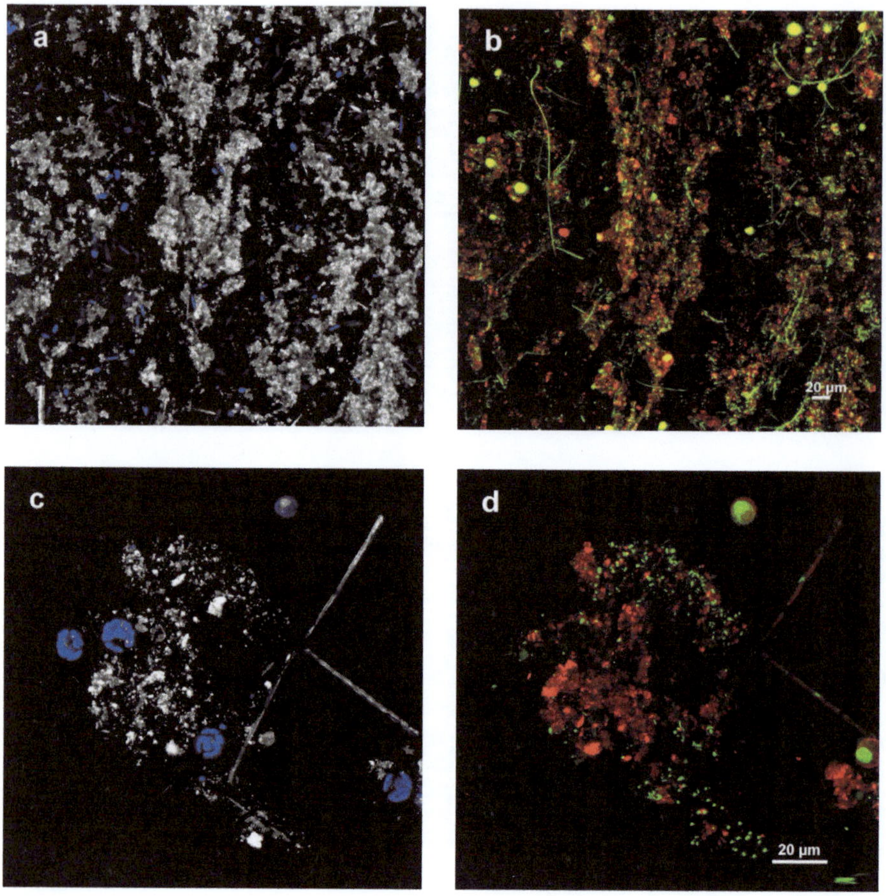

Figure 3 **a, b)** Mature lotic biofilm grown with raw river water in a microcosm after 75 days. Confocal laser scanning micrographs of the same location showing maximum intensity projections of a) reflection signal (white), autofluorescence of chlorophyll (blue) and b) nucleic acid stained bacteria (green), lectin stained glycoconjugates (red). **c, d)** Fresh, unfixed lotic aggregate collected from the river Elbe. Confocal laser scanning micrographs of the same aggregate showing maximum intensity projections of c) reflection signal (white), autofluorescence of chlorophyll (blue) and d) nucleic acid stained bacteria (green), lectin stained glycoconjugates (red).

7. CONCLUSIONS

Microbial biofilm systems play a key role in the interaction of dissolved, colloidal and biological compounds with stationary and mobile interfaces in lotic systems. Despite the knowledge we have to date there are still many questions to be answered. These questions include among others: the identity of bacteria, the activity of specific genes/bacteria, quorum sensing in biofilms, the competition and commensalism between bacteria, the co-operative processes in degradation of contaminants, the identity of EPS, the dynamics of EPS and the complex interaction between bacteria and EPS. In a situation were freshwater is a major issue for many countries the topic “lotic biofilms” may become a very important study area in order to assure drinking water as well as raw water quality and quantity. Finally, the saying "you never step in the same river twice" may indicate the general difficulties in studying lotic systems and might be the reason for this neglected area of microbiological research.

ACKNOWLEDGEMENTS

The study of M. L. P. was supported by a Georg-Forster research fellowship of the Alexander von Humboldt foundation.

REFERENCES

Admiraal, W., Tubbing, G. M. J., and Breebaart, L. (1995). Effects of phytoplankton on metal partitioning in the lower river Rhine. *Water Research* 29, 941-946.

Arp, G., Wedemeyer, N., and Reitner, J. (2001). Fluvial tufa formation in a hard-water creek (Deinschwanger Bach, Franconian Alb, Germany). *Facies* 44, 1-22.

Austin, J. W., Sanders, G., Kay, W. W., and Collinson, S. K. (1998). Thin aggregative fimbriae enhance *Salmonella enteritidis* biofilm formation. *FEMS Microbiology Letters* 162, 295-301.

Baker, J. H. (1984). Factors affecting the bacterial colonization of various surfaces in a river. *Canadian Journal of Microbiology* 30, 511-515.

Battin, T. J. (2000). Hydrodynamics is a major determinant of streambed biofilm activity: From the sediment to the reach scale. *Limnology and Oceanography* 45, 1308-1319.

Böckelmann, U., Manz, W., Neu, T. R., and Szewzyk, U. (2000). Characterization of the microbial community of lotic organic aggregates ("river snow") in the Elbe river of Germany by cultivation and molecular techniques. *FEMS Microbiology Ecology* 33, 157-170.

Böckelmann, U., Manz, W., Neu, T. R., and Szewzyk, U. (2002). A new combined technique of fluorescent in situ hybridization and lectin-binding-analysis (FISH-LBA) for the investigation of lotic microbial aggregates. *Journal of Microbiological Methods* 49, 75-87.

Brümmer, I. H. M., Fehr, W., and Wagner-Döbler, I. (2000). Biofilm community structure in polluted rivers: Abundance of dominant phylogenetic groups over a complete annual cycle. *Applied and Environmental Microbiology* 66, 3078-3082.

Buffle, J. and van Leeuwen, H. P. (1992). "Environmental particles." Lewis Publishers, Boca Raton.

Busscher, H. J. and Mei, H. C. v. d. (2000). Initial microbial adhesion events: mechanisms and implications. *In* "Community structure and co-operation in biofilms" (D. G. Allison, P. Gilbert, H. M. Lappin-Scott, and M. Wilson, Eds.), pp. 25-36. Cambridge University Press, Cambridge.

Calow, P. and Petts, G. E. (1992). "The rivers handbook. Volume 1." Blackwell Scientific Publications, Oxford.

Claret, C. and Fontvieille, D. (1997). Characteristics of biofilm assemblages in two contrasted hydrodynamic and trophic contexts. *Microbial Ecology* 34, 49-57.

Dowell Kearns, K. and Hunter, M. D. (2000). Green algal extracellular products regulate antialgal toxin production in a cyanobacterium. *Environmental Microbiology* 2, 291-297.

Ehrlich, H. L. (1999). Microbes as geologic agents: Their role in mineral formation. *Geomicrobiology Journal* 16, 135-153.

Ferris, F. G., Fyfe, W. S., Witten, T., Schultze, S., and Beveridge, T. J. (1989). Effect of mineral substrate hardness on the population density of epilithic microorganisms in two Ontario rivers. *Canadian Journal of Microbiology* 35, 744-747.

Freytet, P. and Plet, A. (1996). Modern freshwater microbial carbonates: The *Phormidium* stromatolites of southeastern Burgundy (Paris Basin, France). *Facies* 34, 219-238.

Freytet, P. and Verrecchia, E. P. (1998). Freshwater organisms that build stromatolites: A synopsis of biocrystallization by prokaryotic and eukaryotic algae. *Sedimentology* 45, 535-563.

Geesey, G. G., Mutch, R., Costerton, J. W., and Green, R. B. (1978). Sessile bacteria: An important component of the microbial population in small mountain streams. *Limnology and Oceanography* 23, 1214-1223.

Haack, T. K. and McFeters, G. A. (1982). Microbial dynamics of an epilithic mat community in a high alpine stream. *Applied and Environmental Microbiology* 43, 702-707.

Hahn, M. W. and Höfle, M. G. (2001). Grazing of protozoa and its effect on populations of aquatic bacteria. *FEMS Microbiology Ecology* 35, 113-121.

Heyduck-Söller, B. and Fischer, U. (2001). Extracellular cyanobacterial substances inhibit microbial growth. *International Microbiology* 3, 231-234.

Kaebernick, M. and Neilan, B. A. (2001). Ecological and molecular investigations of cyanotoxin production. *FEMS Microbiology Ecology* 35, 1-9.

Kaplan, L. A. and Bott, T. L. (1989). Diel fluctuations in bacterial activity on streambed substrata during vernal algal blooms: Effects of temperature, water chemistry, and habitat. *Limnology and Oceanography* 34, 718-733.

Lawrence, J. R., Korber, D. R., Hoyle, B. D., Costerton, J. W., and Caldwell, D. E. (1991). Optical sectioning of microbial biofilms. *Journal of Bacteriology* 173, 6558-6567.

Lawrence, J. R., Korber, D. R., Wolfaardt, G. M., Caldwell, D. E., and Neu, T. R. (2002). Analytical imaging and microscopy techniques. *In* "Manual of environmental microbiology" (C. J. Hurst, R. L. Crawford, G. R. Knudsen, M. J. McInerney, and L. D. Stetzenbach, Eds.), pp. 39-61. ASM, Washington.

Lawrence, J. R. and Neu, T. R. (1999). Confocal laser scanning microscopy for analysis of microbial biofilms. *Methods in Enzymology* 310, 131-144.

Lawrence, J. R., Neu, T. R., and Swerhone, G. D. W. (1998). Application of multiple parameter imaging for the quantification of algal, bacterial and exopolymer components of microbial biofilms. *Journal of Microbiological Methods* 32, 253-261.

Lawrence, J. R., Scharf, B., Packroff, G., and Neu, T. R. (2002). Microscale evaluation of the effects of grazing by invertebrates with contrasting feeding modes on river biofilm architecture and composition. *Microbial Ecology* 43, 199-207.

Lawrence, J. R., Swerhone, G. D. W., and Neu, T. R. (2000). A simple rotating annular reactor for replicated biofilm studies. *Journal of Microbiological Methods* 42, 215-224.

Lawrence, J. R., Wolfaardt, G., and Neu, T. R. (1998). The study of microbial biofilms by confocal laser scanning microscopy. *In* "Digital image analysis of microbes" (M. H. F. Wilkinson and F. Shut, Eds.), pp. 431-465. Wiley, Chichester.

Leff, L. G. (1994). Stream bacterial ecology : A neglected field? *ASM News* 60, 135-138.

Leff, L. G. and Leff, A. A. (2000). The effect of macroinvertebrates on bacterial distributions in freshwater microcosms. *Archiv für Hydrobiologie* 147, 225-240.

Leff, L. G., McArthur, J. V., and Shimkets, L. J. (1992). Information spiraling: Movement of bacteria and their genes in streams. *Microbial Ecology* 24, 11-24.

Lemke, M. J., Brown, B. J., and Leff, L. G. (1997). The response of three bacterial populations to pollution in a stream. *Microbial Ecology* 34, 224-231.

Lock, M. A. (1993). Attached microbial communities in rivers. *In* "Aquatic microbiology" (T. E. Ford, Ed.), pp. 113-138. Blackwell, Cambridge.

Lünsdorf, H., Brümmer, I., Timmis, K. N., and Wagner-Döbler, I. (1997). Metal selectivity of in situ microcolonies in biofilms of the Elbe river. *Journal of Bacteriology* 179, 31-40.

Manz, W., Wendt-Potthoff, K., Neu, T. R., Szewzyk, U., and Lawrence, J. R. (1999). Phylogenetic composition, spatial structure, and dynamics of lotic bacterial biofilms investigated by fluorescent in situ hybridization and confocal laser scanning microscopy. *Microbial Ecology* 37, 225-237.

Merz-Preiß, M. and Riding, R. (1999). Cyanobacterial tufa calcification in two freshwater streams: Ambient environment, chemical thresholds and biological processes. *Sedimentary Geology* 126, 103-124.

Meyer, J. L. (1994). The microbial loop in flowing waters. *Microbial Ecology* 28, 195-199.

Möller, S., Sternberg, C., Andersen, J. B., Christensen, B. B., Ramos, J. L., Givskov, M., and Molin, S. (1998). In situ gene expression in mixed-culture biofilms: Evidence of metabolic interactions between community members. *Applied and Environmental Microbiology* 64, 721-732.

Neu, T. R. (1994). The challenge to analyse extracellular polymers in biofilms. *In* "Microbial mats" (L. J. Stal and P. Caumette, Eds.), pp. 221-227. Springer, Berlin.

Neu, T. R. (1996). Significance of bacterial surface-active compounds in interaction of bacteria with interfaces. *Microbiological Reviews* 60, 151-166.

Neu, T. R. (2000). In situ cell and glycoconjugate distribution in river snow studied by confocal laser scanning microscopy. *Aquatic Microbial Ecology* 21, 85-95.

Neu, T. R., Kuhlicke, U., and Lawrence, J. R. (2002). Assessment of fluorochromes for two-photon laser scanning microscopy biofilms. *Applied and Environmental Microbiology* 68, 901-909.

Neu, T. R. and Lawrence, J. R. (1997). Development and structure of microbial biofilms in river water studied by confocal laser scanning microscopy. *FEMS Microbiology Ecology* 24, 11-25.

Neu, T. R. and Lawrence, J. R. (1999a). In situ characterization of extracellular polymeric substances (EPS) in biofilm systems. *In* "Microbial extracellular polymeric substances" (J. Wingender, T. R. Neu, and H.-C. Flemming, Eds.), pp. 21-47. Springer, Heidelberg.

Neu, T. R. and Lawrence, J. R. (1999b). Lectin-binding-analysis in biofilm systems. *Methods in Enzymology* 310, 145-152.

Neu, T. R., Swerhone, G. D. W., and Lawrence, J. R. (2001). Assessment of lectin-binding analysis for in situ detection of glycoconjugates in biofilm systems. *Microbiology* 147, 299-313.

Nielsen, P. H. and Jahn, A. (1999). Extraction of EPS. *In* "Microbial extracellular polymeric substances" (J. Wingender, T. R. Neu, and H.-C. Flemming, Eds.), pp. 49-72. Springer, Berlin.

Pitois, F., Jigorel, A., and Bertru, G. (2001). Colonization dynamics of an encrusting cyanobacterial mat in a hard water river (Eaulne, France). *Geomicrobial Journal* 18, 139-155.

Riding, R. E. and Awramik, S. M. (2000). "Microbial sediments." Springer, Berlin.

Sinsabaugh, R. L., Golladay, S. W., and Linkins, A. E. (1991). Comparison of epilithic and epixylic biofilm development in a boreal river. *Freshwater Biology* 25, 179-187.

Staudt, C., Horn, H., Hempel, D. C., Neu, T. R. (2003). Screening of lectins for staining lectin-specific glycoconjugates in the EPS of biofilms. In: Lens P., Moran A. P.,

Mahony T., Stoodley P., O'Flaherty V.(eds) "Biofilms in medicine, industry and environmental biotechnology" IWA Publishing, UK, pp 308-327.

Stock, M. S. and Ward, A. K. (1989). Establishment of a bedrock epilithic community in a small stream: Microbial (algal and bacterial) metabolism and physical structure. *Canadian Journal of Fisheries and Aquatic Sciences* 46, 1874-1883.

Suoniemi, A., Björklöf, K., Haahtela, K., and Romantschuk, M. (1995). Pili of *Pseudomonas syringae* pathovar syringae enhance initiation of bacterial epiphytic colonization of bean. *Microbiology* 141, 497-503.

Sutherland, I. W. (2001). The biofilm matrix - an immobilized but dynamic microbial environment. *Trends in Microbiology* 9, 222-227.

Tay, S. T. L., Hemond, F. H., Krumholz, L. R., Cavanaugh, C. M., and Polz, M. F. (2001). Population dynamics of two toluene degrading bacterial species in a contaminated stream. *Microbial Ecology* 41, 124-131.

Wingender, J., Flemming, H.-C., and Neu, T. R. (1999). What are bacterial extracellular polymeric substances. *In* "Microbial extracellular polymeric substances" (J. Wingender, T. R. Neu, and H.-C. Flemming, Eds.), pp. 1-19. Springer, Berlin.

Winkler, M., Kopf, G., Hauptvogel, C., and Neu, T. R. (1998). Fate of artificial musk fragrances associated with suspended particulate matter (SPM) from the river Elbe (Germany) in comparison to other organic contaminants. *Chemosphere* 37, 1139-1156.

Wolfaardt, G. M., Lawrence, J. R., and Korber, D. (1999). Function of EPS. *In* "Microbial extracellular polymeric substances" (J. Wingender, T. R. Neu, and H.-C. Flemming, Eds.), pp. 171-200. Springer, Berlin.

Wotton, R. S. (1994). "The biology of particles in aquatic systems." Lewis Publishers, Boca Raton.

Chapter 3

ECOSYSTEM FUNCTION, CELL MICRO-CYCLING AND THE STRUCTURE OF TRANSIENT BIOFILMS.

David M. PATERSON[1], Rupert PERKINS[1], Mireille CONSALVEY[1], Graham, J.C. UNDERWOOD[2]
Gatty Marine Laboratory, University of St Andrews
John Tabor Laboratories, University of Essex
dp1@st-andrews.ac.uk

1. INTRODUCTION

The understanding of the complexity of benthic biofilms has progressed rapidly in recent years, as technological advances have allowed analysis of biofilms on more relevant temporal and spatial scales. However, many different forms of biofilms exist and few have been extensively studied. Biofilms form on any submerged or intermittently submerged surface and most follow a known progression of development (or succession) towards an equilibrium state related to environmental conditions. A particularly dynamic sub-section of benthic biofilms are transient films that self-assemble and disperse in response to environmental conditions at the surface of intertidal cohesive or mixed sediment systems (Paterson and Hagerthey 2001). These biofilms are comprised of motile autotrophic cells, given the collective term microphytobenthos, and are largely comprised of diatom, euglenid flagellates and cyanobacteria. A primary characteristic of these organisms is their ability to migrate through sediments in response to environmental cues and their inherent biological rhythms (Admiraal 1984). Transient biofilms have a variable life-span related to the differential pressures of tidal exposure, light conditions, grazing and hydrodynamic forcing. These biofilms are formed from a matrix of cells and associated extracellular polymeric substances (EPS) that many microphytobenthic cells extrude through locomotion (Edgar and Pickett-Heaps, 1984). It is arguable that in terms of temporal and spatial variability, transient biofilms are a particularly complex group. Transient biofilms can form rapidly (within 20 min) (Paterson et al. 1998) and usually disperse shortly before or after the tidal exposure period ends. At their most basic they may comprise a single layer of cells (Paterson et al. 1998) or develop into a coherent matrix of cells and polymeric substances, sometimes several mm in depth. The formation of more sturdy biofilms may lead to microbial mat formation and the system loses its transient nature, being retained over several tidal cycles, or even become semi-permanent on a seasonal basis. An example of these semi-permanent biofilms is the "bubble mat" system described in literature (Yallop et al. 1994).

The ecological importance of transient biofilms has also been reassessed in recent years. Although relatively thin, the films are widespread and centres of very active primary production and biogeochemical activity (Underwood and Kromkamp, 1999). In modern terms, the ecological role of microphytobenthic biofilms can be seen in terms of the "ecosystem services" they provide. Transient biofilms form the basis of the food chain for many coastal species; mediate the flux of nutrients across the sediment/water interface (Nedwell et al. 1999) and are also "ecosystem engineers" enhancing the resistance of the sediment to erosion (Paterson 1997, Black et al. 2002).

Major advances in the understanding of the structure and function of biofilm systems have been made since developments in micro-analytical technology have allowed the examination of biofilm on a scale relevant to the structure and processes, that occur within the film itself. These developments have included the use of micro-sensor systems for oxygen (Revsbech et al. 1980), nutrients and light (Consalvey, 2002) with the analysis of biofilm properties at a 100 μm scale (Taylor and Paterson, 1998; Kelly et al. 2001). In addition, biofilms have now been viewed using a variety of techniques, which enhance our conceptual understanding of their structure, (Richards and Turner, 1984, Paterson, 1995, Defarge et al. 1996), temporal plasticity (Consalvey 2002, Sauer et al. 2002,) and physiological activity (Oxborough et al. 2000, Perkins et al. 2002). A potentially valuable emerging technique is the ability to visualise and measure the photo-physiological activity of individual cells within the biofilm. These investigations have already demonstrated cellular responses under high light conditions (Perkins et al., 2000).

There have been many attempts to define biofilms, usually based on their taxonomy or structure (Neu 1994). Biofilms are often represented as a matrix of relatively stable components, cell consortia, EPS, and space arranged in three-dimensions. The forth dimension, time, is generally considered to vary as successional processes occurring over a scale of days, weeks and years. In permanent biofilms, the primary role of extracellular polymeric material is anchorage. In contrast, EPS production provides the means of locomotion to one of the most important groups of organisms forming transient biofilms, the diatoms (Edgar and Pickett-Heap 1984). The locomotive ability of microphytobenthos represents a possible evolutionary strategy to maximise resource availability (light, space and nutrients) in depositional environments. Cells rapidly congregate at the sediment surface after the retreat of the tides. The traditional view was that cells accumulated at the sediment surface and the relative role and turnover of cells within the mat system was not considered. More limited locomotion is to be expected where a biofilm forms in relation to a surface when attachment is a critical ability. In transient biofilms, however, locomotion is the primary characteristic. Evidence is accumulating, that locomotion of cells within the microphytobenthic biofilm provides a mechanism whereby cells may improve their individual "fitness" and contribute to the overall "efficiency" of the biofilm by the sub-cycling of cells and consortia within the film on a scale of minute to hours (Kromkamp et al. 1998, Perkins et al. 2002). This paper presents evidence in support of the paradigm that the ecosystem functioning of biofilm assemblages is more than the sum of the individual parts (species or consortia). Light availability is a driving force for photosynthetic microbes, and migratory patterns are an adaptation to maximising light utilisation. The temporal variability in biofilm structure, light climate and photosynthetic activity is shown by electron microscopy, light microscopy, light micro-profiling and chlorophyll fluorescence determination of cell and biofilm photophysiology.

2. MATERIALS AND METHODS

Sediment biofilms were collected from a variety of sites around the coast of Europe and the sources are indicated on figure captions. Artificial biofilms were grown in laboratory culture using glass beads and a tidal system with controlled temperature and day/night cycle (after Paterson, 1990).

3. LOW TEMPERATURE SCANNING ELECTRON MICROSCOPY

Samples were obtained for low-temperature scanning electron microscopy (LTSEM) by removal/excision of a 2 x 1 cm strip of biofilm supported on a rigid metal foil planchette. The foil and biofilm was quench frozen in liquid nitrogen (-196°C) and then stored, under liquid nitrogen, until viewed in the chamber of a specially-adapted scanning electron microscope (JOEL 35CF SEM fitted with Oxford Instruments CT 1500B). Samples were viewed while still frozen (-180 °C) and surface water was removed by heat etching (partial freeze drying) following the procedure given in Paterson (1995).

4. LIGHT MICROPROFILING

The light attenuation co-efficient of biofilms from 5 sites were determined. 3 sites on the River Colne, Colchester, Essex, UK (51°50.2' N, 0°59.5' E) were investigated; a euglenid- dominated site at Hythe and diatom-dominated sites at Arlesford Creek and Point Clear. Sediment biofilms from the Eden Estuary, U.K. (56°22'N, 02°51'W) and Sarilhos Pequenos on the Tagus Estuary (38° N, 9° W) were also examined. Light profiles through the sediment biofilms were made using fibre-optic microsensors, constructed after Lassen *et al.* (1992). The fibre optic sensor was inserted into the surface-sediment of a randomly selected sediment core, at a 45° angle to reduce shading and moved down at 100 μm intervals using a micromanipulator. The scalar irradiance at each depth was determined as a percentage of surface incident light calculated from the associated drop in voltage. The light extinction co-efficient (k) was calculated using the Beer-Lambert law:

$$E_z = E_0 \, . \exp(-kz) \tag{1}$$

Where E_z is the light level at depth z, E_0 is the surface incident light level and k is the attenuation coefficient.

5. FLUORESCENCE MEASUREMENTS OF ELECTRON TRANSPORT RATE

Intact sediment cores were collected from Arlesford Creek and the Hythe on the River Colne, Colchester, Essex, UK, using 7.5 cm diameter cores in July 1999. Whereas the Arlesford Creek site is situated midway along the estuarine salinity gradient and supports a diatom dominated biofilm, the Hythe is at the head of the estuary and has lower salinity but higher nutrient concentrations, and supports a biofilm comprised largely of euglenids and cyanobacteria (Underwood et al. 1998; Perkins et al. 2002). Chlorophyll *a* fluorescence images were obtained from surface scrapes (depth 1 cm) carefully removed from the surface of the sediment cores and transferred to a high-resolution fluorescence imaging system (Oxborough and Baker, 1997; Oxborough et al. 2000; Perkins et al. 2002). A detailed description of this system is given in Oxborough and Baker (1997). Values of the quantum efficiency of linear electron transport at PSII (F_q'/F_m') from isolated cells were calculated as the sum of individual pixel values for this parameter (Oxborough and Baker, 1997; Oxborough et al. 2000).

Photosynthetic electron transport rate (ETR) versus light response curves were made on three replicate cultures of *Navicula salinarum,* (Ehr.) Reimann *et* Lewin. grown at a PPFD of 180 mol m^{-2} s^{-1} on a 14 / 10 h light / dark cycle at 18 °C in a growth cabinet using f/2 media (Guillard and Ryther, 1962), salinity 20 and with an added antibiotic cocktail to inhibit bacterial growth (Smith and Underwood, 1998). Data were compared to that for intact diatom-dominated biofilms collected from Alresford Creek in July 1999. Chlorophyll fluorescence was measured using a Xenon PAM fluorometer (Walz, GmbH, Effeltrich, Germany), the saturating pulses used for measurements of F_m' were at a PPFD of 8,600 mol m^{-2} s^{-1} for 300 ms and the non-actinic measuring beam frequency was set at 4 Hz. Chlorophyll fluorescence was defined using a 680 nm bandpass filter (Coherent, Watford, England) (Perkins et al., 2002).

ETR was calculated as the product of F_q'/F_m', photosynthetic photon flux density (PPFD) and the specific light absorption coefficient for chlorophyll *a* (a^*), following Sakshaug *et al.* (1997):

$$\mathrm{ETR} = \frac{F_q'}{F_m'} \cdot \frac{\mathrm{PPFD}}{2} \cdot a^* \tag{2}$$

a^* was determined using an integrating sphere and a cosine corrected Skye quantum sensor. Biofilms were sampled using the lens tissue technique

(Eaton and Moss, 1966). Cells were re-suspended in f/2 media in repli-dishes. This was serially diluted to construct a standard curve of absorption versus chlorophyll *a* concentration (over a range 50 to 800 g l^{-1}), from which *a** could be determined. Cultures of *N. salinarum* were also serially diluted and then treated in the same manner as samples from intact biofilms. Maximum ETR (ETR_{max}) and light utilisation coefficient ( ) were calculated by iteration (Long and Hällgren, 1993), from light response curves (ETR vs. PPFDs of 0 to 830 μmol m^{-2} s^{-1}).

6. RESULTS

Qualitative variation in natural biofilm structure was clearly demonstrated by low-temperature scanning electron microscopy (Fig. 1-2). Natural biofilms varied from isolated cells, to confluent unicellular layers and to more extensive biofilms with several layers of cells (Fig. 1). Comparative analysis of non-transient biofilms showed greater development of the acellular polymeric matrix between the grains of sediment (Fig. 2a-b). The development of the extensive films seemed to be restricted to non-cohesive sediments and could be recreated in the laboratory (Fig 2d). An extensive transient biofilm development on cohesive sediment was found from an intertidal drainage channel from the Severn estuary which had an unusual alignment of nitzschiod and naviculoid cells forming a matrix 500 μm in extent (Fig 1e-f). Artificial biofilms were created which closely resembled the structure of natural films (compare Figs 1 and 2d). Transient biofilms were normally formed as a thin biogenic layer on the surface of the bed (Fig. 2d) which disappeared as cells dispersed among the sediment particles on downward migration. In laboratory cultures, migration ceased after 10-14 d and the surface layer became semi-permanent, and up to 600 μm in depth after 14 d (Fig 2d). In one unusual case, natural biofilm development was found to be discontinuous and related to the surface topography of the sample (Fig 2e-f). In this area, diatoms were found only in the hollows of a very uneven surface. This unusual formation was found on the upper shore of a heavily bioturbated area on intertidal of the Severn estuary.

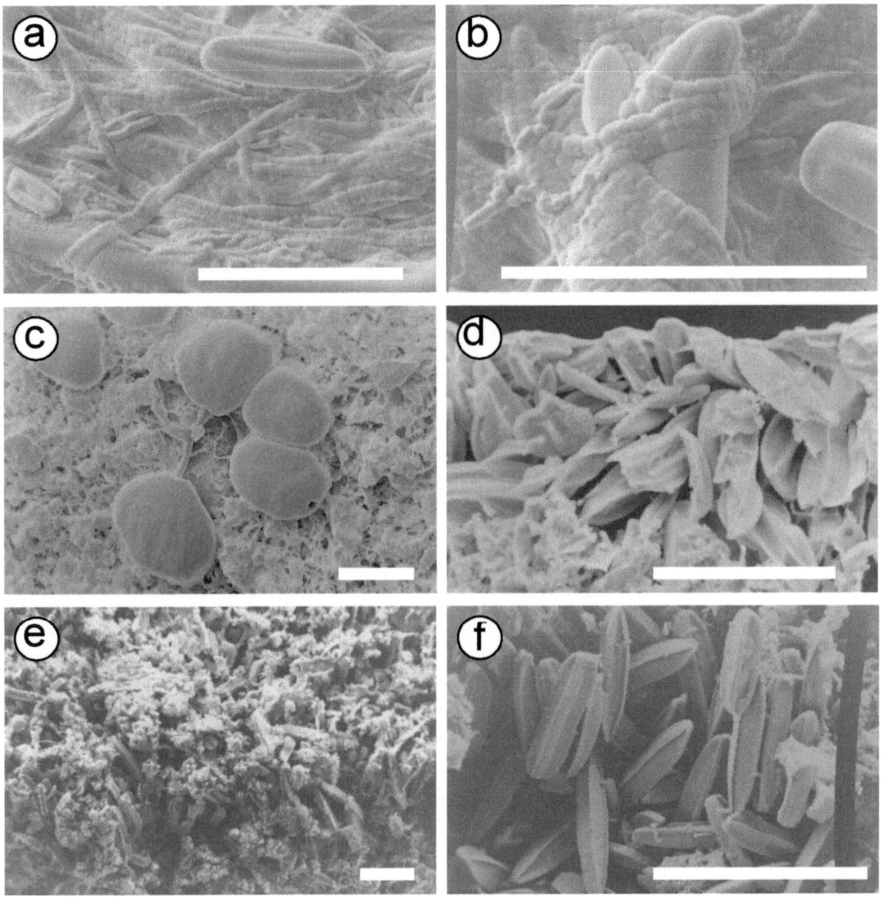

Figure 1: Low-temperature scanning electron micrographs of transient biofilms. a). Mixed assemblage of upper intertidal system dominated by cyanobacteria but still showing active migration patterns. b). Close association between cyanobacteria and diatoms. c). Euglenid-dominated biofilm. d). Cross section of diatom-dominated biofilm showing characteristic depth of several cells. E). Extensive transient biofilm of tidal channel (Severn estuary). F) Detail of unusual upright position of Nitzschiod cells. All bar markers 100 m.

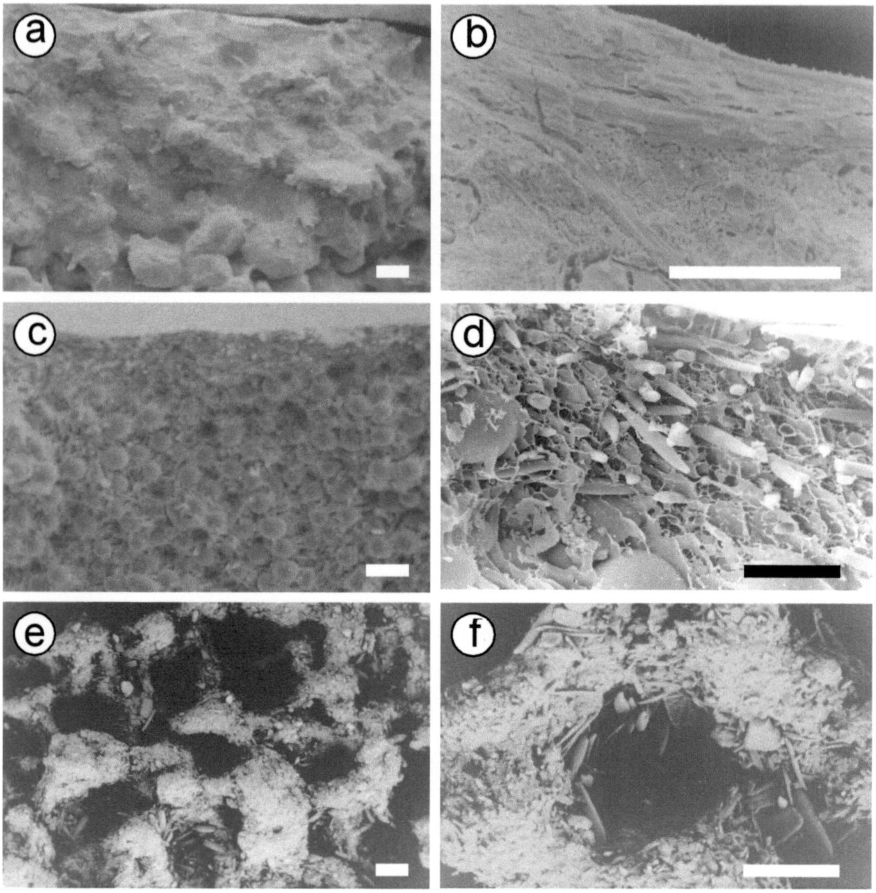

Figure 2: Low temperature scanning electron micrographs of permanent and cultured biofilms. a).Thick cyanobacterial biofilm on the surface of non-cohesive (sandy) sediments. b). Detail of organic matrix of thick permanent biofilm. c). Diatom biofilm cultures on a glass bead substratum. D). Detail of organic matrix developing in glass bead culture. e). Biofilm on the surface of heavily bioturbated sediments. Biofilm, dominated by diatoms, was found only in the hollows. F). Detail of surface hollow filled with diatoms. All bar markers 100 m.

7. LIGHT PROFILES AND BIOFILM TYPE

The transmission of light through different biofilms was examined and the light extinction coefficient (k) calculated. k was determined to significantly vary between biofilm types ($F_{4, 12} = 17.88$, $p < 0.005$) (Table 1). Of the biofilms measured in this study, those dominated by euglenids (Hythe site) were found to have the highest light attenuation co-efficients. Significant differences between diatom-dominated sites were also observed. The Arlesford Creek biofilm had the highest light attenuation coefficient compared to the other diatom-dominated sites. Using these data light, attenuation profiles over sediment depth were constructed (Fig. 3). Cells in biofilms from the Hythe and Arlesford Creek would have experienced almost complete darkness at a depth of 300 μm, whereas cells at the other sites (diatom-dominated) would still have experienced > 4% of the incident PPFD.

Table 1: Light extinction co-efficient of surface biofilms and associated sediments at 5 sites

Site	k (mm^{-1})	± SE
Hythe	21.173	0.7758
Arlesford Creek	15.717	0.5457
Point Clear	9.83	0.6279
Eden	9.61	1.1816
Tagus	11.14	1.6433

8. PHOTOPHYSIOLOGICAL RESPONSES OF BIOFILMS AND CELLS

Images obtained from the high resolution imaging system (Fig 4-6) are of chlorophyll *a* fluorescence exposures (at F', just prior to application of the saturating pulse) obtained at an actinic light level of 220 μmol PPFD m^{-2} s^{-1}. The different community structures of the microphytobenthic biofilms sampled from the Hythe (Fig 4 a-b) and Arlesford Creek (Plate Fig 4c) are clearly visible. The Hythe biofilm was comprised largely of euglenids (Fig 4a) with patches of dense cyanobacterial filaments (Fig. 4b). In contrast, biofilms at Arlesford Creek were dominated by diatoms, principally naviculoids and larger cells such as *Pleurosigma* sp., *Gyrosigma* sp., and *Nitzschia dubia*. This variability in community structure has important consequences when measuring variables such as primary production and electron transport, which vary greatly between taxa and hence biofilms. Such variability is highlighted further by the variation in PSII quantum efficiency (F_q'/F_m') shown (Fig 4c). Individual cells showed different values of F_q'/F_m', despite being imaged at the same PPFD. Values of 0.152 and 0.315 were obtained for two cells of the same species, *Nitzschia dubia*.

The diatom-dominated biofilm obtained from Arlesford Creek varied in community structure over a diel period (Fig 5). Selected fluorescence images (imaged at 220 µmol PPFD m^{-2} s^{-1} as above) of the same biofilm (incubated in a tidal tank system, see Perkins et al. 2002) obtained at 07:00, 13:00, 15:00 and 20:00 h demonstrated variability in biofilm appearance over time. Cells of *Nitzschia dubia* were present over the diel period, whilst smaller naviculoids were most abundant at 07:00 h. Cells of *Pleurosigma* spp. were noted primarily at 13:00 and 20:00 h, with *Gyrosigma* spp. particularly apparent at 20:00 h. At least 10 images were taken at each time to ensure those shown were representative of the biofilm overall.

The transient nature of a microphytobenthic biofilm in response to changes in ambient light level (PPFD) was demonstrated (Fig. 6). Exactly the same surface area of biofilm was exposed to incremental increases in actinic light during a light response curve under the imaging system. Initially the biofilm was comprised of a mixture of small naviculoid cells, with some larger diatoms and euglenids (Fig. 6a, 180 µmol m^{-2} s^{-1}). However at a higher PPFD of 1150 µmol m^{-2} s^{-1}, the surface of the biofilm was dominated by euglenids (Fig. 6b). The cellular composition is clearly shown to vary in response to light climate.

The properties of the biofilm therefore varied depending on surface cell type, light climate and local conditions. The performance of cells in culture and those within a biofim matrix may also vary. This was investigated using a Xenon PAM fluorometer (Walz, Germany). The light response curve for a culture of *Navicula salinarum* grown at 180 µmol m^{-2} s^{-1} was compared to that for an intact diatom-dominated biofilm obtained from Arlesford (Fig. 7). The curve for *N. salinarum* shows a higher initial slope, corresponding to (0.29 ± 0.0065 compared to 0.17 ± 0.0014 µmol e^{-} m^{-2} µg Chl a^{-1} µmol $photons^{-1}$ for *N. salinarum* and the intact biofilm respectively), but saturated at a lower ETR, corresponding to ETR_{max} (4.5 ± 0.57 compared to 6.1 ± 0.42 µmol e^{-} µg Chl a^{-1} s^{-1} for *N. salinarum* and the intact biofilm respectively), compared to the biofilm obtained from the field site (Fig. 7). This suggests a comparatively shade-adapted photophysiology for the cultured cells compared to a light-adapted natural biofilm, with the former grown at comparatively lower light levels than experienced by the natural biofilm *in situ* in July 1999.

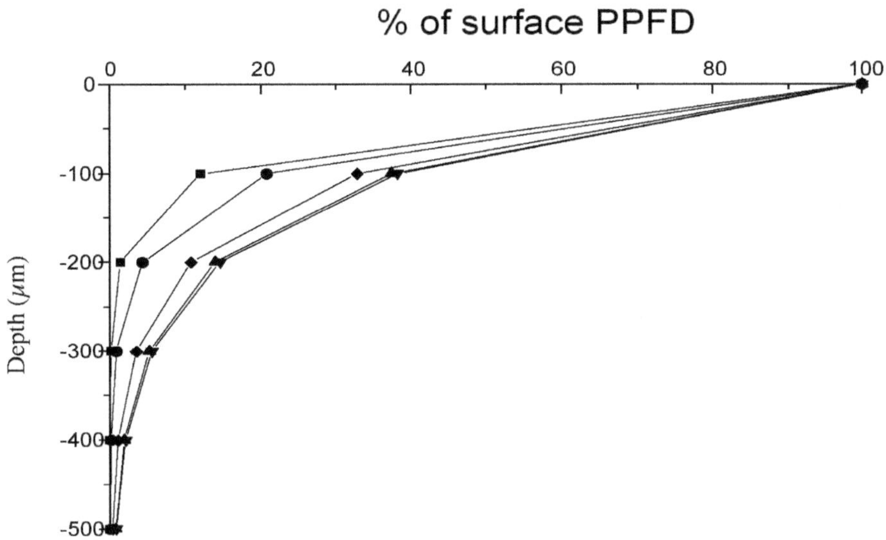

Figure 3: Light attenuation profiles of 5 different biofilms. Hythe (euglenoid) = square; Arlesford Creek (diatom) = circle, Point Clear (diatom) = up triangle, Eden (diatom) = down triangle and Tagus (diatom) = diamond.

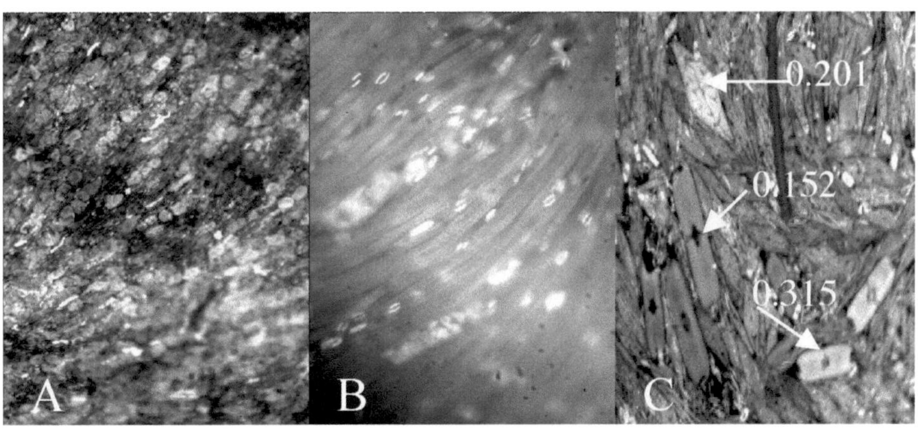

Figure 4: Chlorophyll *a* fluorescence images (*F*', imaged at 220 μmol PPFD m^{-2} s^{-1}) microphytobenthic biofilms dominated by euglenoids (A), cyanobacteria (B) and diatoms (C). PSII quantum efficiencies (F_q'/F_m') are shown for individual diatom cells of a *Pleurosigma* sp. and two *Nitzschia dubia* cells (C). Plates (A) and (B) are of intact biofilms obtained from the Hythe and Plate (C) from Arlesford Creek in July 1999.

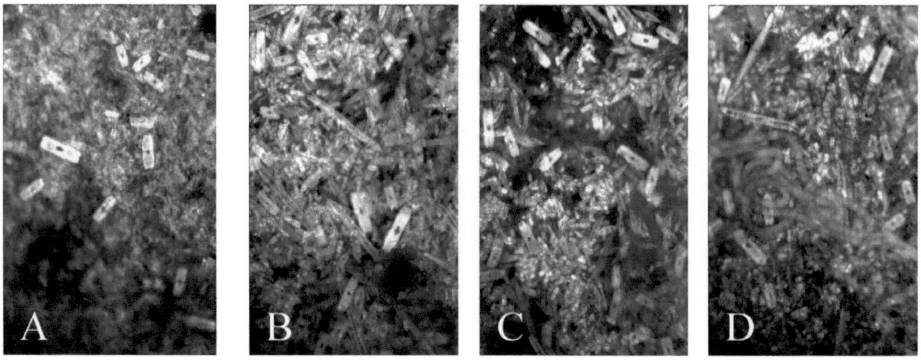

Figure 5: Chlorophyll *a* fluorescence images (*F*', imaged at 220 µmol PPFD m^{-2} s^{-1}) of a diatom dominated biofilm obtained from Arlesford Creek in July 1999. Images were taken at (A) 07:00, (B) 13:00, (C) 17:00 and (D) 20:00 h, from sediment cores incubated in a tidal tank system. At least 10 images were taken at each time to ensure those shown were representative of the biofilm at each time.

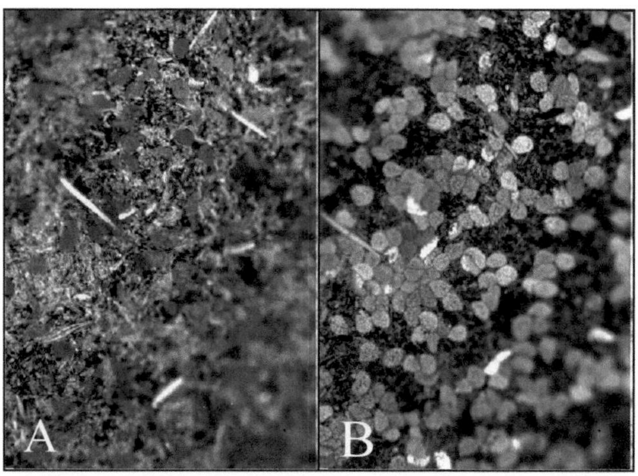

Figure 6: Chlorophyll *a* fluorescence images (*F*') imaged at (A) 180 and (B) 1150 µmol PPFD m^{-2} s^{-1} for a euglenoid dominated biofilm collected from the Hythe in July 1999. Images were taken within 8 minutes of each other. This figure formed part of Fig. 3 in Perkins et al., 2002.

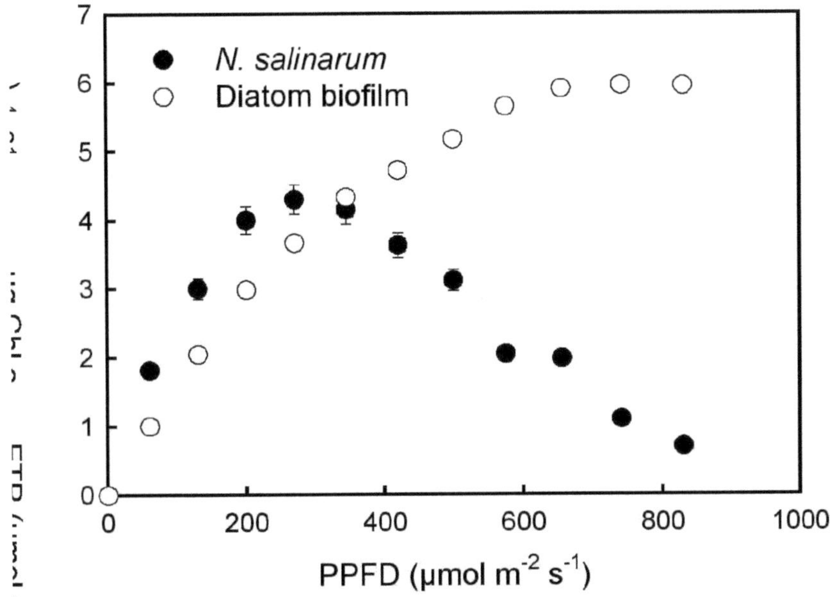

Figure 7: Light response curves of electron transport rate (ETR) versus light level (PPFD) for a culture of *Navicula salinarum* grown at 180 µmol PPFD m^{-2} s^{-1} (n=3, mean ± s.e., closed symbols) and a diatom-dominated biofilm from Arlesford Creek collected in July 1999 (open symbols)

9. DISCUSSION

9.1 Biofilm structure

The structure of several types of intact microphytobenthic biofilms were examined using LTSEM. Variation in the structure, depth and coverage of the biofilm was evident and, as expected, this variation led to differences in light penetration through the biofilm, confirmed by light microsenor measurements. However, it was also evident that some biofilms were highly heterogeneous and that their formation was related to the surface micro relief of the sediments (Fig 2e-f). This has rarely been documented and the mechanisms behind the formation of these unusual structures are not known. However, it is clear that the properties of the film will vary on a scale related to the surface relief, in this case on a 50-200 µm scale. The region of sediment where these biofilms were found had been highly bioturbated but was relatively dry at the time of sampling (during good weather). High light and dry conditions may lead to biofilms developing in the shaded and protected hollows of the uneven surface but this hypothesis would need further investigation.

9.2 Light penetration

The assemblage type affected the penetration of light into the biofilm, with light being attenuated most rapidly in euglenid biofilms (k = 21) at a level more than 2 fold greater than the diatomaceous mat from the Eden Estuary (k = 9). However, the depth of light penetration also varied between diatomaceous biofilms. The variation in light penetration between different biofilms dictates the depth to which cells remain photosynthetically active. Greater k values will limit the photosynthetically active biomass (PAB) to the surface layers of the sediment and lead to a shallower zone of primary productivity. Not only does the depth of light penetration have implications regarding the overall primary productivity of microphytobenthic systems but it also influences the photophysiological state of the cells. It has been suggested that cells are able to alter their position within the sediments optimising light level to avoid photoinhibition/damage therefore resulting in a maximal rate of productivity for the whole biofilm (Kromkamp et al. 1998, Perkins et al. 2002).

9.3 Photophysiological responses of biofilms and cells

High resolution fluorescence imaging revealed a sequential change in diatom taxa at the sediment surface; with a shift towards larger cells at the end of the day. Furthermore short-term changes in taxa at the sediment surface were also observed during a light curve; a shift occurring from diatoms at low light levels to euglenid cells at high light. This demonstrates the plasticity of microphytobenthic biofilms and the ability of cells to respond rapidly to short-term environmental changes in light level. Vertical migrations are widely described for a diverse range of microphytobenthic taxa, the ubiquity of these behaviours suggests that these actions must confer some evolutionary survival advantage (Cohn and Disparti 1994; Kingston 1999). Kingston (1999) experimentally verified the theory that upwards migration maximises photosynthetic capacity and downward migration minimises photoinhibition (through exposures to high light) in *Euglena proxima*. Photoinhibition has rarely been recorded in benthic communities (Blanchard and Cariou-LeGall 1994) and may be attributed to vertical migration. The metabolic cost of a behavioural versus a photophysiological response to high light is unknown and both strategies are likely to be used to complement each other (e.g. Perkins et al., 2002).

A whole suite of factors is involved in the migratory response of diatoms and other microphytobenthos, but the ultimate driving force remains unknown. Due care must be taken with sampling to accommodate any inherent changes in biomass based upon rhythmic migration as well as sub-cycling. The micro-spatial variability in biomass distribution over the surface of the sediment may sometimes be significant. A biofilm may be viewed as more than the sum of its parts, with surface cells shading and perhaps benefiting the cells underneath, but themselves migrating down when necessary, allowing other cells to take their place.

10. CONCLUSION

The functional role of biofilms is recognised as significant in the structuring of coastal ecosystems (Paterson and Hagerthey, 2001). However, is becoming clear that the capabilities of individual cells cannot be used to determine the limits of biofilm function. Evolution has resulted in a system of dynamic turnover which maximise the functional capacities of the biofilm and maximise the extraction of resources from the environment. More research is now required to examine the mechanisms that allow a biofilm to be more than the sum of its parts.

REFERENCES

Admiraal, W.1984. The ecology of estuarine sediment-inhabiting diatoms. In: *Prog. Phycol. Res.* volume 3, (eds, Round. F.E. & Chapman. D.J.). Biopress. pp. 269-322.

Black, K.S., Tolhurst, T.J., Hagerthey, S.E., and Paterson, D.M. 2002. Working with Natural Cohesive Sediments. *J Hydr. Eng.* 128 (1): 1-7.

Blanchard, G. F. and V. Cariou-Le Gall. 1994. Photosynthetic characteristics of microphytobenthos on Marennes-Oleron bay, France: Preliminary results. *J. Exp. Biol. Ecol.* 182: 1-14.

Cohn, S. A. and N. C. Disparti. 1994. Environmental factors influencing diatom cell motility. *J. Phycol.* 30: 818-828.

Consalvey, M. 2002. The structure and function of microphytobenthic biofilms. Ph.D. Thesis. University of St Andrews.

Défarge, C., Trichet, J., Jaunet, A., Robert, M., Tribble, J. and Sansone, F.J. 1996. Texture of microbial sediments revealed by cryo-scanning electron microscopy. *J. Sed.. Res.* 66: 935-947.

Eaton, J.W. & Moss, B. 1966. The estimation of numbers and pigment content in epipelic algal populations. *Limnol. Oceanog.* 11: 584-595.

Edgar, L.A. & Pickett-Heaps, J.D. 1984. Diatom locomotion. In: *Prog. Phycol. Res.* 3: 47-88, Biopress, Bristol, UK.

Guillard, R.R.L. and Ryther, J.H. 1962. Studies of marine planktonic diatoms. I. Cyclotella nana Hustedt, and Detonula confervacea (Cleve) Gran. *Can. J. Mar. Biol.* 27: 191-196

Honeywill, C., Paterson, D.M and Hagerthey, S.E. Determination of microphytobenthic biomass using pulse amplitude modulated minimum fluorescence. *Eur J. of Phycol.* (In press).

Kelly, J., Honeywill, C. and Paterson, D.M. 2001. Microscale analysis of chlorophyll-*a* in cohesive, intertidal sediments: the implications for microphytobenthos distribution.. *Mar. Biol. and Ecol.* 81: 151-162.

Kingston, M. B. 1999. Effect of light on vertical migration and photosynthesis of *Euglena proxima* (Euglenophyta). *J. Phycology* 35: 245-253.

Kromkamp, J., Barranguet, C and Peene, J. 1998. Determination of microphytobenthos PSII quantum efficiency and photosynthetic activity by means of variable chlorophyll fluorescence. *Mar. Ecol. Prog. Ser.* 162: 45-55.

Lassen, G. Ploug, H. & Jorgensen, B.B. 1992. A fibre-optic scalar irradiance microsensor: Application for spectral light measurements in sediments. *FEMS Microbial. Ecol.* 86: 247-254.

Long, S.P. and Hällgren, J-E. 1993. Measurement of CO2 assimilation by plants in the field and laboratory. In: DO Hall, J MO Scurlock HR, Bothar-Nordenkamps RC, Leegood, SPLong (eds). *Photosynthesis and Production in a Changing Environment, a field and laboratory manual.* Chapman and Hall, St. Ives, UK

Nedwell, D. B., Jickells T. D., Trimmer, M. and R. Sanders. 1999. Nutrients in Estuaries. Ed. D. B. Nedwell and D. G. Raffaelli. *Adv. Ecol. Res: Estuaries.* 29: 43-92.

Neu, T.R. 1994. Biofilms and microbial mats. In: *Biostabilisation of sediments.* (Krumbein, W.E., Paterson, D.M. and Stal, L.J., eds), BIS, Carl von Ossietzky Univ. Oldenburg, 9-15.

Oxborough, K. and Baker, N.R. 1997. An instrument capable of imaging chlorophyll a fluorescence from intact leaves at very low irradiance and at cellular and subcellular levels of organization. *Plant Cell Environ.* 20: 1473-1483.

Oxborough, K., Hanlon, A.R.M., Underwood, G.J.C. and Baker N.R. 2000. In vivo estimation of the photosystem II photochemical efficiency of individual microphytobenthic cells using high-resolution imaging of chlorophyll a fluorescence. *Limnol Oceanogr.* 45: 1420-1425.

Paterson, D.M. 1990. The influence of epipelic diatoms on the erodibility of an artificial sediment. In: *Proceedings of the 10th International Symposium on Living and Fossil Diatoms.* (Simola, H. ed) Joensuu, 1988. Koenigstein, 345-355.

Paterson, D.M. 1995. The biogenic structure of early sediment fabric visualised by low-temperature scanning electron microscopy. *J. Geol. Soc.* 152: 131-140.

Paterson, D.M. and Hagerthey, S.E. 2001. Microphytobenthos in contrasting coastal ecosystems: Biology and dynamics. In Ecological Comparisons of Sedimentary shores (K.Reise, ed), *Ecological studies*, 151: 105-125

Paterson, D.M. Yates, M.G., Wiltshire, K.H., McGrorty, S. Miles, A. Eastwood, J.E.A., Blackburn, J. & Davidson, I. 1998. Microbiological mediation of spectral reflectance from intertidal cohesive sediments. *Limnol. Oceanog*. 43(6): 1207-1221.

Paterson, D.M., 1997. Biological mediation of sediment erodibility: ecology and physical dynamics. In: *Cohesive Sediments* (Burt, N., Parker, R. and Watts, J. eds), Wiley and Sons, pp. 215-229.

Perkins, R.G., Oxborough, K., Hanlon, A.R.M., Underwood, J.C. and Baker, N.R. 2002. Can chlorophyll fluorescence be used to estimate the rate of photosynthetic electron transport within microphytobenthic biofilms? *Mar. Ecol. Prog. Ser.* 228: 47 – 56.

Revsbech, N. P., Sørensen, J., Blackburn, T. H., Lomholt, J. P. 1980. Distribution of oxygen in marine sediments measured with microelectrodes. *Limnol. Oceanogr*. 25(4): 03-411.

Richards. R.S. and Turner, R.J. 1984. A comparative study of techniques for the examination of biofilms by scanning electron microscopy. *Water Res.* 18(6): 767-773.

Sakshaug, E., Bricaud, A., Dandonneau, Y., Falkowski, P.G., Kiefer, D.A., Legendre, L. Morel, A., Parslow, J. and Takahashi, M. 1997. Parameters of photosynthesis: definitions, theory and interpretation of results. *J. Plankton Res.* 19: 1637-1670.

Sauer, J., Wenderoth, K., Maier, U.G. and Rhiel, E. 2002. Effects of salinity, light and time on the vertical migration of diatom assemblages. *Diatom Res.* 17(1), 189-203.

Taylor I S & Paterson D M 1998. Microspatial variation in carbohydrate concentrations with depth in the upper millimetres of intertidal cohesive sediments. *Est. Coast. Shelf Sci.* 46: 359-370

Underwood, G.J.C., Phillips, J. and Saunders, K. 1998. Distribution of estuarine benthic diatom species along salinity and nutrient gradients. *Eur J. Phycol.* 33: 173-183

Underwood, G. J. C. & Kromkamp, J. 1999. Primary production by phytoplankton & microphytobenthos in estuaries. *Adv. in Ecol. Res*: *Estuaries*. D. B. Nedwell & D. G. Raffaelli. Acedemic Press, 29: 93-153.

Underwood, G.J.C. & Smith, D.J. 1998. *In situ* measurement of exoploymer production by intertidal epipelic diatom dominated biofilms in the Humber estuary. In: *Sedimentary Processes in the Intertidal Zone.* (Black, K. S., Paterson, D. M., & Cramp, A. eds). Geological Society, London, Special publication. 139: 125-134.

Yallop, M.L. de Winder, B., Paterson, D.M. & Stal, L.J. 1994. Comparative structure primary production and biogenic stabilisation of cohesive and non-cohesive marine sediments inhabited by microphytobenthos. *Est. Coastal & Shelf Sci.*. 39: 565-582.

Chapter 4

STRUCTURE OF MARINE BIOFILMS

Flat Laminated Mats and Modern Marine Stromatolites

John F. STOLZ
Duquesne University
Pittsburgh, PA 15282
stolz@duq.edu

1. INTRODUCTION

Microbial biofilms in marine environments are dynamic systems exhibiting spatial and temporal heterogeneity. Their species composition may range from low to high diversity and change over time. Within the biofilm, populations of individual species are clustered together in microcolonies or bundles of filaments and can be organized into distinct layers. These communities are faced with fluctuating environmental conditions that may change diurnally, tidally, seasonally, and episodically. These changes may be reflected in the overall structure of the biofilm, especially in cases where sedimentary deposits result. In hypersaline lagoons and sabkhas this results in laminated sediments. In carbonate rich waters, where sediment trapping and binding, as well as lithification occur, the end result of these processes is a stromatolite. In this study, two different marine microbial mat systems were compared and contrasted using light and transmission electron microscopy: a flat laminated microbial mat from Laguna Figueroa, Baja California, Mexico, and lithified marine stromatolites from Highborne Cay, Bahamas. This approach, including preliminary image analysis has revealed several unique features.

2. TEM ANALYSIS

Transmission electron microscopy (TEM) has been used to study microbial communities in laminated sediments from hypersaline environments (Stolz 1983; 1984; 1990; 1994; 2000; D'Amelio et al., 1987, 1989, de Wit et al., 1994), and finegrained continental margin sediments (Ransom et al., 1998). More recently, it has been employed to study the microbial role in the sedimentation and lihtification in reef biofilms (Reitner et al., this volume) and modern marine stromatolites (Stolz et al., 2001). TEM especially lends it self to communities dominated by phototrophic bacteria as these organisms can be readily recognized by their chromatophore membranes and light harvesting structures (Stolz, 1991). Ultrastructural details such as bacterial cell wall topology (i.e., Gram negative and Gram positive) and intracellular inclusion bodies can help in the identification of microbial species and provide indications of the organism's physiological state. Although the hydrous nature of microbial biofilms can result in shrinkage artifacts, the overall topology, in particular the microbe-mineral interaction is usually preserved.

For this study two different sets of samples were examined. Flat laminated mats from Laguna Figueroa, Baja California, Mexico were collected in August of 1983 and fixed in 2.5% glutaraldehyde in seawater buffer at the site and kept cold and in the dark. They were subsequently prepared for TEM as described in Stolz (1983). The modern marine stromatolites in this study were collected by R.P. Reid in July of 1999 and fixed in 2.5% glutaraldehyde in seawater. The samples were subsequently prepared as described in Stolz et al. (2001). In both cases, orientation of the samples was maintained. Toluidene blue stained light sections were observed on a Nikon Eclipse 600 microscope. Ultrathin sections were observed on a Philips 201 TEM at 60 kV. Light images were captured using a MTI DC330 CCD camera or Optronics VI470 camera, while TEM negatives were captured using the Kodak DC120 digital camera system. All images were enhanced using Adobe Photoshop. Image analysis was done using Image I (Universal Imaging,West Chester PA).

3. FLAT LAMINATED MATS

The flat laminate mats at Laguna Figueroa, Baja California, Mexico form in a predominantly silicoclastic environment where little carbonate accretion occurs. The field site has been previously described in detail (Margulis et al., 1980; Stolz, 1990). The mats that are involved in the deposition of laminated sediments are dominated by the cyanobacterium *Microcoleus chthonoplastes*. In well developed mats, four distinct layers can be distinguished by color: a surface carotenoid rich layer, a green cyanobacterial layer, a pinkish layer comprised of anoxyphototrophic bacteria, and a salmon-colored bacteriochlorophyll *b*-enriched layer (Stolz, 1990). The mats that were examined in this study were somewhat different in that the third layer was comprised primarily of a population of diatoms in addition to the anoxyphototrophic bacteria. Figure 1 is a toluidene blue stained embedded section through the first 1.2 mm. Within the section, four different communities can be discerned. The surface community was dominated by a population of filamentous cyanobacteria (i.e., *Phormidium* sp.) and a few small diatoms (Figure 1, 2a). This layer was underlain by a community comprised of *Microcoleus chthonoplastes* (Figure 1, 2b). In this particular section there was a gap in the sediment profile, apparently the result of separation during preparation. A community of anoxygenenic phototrophic bacteria and a conspicuous abundance of large diatoms lied beneath the *Microcoleus* layer (Figure 1). Examination at the ultrastructural level revealed that these diatoms were healthy (Figure 2c). The last discernable layer, L-4, was comprised primarily of anoxyphototrophic and heterotrophic bacteria (Figure 2d).

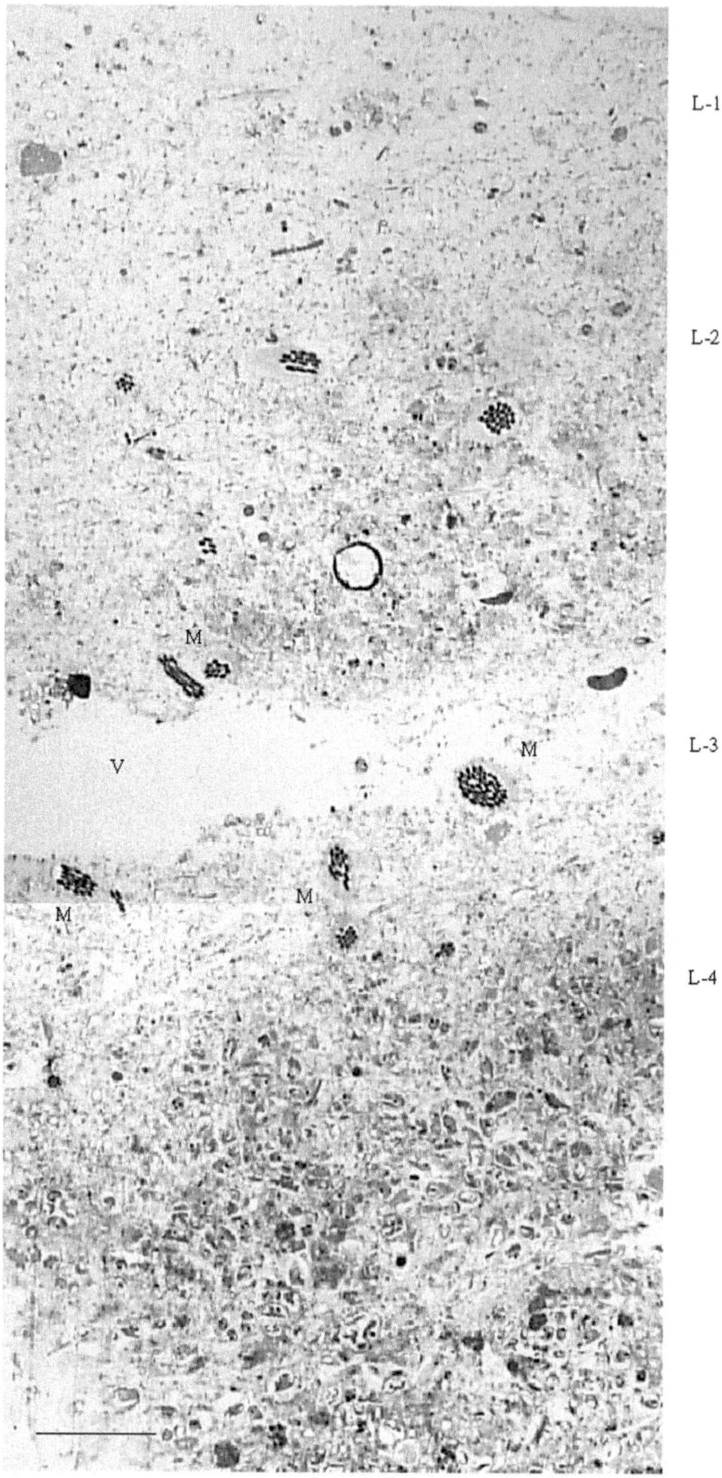
L-1
L-2
M
V
M
L-3
M
M
L-4

Figure 1. (Previous page) Toluidene blue stained light section through the top 1.2 mm of the flat laminated mats at Laguna Figueroa, Baja California, Mexico. The four layers are numbered (L-1, L-2, L-3, L-4). Bundles of *Microcoleus chthonoplastes* (M) can be seen in cross and tangential section surrounding the void (V). The conspicuous abundance of diatoms can be seen in L-3 and L-4. The bar is 100 μm.

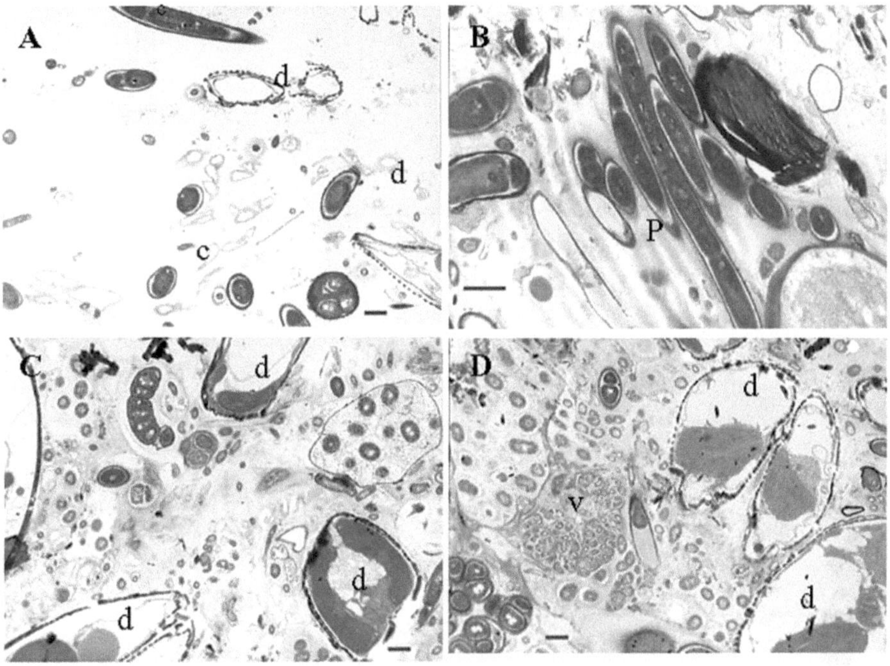

Figure 2. Transmission electron micrographs of the microbial communities in the different layers from the top 2 mm of the flat laminate mat at Laguna Figueroa, Baja California, Mexico. A) L-1, showing filamentous cyanobacteria (c) in cross and longitudinal section and numerous species of small diatoms (d) embedded in EPS, B) L-2, showing a bundle of *Phormidium* sp. (P), sheath material, and small coccoid cyanobacteria, C) L-3, showing several species of large diatoms (d), Gram negative heterotrophic bacteria and anoxyphototrophic bacteria (a), and D) L-4, showing two large diatoms (d), a microcolony of vibriod-shaped Gram negative bacteria (v), and anoxyphototrophic bacteria. All bars are 5 μm.

4. MODERN MARINE STROMATOLITES

The modern marine stromatolites at Highborne Cay, Bahamas thrive in a carbonate rich environment. Three different surface microbial communities have been identified that represent different growth stages (Reid et al., 2000, this volume; Stolz et al., 2001). The first, Type 1, is dominated by the filamentous cyanobacterium *Schizothrix gebeleinii* and is typical of the stromatolite surface during periods of rapid sediment accretion. The second, Type 2, is typified by a surface populated by a community of heterotrophic bacteria embedded in extracellular polymer underlain by filamentous cyanobacteria, and the presence of a micritic crust. The third, Type 3, results from further development of the Type 2 community and the colonization of the near subsurface by the endolithic cyanobacterium *Solentia* sp.. The boring activity of the endolithic cyanobacteria and penecontemporaneous cementation results in fused-grain laminae. The cyclical succession and subsequent burial of these surface communities results over time in the growth and laminated structure of the stromatolite (Reid et al., 2000; Visscher et al., 1998, 2000; Stolz et al., 2001).

A toluidene blue stained light section of the top 1.2 mm of a Type 3 surface community is shown in Figure 3. A thick micritic layer could be clearly seen at the surface (Figure 3A). The ooids just below the surface were colonized by the endolithic cyanobacterium *Solentia* sp. (Macintyre et al., 2000). Notice how the surface face of the ooids have been heavily bored and appear truncated (Figure 3A). Filaments of *S. gebeleinii* were also seen. Their orientation perpendicular to the surface suggests that they were migrating. Closer examination of the micritic layer revealed that it was devoid of cyanobacteria (e.g., *S. gebeleinii*) and contained a population of very small (< 1 μm.) Gram negative bacteria embedded in EPS (Figure 3B and C).

Figure 4 is a toluidene blue stained embedded section through the first 2 mm of a surface community that appeared to be in transition from a Type 3 to a Type 1 surface community. At the surface, carbonate ooids had begun to be trapped by a colonizing community of *S. gebeleinii* (Figure 4, inset 3). Further down section a former surface climax community (i.e., Type 3) could be seen (Figure 4). This layer had the greatest population density and species diversity. In addition to *S. gebeleinii*, filaments of an *Oscillatoria* sp. and a plethora of heterotrophic bacteria were seen (Figure 4, inset 1). Micrite precipitation was abundant (Figure 4, inset 2) and many of the grains in this layer were well-bored and contained the remains of *Solentia* sp.. Still further down section, the presence of unbored ooids, filaments and empty sheaths of

S. gebeleinii, and the scarcity of other species of bacteria suggested that this particular layer was once a Type 1 surface community.

5. COMPARISON

Several striking differences can be discerned between the flat laminated mats and the marine stromatolites. The first, of course, is the difference in sediment matrix. The flat laminated mats have a high percentage of organic material (i.e., sheath, EPS, cells) some chalcocidic minerals (i.e., halite, gypsum) and silicoclastics, and little carbonate. The marine stromatolites on the other hand are composed primarily of carbonate ooids and precipitates (i.e., micrite). The second, is the apparent disparity in overall biomass. An analysis based on percent area occupied by cells in thick sections provided some clues as to the overall community biomass. The surface community in flat laminated mats occupied less than 10%, while in the *Microcoleus* dominated layer, the cells occupied between 75 to 86% area (Figure 1). Remarkably, the diatomatous layer had the highest cell density, with the cells occupying greater than 90% of the total area (Figure 1). The marine stromatolite offered a stark contrast. Analysis of a Type 1 surface community showed that the ooids occupied more than 50% of the total percent area. If the sheath material was also discounted, considering it is mostly water, the total area occupied by cells was less than 12% (for a further discussion of stromatolite exopolymer see Decho et al., this volume). Even more striking was the micritic layer, where cells occupied less than 7% of the total area. The greatest cell densities were found, as expected, in the Type 3 surface community (Figure 4), just below the surface, where cells occupied greater than 37% of the total area. This observation further supports our conclusion that this indeed represents the "climax" community for this microbial ecosytem (Reid et al., 2000 this volume). The third, is species richness. The great species diversity in the flat laminated mats of Laguna Figueroa is readily apparent in light and electron micrographs (Figures 1 and 2). Quantitative data for flat laminated mats is lacking, however, as most studies have focused on specific traits (e.g., desiccation resistance) or trophic groups (i.e., phototrophic bacteria). For the flat mats at Laguna Figueroa, at least 75 species of heterotrophic bacteria (e.g., Gram postive and Gram negative bacteria), 17 species of cyanobacteria, 9 species of anoxyphototrophic bacteria, 5 species of green algae, 70 species of diatoms, 19 species of protists, and 14 species of fungi have been identified by morphology (i.e., light and electron microscopy) or culture (Brown et al., 1985; Stolz, 1990). Neubel and colleagues used three different criteria, morphology, carotenoid composition, and 16S rRNA analysis (i.e., denaturing gradient gel electrotrophoresis, DGGE), to determine species richness in eight different flat laminated microbial mats at Guerrero Negro, Baja California, Mexico (Nuebel et al., 1999). Although the three methods showed a positive linear correlation (i.e., an increase in number of morphological types correlated with an increase in 16S rRNA genes amplified and carotenoid composition) the species richness indices (i.e., Shannon-Weaver indices) did not (Nuebel et al., 1999). Most surprising was the number of 16S rRNA genes that were amplified with respect to the larger number of morphological species observed.

Species richness in modern marine stromatolites from Highborne Cay is dependent on the successional stage. The Type 1 surface community has the lowest diversity, based on morphological analysis, being comprised predominantly of *S. gebeleinii* and a few species of heterotrophic bacteria (Stolz et al., 2001). Morphometric analysis of the *Schizothrix* population in a well-developed Type 1 community (6/97 NS 11), however, indicated at least three different morphotypes (N. Paraska and J.F. Stolz, unpublished result). In addition, five different *Schizothrix* strains have been cultured and shown to be distinct by 16S rRNA sequence analysis (L. Prufert-Bebout, personal communication). These results suggest greater complexity. Type 3 climax community clearly has the greatest species diversity (Figure 3 and 4). Several different species of cyanobacteria have been identified morphologically including species of *Oscillatoria* and *Solentia*. Using MPN incubations, Visscher et al. (1998) showed that the maximum population densities of sulfate-reducing and sulfur-oxidizing bacteria were associated with a Type 3 community that lay 3 to 5 mm below the surface. Anoxygenic phototrophic and denitrifying sulfide-oxidizing bacteria were only detected in subsurface sediments 0.5 to 5 mm below the surface and at their greatest population density was also in the Type 3 community (Visscher et al., 1998). Steppe et al. (2001), taking a molecular approach, identified 21 distinct *NifH* genes (the genes encoding nitrogenase for nitrogen fixation) in sediment cores and enrichment cultures. The clones represent species of heterocystis and non-heterocystis cyanobacteria, and alpha and delta Proteobacteria. Interestingly, none of the *NifH* genes amplified from the enrichment cultures matched sequences obtained from the cores suggesting an even greater diversity (Steppe et al., 2001).

Figure 3 The top 1.2 mm of a Type 3 surface community (7/99 NS 101) from a modern marine stromatolite of Highborne Cay, Bahamas. A) Toluidene blue stained light section showing the thick surface micritic crust (M), and *S. gebeleinii* filaments (S). The ooids closest to the surface appear to be fused to the micritic crust (large arrow). Further down section the side of the ooids facing the surface has been extensively bored and recrystallized (small arrows). B) and C) Transmission electron micrographs of the Gram negative bacteria associated with the micritic crust. Bar in A 100 µm, in B and C 1 µm.

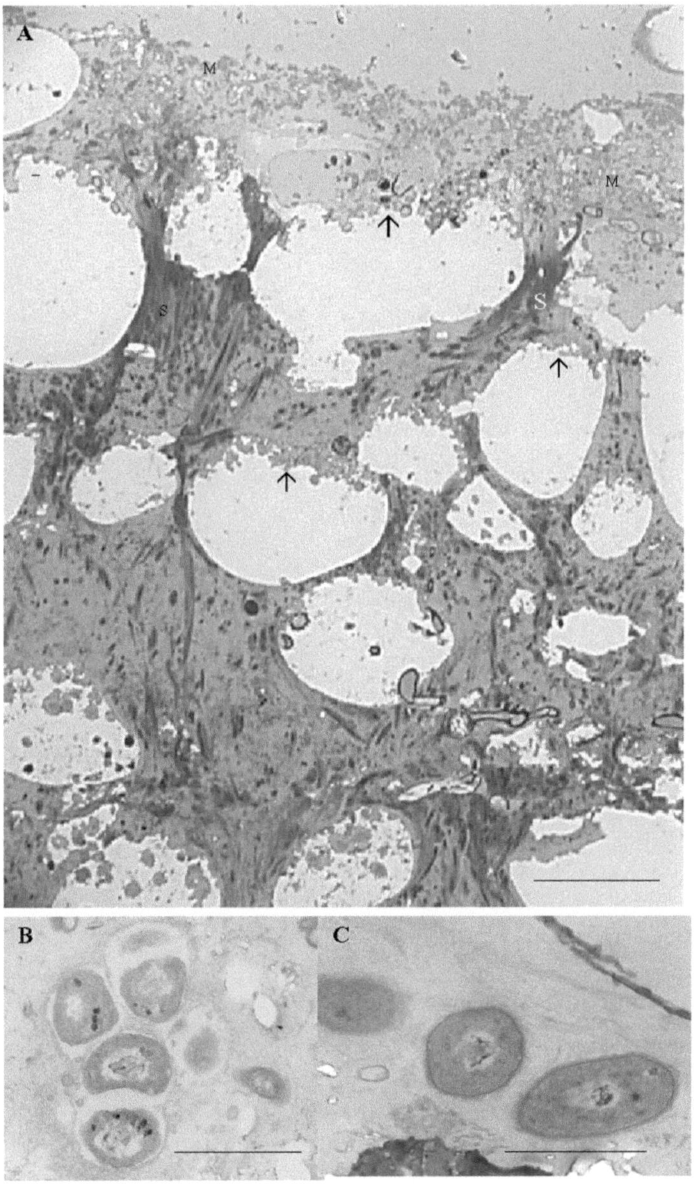

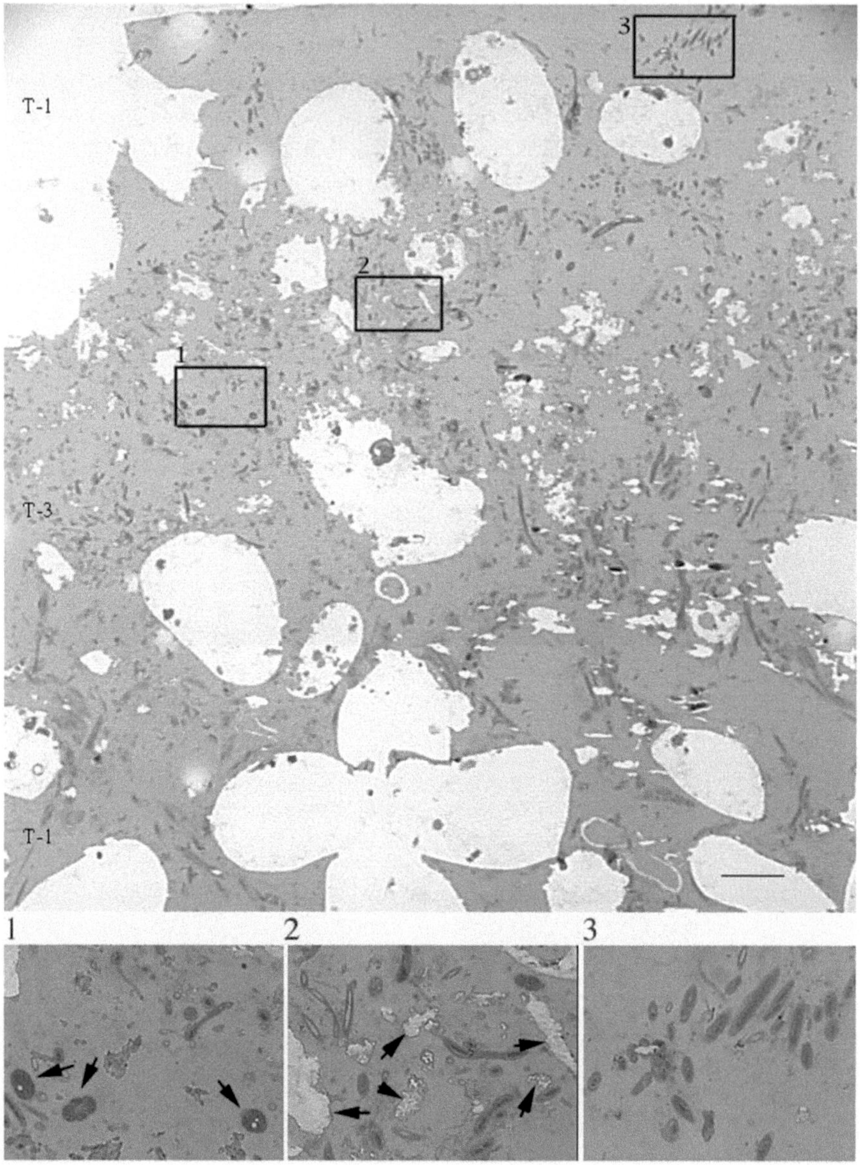

Figure 4. Toluidene blue stained light section through the top 2 mm of a modern marine stromatolite that appears to be in a transition from a Type 3 (T-3) surface community to a Type 1 (T-1)(7/99 8tnb), from Highborne Cay, Bahamas. Note the sparser population density relative to flat laminated mat shown in Figure 1. The insets are enlargements of an area 1) in the climax community showing filamentous of an *Oscillatoria* sp. (arrows), 2) in the climax community showing the presence of micrite precipitation (arrows), and 3) at the surface where filaments of *Schizothrix gebeleinii* have recolonized (beginning of Type 1 community). Bar is 100 μm.

This brief synopsis of flat laminated mats and marine stromatolites has tried to portray the truly complex and dynamic nature of these microbial ecosystems. A head to head comparison of a Type 3 community and the flat laminated mat (Figures 1 and 3) suggests that the oligotrophic waters of the Bahamas supports a less diverse community (at least for the phototrophs). Regardless, an apt analogy for species richness and abundance might be that the flat laminated mat is to a tropical rain forest as Bahamian marine stromatolites are to a temperate forest. Clearly, future studies must address the laminated nature of the mats and seasonal variation, as well as the limitation of each quantitative approach.

6. AKNOWLEDGEMENTS

The author would like to greatly acknowledge the participants of the Research Initiative on Bahamian Stromatolites (RIBS). This work was supported in part by a NSF ROA supplement to OCE9530215 (P. Reid). This is RIBS contribution 15.

REFERENCES

Brown, S., Margulis, L., Ibarra, S., and Siqueiros, D. (1985) Desiccation resistance and contaminations as mechanisms of Gaia. BioSystems 17, 337-360.

D'Amelio, E.D., Cohen, Y., and DesMarais, D.J. (1987) Association of a new type of filamentous purple phototrophic bacterium inside bundles of Microcoleus chthonoplastes in hypersaline cyanobacterial mats. Archives for Microbiology 147, 528-534.

D'Amelio, E.D., Cohen, Y., and DesMarais, D.J. (1989) Comparative functional ultrastructure of two hypersaline submerged cyanobacterial mats: Guerrero Negro, Baja California Sur, Mexico and Solar Lake Sinai, Egypt. In Microbial Mats, Y. Cohen and E. Rosenberg (eds) ASM Press, Washington DC pp 97-113.

DeWit R., Grimalt, J.O., and Hernandez-Marine M., (1994) Morphological and chemical transformations of Microcoleus chthonoplastes during early diagenesis in hypersaline microbial mats. In L.J. Stal and P. Caumette, (eds.), NATO ASI series volume G35, Springer Verlag, Berlin, pp. 69-76.

Margulis, L., Barghoorn, E.S., Giovannoni, S., Chase, D., Banerjee, S., Francis, S., Ashendorf, D., and Stolz, J., 1980. The microbial community at Laguna Figueroa, does it have Precambrian analogues? Precambrian Res 11,93-123.

Macintyre, I.G., Prufert-Bebout, L. and Reid, R.P. (2000) The role of endolithic cyanobacteria in the formation of lithified laminae in Bahamian stromatolites. Sedimentology 47,915-921.

Neubel, U., Garcia-Pichel, F., Kuehl, M., and Muyzer, G. (1999) Quantifying microbial diversity: morphotypes, 16S rRNA genes, and carotenoids of oxygenic phototrophs in microbial mats. Applied and Environmental Microbiology 65, 422-430.

Ransom, B., Bennett, R.H., Baerwald, R., Hulbert, M.H., and Burkett, P.-J. (1998) In situ conditions and interactions between microbes and minerals in fine-grained marine sediments: a TEM microfabric perspective. American Minerologist 84, 183-192.

Reid, R.P., Visscher, P.T., Decho, A.W., Stolz, J.F., Bebout, B.M., Dupraz, C., Macintyre, I.G., Paerl, H. W., Pinckney, J.L., Prufert-Bebout, L., Steppe, T.F., and DesMarais, D.J. (2000). The role of microbes in accretion, lamination and early lithification of modern marine stromatolites. Nature 406,989-992.

Stolz, J.F. (1983) Fine structure of the stratified microbial community at Laguna Figueroa, Baja California, Mexico. I. Methods of in situ study of the laminated sediments. Precambrian Research 20, 479-492.

Stolz, J.F. (1984) Fine structure of the stratified microbial community at Laguna Figueroa, Baja California, Mexico. II. Transmission electron microscopy as a diagnostic tool in studying microbial mats in situ. In Y. Cohen, R.W. Castenholz, and H.O. Halvorson, (eds.), Microbial Mats: Stromatolites, Alan R. Liss Inc., New York, pp. 23-38.

Stolz, J.F. (1990) Distribution of phototrophic microbes in the flat laminated microbial mat at Laguna Figueroa, Baja California, Mexico. BioSystems 23, 345-357.

Stolz J.F. (1991) The ecology of phototrophic bacteria. In Stolz JF (ed) Structure of Phototrophic Prokaryotes, CRC Press, Boca Raton, pp 105-123

Stolz, J.F. (1994) Light and electron microscopy in microbial mat research: an overview. In L.J. Stal and P. Caumette, (eds.), NATO ASI series volume G35, Springer Verlag, Berlin, pp. 173-182.

Stolz, J.F. (2000) The structure of microbial mats and biofilms. In Riding, R.E. and Awramik, S.M. (eds.), Microbial Sediments, Springer Verlag, Berlin, pp. 1-8.

Stolz, J.F., Feinstein, T.N., Salsi, J., Visscher, P.T., and Reid, R.P. (2001). TEM analysis of microbial mediated sedimentation and lithification in modern marine stromatolites. Am. Min. 86, 826-833

Visscher, P.T., Reid, R.P., Bebout, B.M., Hoeft, S.E., MacIntyre, I.G., and Thompson, J.A. Jr. (1998) Formation of lithified micritic laminae in modern marine stromatolites (Bahamas): the role of sulfur cycling. American Minerologist 84, 1482-1493.

Visscher, P.T., Reid, R.P., and Bebout, B.M. (2000) Microscale observations of sulfate reduction: correlation of microbial activity with lithified micritic laminae in modern marine stromatolites. Geology 28, 919-922.

Chapter 5

GROWTH, STRUCTURE AND CALCIFICATION POTENTIAL OF AN ARTIFICIAL CYANOBACTERIAL MAT

Michael KÜHL and Tom FENCHEL
Marine Biological Laboratory, University of Copenhagen
Strandpromenaden 5, DK-3000 Helsingør, Denmark.
mkuhl@zi.ku.dk

Jozef KAZMIERCZAK
Institute of Paleobiology, Polish Academy of Sciences,
Biogeology Division, Twarda 51/55, PL 00-818 Warszawa, Poland

1. INTRODUCTION

Microbial mats growing on and within light-exposed surfaces harbor complex structured microbial communities of both aerobic, microaerophilic and anaerobic microbes, which in their concerted action exhibit almost closed cycles of carbon, sulfur and other essential elements for biological growth and development (Canfield and Des Marais 1993; van Gemerden 1993; Stal 2000). Diatoms and, primarily, cyanobacteria constitute the major primary producers in most microbial mats. Anoxygenic photosynthetic bacteria (both photoautotrophs and photoheterotrophs) are also abundant, but their contribution to the primary production of mats is regarded minor (Canfield and Des Marais 1993), with the exception of certain hot spring mats (Jørgensen and Nelson 1988) and coastal mats of purple bacteria (van Gemerden *et al.* 1989). Due to the absence or very minor presence of higher organisms (macrophytes and animals), microbial mats represent well developed microbially driven ecosystems. In extreme environments, like in hypersaline or geothermal waters, microbial mat communities are relatively stable over time periods >1 year and they can develop into cm to m thick cohesive layers consisting of >90% exopolymer and cells (e.g. Krumbein *et al.* 1977; Jørgensen *et al.* 1983; DesMarais 1995). Coastal mats in temperate environments are thinner and have a more ephemeral occurrence (Stal *et al.* 1985; van Gemerden *et al.* 1989). Intertidal microbial mats play an important role for coastal morphology due to their sediment binding and stabilization properties (Krumbein *et al.* 1994). In special aquatic environments, like alkaline lakes and tropical intertidal and subtidal zones, sediment trapping by microbial mats in combination with calcification leads to formation of conspicuous solid structures like beachrock lining the intertidal of tropical lagoons (Krumbein 1979b), and laminated cushion (Logan 1961; Dravis 1983; Riding *et al.*1991; Reid *et al.* 2000) and dome shaped (Kempe *et al.* 1991) structures interpreted as living stromatolites.

Microbial mats are ideal model systems for studying the interaction and regulation of microbial life and its biodiversity. Furthermore, some microbial mats are regarded as recent analogues of stromatolites, the earliest signs of biotic communities on Earth that are preserved in the fossil record, and which exhibit a striking similarity in both macro- and microstructure to some of the present microbial mats (Grotzinger and Knoll 1999; Schopf and Klein 1992). This has been inferred from (Schopf 2000): i) an impressive structural similarity between preserved microfossils and microorganisms, especially cyanobacteria, present in recent microbial mats, and ii) comparative geological and geochemical studies of fossil and recent microbial mats. Thus, microbial mat studies not only gain fundamental principles and mechanisms of today's microbial ecology, they also allow to infer and test hypotheses on how the earliest ecosystems on Earth may have evolved and functioned. However, many, if not most, extant microbial mats do not lead to stromatolite formation (see e.g. Stal 2000) and extrapolation from studies of recent mats to Precambrian analogues should be done with care.

Most microbial mat studies have been done with natural mats either *in situ* or with samples transferred to the laboratory. Much less effort has been made to induce and study microbial mat growth in the laboratory, in part due to the difficulty to create and maintain the multitude of gradients (light, oxygen, pH, inorganic carbon, nutrients etc.) driving the organization and activity within microbial mats. Giani *et al.* (1989) described a laboratory scale device for growing hypersaline microbial mats, which was later adapted for a number of different model systems for studying intertidal microbial mats (Krumbein *et al.* 1994, pp. 492-497). A mesocosm facility for studying hypersaline microbial mats has been established at the Interuniversity Institute in Eilat, Israel by Yehuda Cohen (see http://www.univ-pau.fr/RECHERCHE/MATBIOPOL/; Abed and Garcia-Pichel 2001), and a controlled greenhouse facility for hypersaline microbial mats has been established by the NASA exobiology section in Palo Alto, California (http://exobiology.nasa.gov/microecobiogeo/html_documents/greenhouse_one.htm; Bebout et al. 2002). A benchtop system for growing artificial microbial mats in defined gradients has been developed by Pringault *et al.* (1996), and has been used extensively to study artificial mats of anoxygenic photosynthetic bacteria (Pringault *et al.* 1999a,b) and cyanobacteria (Pringault and Garcia-Pichel 2000).

The mentioned systems have mostly relied on inoculation with natural mat samples or defined cultures of phototrophic microorganisms and involved elaborate controls of environmental parameters. An alternative and simple approach was invented by Fenchel (1998a), who induced prolific microbial mat growth on top of natural coastal sediment, by experimental removal of fauna. In this chapter we review the development, structure and function of these artificially induced cyanobacterial mats, including an account of their calcification potential.

2. GRAZING AND FORMATION OF ARTIFICIAL CYANOBACTERIAL MATS

The effect of grazing on microbial mats has been studied in both hypersaline and geothermal microbial mats (reviewed by Farmer 1992). The presence of grazers does not always imply destruction of mats (McNamara 1992), and the presence of grazers can significantly affect both the diversity of phototrophs and mat structure.

In hypersaline environments insect (coleopteran and dipteran) larvae and nematodes are predominant grazers of diatoms and cyanobacteria (Gerdes *et al.* 1985; Farmer 1992), especially in mats with heterogeneous surface structures (Gerdes and Krumbein 1987). However, abundant grazing meiofauna and small crustaceans like the harpacticoid copepod *Cletocampus dietersi* seem to have little impact on hypersaline mat distribution (Farmer 1992). Cerithid gastropods (e.g. *Cerithidia californica*), the waterboatman *Trichocorix sp.*, and salt beetles (*Bledius* spp. adults and larvae) are larger conspicuous grazers of hypersaline mats (Javor and Castenholz 1984; Gerdes and Krumbein 1987). Even mat reworking by crabs and grazing fish are reported (Davies 1970; Schwarz *et al.* 1975)

Grazing larvae of the brine fly *Paracoenia turbida* can control mat development in hot springs at <40°C (Wickstrom and Wiegert 1980). Thermophilic ostracods (*Potamocypris* sp.) have been shown to facilitate, by selective grazing, the co-occurrence of two competing cyanobacteria (*Calothrix thermalis* and *Pleurocapsa minor*) in nodular mats in hot springs (Wickstrom and Castenholz 1985). Overgrowth of the nodular mat by more efficiently growing *Synechococcus* and *Oscillatoria* mats was prevented by ostracod grazing at temperatures below 47-51°C, i.e. the upper temperature limit of the ostracod population.

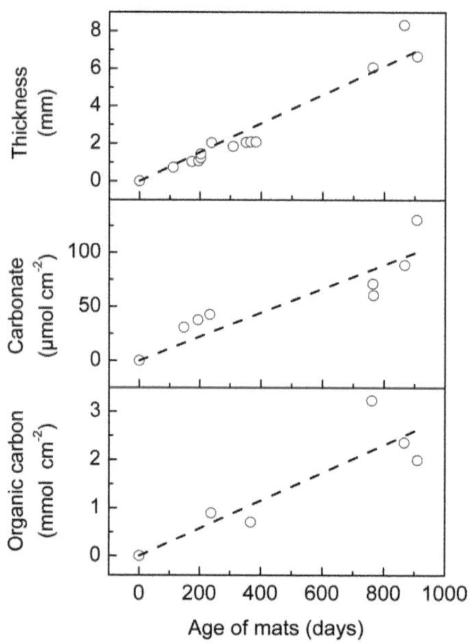

Figure 1. Increment of thickness, carbonate and organic carbon content of the artificial mats as a function of time (from Fenchel and Kühl 2000)

The role of grazers for microbial mat formation and destruction in temperate systems was studied by Fenchel (1998a). Formation of microbial mats was induced by removal of macro- and most meiofauna in coastal sediment by 20h freezing at –20 °C, and subsequent incubation in sterile seawater at constant temperature under a 10h dark-14h light cycle. Within 3-4 weeks after defaunation a cyanobacterial mat was formed on top of the sediment, which over the following 4-5 years has developed further into a complex laminated microbial mat with an accretion rate of ~2-3 mm per year, a carbonate accumulation rate of ~0.3 μmol cm^{-2} d^{-1}, and an accumulation rate of organic carbon of ~2.5 μmol cm^{-2} d^{-1} (Fenchel and Kühl 2000, Fig. 1). The accretion rates of natural hypersaline microbial mats are in the range of 0.5-5 mm per year (Bauld 1981, DesMarais 1995).

The presence of resistant meiofauna and a harpacticoid copepod *Harpacticus* sp. did not affect mat development. However, when fauna (the harpacticoid copepod *Nitocra* sp., hydrobiid snails and polychaetes) was allowed to recolonize ~4 months old mats, the mat was digested within ~2 weeks and the laminated structure changed into bioturbated sediment (Fenchel 1998a, Fig. 2B). It was concluded that normal sediments bear a potential for microbial mat formation (similar to the ones found in extreme habitats), which is, however, only realized when faunal grazing and bioturbation is depressed. These observations are interesting in the context that the origin of metazoan life, and therefore benthic fauna, apparently coincided with the decline of late Precambrian stromatolites (Garret 1970; Walter and Heys 1985), and the fact that most modern stromatolites occur under environmental conditions that restrict metazoan life.

2.1 Changes in structure, biota and diversity

The development of the artificial cyanobacterial mats was followed intensively over the first year of mat development (Fenchel 1998a,b, c). Another detailed investigation was done with 2-2.5 year old mats (Fenchel and Kühl 2000; Kühl and Fenchel 2000). On both occasions a very detailed study of mat structure and composition was done with TEM, and biovolumes were determined by stereological methods.

2.1.1 The first year after defaunation

Within 4 months after defaunation, the mats had developed into a laminated cohesive yellow-brownish matrix of cells and exopolymers, with different phototrophs dominating in different strata (Fenchel 1998b). Calcification was evident from the formation of carbonate precipitates within the cyanobacterial zone 0.2-0.3 mm below the mat surface. Both chlorophyll *a* and bacteriochlorophyll *a* concentrations increased during this period and reached levels of ~140 μg Chl *a* cm^{-2} and 20 μg Bchl *a* cm^{-2} (Fig. 2). A prominent feature was the presence in the upper 3 mm's of the mat of several layers densely populated by filamentous cyanobacteria (mainly two morphotypes of *Pseudanabaena*). Layers with vertical filament orientation were sharply separated from layers where filaments were oriented in parallel to the surface and each other. Maximum volume fractions of filamentous cyanobacteria were ~10-15% (25-30% with slime sheaths) in these layers. The total length of cyanobacterial filaments below 1 cm^2 of mat was estimated to ~14 km at this stage of development. Photosynthetic bacteria were relatively sparse and mainly confined to a layer with *Thiocapsa* colonies ~4-5 mm below the surface. Bacterial biomass accounted for ~2% of the volume in the uppermost 0.2-0.3 mm (~5 x 10^{10}) of the mat and ~0.5% in deeper layers (~5 x 10^9 bacteria ml^{-1}). Bacterial density tended to vary with cyanobacterial density, and bacteria were observed to also colonize the outer and inner surfaces of cyanobacterial sheaths.

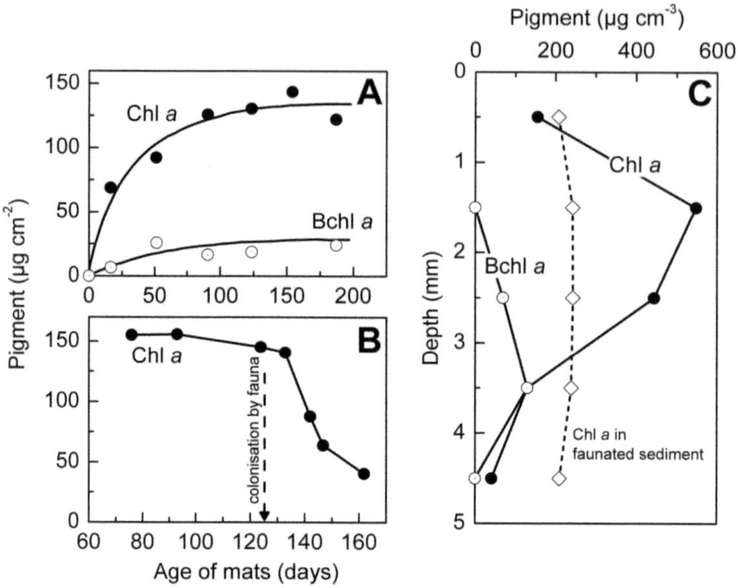

Figure 2. Accumulation of photopigments during mat development (A), and degradation after colonization by fauna (B). Depth distribution of photopigments in an artificial mat and in a faunated sediment (C) (from Fenchel 1998a).

While the mat structure and density of microorganisms was similar to that of natural cyanobacterial mats, the diversity of phototrophs was apparently lower than in natural mats (Nübel *et al.* 1999). Besides the dominating *Pseudanabaena* spp., minor occurrence of coccoid morphotypes and filamentous forms assigned to *Phormidium*, *Microcoleus* and *Calothrix* was found.

2.1.1.1 Later stages of development

Approximately 2-2.5 years after defaunation, *Microcoleus* had disappeared and *Phormidium* and *Calothrix* were more numerous in the upper 1 mm of the mat (Fenchel and Kühl 2000). Three morphotypes of *Pseudanabaena* were still the most abundant cyanobacteria. The stratification observed in young mats was preserved, but zonations in the older mat were more pronounced with a complex texture, and biovolumes were somewhat higher (Fenchel and Kühl 2000). It was estimated that the filamentous cyanobacteria below a cm^2 of mat represented a surface area of ~350 cm^2. The diversity

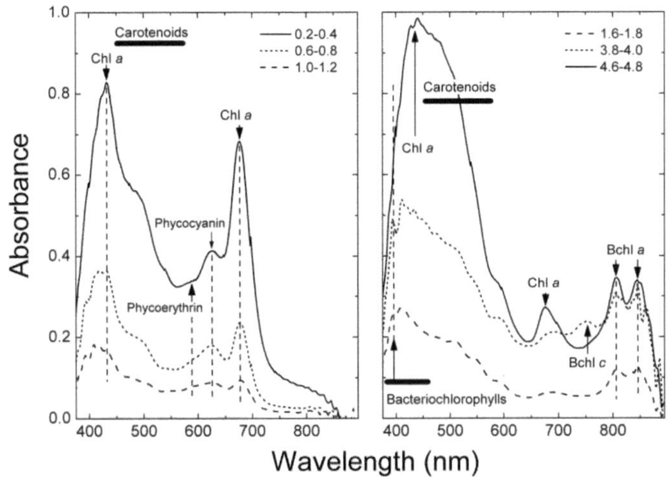

Figure 3. Spectral absorbance of various zones (denoted by depth intervals in mm) as measured by microscope spectrometry on vertical thin sections of an artificial mat (from Kühl and Fenchel 2000).

and abundance of photosynthetic bacterial layers with different photopigments was somewhat increased (Fig. 3). Several carbonate layers were present. A dense 0.2-0.4 mm thick layer of carbonates within the upper mm of the mat, a second less dense carbonate layer 1.5-2 mm below the surface, and sometimes a third layer in-between purple bacteria 3-5 mm below the mat surface. Carbonate precipitates associated with cyanobacterial sheaths were observed. The largest part of the mat was composed of empty cyanobacterial sheaths and exopolymers. The sheaths appear resistant to degradation and their accumulation seems the major cause for the slow increase in mat thickness over time (Fig. 1A). The mats continue to develop in the mesocosms, and most recently conspicuous gelatinous pinnacle structures have developed on the mat surface (Fig. 4).

Besides a stratification of cyanobacteria and anoxygenic phototrophic bacteria (Fig. 3), the mat structure exhibits 3 prominent characteristics found in extant natural mats and often interpreted as preserved in fossil material (Bauld *et al.* 1992): i) a pronounced microfabric of cyanobacterial lamina, where layers with vertical filament orientation are separated from layers with pronounced horizontal filament orientation, ii) abundance of empty cyanobacterial sheaths, which are not mineralized and together with exopolymers constitute the bulk mass of the mats, and iii) presence of distinct layers of carbonate precipitates associated with zones of high photosynthesis and zones of mainly heterotrophic activity.

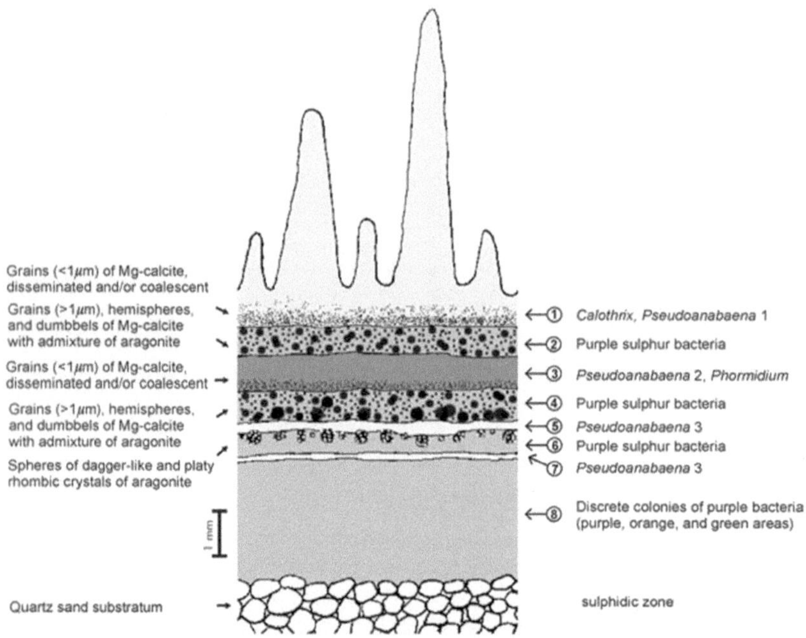

Figure 4. The vertical micro-zonation in a ~3 year old artificial cyanobacterial mat. Different characteristic zonations of phototrophs are numbered and referred to in the text. The location of calcium carbonate precipitates is indicated with arrows. Detailed LM and TEM pictures of the mats in earlier stages of development can be found in Fenchel (1998a, b) and in Fenchel and Kühl (2000). The prominent pinnacles developed within the last 12-18 months and were absent or only minor in the earlier stages of development.

3. MICROENVIRONMENTAL CHARACTERISTICS AND CARBON CYCLING

Besides a high structural similarity, the carbon cycling and microenvironmental characteristics of the artificial mat (Fenchel 1998c, Kühl and Fenchel 2000) are very similar to results obtained in natural marine mats found in hypersaline environments (e.g. Revsbech *et al.* 1983; Jørgensen 1989; Canfield and Des Marais 1993; Des Marais 1995; Wieland and Kühl 2000).

3.1 Physical and chemical microenvironment

Fiber-optic microprobes for field radiance and scalar irradiance (Kühl et al. 1994) were used for measurements of i) depth distribution of chl *a* and phycobilin fluorescence (Thar et *al.* 2001), ii) depth distribution of available light for oxygenic photosynthesis, PAR, (Kühl *et al.* 1997). Electrochemical microsensors for O_2, pH, S_{2-}, and H_2S (Kühl and Revsbech 2001) were used to quantify oxygen and sulfide dynamics in the mats.

Oxygen and sulfur cycling was closely coupled in ~4-12 months old artificial mats (Fenchel 1998c). Oxygen penetration varied from 0 in darkness to ~2.5 mm in light, where photosynthesis in a ~1 mm thick photic zone created O_2 super-saturation within the mat. In darkness, sulfide oxidation could account for all oxygen consumed and mineralization was entirely anaerobic with sulfate as the terminal e--acceptor. The main source of sulfide was in the reduced sediment below the mat but sulfate reduction was also found in the uppermost 1 mm of the mat, corresponding to the photic zone. In light, sulfide was removed by anoxygenic photosynthesis in the mat and sulfide was only present in the deepest mat layers (>3 mm). It was estimated that anoxygenic photosynthesis accounted <10% of primary production.

In ~2-2.5 year old mats, sulfide was only present below the 4-5 mm thick oxic zone in the light, and O_2 penetration in the dark was ~1 mm, allowing for some aerobic mineralization in the uppermost layers of the mat (Fig. 5). pH in the photic zone varied from pH 8.8 in light to pH 8.3 in darkness. Strong attenuation of visible light confined the zone of oxygenic photosynthesis to the upper 3 mm's of the mat. Oxygenic photosynthesis, light attenuation and photopigment fluorescence showed a heterogeneous depth distribution with several peaks corresponding to the different layers of cyanobacteria carbonate precipitates (Fig. 6). Scattering in the matrix of exopolymers, carbonate precipitates, and cells resulted in a local maximum in scalar irradiance in the upper 0.2-0.5 mm of the mat.

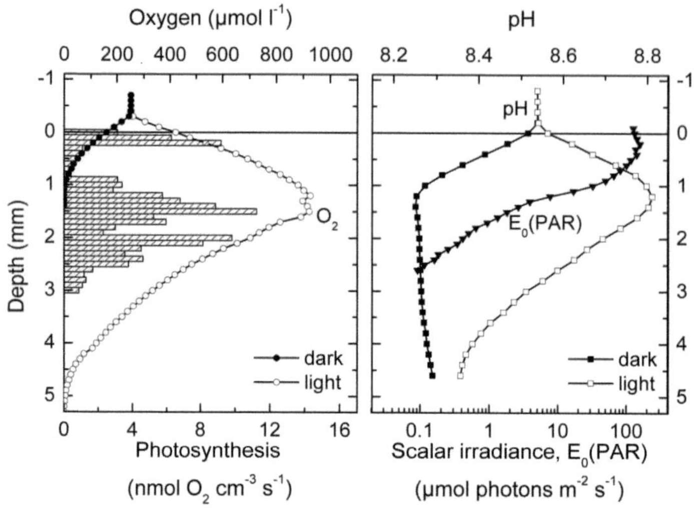

Figure 5. Microprofiles of O2, pH, scalar irradiance (E0(PAR), and gross oxygenic photosynthesis (bars) in a ~2.5 year old 7-8 mm thick artificial mat. Incident irradiance in light was 413 µmol photons m^{-2} s^{-1} (from Kühl and Fenchel 2000).

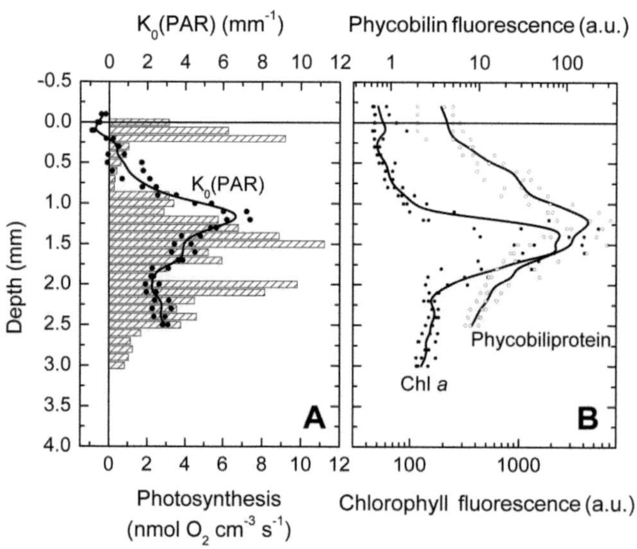

Figure 6. Depth dependence of (A) oxygenic photosynthesis (bars) and scalar irradiance attenuation, and (B) chlorophyll *a* and phycobiliprotein fluorescence in a ~2.5 year old and 7-8 mm thick artificial mat (from Kühl and Fenchel 2000).

3.2 Cycling of carbon and other elements in the mats

Net uptake (1.244 ± 0.305 μmol cm^{-2} h^{-1}) in light and net production (1.235 ± 0.236 μmol cm^{-2} h^{-1}) in dark of inorganic carbon were almost identical in the mat (Fenchel 1998c). This is in accordance with the slow accumulation of organic carbon (estimated to ~10% of net C fixation) within the mats (Fig. 1), and points to an efficient recycling of carbon, sulfur, oxygen and other elements in the mat.

Chemotrophic and photosynthetic sulfur bacteria reoxidized most sulfide produced by sulfate reduction. The dynamics of sulfide was, however, buffered by a pool of intermediate oxidation products like elemental sulfur (S^0) and of iron sulfide (FeS) temporary immobilizing some of the free sulfide. Permanent burial of sulfide as pyrite (FeS_2) was not measured but is probably low in the artificial mats. Carbonate precipitation was estimated to ~1% of inorganic carbon uptake and most probably is stimulated by the high pH of ~9 found in the photic zone during illumination (Fig. 5, see also 4).

Net oxygen fluxes in light and darkness did not balance and were 2-3.5 times lower than the corresponding inorganic carbon fluxes. In the light, oxygen is also used to reoxidize i) reduced pools of carbon and sulfur accumulating during darkness, and ii) reduced storage products accumulating in the cyanobacteria during illumination periods. Furthermore, oxygen respiration within the photic zone is also stimulated in light due to excretion of photosynthates by the cyanobacteria, which are surrounded by a high density of heterotrophic bacteria. Whether this apparent cross-feeding of heterotrophs, often observed in mats (e.g. Canfield and Des Marais 1993; Kühl *et al.* 1996; Paerl *et al.* 2000), also results in significant CO_2 transfer to the cyanobacteria is unknown, but together with carbonate precipitation in the vicinity of the cyanobacteria this could help alleviate inorganic carbon limitation induced by the high pH in the photic zone.

4. CALCIFICATION POTENTIAL OF THE ARTIFICIAL CYANOBACTERIAL MATS

The mechanism of calcification in cyanobacterial mats, and the use of such information as a key to elucidate the origin of ancient sedimentary structures known as stromatolites, and of other carbonate deposits interpreted as products of calcified benthic cyanobacteria (Kazmierczak *et al.* 1996; Riding 2000), are still poorly understood and controversial (Krumbein 1979a; Ginsburg 1991; Grotzinger and Knoll 1999; Reid et al. 2000). Hence, the possibility to study $CaCO_3$ precipitates in well-characterized artificial cyanobacterial mats offers new insights to the origin of $CaCO_3$ deposits, but also to the genesis of their fossil counterparts.

4.1 Distribution and morphology of $CaCO_3$ precipitates

First $CaCO_3$ precipitates appeared in the mats ~2 months after defaunation of the original sediment (Fenchel 1998a). After 4 months $CaCO_3$ deposits occurred as 100-200 μm large white granules, which in ~1 year old mats had grown together to form an almost continuous, but perforated, 100-200 μm thick plate (cf. Fig. 1F in Fenchel 1998a). After ~2 years growth, a second carbonate layer had been generated in the mats (Fenchel and Kühl 2000). The carbonate precipitates were investigated with cryo-microtome sectioning techniques, scanning electron microscopy (SEM), X-ray diffractometry (XRD), electron microprobe analysis (EDS), and stable carbon isotope ratio measurements (^{13}C). The basis for these studies were ~2 year old mats. Selected control analyses were carried out on carbonates in ~3 year old mats.

The results of morphological and mineralogical investigations of the $CaCO_3$ particles identified in the mat vertical section are summarized in Fig. 4. They are based on cryo-microtome sections (Fig. 7A) and SEM images of the $CaCO_3$ precipitates occurring in particular biozones (Figs. 7B-E, 9, and 10). The distribution of $CaCO_3$ precipitates within the mats was non-random and associated to particular microzones (numbered in Fig. 4). Several carbonate layers could be distinguished in the ~7 mm thick mat beginning from their pinnacled uppermost biozone down to the sandy sulfidic substratum.

At the basis of the uppermost biozone composed of *Calothrix* and *Pseudanabaena* 1 (zone 1 in Fig. 4; Fig. 7 A1) an almost continuous dense layer of spherical/subspherical submicron-sized $CaCO_3$ grains occurred. The grains were distributed both on the slime sheaths of living cyanobacterial filaments and within the extracellular polymers (EPS) surrounding the sheaths. The 50-750 nm-sized particles occurred either as irregular individual bodies or as loosely coalesced chain-like groups (Fig. 7B). Larger clusters of similar particles were occasionally observed (Fig. 7C). X-ray diffractometry showed that this carbonate layer was composed exclusively of high Mg-calcite enclosing 15-17 mole% $MgCO_3$ (Fig. 8, curve *a*). Interestingly, electron microprobe analysis (EDS) revealed that in the ~3 years old mats the same layer now also contained a mixture of amorphous magnesium silicates.

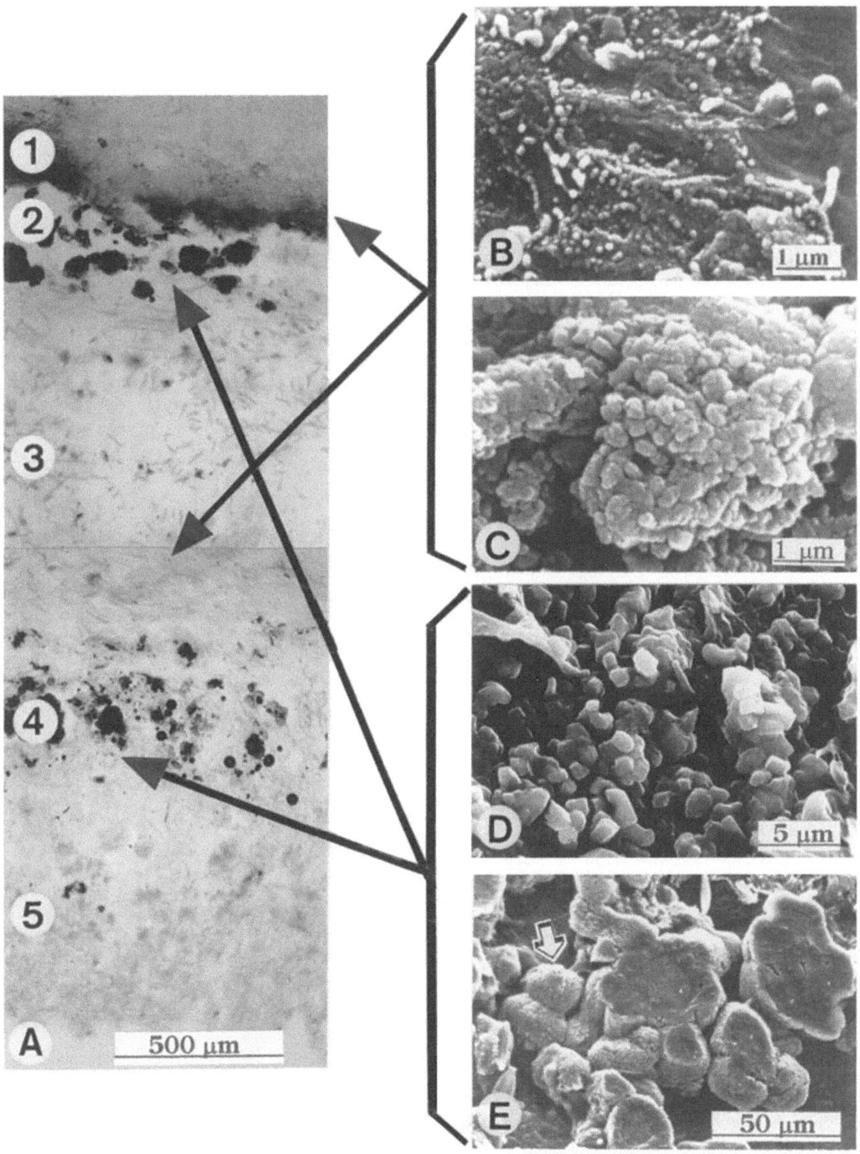

Figure 7. (A) Vertical cryotome section of a ~2 year old mat showing the location of $CaCO_3$-rich layers (arrows). Numbers indicate particular layers corresponding to those marked in Fig. 4. (B, C) Submicrometer-sized Mg-calcite grains dispersed (B) or forming clusters (C) at the bottom of the living cyanobacterial layer (zones 1 and 3). (D) Larger (>1 µm) anhedral and subhedral grains of Mg-calcite from zones 2 and 4. (E) Hemispheres and dumbbells (arrow) of Mg-calcite with admixture of aragonite found in zones 2 and 4.

In the underlying biozone composed of purple sulfur bacteria and heterotrophic bacteria (zone 2 in Fig. 4) and separated from the top cyanobacterial zone by a sharp boundary, a carbonate layer with much larger anhedral and subhedral 2-4 µm sized $CaCO_3$ grains occurred (Fig. 7 A2). The grains showed a tendency to develop crystal faces (Fig. 7D), button-like 20-110 µm large hemispheres (Fig. 7E; Fig. 9A-B), as well as 20-50 µm long peanut- and dumbbell-like bodies (Fig. 7E; Fig. 9A, C, D). The hemispheres were often coalesced into larger plate-like structures. X-ray diffractometry of the layer showed up to 90% of high Mg-calcite (with ~15 mole % $MgCO_3$) and ~10% aragonite (Fig. 8, curve *b*). The carbonate particles in this zone were often entangled in felt-like remnants of strongly degraded cyanobacterial sheaths.

A distinct deep green colored cyanobacterial layer composed of filaments of *Pseudanabaena* 2 and *Phormidium* (zone 3 in Fig. 4; Figs. 7A, 9) was found in deeper mat layers. At the base of this zone a carbonate layer comprised of ultra small high Mg-calcite particles occurred on the cyanobacterial sheaths and in the EPS. The particles were practically identical with those from the base of the uppermost layer, although the density of their distribution was much lower (compare Figs. 7A, 9).

In a deeper zone with purple sulfur bacteria (zone 4 in Fig. 4), a layer of relatively large carbonate particles occurred (Figs. 7A, 9). The particles comprised >1 µm large anhedral and subhedral grains, many with well-developed crystal faces, button-like hemispheres and dumbbell-like bodies. There were many imperfect peanut shaped particles, which were particularly numerous in places rich in felt-like remnants of degraded cyanobacterial sheaths (Fig. 9C, D). Like in zone 2, the carbonate particles in this layer were composed of high Mg-calcite with admixture of aragonite.

The deepest located carbonate bodies occurred in a zone of purple sulfur bacteria (zone 6 in Fig. 4) located between two thin zones composed of *Pseudanabaena* 3 (zones 5 and 7 in Fig. 4). The carbonate precipitates were 50-120 µm wide spherical and sub-globular aggregates composed of i) dagger-like crystals (Fig. 10A, B), which resembled trigonal prism crystallites described by Given and Wilkinson (1985), or ii) sub-globular aggregates of platy-rhombic crystals bearing a character of trigonal prisms with rhombic termination (Fig. 10C, D). Aggregates of dagger-like crystals and rhombic plates were relatively rare and most of them were aragonitic, sometimes with small amounts of high Mg-calcite.

4.2 The origin of calcium carbonate precipitates

Saturation indices calculated with respect to calcite, aragonite, and dolomite showed very high supersaturation values in the uppermost 2-3 mm's of the mat. The high values were due to high pH and carbonate alkalinity levels generated by microbial metabolism in the mats and to increased concentration of Ca^{2+} and Mg^{2+} liberated from decaying Ca- and Mg-enriched cyanobacterial sheaths and exopolymers. The mode of occurrence of calcium carbonate precipitates in the studied mats indicates two different processes involved in their formation within two different microbial zones:

4.2.1 Calcium carbonate precipitation in vicinity of living cyanobacteria.

Precipitation of ultra small high Mg-calcite grains was found in the cyanobacterial layers of biozones 1 and 3 (Fig. 4). The nucleation and precipitation of calcium carbonate deposits on and/or within the mucus sheaths surrounding living cyanobacteria is a spontaneous process resulting from the high pH induced by photoassimilatory uptake of CO_2 and/or HCO_3^-, and the resulting increase in saturation level with respect to the solubility product of common calcium carbonate minerals (Pentecost and Bauld 1988; Merz 1992; Merz and Zankl 1993; Ferris *et al.* 1994, 1997; Schultze-Lam *et al.* 1996).

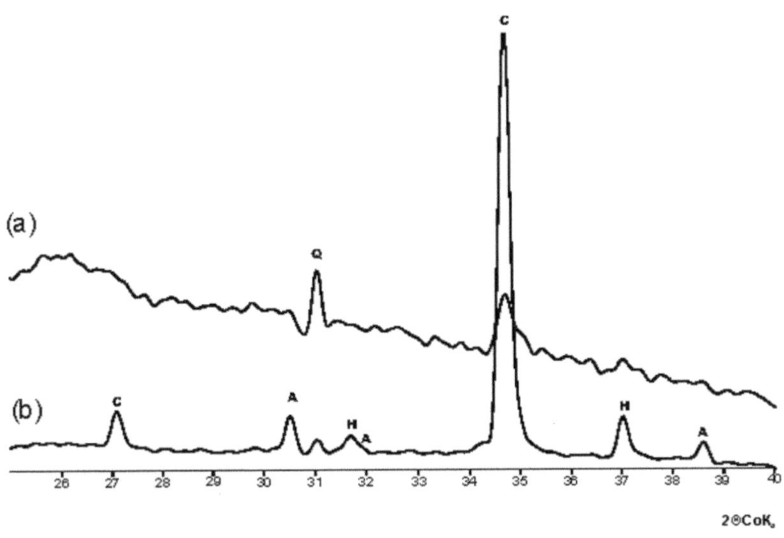

Figure 8. X-ray diffraction spectra of mineral components from mat zones 1 (a) and 2 (b). C - Mg-calcite, A - aragonite, H - halite, Q - quartz (sand grains).

The presence of ultra small high Mg-calcite particles within the two active cyanobacterial layers is related to the high photosynthetic rates observed within these zones. The $CaCO_3$ saturation level within cyanobacterial mats and surrounding waters seems to be the major factor controlling the calcification potential of microbial mats (Simkiss 1986; Kempe and Kazmierczak 1990). Table 1 shows saturation indices (SI) calculated for the cultured mats and the original sediment. The calculations indicate that both the overlaying water and the mat interiors down to the depth of calcium carbonate precipitation (~2.5 mm) were extremely supersaturated with respect to the solubility product of calcite, aragonite and dolomite. The supersaturation was much lower in the original sediments from Nivå Bay, Øresund (Table 1), mostly due to its significantly lower pH, as compared with the apparent pH increase in the culture medium during the mat growth. SI values measured in various natural settings sustaining calcifying benthic cyanobacterial mats (e.g. Kempe and Kazmierczak 1990; Merz-Preiß] and Riding 1999) have shown that SI >0.8 is the environmental saturation threshold necessary for inducing $CaCO_3$ precipitation (calcite or aragonite) by living cyanobacterial mats. The $CaCO_3$ saturation in the natural environment of the studied mats (as in the majority of other "normal"marine environments) is below this value and therefore cannot support their *in vivo* calcification.

Highly viscous media are known to influence the kinetics of calcium carbonate crystallization by slowing the ion diffusion rate (Given and Wilkinson 1985). Calcite precipitation instead of aragonite is favored in such a situation (Given and Wilkinson 1985; Buczynski and Chafetz 1991). The high viscosity of the exopolymers can explain the extremely small size and spherical shapes of high Mg-calcite grains precipitated in the active cyanobacterial layers *in vivo*. Amorphous $CaCO_3$ nanoparticles in more liquid media are rather unstable and quickly transform into macrocrystals (Cölfen and Antonietti 1998; Cölfen 2001).

4.2.2 Calcium carbonate precipitation associated with degrading cyanobacteria

The second mode of calcium carbonate precipitation was associated with degradation of cyanobacterial biomass in zones containing purple bacteria and heterotrophic bacteria (Fig. 4). The larger dimensions of the carbonate particles in this zone and their mixed high Mg-calcite/aragonite mineralogy, is probably related to a less viscous precipitation medium, where amorphous precursor particles transformed faster into superstructured grains composed of ordered nanocrystals (Cölfen 2001; Cölfen and Antonietti 1998). The admixture of aragonite in the high Mg-calcite that formed the bulk of the grains in the two uppermost purple zones, and particularly the presence of pure aragonitic bodies in zone 6, is in accordance with a less viscous precipitation microenvironment as has been postulated for such grain morphologies by Buczynski and Chafetz (1991).

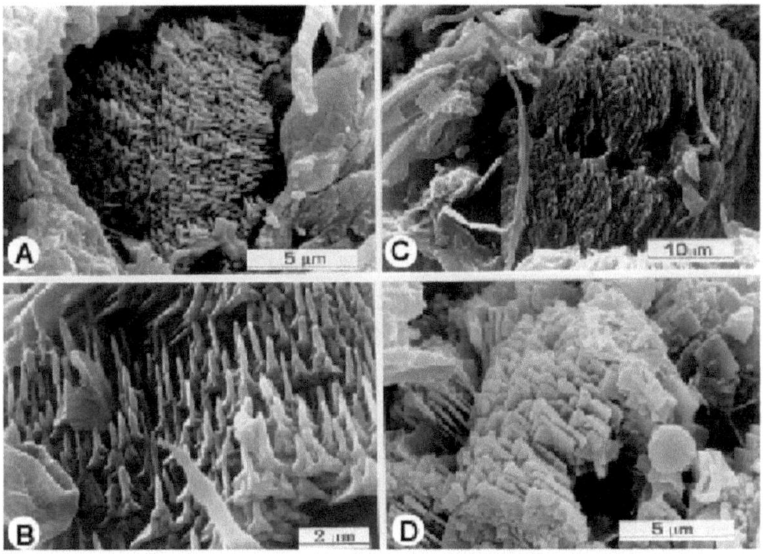

Figure 9. (A) Examples of coalesced Mg-calcite/aragonite hemispheres associated with peanut shaped bodies (arrow); note also numerous frustules of diatoms (*Navicula*). (B) A group of hemispheres showing significant size differences. (C, D) - Examples of imperfect Mg-calcite/aragonite peanut shaped particles irregularly distributed within felt-like remnants of cyanobacterial sheaths (C, from zone 2; D, from zone 4 in Fig. 1). All from a ~2 year old mat.

Bacterial mediation has been suggested instrumental in the formation of $CaCO_3$ precipitates obtained from cyanobacterial mats decaying in calcium rich solutions (Buczynski and Chafetz 1991; Chafetz and Buczynski 1992). Particle morphologies produced in these experiments are strikingly similar, if not identical, to those generated in our artificial mats. Very similar $CaCO_3$ particles have also been found in cultures of heterotrophic bacteria growing in media with calcium carbonate supersaturation (Morita 1980; Riege *et al.* 1991; Rivadeneyra *et al.* 1998; Gonzáles-Muñoz *et al.* 2000; Knorre and Krumbein 2000; Warren *et al.* 2001). However, neither of these studies provided direct evidence for an active participation of bacteria in $CaCO_3$ precipitation, and a whole spectrum of almost identical $CaCO_3$ morphs was obtained from sterile solutions supersaturated with calcium carbonate (Tai and Chen 1998); in particular when polymers were added (Cölfen and Antonietti 1998; Cölfen 2001).

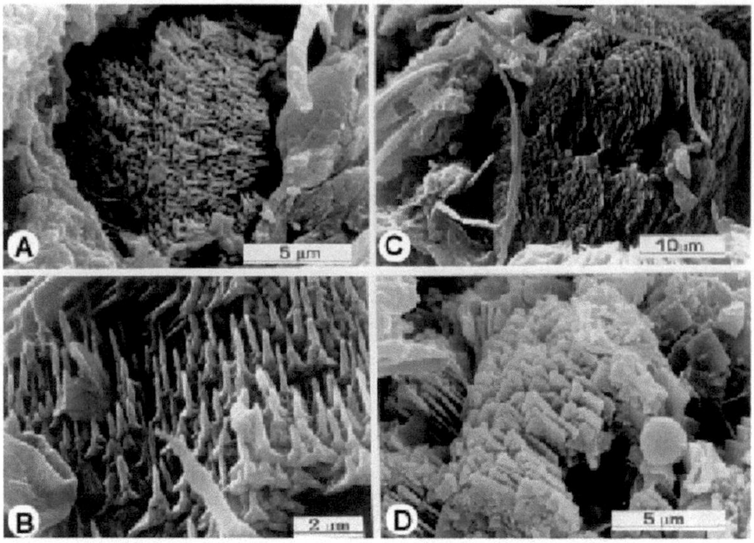

Figure 10. (A, B) Aggregates of dagger-like aragonite crystals. (C, D) Aggregates of platelet rhombic aragonite crystals. All from zone 6 (see Fig. 4) of a ~3 year old mat.

Thus, the role of bacteria in $CaCO_3$ precipitation in the mat could be to supply the supersaturated medium with polymers effective in crystallization control, and, in some cases, to serve as a nucleation centers for the $CaCO_3$ crystals. Furthermore, bacteriolysis of cyanobacterial biomass may significantly increase calcium saturation levels inside mats by releasing Ca^{2+} and Mg^{2+} cations concentrated in cyanobacterial sheaths and EPS (Somers and Brown 1978; Amemiya and Nakayama 1984; Decho 1990). This process may be of key importance in $CaCO_3$ precipitation both in lacustrine and marine cyanobacterial mats (Kazmierczak and Kempe 1998; Stal 2000; Paerl *et al.* 2001). The existence of two different $CaCO_3$ precipitating environments in the artificial mats was also evident from the stable carbon isotope analysis of the various $CaCO_3$ layers. Layers generated in the living cyanobacterial zones showed ^{13}C values ~2‰ heavier than in layers of decaying cyanobacterial biomass. In accordance with the stable carbon isotope fractionation within living cyanobacterial mats, i.e. preferential photosynthetic uptake of ^{12}C from the inorganic carbon pool (Calder and Parker 1973; Miller et al. 1990), this confirms that the carbonate species used for $CaCO_3$ precipitation must have been relatively enriched in ^{13}C. In turn, the relatively lighter ^{13}C values characterizing the $CaCO_3$ generated in the zones with purple and heterotrophic bacteria suggest input to the inorganic carbon pool of lighter carbon from respiration, i.e. remineralization of cyanobacterial biomass.

Table 1. Hydrochemistry and the carbonate system in the artificial mat (A and B) and in the Nivå Bay, Øresund, Denmark (the origin of precursor material for the artificial mats).

Artificial mats

A) Calculated for S = 21‰ and T = 20°C Ca: 6.0; Mg: 31.8; K: 5.16; Na: 270.9; SO_4: 15.6 (all mmol/l)

	Day above mat	Day 1 mm depth	Day 2 mm depth	Night 1 mm depth	Night 2 mm depth
PH	8.54	8.82	8.73	8.25	8.26
A_c	9.60	10.10	6.58	10.10	6.58
$SI_{Calcite}$	1.48	1.73	1.48	1.18	1.00
$SI_{Aragonite}$	1.33	1.58	1.33	1.03	8.85
$SI_{Dolomite}$	3.79	4.27	3.78	3.11	2.76
PCO_2	3.04	3.35	3.43	2.72	2.92

B) Calculated for S = 27‰ and T = 20°C Ca: 12.6; Mg: 39.2; K: 4.57; Na: 259.4; SO_4: 14.6 (all mmol/l)

	Day above mat	Day 1 mm depth	Day 2 mm depth	Night 1 mm depth	Night 2 mm depth
PH	8.54	8.82	8.73	8.25	8.26
A_c	9.60	10.10	6.58	10.10	6.58
$SI_{Calcite}$	1.77	2.01	1.77	1.48	1.30
$SI_{Aragonite}$	1.62	1.87	1.62	1.33	1.45
$SI_{Dolomite}$	4.14	4.63	4.13	3.48	3.12
PCO_2	3.07	3.38	3.46	2.75	2.95

Nivå Bay

S = 22 ‰ and T = 15°C; Ca: 6.3; Mg: 33.3; K: 6.3; Na: 295.4; SO_4: 17.6 (all mmol/l)

A_c	8.49	mmol/l	$SI_{Calcite}$	0.67
PH	7.80		$SI_{Aragonite}$	0.52
PCO_2	2.35		$SI_{Dolomite}$	2.00

S = salinity; T = temperature; Ac = carbonate alkalinity (mmol/l); pCO_2 = partial pressure of CO_2. Saturation Indices (SI) were calculated according to the formula:
$SI = \log ([Ca^{2+}] \times [CO_3^{2-}]/K)$ with the software PHREEQE (Parkhurst et al. 1980).

5. SUMMARY

The induction of prolific microbial mat growth by removal of fauna shows that grazing and bioturbation are major controlling factors preventing microbial mat development in the majority of extant benthic environments. The integrity and structure of microbial mats is regulated by strong gradients of light and essential chemicals and by hydrodynamic factors affecting diffusive fluxes and boundary layers. When kept under constant temperature, flow, salinity, and light-dark periods, complex laminations of phototrophic microorganisms and EPS developed in the mat. The EPS structure in combination with the metabolic activity of phototrophs and heterotrophs lead to calcification within certain microzones, each with characteristic sizes and shapes of carbonate. Faunal grazing and especially bioturbation destroys and prevents formation of such patterns.

Artificial cyanobacterial mats induced by removal of fauna represent a simple to use model system for studying the microbial ecology and biogeochemistry of microbial mat formation and development. The mats exhibit the structure and microenvironment of natural mats, and carbon, sulfur and oxygen cycling within the artificial mats is similar to natural ones. Future studies should focus on how the boundary conditions for incubation after removal of fauna affect the composition and structure of developing microbial mats. Also the distribution and dynamics of intermediate sulfide oxidation products and of iron and nitrogen species warrant more investigation. Especially, the microenvironmental controls of calcification potential in the mats under different boundary conditions should be investigated in more detail; e.g. with Ca^{2+} microsensors, which have proven valuable tools in studies of calcifying symbioses (DeBeer *et al.* 2000; Köhler-Rink and Kühl 2000) and biofilms (Hartley *et al.* 1996). The described mats are still actively growing, and we would be happy to share moderate amounts of samples with interested colleagues.

6. ACKNOWLEDGEMENTS

Our studies were funded by the Danish Natural Science Research Council (MK and TF) and the Foundation for Polish Science (JK). Thanks are extended to I. Duun, A. Glud, S. Kempe, C. Kulicki, M. Kuzniarski, B. Lacka, and K. Malkowski for technical assistance and discussions. We thank Wolfgang Krumbein for hosting an excellent workshop on microbial mats and for taking the initiative to this book.

REFERENCES

Abed, R.M.M. and Garcia-Pichel, F. (2001) Long-term compositional changes after transplant in a microbial mat cyanobacterial community revealed using a polyphasic approach. *Environ. Microbiol.* 3, 53-62.

Amemiya,Y. and Nakayama, O. (1984) The chemical composition and metal adsorption capacity of the sheath material isolated from *Microcystis*, Cyanobacteria. *Jap. J. Limnol.* 45, 187-193.

Bauld, J. (1981) Occurrence of benthic microbial mats in saline lakes. *Hydrobiologia* 81, 87-111.

Bauld, J., D'Amelio, E., and Farmer, J.D. (1992) Modern microbial mats. In: Schopf, J.W. and Klein, C. (eds.), *The Proterozoic Biosphere*. Cambridge Univ. Press, New York, pp. 261-269.

Bebout, B. M., Carpenter, S., Des Marais, D. J., and 12 co-authors (2002) Long-term manipulations of intact microbial mat communities in a greenhouse collaboratory: Simulating Earths present and past field environments. Astrobiology 2, 383-402.

Buczynski, C. and Chafetz, H.S. (1991) Habit of bacterially induced precipitates of calcium carbonate and the influence of medium viscosity on mineralogy. *J. Sed. Petrol.* 61, 226-233.

Calder, J.A. and Parker, P.L. (1973) Geochemical implications of induced changes in C13 fractionation by blue-green algae. *Geochim. Cosmochim. Acta* 37, 133-140.

Canfield, D.E. and DesMarais, D.J. (1993) Biogeochemical cycles of carbon, sulfur, and free oxygen in a microbial mat. *Geochim. Cosmochim. Acta* 57, 3971-3984.

Chafetz, H.S. and Buczynski, C. (1992) Bacterially induced lithification of microbial mats. *Palaios* 7, 277-293.

Cölfen, H. (2001) Double-hydrophilic block copolymers: Synthesis and application as novel surfactants and crystal growth modifiers. *Macromol. Rapid Commun.* 22, 219-252.

Cölfen, H. and Antonietti, M. (1998) Crystal design of calcium carbonate microparticles using double-hydrophilic block copolymers. *Langmuir* 4, 582-589.

Davies, G.R. (1970) Algal-laminated sediments, Gladstone Embayment, Shark Bay, Western Australia. *Am. Soc. Petr. Geol. Mem.* 13, 169-205.

Decho, A. (1990) Microbial exopolymer secretions in ocean environments: their role(s) in foodwebs and marine processes. *Oceanogr. Mar. Biol. Annu. Rev.* 28, 73-154.

DeBeer, D., Kühl, M., Stambler, N., and Vaki, L. (2000) A microsensor study of light enhanced Ca^{2+} uptake and photosynthesis in the reef-building coral *Favia* sp. *Mar. Ecol. Progr. Ser.* 194, 75-85.

Des Marais, D.J. (1995) The biogeochemistry of hypersaline microbial mats. *Adv. Microb. Ecol.* 14, 251-274.

Dravis, J.J. (1983) Hardened subtidal stromatolites, Bahamas. *Science* 219, 385-387.

Farmer, J. (1992) Grazing and bioturbation in modern microbial mats. In: Schopf, J.W. and Klein, C. (eds.), *The proterozoic biosphere*. Cambridge Univ. Press, New York, pp. 295-297.

Fenchel, T. (1998a) Formation of laminated cyanobacterial mats in the absence of benthic fauna. *Aq. Microb. Ecol.* 14, 235-240.

Fenchel, T. (1998b) Artificial cyanobacterial mats: structure and composition of the biota. *Aq. Microb. Ecol.* 14, 241-251.

Fenchel, T. (1998c) Artificial cyanobacterial mats: cycling of C, O, and S. *Aq. Microb. Ecol.* 14, 253-259.

Fenchel, T. and Kühl, M. (2000) Artificial cyanobacterial mats: growth, structure, and vertical zonation patterns. *Microb. Ecol.* 40, 85-93.

Ferris, F.G., Wiese, R.G., and Fyfe, W.S. (1994) Precipitation of carbonate minerals by microorganisms: Implications for silicate weathering and the global carbon dioxide budget. *Geomicrobiol. J.* 12, 1-13.

Ferris, F.G., Thompson, J.B., and Beveridge, T.J. (1997) Modern freshwater microbialites from Kelly Lake, British Columbia, Canada. *Palaios* 12, 213-219.

Garret, P. (1970) Phanerozoic stromatolites: noncompetitive restriction by grazing and burrowing animals. *Science* 169, 171-173.

van Gemerden, H., Tughan, C.S., de Wit, R., and Herbert, R.A. (1989) Laminated microbial ecosystems on sheltered beaches in Scapa Flow, Orkney Islands. *FEMS Microbiol. Ecol.* 62, 87-102.

van Gemerden, H. (1993) Microbial mats: a joint venture. *Mar. Geol.* 113, 3-25.

Gerdes, G., Spira, J., and Dimentman C. (1985) The fauna of the Gavish Sabkha and the Solar Lake – a comparative study. In: G.M. Friedman and W.E. Krumbein (eds.), *Hypersaline Ecosystems: The Gavish Sabkha*, Springer, New York, pp. 322-345.

Gerdes, G. and Krumbein, W.E. (1987) Biolaminated deposits. In: S. Bhattacharji, G.M. Friedman, H.J. Neugebauer, and A. Seilacher (eds.), *Lecture notes in Earth sciences*, Springer, Heidelberg, Vol. 9, 1-183.

Giani, D., Seeler, J., Giani, L., and Krumbein, W.E. (1989) Microbial mats and physicochemistry in a saltern in the Bretagne (France) and in a laboratory scale saltern model. *FEMS Microbiol. Ecol.* 62, 151-162.

Ginsburg, R.N. (1991) Controversies about stromatolites: Vices and virtues. In: D.W. Müller, J.A. MacKenzie, and H. Weissert (eds.), *Controversies in Modern Geology*, Academic Press, London, pp. 25-36.

Given, R.K. and Wilkinson, B.H. (1985) Kinetic control of morphology, composition, and mineralogy of abiotic sedimentary carbonates. *J. Sed. Petrology* 55, 109-119.

González-Muñoz, M.T., Ben Chekroun, K., Ben Aboud, A., Arias, J.M., and Rodriguez-Gallego, M. (2000) Bacterially induced Mg-calcite formation: role of Mg^{2+} in development of crystal morphology. *J. Sed. Res.* 70, 559-564.

Grotzinger, J.P. and Knoll, A.H. (1999) Stromatolites in Precambrian carbonates: evolutionary mileposts or environmental dipstics? *Ann. Rev. Earth Planet. Sci.* 27, 313-358.

Hartley, A.M., House, W.A., Leadbeater, B.S.C., and Callow, M.E. (1996) The use of microelectrodes to study the precipitation of calcite upon algal biofilms. *J.Coll. Interf. Sci.* 183, 498-505.

Javor, B.J. and Castenholz, R.W. (1984) Invertebrate grazers of microbial mats, Laguna Guerrero Negro, Mexico. In: Y. Cohen, R.W. Castenholz, and H.A. Halvorson (eds.), *Microbial Mats: Stromatolites*. Alan R. Liss, New York, pp. 85-94.

Jørgensen, B.B., Revsbech, N.P., and Cohen, Y. (1983) Photosynthesis and structure of benthic microbial mats: microelectrode and SEM studies of four cyanobacterial communities. *Limnol. Oceanogr.* 28, 1075-1093.

Jørgensen, B.B. and Nelson, D.C. (1988) Bacterial zonation, photosynthesis, and spectral light distribution in hot spring microbial mats of Iceland. *Microb. Ecol.* 16, 133-147.

Jørgensen, B.B. (1989) Light penetration, absorption, and action spectra in cyanobacterial mats. In: Y. Cohen and E. Rosenberg (eds.), *Microbial Mats: Physiological Ecology of Benthic Microbial Communities*. Am. Soc. Microbiol., Washington D.C., pp. 123-137.

Kazmierczak, J. and Kempe, S. (1998) Bacterial spherulites associated with coccoid cyanobacterial microbialites from Lake Van, Turkey: Clues for the origin of ooids. In: J.C. Canaveras, M.A. Garcia del Cura, and J. Soria (eds.), *Abstracts of the 15th International Sedimentological Congress*, Universidad de Alicante Press, Alicante, pp. 465-466.

Kazmierczak, J., Coleman, M.L., Gruszczynski, M., and Kempe, S. (1996) Cyanobacterial key to the genesis of micritic and peloidal limestones in ancient seas. *Acta Palaeont. Pol.* 41, 319-338.

Kempe, S. and Kazmierczak, J. (1990) Calcium carbonate supersaturation and the formation of *in situ* calcified stromatolites. In: V. Ittekkot, S. Kempe, W. Michaelis, and A. Spitzy (eds.), *Facets of Modern Biogeochemistry*, Springer, Berlin, pp. 255-278.

Kempe, S., Kazmierczak, J., Landmann, G., Konuk, T., Reimer, A., and Lipp, A. (1991) Largest known microbialites discovered in Lake Van, Turkey. *Nature (London)* 349, 605-608.

Knorre, H.v. and Krumbein, W.E. (2000) Bacterial calcification. In: R.E. Riding and S.M. Awramik (eds.), *Microbial Sediments*, Springer-Verlag, Berlin, pp. 25-31.

Köhler-Rink, S. and Kühl, M. (2000) Microsensor analysis of photosynthesis and respiration in larger symbiotic foraminifera. 1. The physico-chemical microenvironment of *Amphistegina lobifera*, *Amphisorus hemprichii* and *Marginopora vertebralis*. *Mar. Biol.* 137, 473-486.

Krumbein, W.E., Cohen, Y., and Shilo, M. (1977) Solar Lake (Sinai). 4. Stromatolithic cyanobacterial mats. *Limnol. Oceanogr.* 22, 635-656.

Krumbein, W.E. (1979a) Calcification by bacteria and algae. In: P.A. Trudinger and D.J. Swaine (eds.), *Biogeochemical Cycling of Mineral-Forming Elements*, Elsevier, Amsterdam, pp. 47-68.

Krumbein, W.E. (1979b) Photolithotrophic and chemoorganotrophic activity of bacteria and algae as related to Beachrock formation and degradation (Gulf of Aqaba, Sinai). *Geomicrobiol. J.* 1, 139-203.

Krumbein, W.E., Paterson, D.M., and Stal, L.J. (1994) *Biostabilization of Sediments*. BIS-Verlag, Oldenburg.

Kühl, M., Lassen, C., and Jørgensen, B.B. (1994). Optical properties of microbial mats: light measurements with fiber-optic microprobes. In L. J. Stal and P. Caumette (eds.), *Microbial Mats: Structure, Development and Environmental Significance*. Springer, Berlin, pp. 149-167.

Kühl, M., Glud, R.N., Ploug, H., and Ramsing, N.B. (1996) Microenvironmental control of photosynthesis and photosynthesis-coupled respiration in an epilithic cyanobacterial biofilm. *J. Phycol.* 32, 799-812.

Kühl, M., Lassen, C., and Revsbech, N.P. (1997). A simple light meter for measurements of PAR (400-700 nm) with fiber-optic microprobes: application for P vs. I measurements in microbenthic communities. *Aq. Microb. Ecol.*. 13, 197-207.

Kühl, M. and Fenchel, T. (2000) Bio-optical characteristics and the vertical distribution of photosynthetic pigments and photosynthesis in an artificial cyanobacterial mat. *Microb. Ecol.* 40, 94-103.

Kühl, M. and N.P. Revsbech. (2001) Biogeochemical microsensors for boundary layer studies. In: B.P. Boudreau and B. B. Jørgensen (eds.), *The Benthic Boundary Layer.* Oxford University Press, New York, pp. 180-210.

Logan, B.W. (1961). Cryptozoon and associate stromatolites from the recent of Shark Bay, Western Australia. *J. Geol.* 69, 517-533.

McNamara, K. (1992) *Stromatolites*. Western Australian Museum, Perth.

Merz, M. (1992) The biology of carbonate precipitation by cyanobacteria. *Facies* 26, 81-102.

Merz, M. and Zankl, H. (1993) The influence of the sheath on carbonate precipitation by Cyanobacteria. In: F. Baratollo, P. De Castro and M. Parente (eds.), Studies on Fossil Benthic Algae. *Bolletino della Societa Palaeontologica Italiana*, Special Volume 1, pp. 325-331.

Merz-Preiss, M. and Riding, R. (1999) Cyanobacterial tufa calcification in two freshwater streams: ambient environment, chemical thresholds and biological processes. *Sediment. Geol.* 126, 103-124.

Miller, A.G., Espie, G.S., and Canvin, D.T. (1990) Physiological aspects of CO_2 and HCO_3^- transport by cyanobacteria: a review. *Can. J. Bot.* 68, 1291-1302.

Morita, R.Y. (1980) Calcite precipitation by marine bacteria. *Geomicrobiol. J.* 2, 63-82.

Nübel, U., Garcia-Pichel, F., Kühl, M., and Muyzer, G. (1999) Quantifying microbial diversity: Morphotypes, 16S rRNA genes, and carotenoids of oxygenic phototrophs in microbial mats. *Appl. Environ. Microbiol.* 65, 422-430.

Paerl, H.W., Pinckney, J.L., and Steppe, T. F. (2000) Cyanobacterial-bacterial mat consortia: examining the functional unit of microbial survival and growth in extreme environments. *Environ. Microbiol.* 2, 11-26.

Paerl, H.W., Steppe, T.F., and Reid, R.P. (2001) Bacterially mediated precipitation in marine stromatolites. *Environ. Microbiol.* 3, 123-30.

Parkhurst, D.L., Thorstenson, D.C., and Plummer, L.N. (1980) PHREEQE – A computer program for geochemical calculations. *U.S. Geol. Surv. Wat. Res. Investig.* 80-96, 1-210.

Pentecost, A. and Bauld, J. (1988) Nucleation of calcite on the sheaths of cyanobacteria using a simple diffusion cell. *Geomicrobiol. J.* 6, 129-135.

Pringault, O. and Garcia-Pichel, F. (2000) Monitoring of oxygenic and anoxygenic photosynthesis in a unicyanobacterial biofilm, grown in a benthic gradient chamber. *FEMS Microbiol. Ecol.* 33, 251-258.

Pringault, O., de Wit, R., and Caumette, P. (1996) A benthic gradient chamber for culturing phototrophic sulfur bacteria on reconstituted sediments. *FEMS Microbiol. Ecol.* 20, 237-250.

Pringault, O., de Wit, R., and Kühl, M. (1999a) A microsensor study of the interaction between purple sulfur and green bacteria in experimental benthic gradients. *Microb. Ecol.* 37, 173-184.

Pringault, O., Epping, E., Guyoneaud R., Khalili, A., and Kühl, M. (1999b) Dynamics of anoxygenic photosynthesis in an experimental green sulphur bacteria biofilm. *Environ. Microbiol.* 1, 295-305.

Reid, R.P., Visscher, P.T., Decho, A.W., Stolz, J.F., Bebout, B.M., Dupraz, C., Macintyre, I.G., Paerl, H.W., Pinkney, J.L., Prufert-Bebout, L., Steppe, T.F., and DesMarais, D.J. (2000) The role of microbes in accretion, lamination and early lithification of modern stromatolites. *Nature* (London) 406, 989-992.

Revsbech, N.P., Jørgensen, B.B., Blackburn, T.H., and Cohen, Y. (1983) Microelectrode studies of the photosynthesis and O_2, H_2S, and pH profiles of a microbial mat. *Limnol. Oceanogr.* 28, 1062-1074.

Riege, H., Gerdes, G., and Krumbein, W.E. (1991) Contribution of heterotrophic bacteria to the formation of $CaCO_3$-aggregates in hypersaline microbial mats. *Kieler Meeresforsch.* Sonderh. 8, 168-172.

Riding, R. (2000) Microbial carbonates: the geological record of calcified bacterial-algal mats and biofilms. *Sedimentology* 47 (Suppl.), 179-214.

Riding, R., Awramik, S.M., Winsborough, B.M., Griffin, K.M., and Dill, R.F. (1991) Bahamian giant stromatolites - microbial composition of surface mats. *Geol. Mag.* 128, 227-234.

Rivadeneyra, M.A., Delgado, G., Ramos-Cormenzana, A., and Delgado, R. (1998) Biomineralization of carbonates by *Halomonas euryhalina* in solid and liquid media with different salinities: crystal formation sequence. *Res. Microbiol.* 149, 277-287.

Schopf, J.W. and Klein, C. (eds.) (1992) *The Proterozoic Biosphere*. Cambridge Univ. Press, Cambridge.

Schopf, J.W. (2000) The fossil record: tracing the roots of the cyanobacterial lineage. In: B.A. Whitton and M. Potts (eds.), *The Ecology of Cyanobacteria,* Kluwer Acad. Publ., Dordrecht, pp. 13-35.

Schultze-Lam, S., Ferris, F.G., Sherwood-Lollar, B., and Gerits, J.P. (1996) Ultrastructure and seasonal growth patterns of microbial mats in a temperate climate saline-alkaline lake: Goodenough Lake, British Columbia, Canada. *Can. J. Microbiol.* 42, 147-161.

Schwartz, H.-U., Einselle, G., and Herm, D. (1975) Quartz-sandy, grazing contoured stromatolites from coastal embayments of Mauritania, West Africa. *Sedimentology* 22, 529-561.

Simkiss, K. (1986) The processes of biomineralization in lower plants and animals. In: B.S.C Leadbeater and R. Riding (eds.), *Biomineralization in Lower Plants and Animals*, Oxford Univ. Press, Oxford, pp.19-36.

Somers, G.F. and Brown, M. (1978) The affinity of trichomes of blue-green algae for calcium ions. *Estuaries* 1, 17-28.

Stal, L.J., van Gemerden, H., and Krumbein, W.E. (1985) Structure and development of a benthic marine microbial mat. *FEMS Microbiol. Ecol.* 31, 111-125.

Stal, L.J. (2000) Cyanobacterial mats and stromatolites. In: B. A. Whitton & M. Potts (eds.), *The Ecology of Cyanobacteria,* Kluver Acad. Publ., Dordrecht, pp. 61-120.

Tai, C.Y.and Chen, F.-B. (1998) Polymorphism of $CaCO_3$ precipitated in a constant-composition environment. *AIChE J.* 44, 1790-1798.

Thar, R., Kühl, M., and Holst, G. (2001) A fiber-optic fluorometer for microscale mapping of photosynthetic pigments in microbial communities. *Appl. Environ. Microbiol.* 67, 2823-2828.

Walter, M.R. and Heys, G.R. (1985) Links between the rise of metazoa and the decline of stromatolites. *Precambrian Res.* 29, 149-174.

Warren, L.A., Maurice, P.A., Parmar, N., and Ferris, F.G. (2001) Microbially mediated calcium carbonate precipitation: Implications for interpreting calcite precipitation and for solid-phase capture of inorganic contaminants. *Geomicrobiol. J.* 18, 91-115.

Wickstrom, C.E. and Wiegert, R.E. (1980) Response of thermal algal-bacterial mat to grazing by brine flies. *Microb. Ecol.* 6, 313-315.

Wickstrom, C.E and Castenholz R.W. (1985) Dynamics of cyanobacterial and ostracod interactions in an Oregon hot spring. *Ecology* 66, 1024-1041.

Wieland, A. and Kühl, M. (2000) Short-term temperature effects on oxygen and sulfide cycling in a hyper-saline cyanobacterial mat (Solar Lake, Egypt). *Mar. Ecol. Progr. Ser.* 197, 87-102.

Chapter 6

MICROBIAL PROCESSES FORMING MARINE STROMATOLITES

Microbe-Mineral Interactions with a Three-Billion-Year Rock Record

Pamela REID, C.D. DUPRAZ
Rosenstiel School of Marine and Atmospheric Science
University of Miami
4600 Rickenbacker Causeway
Miami FL 33149 USA
preid@rsmas.miami.edu

P.T. VISSCHER
University of Connecticut
1084 Shennecosset Road
Groton CT 06340 USA

D.Y. SUMNER
University of California
One Shields Avenue
Davis CA 95616 USA

1. INTRODUCTION

Research in the burgeoning field of geomicrobiology reveals an "intimate juxtaposition and interdependence" of microbes and minerals that we are only beginning to appreciate (Skinner 1997, p. 1). Future studies of microbe-mineral interactions are likely to lead to major advances in our understanding of such fundamental issues as the dynamics of sedimentation, the flow of energy and matter through the biosphere, and the evolution of life on Earth (Nealson 2000, Nealson and Stahl 1997, Hazen 2001)

An ideal model system in which to study microbe-mineral interactions and ecological principles governing these interactions is modern marine stromatolites. Modern marine stromatolites are living examples of one of earth's oldest and most persistent widespread ecosystems. Layered deposits of calcium carbonate known as stromatolites first appeared in the geological record at least three and a half billion years ago (Grotzinger and Knoll 1999, Hofmann et al. 1999). Stromatolites are neither biotic fossils nor abiotic structures. Rather, they represent the complex interactions of microbes, minerals, and the environment (Walter 1994, Grotzinger and Knoll 1999). For almost 80% of Earth's history, stromatoliteforming microbial communities played a major role in regulating sedimentation and global cycles of major elements via production and decomposition of organic matter, trapping and binding of sediment, and precipitation of calcium carbonate.

Modern stromatolites in Exuma Cays, Bahamas, offer a unique opportunity to investigate interactions between microbes and minerals in the highly successful stromatolite ecosystem. In this paper, we examine microbial processes forming Exuma stromatolites. We review what is currently known and identify unanswered questions and future research directions. Our goal is to demonstrate that a thorough understanding of microbe-mineral interactions in modern marine stromatolites will have broad implications in a wide variety of fields, including ecology, biogeochemistry, sedimentology, and paleobiology, and will improve our ability to interpret the fossil record of ancient ecosystems.

2. STROMATOLITE GROWTH

Stromatolites in Exuma Cays, Bahamas (Dravis 1993, Dill et al. 1986, Reid and Browne 1991, Reid et al. 1995, Reid et al. 1999) are the only known examples of stromatolites presently forming in open marine environments equivalent to those of many Precambrian platforms. Our recent research results (e.g. Reid et al. 2000) show that growth of these stromatolites results from successive episodes of sediment accretion and lithification of microbial mats. Periods of rapid accretion, during which stromatolite surfaces are dominated by "pioneer" communities of motile filamentous cyanobacteria, alternate with hiatal intervals. Hiatal periods are characterized by development of surface films with abundant aerobic and anaerobic heterotrophic bacteria, which form thin crusts of microcrystalline carbonate. During prolonged hiatal periods, "climax" communities develop, which include endolithic coccoid cyanobacteria. These coccoids fuse sediment grains, forming thicker lithified laminae. Preservation of lithified layers at depth creates millimeter-scale lamination.

This growth scenario is based on extensive field and laboratory-based studies using a wide range of geological and microbiological techniques. These studies revealed that surfaces of Exuma stromatolites are covered with cyanobacterial mats, which show distinct variations in microbial community structure and composition. Three mat types, representing a continuum of growth stages are recognized (Figure 1).

2.1 Type 1 mats

About 70% of surface mats consist of a sparse population of the filamentous cyanobacterium *Schizothrix* sp. within a mucilaginous exopolymer matrix (Stolz et al. 2001). *Schizothrix* filaments are generally vertically oriented and are entwined around carbonate sand grains (Figures 1a, 1b). These mats are "pioneer" communities (Stal et al. 1985), which dominate during periods of rapid sediment accretion of up to one-grain layer per day. Populations of diatoms and other eukaryotes are rarely found in these accreting mats (Golubic and Browne 1996, Pinckney and Reid 1997), and, contrary to previous reports (Awramik and Riding 1988, Riding 1994), eukaryotic organisms are not required for the trapping and binding of these coarse-grained sediments.

2.2 Type 2 mats

Approximately 15% of mats show development of surface films of amorphous exopolymer and bacterial cells; these surface films are referred to in this paper as "biofilms". The biofilms are calcified and appear as thin crusts (~20-60 μm thick) of microcrystalline carbonate (micrite) at the uppermost mat surface (Figure 1c). A sparse to moderately dense population of *Schizothrix* lies below the surface films (Figure 1c). *Schizothrix* filaments are also present, but are not abundant, in the surface films, which are comprised mainly of copious amounts of amorphous exopolymer, metabolically diverse heterotrophic microorganisms and aragonite needles (Visscher et al. 1998, 1999; Stolz et al. 2001; Stolz this volume) Spherical aggregates of aragonite needles, 2 to 5 μm in diameter, are embedded in the exopolymer matrix (Figure 1e). Bacteria are abundant and are commonly observed at the edges of the aragonite spherules (Decho and Kawaguichi 1999; Paerl et al. 2001). This mat type represents a "more mature" (Stal et al. 1985, Van Gemerden 1993) surface community and develops during hiatal periods when sediment accretion ceases and mats begin to lithify. Initial mesocosm manipulations suggest that continuous surface biofilms form in a matter of days.

2.3 Type 3 mats

The remaining 15% of mats are characterized by an abundant population of the coccoid cyanobacterium *Solentia* sp. and randomly oriented *Schizothrix* filaments below a calcified biofilm (Figures 1f-1h). This *Solentia*-rich mat type represents the "climax" community of the stromatolite system. *Solentia* is an endolith that bores into carbonate sand grains. These bored grains appear grey when viewed in plane polarized light under a petrographic microscope (Figure 1f), contrasting with the golden brown coloration of unbored grains (Figures 1a, 1c). In contrast to the conventional view that microboring is a destructive process (Golubic and Browne 1996; Perry 1998), the microboring and infilling process associated with *Solentia* activity in these mats is an important constructive process. This process fuses grains at point contacts to create laterally cohesive carbonate crusts ("welded" grains, Figure 1h) (Macintyre et al. 2000). Field and laboratory studies show that layers of fused microbored grains are formed in periods of weeks to months (Macintyre et al. 2000). As *Solentia* is a photosynthetic microorganism, such prolonged periods of microboring activity can only be sustained when this population remains at the surface during long hiatal periods. Longer hiatal periods can result in development of eukaryotic algal communities, which do not form laminated structures (Steneck et al. 1998, Golubic and Browne 1996).

2.4 Subsurface structure

The laminated stromatolitic fabric records a chronology of former surface mats (Figure 2). Although lamination is readily apparent in hand samples (Figure 2a), it has a subtle expression in petrographic thin sections. Detailed observations show, however, that lithified layers have two distinct petrographic appearances (Figure 2b). These laminae correspond to (1) thin crusts of microcrystalline carbonate (micrite), 10-60 μm thick (Figures 2b, 2c); and (2) layers of fused, microbored grains infested with *Solentia* sp.; these layers are 1-2 mm thick (Figures 2b, 2d) and underlie micritic crusts. Light microscopy combined with scanning electron microscopy shows that the thin crusts are identical in thickness, composition, and texture to the calcified biofilms described above. They are also similar in thickness to micritic laminae in many ancient stromatolites (Walter 1983, Bertrand-Sarfati 1976).

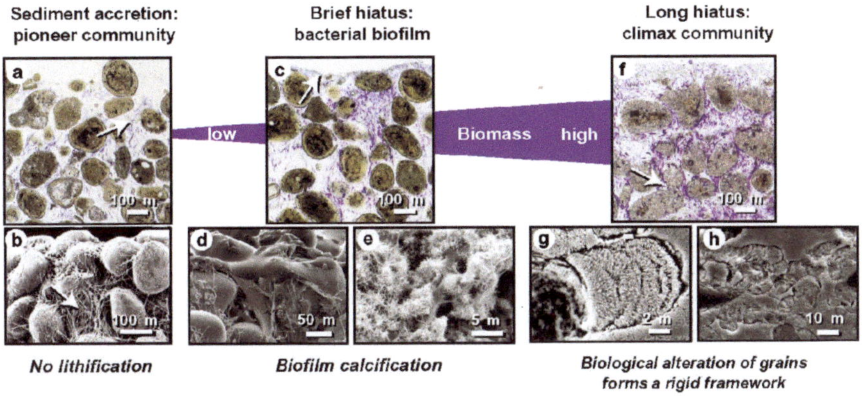

Figure 1. Surface mats shown as a response to intermittent sedimentation. a,b, Type 1 mats; filamentous cyanobacteria (arrows) bind carbonate sand grains. c,d,e, Type 2 mats; a continuous sheet of amorphous exopolymer with abundant heterotrophic bacteria drapes the surface (a, arrow; d); aragonite needles precipitate within this film (e). f,g,h, Type 3 mats; a surface biofilm overlies filamentous cyanobacteria and endolithinfested grains, which appear gray and are fused (arrow, f). Banded pattern of fibrous aragonite in bore holes (g) indicates progressive infilling. Precipitation in tunnels that cross between grains leads to welding (h). (a,c,f) Petrograhic thin sections, plane polarized light. (b,d,e,g,h) Scanning electron microscope images

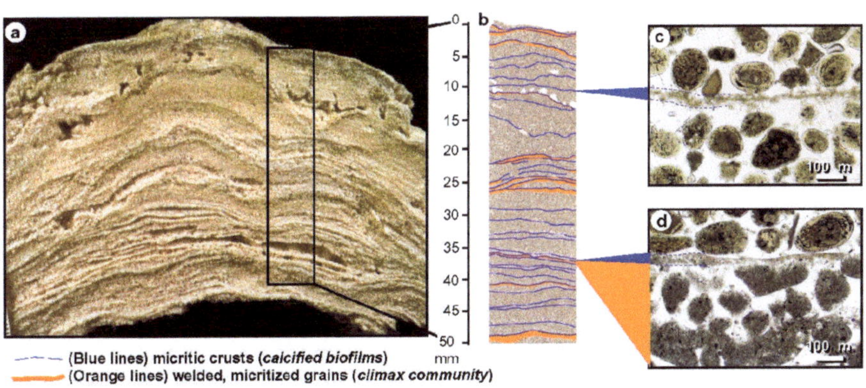

Figure 2. Subsurface distribution of lithified layers, which form at 1-2 mm intervals. a, Water-washed vertical section showing lithified laminae, which stand out in relief. b, Low magnification thin-section photomicrograph of boxed area in (a) showing the distribution of lithified layers. c, Micritic crust, equivalent to the blue lines in (b). d, Layer of microbored, fused grains, equivalent to the orange lines in (b), underlying a micritic crust (black dashed line).

In addition, the microstructure of the layers of fused, microbored grains is identical to that formed by the climax community described above. Analyses of the distribution of these layers indicate that micritic crusts, representing mature biofilm communities, occur at 1 to 2 mm intervals, whereas layers of fused grains, representing climax communities, occur at 2 to 3 mm and 3 to 4 mm intervals (Figure 2) (Reid et al. 2000).

3. CONTROLS OF STROMATOLITE MORPHOGENESIS

Our recent research results, as summarized above, are the first to document a set of microbemineral interactions resulting in the growth of lithified, laminated carbonate buildups in a modern environment. Theses studies indicate that Exuma stromatolites are the net result of interactions between microbes and minerals in three distinct mat types (Figure 3). These mat types represent a spectrum of community development and include (1) a "pioneer" stage of motile cyanobacteria; (2) more mature mats having continuous surface biofilms with a more substantial heterotrophic population; and (3) mats developed to a stage where surface carbonate grains have been invaded by endolithic cyanobacteria and grains are fused. Processes of trapping and binding of sediment by the pioneer communities and carbonate precipitation in biofilm and endolithic communities result in stromatolite accretion. Cycling between microbial communities, with accompanying changes in accretion style, leads to lamination, a fundamental feature of stromatolites through time. Iteration of lamination is expressed as stromatolite morphology (Figure 3).

Currently, the dynamics of the microbial system and the factors that regulate community succession and stromatolite morphogenesis are unconstrained (Figure 3). On first consideration, it may seem likely that successions between pioneer communities, which trap and bind sediment, and biofilm and climax communities, which precipitate calcium carbonate, are controlled primarily by sediment supply. Initial trapping of sediment is, however, dependent on adhesion of grains to a microbial mat. Thus, cessation of sediment accretion could result for two different reasons: (1) a lack of sediment influx (an environmental control) or (2) sediment adhesion could be inhibited by accumulation of exopolymer that is not sticky enough to "trap" grains (a biological control). In other words, hiatuses in sediment accretion could conceivably reflect conditions of mat biology! Microbial communities may, therefore, be actively controlling stromatolite morphogenesis.

Our recent results provide critical insight into microbial functional groups and major processes that are involved in stromatolite growth. Specific mechanisms and rates of these processes, however, remain largely unknown, and at present we can only speculate on intrinsic and extrinsic factors that control stromatolite morphogenesis. Future advances in understanding stromatolite growth will require detailed knowledge of accretion mechanisms, microbial and environmental factors that regulate accretion, and quantitative models that link accretion to morphology.

3.1 Stromatolite accretion

3.1.1 Sediment Trapping and Binding

Exact mechanisms of trapping and binding by pioneer communities of filamentous cyanobacteria are not known. Initial observations suggest, however, that "trapping" occurs when grains adhere to sticky exopolymer at the surface of a mat (unpublished video recordings). These grains are subsequently "bound" when they are entwined by filamentous sheaths as cyanobacteria move upward to the mat surface. Water turbidity fluctuates, but sediment is typically abundant due to tidal cycles, frequent high winds and frequent burial events. The factor that is considered most likely to impact initial trapping is "stickiness" of exopolymer (Decho, this volume); additional factors that may be important in the binding of sediment are cyanobacterial growth rates, motility and response to light, substrate availability or chemical cues. In addition, erosive events, such as storms, may have a negative impact on accretion; erodibility will reflect the cohesive properties of the "bound" sediment.

3.1.2 Carbonate Precipitation

Carbonate precipitation in Exuma stromatolites is primarily associated with biofilm and climax communities. In biofilm communities, precipitation occurs within surface biofilms, forming thin micritic crusts. In climax communities, additional precipitation occurs in endolithic borings, resulting in layers of fused grains. Heterotrophic degradation of exopolymer by sulfate reducing bacteria has been identified as a major precipitation process forming micritic crusts, as evidenced by microscale observations that high rates of sulfate reduction coincide with micritic crusts (Figure 4) (Visscher et al. 1998, 2000). Precipitation within microborings has not yet been investigated, but observations of organic matter in some boreholes (Kawaguchi and Decho 2000), together with high sulfate reduction activity in these layers (Visscher et al. 2000) suggests that, as in biofilm layers, heterotrophic activity may be important in the precipitation process. In addition, although our previous work has emphasized exopolymer degradation in precipitation, recent work suggests that low molecular weight dissolved organic compounds (DOC) may play an important role in the precipitation process.

Carbonate precipitation in stromatolites may, therefore, be impacted by many factors that concurrently affect microbial carbon and sulfur cycling, including community composition, light, flow, and availability of oxygen and carbon. The carbonate saturation state of seawater is an additional extrinsic factor that can affect precipitation. Previous studies have linked extensive cementation in the Exuma Cays region to supersaturation resulting from degassing (loss of $CO2$) and increase in temperature as cool oceanic water from Exuma Sound moves onto the Bahama Bank (e.g. Dill 1991). The distribution of stromatolites along the ocean-facing margins of Exuma Sound (Reid et al. 1995) and their apparent absence on Bank interiors suggest that ocean chemistry may indeed play a role in stromatolite development.

3.2 Stromatolite lamination

Lamination in modern marine stromatolites represents successive intervals of sediment accretion and carbonate precipitation. As discussed above, microbial factors that control the "stickiness" of exopolymer in a filamentous cyanobacterial mat may be important determinants of accretion. Therefore, biological processes controlling community succession, such as sulfide and light cues, microbial growth rates, relative rates of photosynthesis and respiration (aerobic and anaerobic), exopolymer production, etc. may be critical controls on lamination.

A fundamental feature of lamination is episodic accretion. The distance between lamina represents the frequency of events (external disturbance events or microbial growth factors) associated with layer formation. In modern marine stromatolites, micritic crusts, representing mature biofilm communities, occur at 1 to 2 mm intervals, whereas layers of fused grains, representing climax communities, occur at 2 to 3 mm and 3 to 4 mm intervals (Figure 2) (Reid et al. 2000). Statistical analyses of lamina spacing are needed to define long-term patterns of mat development and community succession. Differentiating between lamination styles that are cyclical or random will help to constrain factors regulating mat succession and resultant stromatolite accretion.

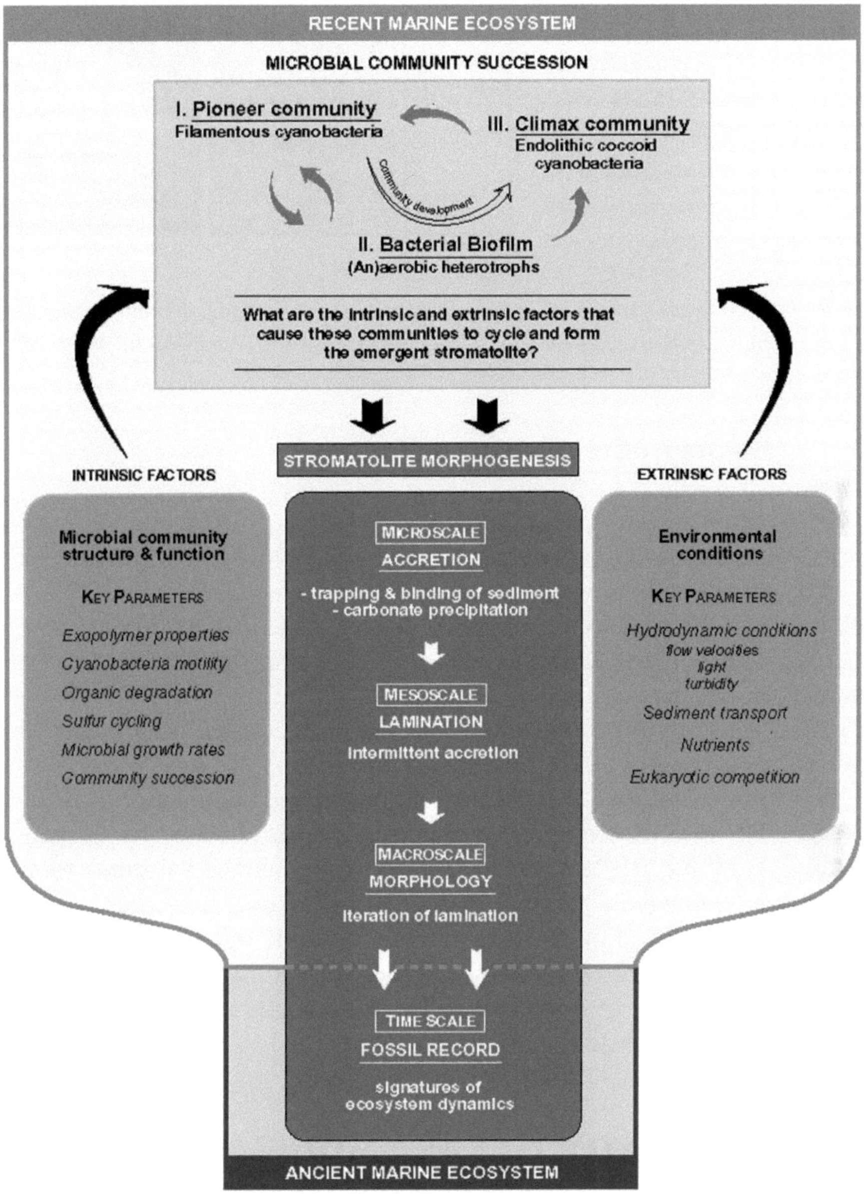

Figure 3. Microbial-environmental interactions leading to stromatolite morphogenesis.

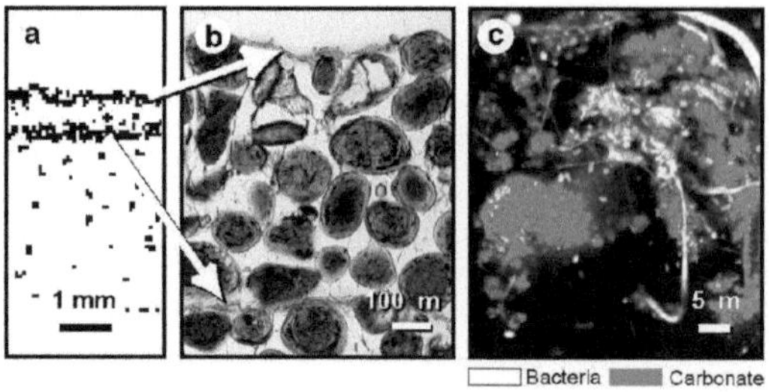

Figure 4. Bacterial activity related to carbonate precipitation. a, Sulfate reduction mapped using the 35S-Ag technique; black dots indicate zones of high activity. b, Thin section of mat analyzed in (a) showing a correspondence between high sulfate reduction activity and micritic crusts (arrows). c, Scanning laser confocal microscope image of a micritic crust showing an intimate association between bacteria (white) and carbonate precipitates (grey). Filament in upper left is *Schizothrix* sp.

3.3 Stromatolite morphology

Recent studies have used mathematical models to simulate stromatolite morphogenesis (e.g. Grotzinger and Rothman 1996, Grotzinger and Knoll 1999). These efforts have focused primarily on surface growth processes that could occur in the absence of microbial activity. The models produce layered structures similar to stromatolites by incorporating vertical accretion (sediment influx or microbial growth), lateral diffusion of loose sediment or microbes, surface normal growth (mineral precipitation or microbial growth), and random noise. Although these models may capture some aspects of microbial mat growth, they do not include population or reaction dynamics intrinsic to ecological interactions. As articulated by Grotzinger and coworkers, the challenge is to identify morphological features that can be produced exclusively through biological behavioral responses or related nonlinear dynamics. Future insights into modern microbial mat growth processes will allow development of models that are capable of expressing unique characteristics of microbial mat dynamics and can predict effects of microbial processes on stromatolite morphology.

4. STROMATOLITES AS A MODEL SYSTEM

As indicated above, stromatolite morphogenesis encompasses a wide range of fundamentally important ecological, biogeochemical, and sedimentological processes and principles. Modern marine stromatolites are an ideal model system for investigating these topical issues because of the relative simplicity of the stromatolite system, i.e. a relatively small number of organisms participate in the development of a highly organized ecosystem (Figure 3). This simplicity enables individual components to be sampled, experimentally manipulated, and modeled, such that the sensitivity of the system to internal and external perturbation can be evaluated and biogeochemical cycling and mineral formation can be quantified. The formation of three distinct mineral "products" by three mat types in a spectrum of community succession is particularly beneficial because these "end points" enable population dynamics to be monitored.

Examination of the stromatolite system using interdisciplinary approaches will allow fundamental ecological questions to be addressed. For example, what adaptations have allowed the stromatolite ecosystem to persist for over 3 billion years? How do microbes respond to environmental perturbations? What extent do microbes manipulate or "engineer" their own environment? The potential for stromatolite-forming communities to "self organize" to achieve a common goal of ecological stability and persistence is of particular significance. Self-perpetuating episodes of community succession may be a critical adaptive strategy that has ensured the long-term survival of this highly successful ecosystem.

It is intriguing to consider that stromatolites, which represent earth's earliest biofilms, may represent the simplest organized system of calcification. The spatially- (and temporally-) ordered interactions between microbial components within biofilms may evoke and regulate precipitation in a relatively organized manner— "one that bears functional analogy to primitive shells and bones" (Decho, this volume). This posits an important question regarding the microbial communities involved in precipitation: are these communities (e.g. sulfate reducers and cyanobacteria) actively coordinating their metabolic activities (production and consumption of carbon) to effect a layered precipitation of CaCO3? Such metabolic coordination would likely involve chemical signaling processes, called "quorum sensing", within and/or between different microbial groups (Fuqua et al., 1996; Decho, 1999). An ecological advantage of metabolic cooperation leading to organized precipitation of micritic layers (laminae) would be to physically stabilize the overall system.

Modern marine stromatolites are also ideal systems in which to study biogeochemical cycling. As in non-lithifying microbial mats, microbial populations and processes within the surface mats of stromatolites exhibit vertical compartmentalization with a high degree of horizontal homogeneity. In contrast to non-lithifying mats, each of the stromatolitebuilding communities has a unique geochemical signature. In addition, the stromatolite mats are associated with lower biomass and display correspondingly lower metabolic rates than typical non-lithifying mats (Visscher et al. 2001; Decho et al. submitted). The trapping and binding of sediment in stromatolite mats provides a "geophysical dilution" of biochemical processes, and hence a stretching of spatial scales. As a result of these combined features, geochemical parameters of the stromatolite ecosystem are relatively easy to study. Sampling and measurement of individual strata are simplified and, most significantly, mathematical models that can adequately describe ecosystem interactions and changes through time can be developed.

We predict that the biological properties of the stromatolite system interact with physicochemical environment to regulate biogeochemical cycling. To evaluate this hypothesis, elemental budgets within stromatolites and the factors that control the balance of elemental exchange within stromatolites and the external environment must be determined. These measurements, combined with combined with knowledge of microbial community composition and activity, will reveal fundamental links between ecosystem function and organismal ecology by identifying feedbacks between the cycling of elements, population dynamics, and shifts in community structure.

Understanding pathways of biogeochemical cycling instromatolite-forming communities will also have a fundamental impact in studies of biofouling, biocorrosion, and biostabilization of sediments. Biogenic stabilization of sediment, for example, is important in coastal sediments worldwide (e.g. Krumbein 1994). Microbial mats typically function to reduce sediment transport. Community structure and function is, in turn, influenced by sediment type (Yallop et al. 1994). In addition to contributing to sediment cohesion, microbial mats affect water quality when they are eroded. By injecting high quality organic matter into the water column, eroded mats can play a central role in the energy flow of the coastal marine ecosystem (Grant and Emerson 1994). At present, microbe-mineral interactions controlling biogenic stabilization are poorly understood. New approaches, such as those being developed in studies of stromatolite morphogenesis, will show how specific chemical components of exopolymer influence sediment cohesiveness and how the composition of exopolymer varies in response to environmental parameters.

5. APPLICATION OF THE ROCK RECORD

Can modern marine stromatolites serve as a key to recognizing and interpreting biological and environmental signatures in the rock record? We answer, emphatically, yes. A rigorous understanding of present-day interactions between microbes, minerals and the environment is of critical importance for establishing criteria that identify microbial and environmental influences on stromatolite growth and for developing process models that "accurately describe stromatolite accretion dynamics"

(Grotzinger and Knoll 1999, p. 316). Processes of carbonate precipitation and sediment trapping and binding, which form modern marine stromatolites, are the primary accretion mechanisms in stromatolites through time and in other microbial deposits.

A defining feature of stromatolites in the rock record is lamination. Previous studies of ancient stromatolites have typically attributed lamination to variations in rates of sediment supply (extrinsic control) (e.g. Seong-Joo and Golubic, 1999, 2000; Seong-Joo, et al. 2000). The possibility that intrinsic biologic factors may regulate lamination in modern marine stromatolites, as previously discussed, has important implications for the interpretation of ancient ecosystems. If "stickiness" of a mat is indeed a critical factor in initial adhesion of sediment to the stromatolite surface, then lamination may record information on mat characteristics, such as species abundances and the composition of exopolymer. This interpretation has significantly different implications for stromatolite growth rates and dynamics of the microbial-sediment system than a strictly abiotic control of lamination related to sedimentation events.

The example above demonstrates the fundamental importance of differentiating between biological and environmental influences in stromatolite morphogenesis. At present, effects of environmental processes on stromatolite growth are much better understood than biological influences (e.g. Grotzinger and Knoll 1999). Until the range of microbial influences on carbonate sedimentation is established in modern environments, the role of microbial communities in the accretion and lithification of ancient stromatolites will remain ambiguous. Differentiating between biological and environmental influences on stromatolite morphogenesis is of critical importance in interpreting the rock record and understanding the early evolution of life.

ACKNOWLEDGEMENTS

Research support was provided by U.S. National Science Foundation. Research Initiative on Bahamian Stromatolites (RIBS) Contribution #17.

REFERENCES

Awramik, S.M. and Riding, R. (1988) Role of algal eukaryotes in subtidal columnar stromatolite formation. *Proceeding of National Academy of Science*, USA, 85, 1327-1329.

Bertrand-Sarfati, J. (1976) An attempt to classify late Precambrian stromatolite microstructures. In: Walter, M.R. (ed.), *Stromatolites* Developments in Sedimentology, Elsevier Scientific Publishing Company, 20, 251-259.

Decho A.W., and Kawaguchi, T. (1999) Confocal imaging of in situ natural microbial communities and their extracellular polymeric secretions using Nanoplast resin. *BioTechniques* 27, 1246-1252.

Decho, A.W. (1999) Chemical communications within microbial biofilms: chemotaxis and quorum sensing in bacteria, in Wingender, J., New, T.R., Flemming, H.C. (eds.), *Microbial Extracellular Polymeric Substances*, Springer-Verlag, New York, pp. 155-170.

Decho, A.W. (2002) The extracellular polymeric (EPS) matrix and calcification within modern marine stromatolite, in Krumbein, W.E., Paterson, D.M., and Zavarzin, G.A. (eds), *Fossils and Recent Biofilms- a natural history of life on Earth*, Kluwer Academic Publishers, Dordrecht, pp. xx-xx.

Decho, A.W., Visscher, P.T. and Reid, R.P. (submitted) Cycling and production of natural microbial exopolymers (EPS) within a marine stromatolite. Aquatic Microbial Ecology.

Dill, R.F. (1991) Subtidal stromatolites, ooids and crusted-lime muds at the Great Bahama Bank Margin, in Osborne, R.H. (ed.), *From Shoreline to Abyss*, SEPM Special Publication 46, 147-171, Tulsa.

Dill, R.F., Shinn, E.A., Jones, A.T., Kelly, K. and Steinen, R.P. (1986) Giant subtidal stromatolites forming in normal salinity water. *Nature* 324, 55-58.

Dravis, J.J. (1983): Hardened subtidal stromatolites, Bahamas, *Science*, 219, 385-386.

Fuqua, C., Winans, S.C., and Greenberg, E.P. (1996) Census and consensus in bacterial ecosystems: the LuxRLuxI family of quorum sensing transcriptional regulators, *Ann. Rev. Microbiology* 50, 727-751.

Golubic, S., and Browne, K.M. (1996) *Schizothrix gebeleinii* sp. nova builds subtidal stromatolites, Lee Stocking Island, *Algological Studies* 83, 273-290.

Grant J. and Emerson, C. (1994) Resuspension and stabilization of sediments with microbial biofilms: implications for benthic-pelagic coupling, in Krumbein, W.E., Paterson, D.M. and Stal, L.J. (eds.), *Biostabilization of Sediments,* Bibliotheks und Informationssystem der Universitaet Oldenburg, Oldenburg, pp. 121-134.

Grotzinger, J. P. and Rothman, D. H. (1996) An abiotic model for stromatolite morphogenesis, *Nature* 383, 423-425.

Grotzinger, J.P. and Knoll, A.H. (1999) Stromatolites in Precambrian carbonates: evolutionary mileposts or environmental dipsticks? *Annu. Rev. Earth Planet. Sci.* 27, 313-358.

Hazen, R.M. (2001) Life's rocky start, *Scientitic American* April 2001, 76-85.

Hofman, H.J., Grey, A.H., Hickman, A.H. and Thorpe, R.I. (1999) Origin of 3.45 Ga coniform stromatolites in Warrawoona Group, Western Australia. *Geological Society of America Bulletin* 111, 1256-1262.

Kawaguchi, T. and A.W. Decho (2000) Biochemical characterization of cyanobacterial extracellular polymers (EPS) from modern marine stromatolites, *Preparative Biochemistry and BioTechnology* 30, 321-330.

Krumbein, W.E. (1994) The year of the slime, in Krumbein, W.E., Paterson, D.M. and Stal, L.J. (eds.), *Biostabilization of Sediments*, Bibliotheks und Informationssystem der Universitaet Oldenburg, Oldenburg. pp. 1-7.

Macintyre, I.G., Prufert-Bebout, L. and Reid, R.P. (2000) The role of endolithic cyanobacteria in the formation of lithified laminae in Bahamian stromatolites, *Sedimentology* 47, 915-921.

Nealson, K.H. (2000) Crystal ball: the future of biocomplexity, *Environmental Microbiology* 2, 3-4.

Nealson, K.H. and Stahl, D.A. (1997) Microorganisms and biogeochemical cycles: what can we learn from layered microbial communities? in Banfield, J.F and Nealson, K.H. (eds.), *Geomicrobiology: Interactions between Microbes and Minerals*, Mineralogical Society of America, Reviews in Mineralogy 35, 5-34.

Paerl, H.W., Steppe, T.F., Reid, R.P. (2001) Bacterially-mediated precipitation in marine stromatolites, *Environmental Microbiology* 3, 123-130

Perry, C. T. (1998) Grain susceptibility to the effects of microboring: implications for the presevation of skeletal carbonates, *Sedimentology* 45, 39-51.

Pinckney, J.L. and Reid, R.P. (1997) Productivity and community composition of stromatolitic microbial mats in the Exuma Cays, Bahamas, *Facies* 36, 204-207.

Reid, R.P. and Browne, K.M. (1991) Intertidal stromatolites in a fringing Holocene reef complex in the Bahamas, *Geology* 19, 15-18.

Reid, R.P., MacIntyre, I.G., Browne, K.M., Steneck, R.S. and Miller, T. (1995) Modern marine stromatolites in the Exuma Cays, Bahamas: uncommonly common, Facies 33, 1-18.

Reid, R.P., Macintyre, I.G., and Steneck, R.S. (1999) A microbialite/algal ridge fringing reef complex, Highborne Cay, Bahamas, *Atoll Research Bulletin* 466, 1-18.

Reid, R.P., Visscher, P.T., Decho, A.W., Stolz, J.K., Bebout, B.M., Dupraz, C., Mactintyre, I.G., Paerl, H.W., Pinckney, J.L., Prufert-Bebout, L., Steppe, T.F., and DesMarais, D.J. (2000) The role of microbes in accretion, lamination and early lithification of modern marine stromatolites, *Nature* 406, 989-992.

Riding, R. (1994) Stromatolite survival and change: the significance of Shark Bay and Lee Stocking Island subtidal columns, in Krumbein, W.E., Paterson, D.M. and Stal., L.J. (eds.), *Biostabilization of Sediments*, Bibliotheks und Informationssystem der Universitaet Oldenburg, Oldenburg, 183-202.

Seong-Joo, L. and Golubic, S. (1999) Early cyanobacterial fossil record: preservation, palaeoenvironments and identification, *European J. Phycology* 34, 339-348.

Seong-Joo, L. and Golubic, S. (2000) Biological and mineral components of an ancient stromatolite; Gaoyuzhuang Formation, Mesoproterozoic of China, *SEPM Spec. Pub.* 67, 92-102.

Seong-Joo, L., Browne, K.M. and Golubic, S. (2000) On stromatolite lamination, in:Riding, R.E. and Awramik, S.M. (eds.), *Microbial Sediments*, Springer Verlag, New York, pp. 16-25.

Skinner, H.C.W. (1997) Preface, in Banfield, J.F. and Nealson, K.H. (eds.), *Geomicrobiology: Interactions Between Microbes and Minerals*, Mineralogical Society of America, Reviews in Mineralogy 35, 1-4.

Stal, L.J., van Gemerden, H. and W.E. Krumbein. (1985) Structure and development of a benthic microbial mat. *FEMS Microbiology Ecology* 31, 111-125.

Steneck, R. S., Miller, T.E., Reid, R. P. and Macintyre, I.G. (1998) Ecological controls on stromatolite development in a modern reef environment: a test of the ecological refuge paradigm, *Carbonates and Evaporites* 13, 48-65.

Stolz, J.F, Feinstein, T.N., Salsi, J. and Reid, R.P. (2001, in press) Microbial role in sedimentation and lithification in a modern marine stromatolite: a TEM perspective, American *Mineralogist* 86.

Stolz, J.F. (2002) Structure in marine biofilms, in Krumbein, W.E., Paterson, D.M., and Zavarzin, G.A. (eds), *Fossils and Recent Biofilms- a natural history of life on Earth*, Kluwer Academic Publishers, Dordrecht, pp. xx-xx.

Van Gemerden, H. (1993) Microbial mats: a joint venture, *Marine Geology* 113, 3-25.

Visscher, P.T., Gritzer, R.F. and Leadbetter, E.R. (1999) Low-molecular weight sulfonates: a major substrate for sulfate reducers in marine microbial mats, *Applied and Environmental Microbiology* 65, 3272-3278.

Visscher, P.T., Hoeft, S.E., Surgeon, T.M.L., Rogers, D.R., Bebout, B.M., Thompson, J.S. Jr., Reid, R.P. (2001) Microelectrode measurements in stromatolites: unraveling the Earth's past? in Taillefert, M., and Rozan, T. (eds.), ACS Symposium Series 220, *Environmental Electrochemical Analyses of Trace Metal Biogeochemistry* (in press).

Visscher, P.T., Reid, R.P. and Bebout, B.M. (2000) Microscale observations of sulfate reduction: correlation of microbial activity with lithified micritic laminae in modern marine stromatolites, *Geology* 28, 919-922.

Visscher, P.T., Reid, R.P., Bebout, B.M., Hoeft, S.E., Macintyre, I.G. and Thompson, J.Jr. (1998) Formation of lithified micritic laminae in modern marine stromatolites (Bahamas): the role of sulfur cycling, *American Mineralogist* 83, 1482-1491.

Walter, M.R. (1983) Archean stromatolites: evidence of the earth's earliest benthos. In: Schopf, J.W., (ed.), *Earth's Earliest Biosphere,* Princeton University Press, Princeton, N.J., pp. 187-213.

Walter, M.R. (1994) Stromatolites: the main geological source of information on the evolution of the early benthos, in Bengtson, S. (ed.), *Early Life on Earth*, Nobel Symposium 84, 270-286.

Yallop M.L., De Winder B. and Paterson, D.M. (1994) Microbially mediated processes in tide influenced deposits and their importance in stabilization and diagenesis of sediments: Texel survey, in Krumbein, W.E., Paterson, D.M. and Stal, L.J. (eds.), *Biostabilization of Sediments*, Bibliotheks und Informationssystem der Universitaet Oldenburg, Oldenburg, pp. 327-338.

Chapter 7

NITROGEN CYCLING IN MARINE CYANOBACTERIAL MATS

Lucas J. STAL
Netherlands Institute of Ecology
Department of Marine Microbiology
P.O. Box 140
NL-4400 AC Yerseke
The Netherlands
stal@cemo.nioo.knaw.nl

1. INTRODUCTION

Cyanobacteria are a diverse group of photoautotrophic prokaryotes. Many species possess a versatile metabolism but their dominant physiology is a plant-type oxygenic photosynthesis, using two photosystems (PS I and PS II) connected in series, allowing them to utilize light as the only source of energy (Bryant, 1994; Whitton & Potts, 2000). Just like plants, cyanobacteria use H_2O as the electron donor and evolve O_2. CO_2 serves as the carbon source and is fixed through the reductive pentose-phosphate pathway (Calvin cycle). It is partly used for the synthesis of structural cell material while another part is diverted to a glycogen-like storage polymer. This polysaccharide is mobilized in the dark and energy is generated through aerobic respiration. To some extend, dark metabolism allows even growth of the cyanobacterium. Under anaerobic conditions several species have been shown to generate energy by fermentation of glycogen (Stal & Moezelaar, 1997). Under anaerobic conditions in the light and in the presence of sulfide, several cyanobacteria are able to switch to anoxygenic photosynthesis, involving only PS I (Cohen *et al.*, 1986).

Virtually all cyanobacteria are able to use NO_3^- as source of N. NO_3^- is considered to prevail in the mostly aerobic environments in which cyanobacteria thrive. Its uptake and subsequent reduction to NH_4^+ and assimilation in the cell is closely associated with photosynthesis (Flores & Herrero, 1994). Under anaerobic conditions NH_4^+ is probably the preferred source of N, which can be taken up passively as NH_3. However, at very low concentration of NH_4^+, cyanobacteria may induce an active transport of this ion. Other sources of bound N that may be used by cyanobacteria are NO_2^-, urea or amino acids.

A property that is shared by many, though not all, cyanobacteria is the capacity of fixing N_2. The possibility of utilizing N_2, which is ubiquitously present in the atmosphere, is uniquely reserved to some specialized prokaryotes (Bacteria and Archaea) that possess a highly conserved enzyme, nitrogenase. Nitrogenase catalyses the reaction:

$$N_2 + 8[H] + 16\ ATP \bullet 2\ NH_3 + H_2 + 16\ ADP + 16\ P_i \qquad (1)$$

Hence, the fixation of N_2 is at the expense of a considerable amount of reducing equivalents and energy. Cyanobacteria cover this demand through photosynthesis but need to protect the extremely oxygen sensitive nitrogenase (Fay, 1992). With respect to the protection of nitrogenase from O_2 inactivation, three groups of cyanobacteria can be distinguished (Bergman *et al.*, 1997; Stal, 1995). (1) Heterocystous cyanobacteria confine nitrogenase to specialized cells, the heterocysts, which are unable to carry out oxygenic photosynthesis and which are essentially anoxic. (2) Filamentous and unicellular non-heterocystous cyanobacteria that are capable of aerobic N_2 fixation. Their strategy to protect nitrogenase from O_2 is still an enigma. (3) Filamentous and unicellular cyanobacteria that avoid O_2 and fix N_2 only under anaerobic conditions.

Cyanobacteria are among the least exacting organisms with respect to nutritional demands, particularly the N_2-fixing species. These organisms have therefore occupied almost all types of illuminated environments on earth, including extreme environments such as hot- and cold deserts, hot springs, sulfureta and hypersaline ponds and lakes. Cyanobacteria form often benthic communities, known as microbial or cyanobacterial mats. These are complex microbial ecosystems of which fossil examples (stromatolites) have shown that they existed as long as 3.5 billion years ago and are considered as the earliest signs of life on earth (Stal, 2000).

Cyanobacterial mats are usually characterized by their layered structure. The typical structure of microbial mats such as can be found in intertidal marine sediments all over the world is as follows. The layer of cyanobacteria is often covered by a thin layer of sand. Sometimes this top layer is pigmented due to empty sheaths of cyanobacteria and this may serve to protect them from high light intensities or UV irradiance (Garcia-Pichel & Castenholz, 1991). Also diatoms my occur in the top layer and cyanobacteria may migrate in order to occupy the layer with optimum light conditions. The organic matter that is produced by the photosynthetic activity of the cyanobacteria is degraded by a variety of heterotrophic bacteria, but the end-oxidation of low-molecular weight compounds is predominantly carried out by the sulfate-reducing bacteria. The sulfide which is produced by them can be used as electron donor for anoxygenic photosynthesis by the anaerobic purple sulfur bacteria, that form often a distinct layer beneath the cyanobacteria (Stal *et al.*, 1985; van Gemerden, 1993). Sulfide can also be oxidized by aerobic colorless sulfur bacteria. Only the phototrophic micro-organisms occur stratified in different layers. The chemotrophic organisms can be found throughout the mat without producing a typical layer. As a result of the metabolic activity of the phototrophic component of the mat, steep and fluctuating gradients of O_2, H_2S and pH are formed (Revsbech *et al.*, 1983). In the light, the mat may become highly supersaturated with O_2, while in the dark it may become anoxic up to the mat surface. The gradient of H_2S is just opposite of that of O_2. Both components may only co-exist at low concentrations at the anaerobic-aerobic transition layer (Krumbein *et al.*, 1979).

Primary productivity in the marine environment is generally limited by N. The formation of a cyanobacterial mat in marine intertidal sediments is therefore often associated with N_2 fixation. Although in many cases cyanobacteria are the predominant N_2-fixing organisms in these mats also other bacteria in the mat may fix N_2. Moreover, not all mat-forming cyanobacteria are N_2 fixers. While the research has been mostly focused on the N_2-fixing cyanobacteria, amazingly few investigators have addressed the other aspects of the N-cycle in cyanobacterial mats. This chapter reviews what is known and discusses the importance of the various steps of the N cycle for the N budget of marine cyanobacterial mats.

2. THE NITROGEN CYCLE

Nitrogen occurs in nature in great variety of redox-states from its most reduced form, ammonia (NH_3, redox number –3) to its most oxidised form, nitrate (NO_3^-, redox number +5). Nitrogen is therefore subject in nature to many biological transformations, together forming the biological nitrogen cycle (Fig. 1) (Herbert, 1999). About 7 – 10% of biomass is composed of nitrogen and, after carbon, this element is therefore quantitatively most important. Nitrogen occurs in living matter in its reduced form. Organisms can acquire nitrogen in different ways. Many heterotrophic organisms assimilate organic nitrogen, e.g. amino acids or take up ammonium. Particularly autotrophic organisms may take up nitrate which is subsequently assimilated after reducing it to ammonium (assimilatory nitrate/nitrite reduction). Only a few specialized micro-organisms are capable of using dinitrogen (N_2). These organisms are without exception prokaryotes (Bacteria or Archaea) that are able to synthesize the enzyme complex nitrogenase. Nitrogenase reduces N_2 to ammonium, which is subsequently assimilated. Many Eukarya live symbiotically with N_2-fixing prokaryotes and as such can also grow at the expense of N_2.

One of the first steps in the decomposition of organic matter after polymers have been hydrolyzed to low-molecular compounds is ammonification, by which the (reduced) nitrogen is removed and liberated as ammonium. Ammonium represents an energy-rich compound and ammonium-oxidizing bacteria live at the expense of the oxidation of ammonia to nitrite and a second group of bacteria, the nitrite-oxidizing bacteria, oxidize nitrite further to nitrate. Both groups of bacteria are aerobes and use O_2 as terminal electron acceptor. Moreover, these ammonium-oxidizing bacteria are autotrophic, i.e. they fix CO_2 and use this as the sole carbon source. Heterotrophic ammonium oxidation occurs among pro- and eukaryotes but is quantitatively not important for the nitrogen cycle (Kuenen & Robertson, 1988). Some micro-organisms are capable of anaerobic ammonium oxidation.

Anaerobic ammonium oxidation (anammox) is a recently discovered part of the biological nitrogen cycle (Jetten *et al.*, 1999). This process involves the conversion of ammonium and nitrite to dinitrogen gas. Ammonium is oxidized using nitrite as the electron acceptor ($NO_2^- + NH_4^+ \quad N_2 + 2H_2O$). Also aerobic ammonium oxidizers have been shown to be capable of denitrification under low concentrations of dissolved oxygen by reducing nitrite or NO_2 to N_2 (Poth, 1986; Schmidt & Bock, 1997). However, these organisms prefer O_2 as the terminal electron acceptor and denitrification is 25-fold lower than the novel anammox reaction. It was recently shown that the organism carrying out anammox is a new, autotrophic member of the order *Planctomycetales* (Strous *et al.*, 1999). This organism has been found in a wastewater treatment plant with high ammonium concentrations. It is not known whether this process occurs in microbial mats. However, the occurrence of an aerobic – anaerobic transition in microbial mats and the importance of ammonium rather than nitrate, might present an environment in which anammox could occur. Since denitrification from nitrate does not seem to play a major role in cyanobacterial mats and the fate of the fixed N_2 is unknown it will be worthwhile to investigate this possibility.

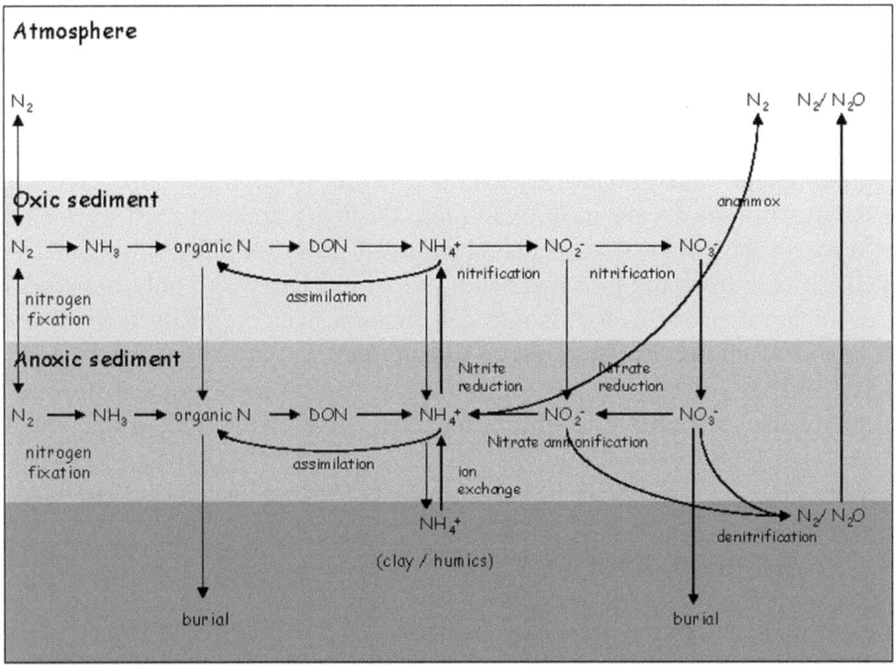

Figure 1. The nitrogen cycle in a marine cyanobacterial mat

As mentioned above, nitrate can be used through the assimilatory nitrate reduction pathway and as such be used as source of nitrogen. Nitrite is an intermediate in assimilatory nitrate reduction and when available in the environment can also itself be used as source of nitrogen. Only organisms that possess the assimilatory nitrate reduction pathway and the appropriate uptake system can use nitrate or nitrite as source of nitrogen. This pathway is particularly common in the cyanobacteria and eukaryotic algae. Nitrate reduction requires a large amount of low-potential electrons and also ist uptake and assimilation demands energy. Therefore, nitrate represents a source of nitrogen that is only marginally lower in energetic cost than N_2. The other fate of nitrate is dissimilatory reduction. The first step is the dissimilatory reduction of nitrate to nitrite. Nitrite can either be reduced to ammonia (nitrate ammonification) or via N_2O to N_2 (denitrification). Nitrate reduction and denitrification is a form of anaerobic respiration in which nitrate, nitrite or one of the intermediates serve as terminal electron acceptor. Nitrate reduction may also occur uncoupled of respiratory electron transport and simply serve as electron sink as is mostly the case with nitrate ammonification. However, some micro-organisms are known that couple the reduction of nitrate to ammonia to terminal electron transport and therefore is associated with energy generation. Not many studies of nitrate ammonification have been undertaken and it is unknown whether this process is important in microbial mats. It has been shown that the organisms involved are heterotrophic, fermentative bacteria occurring in numbers of 10_6-10_7 g dry weight of sediment in certain coastal surface sediments (Herbert, 1999).

It may be clear from the description above that nitrogen is subject to rapid biological transformations. Several essential steps in the nitrogen cycle are catalyzed exclusively by prokaryotes. Denitrification is such a process that leads to the conversion of bound nitrogen to molecular dinitrogen (N_2) which escapes from the biosphere into the atmosphere. The only biological process that reverses this loss is nitrogen fixation, which equally is a process unique to a small but diverse group of prokaryotes.

The nitrogen cycle has also a geological component. Organic nitrogen and NO_3^- may be subject to burial and NH_4^+ may be bound to clays and humics.

3. NITROGEN FIXATION

Stal *et al.* (1985) have investigated the structure and development of cyanobacterial mats of the North Sea island Mellum (Wadden Sea, Germany). These mats possessed high nitrogenase activity (Stal *et al.*, 1984). The mats of this desert island developed during spring and summer on an extended intertidal sand flat, in the upper littoral. Three different types of mats were distinguished, representing three different stages of development. Site 1 was considered as an initial stage of development of a cyanobacterial mat. There were no macroscopic structures, typical for mature cyanobacterial mats. The cyanobacteria usually occurred underneath a layer of sand, and the colonized areas could be recognized by the smooth flat surface of the sediment, while neighboring un-colonized sediments showed ripples that were formed by water running of the sediment when the tide is on the ebb. This was taken as evidence that the cyanobacteria were able to fix the sand grains and increased the erosion threshold. Site 2 was characterized by macroscopic sediment structures produced by continued sediment trapping by cyanobacterial mats in consecutive years. Consequently, these mats were found much higher in the intertidal flat and in contrast to site 1, site 2 was not submerged at every high tide. Site 3 was located in the upper littoral where matured cyanobacterial mats occurred that were only irregularly flooded (mainly at spring tide). These mats also contained some rooted littoral vegetation and could be considered as an intermediate stage to the development of a salt marsh.

The dominant cyanobacterium in the mats of site 1 was identified by Stal & Krumbein (1985) as *Oscillatoria limosa* on the basis of the morphological description of Geitler (1932). This organism possessed a thin sheath and it was motile by gliding (and rotating along its length axis). However, it was already emphasized that this morphotype could also easily be identified as *Lyngbya aestuarii*. On the basis of 16S rDNA sequence analyses it has become clear that *O. limosa* (strain 23) is identical to *L. aestuarii* (Ernst, Wollenzien & Stal, unpublished). As has been reported for several natural occurring and isolated *L. aestuarii* (Bergman *et al.*, 1997), also *O. limosa* (strain 23) is capable of N_2 fixation under aerobic and photoautotrophic conditions (Stal & Krumbein, 1981). Site 1 rarely was inhabited by the cosmopolitan mat-forming cyanobacterium *Microcoleus chthonoplastes*, which was the dominant species in the sites 2 and 3. Both sites also usually contained considerable numbers of *L. aestuarii* (*O. limosa*, strain 23) and the presence of this organism was correlated with the level of nitrogenase activity in these mats.

Table 1. Nitrogenase activity and increase of chlorophyll *a* content in microbial mats at 3 sites on the intertidal flat of the island of Mellum, Germany

	Site 1	Site 2	Site 3
	Cyanobacterial biomass: chlorophyll *a* (mg m-2)		
July	8	154	292
October	111	326	536
Increase Chl	103	172	244
Growth rate (d-1)	0.030	0.008	0.007
	ARA (µmol mg-1 h-1)		
N2-ase activity July	5.4	1.3	2.2
Average N2-ase activity	1.8	1.4	0.7
	ARA (µmol m-2 h-1)		
Average N2-ase activity	106	330	298
	Cyanobacterial biomass: chlorophyll *a* (mg m-2)		
Increase Chl (calc)	69	214	193

The mats of Mellum showed a more or less exponential increase in the content of chlorophyll *a* during the period July to October (Stal *et al.*, 1985). The absolute amounts of chlorophyll *a* were lowest at site 1 and highest at site 3, with an intermediate value at site 2. The same was true for the absolute increase in chlorophyll *a* (Table 1). However, when the relative increase of chlorophyll *a* was considered (specific growth rate), it was clear that site 1 revealed by far the highest growth rate, which was approximately 4-fold higher than measured for the sites 2 and 3. The much higher growth rate of the mat at site 1 was also reflected in the measured nitrogenase activity, which was 3-4-fold higher than in the sites 2 and 3. This was particularly apparent from the measurements of N_2-ase activity at the onset of mat development, but also when the chlorophyll-normalized activity was averaged over the whole growth period (Table 1). When N_2-ase activity was expressed as per mat surface the highest absolute values were of course obtained in the mats with the highest biomass at the sites 2 and 3.

Table 2. Photosynthesis and nitrogen fixation at 2 sites of the microbial mats of the island of Mellum, Germany

Parameter	Site 1	Site 2
Integrated primary production (O2) (mmol L-1 24 h)	315	1070
(C) (g d-1 m-2)	1.5	5.2
(C) (g mg-1(Chla) d-1)	0.2	0.03
(C) d-1	0.020	0.003
N required for C fixation (mmol)	6.25	21.67
N fixed (mmol)	1.27	3.96
Excess C fixed (%)	80	82

The measured N_2-ase activities were used to calculate the increase in biomass that it would have allowed. Acetylene reduction activity (ARA) was converted in N_2 fixation using a conversion factor of 4 (Jensen & Cox, 1983). It was further assumed that C:N = 20, C:Chl = 100 (Ortega-Calvo & Stal, 1991) and that N_2 fixation occurred 12 h d_{-1} at the average rate. Using these assumptions it was calculated that the measured rate of N_2 fixation would allow a growth of chlorophyll *a* of 69, 214 and 193 mg m_{-2} for the period July – October at the sites 1, 2 and 3, respectively. This compared well with the measured amounts of respectively103, 172 and 244 mg m_{-2} chlorophyll *a*. The mean increase for all sites was 173 mg m_{-2} while this was 159 mg m_{-2} when calculated from N_2 fixation.

Villbrandt *et al.* (1990) investigated photosynthesis and N_2 fixation jointly in the cyanobacterial mats of Mellum. From their results it was calculated that just like N_2 fixation, primary production was almost 4-fold higher at site 2 compared to site 1. However, when these values were normalized to chlorophyll *a*, the highest specific rates of photosynthesis were calculated for site 1, which were about 7-fold higher than at site 2 (Table 2). These calculations were done on the basis of measurements of primary productivity, using the O_2 microsensor technique (Revsbech & Jørgensen, 1983), converting O_2 production to CO_2 fixation using a conversion factor of 0.8 and a porosity of 50%. Photosynthesis was integrated over the top 1 mm, which represented the whole photosynthetically active mat (Villbrandt *et al.*, 1990). The specific growth rates calculated from these photosynthesis data, 0.020 and 0.003 d_{-1} for site 1 and 2, respectively (Table 2), matched those calculated on the basis of the increase of chlorophyll *a* remarkably well (respectively 0.030 and 0.008 d_{-1}

for site 1 and 2, table 1). If the observed C fixation would have been used for the synthesis of structural cell material, 6.25 and 21.67 mmol N (d^{-1} m^{-2}) would have been needed for sites 1 and 2, respectively. However, only 1.27 and 3.96 mmol N were really fixed at the sites 1 and 2, respectively. This means that for both sites C fixation was about 80% in excess of N fixation. These calculations assume that the mats did not receive any external input of N, except N_2, did not loose N by e.g. denitrification and not taking into account the N required for non-cyanobacterial biomass (anoxygenic phototrophic bacteria, chemotrophic bacteria).

It is likely that the excess C that is fixed in these cyanobacterial mats was excreted as extracellular polysaccharides, as sheath material and as colloidal carbohydrate. It is known that a substantial amount of microbial mats is composed of carbohydrates that form a gel in which the micro-organisms are embedded. It is conceived that these carbohydrates are important to render stability to the sediment and result in the increase of erosion threshold. This has been studied particularly well in diatom biofilms in intertidal mudflats and it is interesting to note that in these ecosystems it has also been observed that up to 80% of photosynthetically fixed C is excreted as extracellular polysaccharides (Goto *et al.*, 1999; Middelburg *et al.*, 2000).

Amounts of N fixed in a selected number of benthic marine intertidal cyanobacterial communities ranged from 0.8 – 15 g m^{-2} $year^{-1}$ (Table 3). It should be noted that the systems with the highest rates often were dominated by heterocystous cyanobacteria. The fact that the measured rates varied over more than one order of magnitude may also be caused by the fact that the extrapolation of N_2 fixation on an annual basis were often made on a limited number of measurements. Nitrogen fixation in cyanobacterial mats greatly depend on the ambient temperature and irradiance, factors that fluctuate not only seasonally but also on a daily basis (Carpenter *et al.*, 1978).

Table 3. Annual rates of N2 fixation in a selected number of marine intertidal cyanobacterial mats.

System	N2-fixation rate (g N m-2 year-1)	Reference
Island of Mellum, Germany	0.8 – 1.5	Stal *et al.* (1984)
Tomales Bay, California (outer bay)	1.0	Joye & Paerl (1993b)
Tomales Bay, California (inner bay)	1.5	Joye & Paerl (1993b)
Sippewisset salt marsh, Massachusetts	2.3	Carpenter *et al.* (1978)
Colne Point marsh UK	6.0	Nedwell & Azni b Abd Aziz (1980)
Baja California, Mexico, Lyngbya mat	7.1 – 10.5	Stal, unpublished
Hiddensee Island, Baltic Sea	7.6	Savela (1983)
Bank End salt marsh, UK	10.1	Jones (1974)
Baja California, Mexico, Calothrix mat	12.2	Stal, unpublished
Flax Pond salt marsh, New York	13.4	Whitney *et al.* (1975)
Island of Schiermonnikoog, Netherlands	15.0	Staal and Stal, unpublished

4. DENITRIFICATION

Denitrification is the dissimilatory reduction of nitrate to a gaseous product, usually dinitrogen (N_2), but also nitrous oxide (N_2O). By definition, the first step, the reduction of nitrate to nitrite, is not denitrification. Dissimilatory nitrate reduction and denitrification (NO_3^- NO_2^- [NO] N_2O N_2) is a form of anaerobic respiration, in which nitrate or any of the intermediates is used as terminal electron acceptor for respiratory electron transport (Kuenen & Robertson, 1988). A large variety of different microorganisms are capable of carrying out the whole sequence of the reduction of nitrate to N_2 or only certain steps. As a rule, these bacteria are facultative anaerobes, i.e. they will preferentially respire at the expense of O_2 when available. Most denitrifying bacteria are heterotrophs, i.e. depend on an organic substrate, but also some chemoautotrophic bacteria such as the sulfide-oxidizing *Thiobacillus* or ammonium-oxidizing *Nitrosomonas* are known to denitrify. Denitrification is considered as the key process in nature that causes a loss of bound nitrogen from the biosphere. Biological nitrogen fixation is considered as the only process that can counteract this loss. It is often assumed that denitrification is also the main loss factor of bound nitrogen in N_2-fixing cyanobacterial mats. However, remarkably few investigations have addressed this aspect.

Azni b Abd Aziz & Nedwell (1986) investigated the nitrogen cycle of a salt marsh in Colne Point, UK. Part of the salt marsh was covered by cyanobacterial mats. Nitrogen fixation was particularly associated with these cyanobacterial mats, although the authors did not report which species were present in these mats. They found only low rates of *in situ* denitrification and the lowest rates were associated with the cyanobacterial mat. The capacity of denitrification was nevertheless high but the *in situ* rates were limited by the availability of nitrate. Only a small portion of the NH_4^+ that was formed during breakdown of organic matter was nitrified to NO_3^-. Moreover, most of the NO_3^- that was present was reduced to NH_4^+, rather than being denitrified. Hence, in the Colne Point salt marsh very little N was lost by denitrification. The small losses of N from this salt marsh were mostly due to wash-off during the tide. These losses were compensated by N_2 fixation, by which the salt marsh as a whole showed a balanced nitrogen budget. Joye & Paerl (1994) investigated in detail nitrogen cycling in microbial mats of Tomales Bay, California, USA. Their results agree largely with those of Azni b Abd Aziz & Nedwell (1986) that nitrogen-fixing cyanobacterial mats are sources rather than sinks of combined nitrogen. (Joye & Paerl, 1994) reported that N_2 fixation and denitrification were 6 – 79 mg N m_{-2} d_{-1} and 1 – 10 mg N m_{-2} d_{-1}, respectively. Denitrification was 6 – 20% of N_2 fixation. While N_2 fixation exceeded denitrification during the light, in the dark these processes were more or less in balance. Hence, in order to determine the net N input, it was necessary to integrate both N_2 fixation and denitrification over 24-h periods.

The mats of the Tomales Bay were composed of heterocystous and non-heterocystous cyanobacteria, representatives of both are known to fix N_2. The mats of the salt marsh contained both heterocystous and non-heterocystous cyanobacteria ('inner bay') and fixed 1.5 g N m_{-2} $year_{-1}$, while the mudflats mats ('outer bay') fixed somewhat less, 1 g N m^{-2} $year^{-1}$ (Joye & Paerl, 1993b). These values corresponded very well with the values of 0.8 – 1.5 g m^{-2} $year^{-1}$ reported by Stal *et al.* (1984) for the Mellum mats. The values given for C fixation in the mats of Tomales Bay were 11.1 and 14.3 g C m^{-2} $year^{-1}$ for the inner- and outer bay, respectively. These values were 24 and 61% (respectively for the inner- and outer bay mats) in excess of what was required for balanced growth, based on the measured N_2 fixation and the Redfield ratio of 6.6. This is much less than what was calculated for the mats of Mellum. However, Joye & Paerl (1993b) used $_{14}C$ incorporation, while the calculations for the Mellum mats were based on measurements of O_2 evolution. Measurements based on $_{14}C$ incorporation do not account for losses during the incubation, which may in part explain the lower excess of C fixation. The excess C fixation suggests a shortage of N in these cyanobacterial mats. Denitrification in these mats is usually low or not detectable, however the capacity of denitrification is large (Joye & Paerl, 1993a). The salt marsh mats ('upper bay') do not show denitrification when no additions were made, however, when nitrate was added denitrification could reach up to 25% of the rate of N_2 fixation. This high capacity of denitrification is used during run-off events after rainfall. Rain is a source of N, which caused N_2 fixation to decrease and denitrification to increase. Run-off events turned the mats into a sink for N, rather than a source and denitrification became limited by organic carbon, rather than nitrate. On an annual basis, denitrification in the mats of Tomales Bay amount only 15% of N_2 fixation, and, hence, these mats were a source of fixed nitrogen. However, since the mats occur only in the inter tidal area, they contribute only little to the N budget of the whole bay, which acts as a sink for N.

Denitrification is regulated by the availability of NO_3^- and organic matter, and by O_2. Sources of nitrate are (1) *in situ* nitrification, (2) diffusion from the overlying water column and (3) advection of NO_3^- rich ground- or porewater. Nitrification is possible when NH_4^+ and O_2 is available. While NH_4^+ is usually present in appreciable amounts in the porewater, O_2 is available only during the light, when photosynthesis occurs. However, this O_2 will inhibit denitrification and the NO_3^- that is formed must diffuse to the oxic-anoxic transition layer. Cyanobacterial mats in inter tidal areas are most of the time emersed. Hence, diffusion of NO_3^- from the overlying water most of the time will not occur. Also seawater usually has a low concentration of NO_3^-. However, when mats are located close by a river mouth, higher concentrations of NO_3^- are likely, as is the case with run-off after rainfall (Joye & Paerl, 1994). In such cases, the overlying water may be a source of NO_3^-. Ground- and porewater in cyanobacterial mats are usually very low in NO_3^- and are not likely to serve as a source for denitrification. Even if NO_3^- would be present, it is questionable if it is available for denitrifying bacteria. In the light, NO_3^- is rapidly taken up by the phototrophs (Rysgaard *et al.*, 1996) and, as has been shown by Azni b Abd Aziz & Nedwell (1986), nitrate-ammonification is probably more important than denitrification. Heterotrophic denitrifyers require a source of organic matter as substrate. While microbial mats are usually very rich in organic matter, its availability for denitrification is questionable. A considerable part of the extracellular organic matter is composed of recalcitrant polymeric material, which is not readily available for microbial degradation. The low-molecular organic matter is produced by extrusion of osmoregulatory solutes after an osmotic down shock (rainfall), excretion of glycolate as a result of photorespiration in the light and a variety of fermentation products produced in the dark (Stal & Moezelaar, 1997). Considering the low concentrations of NO_3^- ($< 1\ \mu M$) in the porewater and very high concentrations of SO_4^{2-} in seawater (28 mM), it is not very likely that denitrifying bacteria will succeed to decompose any of these compounds and will loose the competition from sulfatereducing bacteria. In fact, the sulfide that is produced by the latter is known to inhibit denitrification (Sørensen, 1987). Autotrophic denitrifying sulfide-oxidizing bacteria may be important in the dark or in the aphotic zone of the mat, provided that nitrate is available. In the photic zone in the light these bacteria will loose the competition from anoxygenic phototrophic bacteria (Visscher *et al.*, 1992b). Cyanobacterial mats may contain large numbers of colorless sulfide oxidizing bacteria (Visscher *et al.*, 1992a) and some of these species are known to possess the capacity of denitrification. Nothing is known about autotrophic denitrifying ammonium oxidizers in cyanobacterial mats. These organisms might play a role since they do not depend on nitrate as electron acceptor but nitrite, which they provide themselves with.

O_2 is the third regulatory factor for denitrification. In the presence of O_2 denitrifying bacteria will prefer aerobic respiration rather than denitrify. Only at low O_2 concentrations or anoxia, denitrification will take place, although also reports of aerobic denitrification exist. Cyanobacterial mats are characterized by steep and fluctuating gradients of O_2. In the light, the mat will become supersaturated with O_2 due to photosynthesis, while in the dark the mat will quickly become anoxic, as a result of respiratory activity and diffusion limitation. Hence, the suboxic transition zone, where denitrification can be expected, is shifting up and down in the sediment, keeping pace with the light impinging on the mat surface. It is unlikely that denitrifying bacteria follow the suboxic transition zone, but rather are distributed throughout the mat, either carrying out aerobic respiration or denitrification. This would require that the organisms maintain the capacity of denitrification constitutive. Indeed it has been observed that denitrification is much higher during night time (Joye & Paerl, 1994).

Hence, cyanobacterial mat photosynthesis may affect denitrification negatively in several ways. It increases porewater O_2 concentration and increases the depth of the aerobic zone, and thereby the distance that NO_3^- must travel into the denitrifying community. Also, increased photosynthesis will lead to increased uptake and assimilation of NO_3^- by the cyanobacteria.

Many marine cyanobacterial mats show high rates of N_2 fixation, indicating the lack of other sources of N. However, a continued fixation of N_2 would eventually result in an accumulation of combined nitrogen, unless losses occur. The few investigations that measured denitrifcation and N_2 fixation simultaneously, have demonstrated that the former is very small compared to the latter or even absent. It is possible that denitrification occurs during a different period of the year than N_2 fixation, but the reports in the literature are not consistent with regard to this (Herbert, 1999). Some reports show that denitrification occurs with the same seasonality as N_2 fixation, whereas others do not and e.g. find denitrification to be important during the winter period. Of course, fixed N will be recycled in the mat and become available to other, non-diazotrophic micro-organisms. However, the total biomass of an active microbial mat is finite as is this sink of N. The third option is that fixed N is lost in the overlying water. This could happen only during immersion of the mat at high tide. Mats are sloughed off and eroded away during high tide, particularly due to strong currents and wave action. Hence, there is certainly a flux of combined N into the water. In addition, there may be a geological component in which N is stored. None of these possibilities have been investigated or quantified in sufficient detail. Last but perhaps not least is the possibility that anaerobic ammonium oxidation represents a loss factor in microbial mats.

5. NITRIFICATION

There are hardly any studies of nitrification in marine cyanobacterial mats, but the scant information available show that this process is of minor importance. Azni b Abd Aziz & Nedwell (1986) calculated that for a cyanobacterial mat of the Colne salt marsh (UK) only 16.4% of the NH_4^+ produced was nitrified and that this amount was very small compared to N_2 fixation. Nitrification is the oxidation of NH_4^+ to NO_3^- which occurs in two stages with NO_2^- as intermediate. The process is carried out by two groups of obligate aerobic and autotrophic bacteria. The ammonium-oxidizing bacteria produce NO_2^-, which in turn is oxidized to NO_3^- by nitrite-oxidizing bacteria. In shallow coastal sediments *Nitrosomonas* spp. and *Nitrobacter* spp. are the dominant species that respectively oxidize NH_4^+ and NO_2^- (Herbert, 1999). Nitrification depends on a number of environmental factors such as the temperature, pH and salinity. Furthermore, the concentrations of NH_4^+, O_2 and CO_2 are important. Because of the relative low energy gain from the oxidation of NH_4^+ and the high cost of CO_2 fixation, growth of nitrifying bacteria is usually slow and the yield is low. Nitrifying bacteria fix CO_2 through the Calvin cycle. Theoretically, a small number of nitrifying bacteria may oxidize a large amounts of NH_4^+.

In a cyanobacterial mat, nitrifying bacteria must compete with cyanobacteria for NH_4^+ and CO_2. Because nitrification depends on O_2 it can only take place in the light, when cyanobacteria actively photosynthesise and fix CO_2 and assimilate NH_4^+. Photosynthesis also results in an alkalinization and the pH may reach values well over 10. Moreover, the cyanobacterial mat will supersaturate with O_2, reaching easily 3 times air saturation. Such high O_2 concentrations may be inhibitory to nitrifying bacteria (Henriksen & Kemp, 1988). On the other hand, in the dark the mat turns anoxic. Ammonium oxidisers are known to denitrify under low O_2 concentrations, using NO_2^- as electron acceptor and producing N_2O (Hynes & Knowles, 1984; Jørgensen *et al.*, 1984). However, the rates at which ammonium is anaerobically oxidized by these organisms is slow. Much faster is a newly discovered anaerobic ammonium oxidation, known as anammox (anaerobic ammonium oxidation) (Jetten *et al.*, 1999). The bacterium that carries out this reaction oxidizes NH_4^+ to N_2, while reducing NO_2^- just as well to N_2. This process does not depend on nitrate but rather on nitrite, the first product of aerobic ammonium oxidation, which could occur in the light. During the subsequent anaerobic dark period the anammox reaction could then oxidize NH_4^+ and reduce NO_2^- both to N_2. This would be consistent with the relatively high concentrations of NH_4^+ (50 μM) and the low NO_2^- (<0.5 μM) found in the mats of Mellum (Stal *et al.*, 1984). Hence, if at all, ammonium oxidation in cyanobacterial mats may be particularly confined to the dark or to the anoxic part of the mat. If this is the case remains to be seen.

6. DIVERSITY OF NITROGEN-FIXING ORGANISMS

Nitrogen fixation in cyanobacterial mats is often attributed to the cyanobacteria. Even though nitrogen fixation in cyanobacterial mats is often light stimulated or occurs mainly during the day, the conclusion that the cyanobacteria are the major N_2-fixing agents does no give justice to the complexity of the mat community (Steppe *et al.*, 1996). Besides the cyanobacteria that form the most conspicuous top layer of the mat, very often anoxygenic phototrophic bacteria are present as well. In marine microbial mats, sulfate-reducing bacteria form a key group of bacteria, responsible for the decomposition of organic matter, providing sulfide for the anoxygenic phototrophic bacteria and for colorless sulfide-oxidizing bacteria. Also the sulfate-reducing bacteria are responsible maintaining anoxia in the mat, providing an environment for a variety of anaerobic or facultative anaerobic heterotrophic bacteria, involved in the degradation of complex organic matter. All of these groups of micro-organisms comprise diazotrophic representatives. The stimulation of N_2 fixation by light may be attributed by anoxygenic phototrophic bacteria the more so as at the same time activity is also stimulated by low O_2 or the inhibition of oxygenic photosynthesis. Also, stimulation of N_2 fixation by light can be explained by exudates of cyanobacteria that fuel heterotrophic diazotrophic bacteria. Moreover, in many mats of non-heterocystous cyanobacteria, nitrogen fixation is confined to the night. Since in culture, many non-heterocystous diazotrophic cyanobacteria also fix N_2 during the dark period when grown under an alternating lightdark regime (Bergman *et al.*, 1997), fixation of N_2 at night does not prove that cyanobacteria are not involved, but still heterotrophic bacteria may do the job.

Many cyanobacterial mats are formed by the cosmopolitan non-heterocystous cyanobacterium *Microcoleus chthonoplastes* (Stal, 2000). This species is easily recognizable by its occurrence in bundles of trichomes, enveloped by a common polysaccharide sheath. Since N_2-fixing activity has often been found associated with these mats, it was originally proposed that bundle formation was a morphological adaptation that would allow this organism to fix N_2 and to protect nitrogenase from oxygen inactivation. It is now clear that *M. chthonoplastes* is unable to fix N_2. Zehr *et al.* (1995) did not detect any cyanobacterial *nifH* sequence in mats dominated by *M. chthonoplastes* but only heterotrophic *nifH* genes. Steppe *et al.* (1996) were unable to detect *nifH* genes in 4 cultures of *M. chthonoplastes* and these strains were unable to grow in the absence of combined N and did not reduce acetylene. *M. chthonoplastes* strains isolated from a variety of different cyanobacterial mats from all over the world form a tight phylogenetic cluster and 16S rDNA sequences differ hardly among these isolates (Garcia-Pichel *et al.*, 1996). None of these strains were capable of diazotrophic growth or acetylene reduction (Villbrandt & Stal, unpublished results). The aerobic N_2-fixing cyanobacterium identified by Pearson *et al.* (1979) as *M. chthonoplastes*, was later re-assigned as *Symploca* sp.(Janson *et al.*, 1998). Steppe *et al.* (1996) detected several non-cyanobacterial *nifH* genes associated with their non-axenic cultures of *M. chthonoplastes*, which were also found in the mats of Tomales Bay, California and in a cryptobiotic soil crust in Utah. These authors conceived that heterotrophic bacteria formed diazotrophic consortia with *M. chthonoplastes*. The heterotrophic bacterial component of these consortia would receive organic matter from the cyanobacterium. This could be through photorespiratory glycolate excretion in the light, fermentation products in the dark (Stal & Moezelaar, 1997) or even through the decomposition of extracellular polymeric substances. The bacteria would then provide the cyanobacterium with fixed N. The bacteria could also return CO_2 to the cyanobacterium. There may further be an exchange of other growth factors between the two components of the microbial consortium.

The *nifH* gene sequences of cyanobacteria are sufficient variable to distinguish cyanobacterial genes from other Bacteria and Archaea and also to distinguish heterocystous from non-heterocystous cyanobacteria (Ben-Porath & Zehr, 1994). The sequences of heterocytsous cyanobacteria form a coherent cluster (Zehr *et al.*, 1997). Also *Gloeothece* and *Synechococcus*, and *Xenococcus* and *Myxosarcina* cluster, but the *nifH* sequences of non-heterocystous filamentous cyanobacteria are not closely related. The phylogeny of the *nifH* sequences are largely consistent with 16S rDNA. Mats that contain obvious N_2-fixing cyanobacteria such as heterocystous cyanobacteria do contain cyanobacterial *nifH* gene sequences (Steppe *et al.*, 1996). Zehr *et al.* (1995) found for instance that when the cyanobacterial mat was dominated by *Lyngbya* sp. contained *nifH* sequences that corresponded with those from *L. lagerheimeii*.

The *nifH* genes that were found by Zehr *et al.* (1995) in the mats dominated by *M. chthonoplastes* were most closely related to anaerobic bacteria such as the anoxygenic phototrophic purple sulfur bacterium *Chromatium* and the sulfate-reducing bacterium *Desulfovibrio*. Other sequences were from unknown bacteria related to sequences obtained from seagrass beds. A few sequences were related to the •-proteobacteria (containing *Azotobacter* and *Klebsiella*). The mats contained a high diversity of *nifH* genes but their relative abundance is not known and thus far it remains to be investigated whether these genes are expressed in the environment.

Mats dominated by *M. chthonoplastes* usually contain low nitrogenase activity without showing a peculiar pattern over a 24-h day (Steppe *et al.*, 1996). As argued above, heterotrophic bacteria associated with the cyanobacteria are responsible for nitrogen fixation in these mats. Mats of the non-heterocystous cyanobacterium *Lyngbya* sp. or the heterocystous *Calothrix* sp. show much higher nitrogen-fixing capacities that are clearly attributed to the cyanobacteria. These mats show completely different diurnal patterns of nitrogenase activity (Stal, 1995). The non-heterocystous species fix N_2 predominantly during the night, whereas the heterocystous species confine nitrogenase activity completely to the daytime. It is not fully known what factors determine when a heterocystous cyanobacterium or a non-heterocystous species becomes dominant. The majority of marine cyanobacterial mats are built by non-heterocystous species, notably *Lyngbya – Oscillatoria* types and mats of heterocystous species are relatively rare. Now, heterocystous species are in fact rare in the marine environment, but notwithstanding, these species seem best equipped for diazotrophic photosynthetic growth. The spatial separation of the incompatible processes of oxygenic photosynthesis and nitrogen fixation in two different cell types, respectively vegetative cells and heterocysts, is very efficient (Stal *et al.*, 1994). Based on research carried out in diazotrophic mats in Baja California, Mexico, Stal (1995) conceived that heterocystous cyanobacteria were out-competed in mats with extreme dynamics of O_2 (very high super saturation in the light and anoxic conditions in the dark) and perhaps the presence of free sulfide. But also high salinity may be a factor that acts against heterocystous species.

Recently mats have been discovered on the North Sea beaches of the Wadden islands in the Netherlands that contain a high diversity of cyanobacteria, including heterocystous and non-heterocystous N_2-fixing cyanobacteria. These mats are influenced by both freshwater and seawater and are further subject to extreme changes in environmental conditions.

Preliminary results have shown that different species become active in N_2 fixation at different times and besides heterocystous and non-heterocystous cyanobacteria, probably also other bacteria are involved (Stal, unpublished results). The high diversity of N_2-fixing bacteria in these mats seems to function as an adaptation to cope with highly unstable environmental conditions. Together they achieve an unprecedented high rate of N_2 fixation (Table 3).

7. CONCLUDING REMARKS

Virtual all marine cyanobacterial mats show very high rates of N_2 fixation, which is probably inherent to the low concentrations of combined N in seawater and to the high biomass of which these mats are usually built of. Not in all cases are cyanobacteria the N_2-fixing organisms. For instance, N_2 fixation in mats that are exclusively built by *Microcoleus chthonoplastes* is most likely due to other (heterotrophic) bacteria associated with them. Mats which are dominated by *Lyngbya – Oscillatoria* type of cyanobacteria often show very high rates of N_2 fixation which can be attributed to them. Marine mats built by heterocystous cyanobacteria are rare, but they show generally the highest rates of N_2 fixation. Mats may also be composed of a variety of different diazotrophic micro-organisms. High diversity of diazotrophic microorganisms may render a great ecological stability to these mats. Very little information is available on other aspects of the N_2 cycle in cyanobacterial mats. Measured rates of denitrification and (aerobic) nitrification are very low compared to N_2 fixation. Hence, the fate of the fixed N_2 in cyanobacterial mats remains as yet unknown.

ACKNOWLEDGEMENTS

This is publication 3123 of NIOO-CEMO.

REFERENCES

Azni b Abd Aziz S, Nedwell DB. 1986. The nitrogen cycle of an east coast, U.K. saltmarsh: II. Nitrogen fixation, nitrification, denitrification, tidal exchange. *Estuarine Coastal and Shelf Science* 22: 689-704.

Ben-Porath J, Zehr JP. 1994. Detection and characterization of cyanobacterial *nifH* genes. *Applied and Environmental Microbiology* 60: 880-887.

Bergman B, Gallon JR, Rai AN, Stal LJ. 1997. N2 fixation by non-heterocystous cyanobacteria. *FEMS Microbiology Reviews* 19: 139-185.

Bryant DA. 1994. *The molecular biology of cyanobacteria*. Dordrecht: Kluwer Academic Publishers.

Carpenter EJ, Van Raalte CD, Valiela I. 1978. Nitrogen fixation by algae in a Massachusetts salt marsh. *Limnology and Oceanography* 23: 318-327.

Cohen Y, Jørgensen BB, Revsbech NP, Poplawski R. 1986. Adaptation to hydrogen sulfide of oxygenic and anoxygenic photosynthesis among cyanobacteria. *Applied and Environmental Microbiology* 51: 398-407.

Fay P. 1992. Oxygen relations of nitrogen fixation in cyanobacteria. *Microbiological Reviews* 56: 340-373.

Flores E, Herrero A. 1994. Assimilatory nitrogen metabolism and its regulation. *The Molecular Biology of Cyanobacteria* 487-517.

Garcia-Pichel F, Castenholz RW. 1991. Characterization and biological implications of scytonemin, a cyanobacterial sheath pigment. *Journal of Phycology* 27: 395-409.

Garcia-Pichel F, Prufert-Bebout L, Muyzer G. 1996. Phenotypic and phylogenetic analyses show *Microcoleus chthonoplastes* to be a cosmopolitan cyanobacterium. *Applied and Environmental Microbiology* 62: 3284-3291.

Geitler L. 1932. Cyanophycea. *Rabenhorst' Kryptogamen Flora* 14: 1-1196.

Goto N, Kawamura T, Mitamura O, Terai H. 1999. Importance of extracellular organic carbon production in the total primary production by tidal-flat diatoms in comparison to phytoplankton. *Marine Ecology Progress Series* 190: 289-295.

Henriksen K, Kemp WM. 1988. Nitrification in estuarine and coastal marine sediments. In: Blackburn H, Sørensen J, eds. *Nitrogen cycling in coastal marine environments*. New York: John Wiley and Sons, 207-249.

Herbert RA. 1999. Nitrogen cycling in coastal marine ecosystems. *FEMS Microbiology Reviews* 23: 563-590.

Hynes RK, Knowles R. 1984. Production of nitrous oxide by *Nitrosomonas europaea*: effects of acetylene, pH and oxygen. *Canadian Journal of Microbiology* 30: 1397-1404.

Janson S, Matveyev A, Bergman B. 1998. The presence and expression of *hetR* in the non-heterocystous cyanobacterium *Symploca* PCC 8002. *FEMS Microbiology Letters* 168: 173-179.

Jensen BB, Cox RP. 1983. Direct measurements of steady-state kinetics of cyanobacterial N2 uptake by membrane-leak mass spectrometry and comparisons between nitrogen fixation and acetylene reduction. *Applied and Environmental Microbiology* 45: 1331-1337.

Jetten MSM, Strous M, van de Pas-Schoonen KT, Schalk J, van Dongen UGJM, Van de Graaf AA, Logemann S, Muyzer G, Van Loosdrecht MCM, Kuenen JG. 1999. The anaerobic oxidation of ammonium. *FEMS Microbiology Reviews* 22: 421-437.

Jones K. 1974. Nitrogen fixation in a salt marsh. *Journal of Ecology* 62: 553-565.

Joye SB, Paerl HW. 1993a. Contemporaneous nitrogen fixation and denitrification in intertidal microbial mats - rapid response to runoff events. *Marine Ecology Progress Series* 94: 267-274.

Joye SB, Paerl HW. 1993b. Nitrogen fixation and denitrification in the intertidal and subtidal environments of Tomales Bay, California. In: Oremland RS, ed. *Biogeochemistry of global change. Radiatively active trace gases*. New York: Chapman & Hall, 633-653.

Joye SB, Paerl HW. 1994. Nitrogen cycling in microbial mats: rates and patterns of denitrification and nitrogen fixation. *Marine Biology* 119: 285-295.

Jørgensen KS, Jensen HB, Sorensen J. 1984. Nitrous oxide production from nitrification and denitrification in marine sediment at low oxygen concentrations. *Canadian Journal of Microbiology* 30: 1073-1078.

Krumbein WE, Buchholz H, Franke P, Giani D, Giele C, Wonneberger K. 1979. O2 and H2S coexistence in stromatolites. *Naturwissenschaften* 66: 381-389.

Kuenen JG, Robertson LA. 1988. Ecology of nitrification and denitrification. *The Nitrogen and Sulphur Cycles* 161-218.

Middelburg JJ, Barranguet C, Boschker HTS, Herman PMJ, Moens T, Heip CHR. 2000. The fate of intertidal microphytobenthos carbon: An in situ 13C-labeling study. *Limnology and Oceanography* 45: 1224-1234.

Nedwell DB, Azni b Abd Aziz S. 1980. Heterotrophic nitrogen fixation in an intertidal salt marsh sediment. *Estuarine Coastal and Shelf Science* 10: 699-702.

Ortega-Calvo JJ, Stal LJ. 1991. Diazotrophic growth of the unicellular cyanobacterium *Gloeothece* sp. PCC 6909 in continuous culture. *Journal of General Microbiology* 137: 1789-1797.

Pearson HW, Howsley R, Kjeldsen CK, Walsby AE. 1979. Aerobic nitrogenase activity associated with a non-heterocystous filamentous cyanobacterium. *FEMS Microbiology Letters* 5: 163-169.

Poth M. 1986. Dinitrogen gas production from nitrite by a *Nitrosomonas* isolate. *Applied and Environmental Microbiology* 52: 957-959.

Revsbech NP, Jørgensen BB. 1983. Photosynthesis of benthic microflora measured with high spatial resolution by the oxygen microprofile method: Capabilities and limitations of the method. *Limnology and Oceanography* 28: 749-756.

Revsbech NP, Jørgensen BB, Blackburn TH, Cohen Y. 1983. Microelectrode studies of the photosynthesis and O2, H2S and pH profiles of a microbial mat. *Limnology and Oceanography* 28: 1062-1074.

Rysgaard S, Risgaard-Petersen N, Sloth NP. 1996. Nitrification, denitrification, and nitrate ammonification in sediments of two coastal lagoons in Southern France. *Hydrobiologia* 329: 133-141.

Savela K. 1983. Nitrogen fixation by the blue-green alga *Calothrix scopulorum* in coastal waters of the Baltic. *Annals of Botany Fennica* 20: 399-405.

Schmidt I, Bock E. 1997. Anaerobic ammonia oxidation with nitrogen dioxide by *NItrosomonas eutropha*. *Archives of Microbiology* 167: 106-111.

Sørensen J. 1987. Nitrate reduction in marine sediment: pathways and interactions with iron and sulfur cycling. *Geomicrobiology Journal* 5: 401-421.

Stal LJ. 1995. Physiological ecology of cyanobacteria in microbial mats and other communities. *New Phytologist* 131: 1-32.

Stal LJ. 2000. Cyanobacterial mats and stromatolites. In: Whitton BA, Potts M, eds. *The Ecology of Cyanobacteria*. Dordrecht: Kluwer, 61-120.

Stal LJ, Grossberger S, Krumbein WE. 1984. Nitrogen fixation associated with the cyanobacterial mat of a marine laminated microbial ecosystem. *Marine Biology* 82: 217-224.

Stal LJ, Krumbein WE. 1981. Aerobic nitrogen fixation in pure cultures of a benthic marine *Oscillatoria* (cyanobacteria). *FEMS Microbiology Letters* 11: 295-298.

Stal LJ, Krumbein WE. 1985. Isolation and characterization of cyanobacteria from a marine microbial mat. *Botanica Marina* 28: 351-365.

Stal LJ, Moezelaar R. 1997. Fermentation in cyanobacteria. *FEMS Microbiology Reviews* 21: 179-211.

Stal LJ, Paerl HW, Bebout B, Villbrandt M. 1994. Heterocystous versus non-heterocystous cyanobacteria in microbial mats. In: Stal LJ, Caumette P, eds. *Microbial mats.Structure, development and environmental significance*. Heidelberg: Springer Verlag, 403-414.

Stal LJ, van Gemerden H, Krumbein WE. 1985. Structure and development of a benthic marine microbial mat. *FEMS Microbiology Ecology* 31: 111-125.

Steppe TF, Olson JB, Paerl HW, Litaker RW, Belnap J. 1996. Consortial N2 fixation: a strategy for meting nitrogen requirements of marine and terrestrial cyanobacterial mats. *FEMS Microbiology Ecology* 21: 149-156.

Strous M, Fuerst JA, Kramer EHM, Logemann S, Muyzer G, van de Pas-Schoonen KT, Webb R, Kuenen JG, Jetten MSM. 1999. Missing lithotroph identified as new planctomycete. *Nature* 400: 446-449.

van Gemerden H. 1993. Microbial mats: A joint venture. *Marine Geology* 113: 3-25.

Villbrandt M, Stal LJ, Krumbein WE. 1990. Interactions between nitrogen fixation and oxygenic photosynthesis in a marine cyanobacterial mat. *FEMS Microbiology Ecology* 74: 59-72.

Visscher PT, Prins RA, van Gemerden H. 1992a. Rates of sulfate reduction and thiosulfate consumption in a marine microbial mat. *FEMS Microbiology Ecology* 86: 283-293.

Visscher PT, van den Ende FP, Schaub BEM, van Gemerden H . 1992b. Competition between anoxygenic phototrophic bacteria and colorless sulfur bacteria in a microbial mat. *FEMS Microbiology Ecology* 101: 51-58.

Whitney DE, Woodwell GM, Howarth RW. 1975. Nitrogen fixation in Flax Pond: A Long Island salt marsh. *Limnology and Oceanography* 20: 640-643.

Whitton BA, Potts M. 2000. *The ecology of cyanobacteria. Their diversity in time and space.* Dordrecht: Kluwer Academic Publishers.

Zehr JP, Mellon M, Braun S, Litaker W, Steppe T, Paerl HW. 1995. Diversity of heterotrophic nitrogen fixation genes in a marine cyanobacterial mat. *Applied and Environmental Microbiology* 61: 2527-2532.

Zehr JP, Mellon MT, Hiorns WD. 1997. Phylogeny of cyanobacterial *nif*H genes: evolutionary implications and potential applications to natural assemblages. *Microbiology* 143: 1443-1450.

Chapter 8

DIVERSITY OF CYANO-BACTERIAL MATS

Georgii.A. ZAVARZIN
Institute of microbiology RAS,
117812, Moscow, Russia
gzavarzin@h-w-k.de

1. INTRODUCTION: CYANO-BACTERIAL MAT AS COMPLETE BIOTIC SYSTEMS

Sustainable existence of biotic systems depends on their fitness to environment, which means minimal changes caused by their activity in habitat. Otherwise changed environment will call to existence another biocoenosis fitted to it, as it occurs during succession. This statement is to some extent opposite to the Gaian idea of reshaping environment by biota to its own benefits. Cyano-bacterial community is the most sustainable biotic system, that dominated Earth from Early Proterozoic to the end of Neoproterozoic revolution and still survives in habitats from which it is not expelled by its successors (Sergeev at all. 1998). Throughout this time cyano-bacterial community transformed the Earth surface to what represents now the biosphere but despite all changes was able to survive thanks to close combining productive cyanobacterial and destructive bacterial components of the biogeochemical cycles. This is a reason why both groups of prokaryotes, photoautotrophic and organotrophic, should be included in the definition of cyano-bacterial community, being designated by the hyphen (Stal, Caumette, 1993). Cyanobacterial productive component depends on light, CO_2, and phosphorus with C:P ratio 100. Equimolar ratio of photoautotrophic assimilation $CO_2:O_2:C_{org}=1:1:1$ represents the main driving force of C_{org} cycle in the biosphere (fig.1). Bioticaly mediated cycle of O_2 causes the oxidative transformation of the biosphere due to the inability of decomposing organotrophs to utilize all C_{org} produced. Part of C_{org} is buried during the sedimentary process as kerogen, representing undecomposed mortmass together with amorphous humic substances (Tyson, 1999). Equimolar quantity of O_2 is liberated and involved in oxidative branches of

S and Fe cycles driven exclusively by prokaryotes with production and deposition of inert Fe-oxides and oceanic sulphates as O_2 sink (Schidlowski, 1993). Partially biologically mediated is inorganic CO_2-sink driven by Ca-carbonate-silicate weathering chain (for bibliog. see Schwartzman, 1999) leading to formation of carbonate deposits and production of clays as sustainable residual substances. Mortmass and kerogen are buried in clays during the sedimentary processes. Carbonate deposits of Proterozoic are obviously linked to cyanobacterial communities by formation of stromatolites. Geochemical calcium cycle is responsible for the neutrophilic conditions on the Earth and it is amplified by biologically mediated reactions, in Proterozoic – driven by microbes exclusively (Zavarzin, 2002).

Overall trophic system of cyanobacterial community is given as an oriented graph of universal trophic network for prokaryotic osmotrophic communities (fig.2). Functional groups of bacteria are given in boxes, substances—in ovals, interactions and flows—as arrows. Destructive loop is closed by aerobes, utilizing O_2 and preventing its toxicity to anaerobes. Aerobes from the very beginning of cyano-bacterial community should be adapted to the oxidative stress. During the night community is under O_2-depleted conditions and destruction is driven by anaerobes. About 20% of carbon comes into the system during transformation of biomass into mortmass by autolysis. Prokaryotic systems are limited in destruction of the viable cells (see Zavarzina 1958, 1961 describing the first algophage); this limitation for decomposition of particulate organic matter was overcome with radiation of Eukarya during Neoproterozoic. Anaerobic decomposition occurs via sequential trophic steps. The key step for the anaerobic destructive branch is the activity of the secondary anaerobes, consuming H_2 and acetate by activity of three main groups: methanogens, sulfate reducers, and anoxygenic phototrophs. Only the first group can complete the destruction both for H_2 and acetate pathways without external forces as inflow of light or sulfate. Gaseous CH_4, if formed, escapes from the cyano-bacterial community. Trophic interactions within communities conform to the rule that phylogenetically distant organisms interact most closely (Zavarzin, 1989).

From evolutionist point of view in cyano-bacterial community organisms of different origin cooperate to construct an entity. Functionally complementing characters make important only those phenotypic traits, which serve to the common goal - sustainability of the whole. Opposite situation leads to instability of community and its substitution by something else. That is natural selection for existing, which has no antithese. New organisms should be adapted to already existing cooperative system (Zavarzin, 2000). Competition in Darwinian sense occurs within the boxes of fig.2. It was noticed earlier (Zavarzin et all. 1990), that the main most numerous phylogenetic clades of bacteria began close to the point of branching of cyanobacteria. This statement has additional reasoning: phylogenetic tree comes out from cyano-bacterial community because everything new needs to fit to biogeochemical cycles first established by this communit

2. ARCHITECTURE OF CYANO-BACTERIAL MATS

Uniform architecture of cyano-bacterial mats is their most impressive character driving to the field of parahystology. There are two main types of structures: one designated by Issatchenko (1934) after Pallas as "leather" and another as "felt". The first regards mostly cyanobacterial mats where cells are imbedded into colloid EPS gel resembling approximately 1% agar consistence with very low water permeability and high mechanical resistance. "Felt" is related mainly to dense algal benthic communities of interwoven filaments with free water space between build by algae like *Vausheria, Cladophora*, or *Nitella.* Some benthic communities of cyanobacteria, e.g. *Spirulina,* are of that type and live in steady water. A term "mat" is usually applied to the coherent continuous covers, but chaotic usage of this term occurs. A term "biofilm" is used for various kinds of attached continuous microbial covers, both subaqual and subaerial. Dominating organism, which determines the structure of community, is designated as edificatory (from the *edifice* - building). L.Stal (2000) gave most comprehensive comparative analysis of cyanobacterial mats. Similarity of the spatial structure of mats in spite of the diversity of habitats indicates their submission to general principles, underlying the system.

Requirements forcyanobacterial mat are rather obvious. The first principle is optimal utilization of light, which predicts the flat surface with maximal chlorophyll content per square. Cyanobacterial mat has chlorophyll density close to that of plant community and that means that chlorophyll cover of the land surface could be approximately the same during the history of biosphere and limited by the humidity of subaerial surfaces. It is well known that photosynthetically active radiation (PHAR) penetrates one-few millimeters of mat; light scattering pattern from minerals might lead to the maximum of light at some depth within this layer (Jorgensen, 1989). Full light harvesting implies self-shadowing. Growth and gliding movement of the cyanobacteria leads to renovation of the surface, while older cells remain below. This is the main mechanism for layered structure of cyanobacterial community. The second principle is mechanical stability of the mat, which leads to the stabilization needed for the upper layer of mat and here either coherent EPS or woven fabric by trichomes or both make cyano-bacterial community stable to the mechanical stresses. When the surface of mat is covered by chemical precipitate or suspension of minerals with approximate thickness of 1 mm young members of cyanobacterial community glide out and spread on the surface, while other remain buried. Precipitation was used by Orleansky (1999) for laboratory modeling stromatolites of different shape and texture as organo-sedimentary structures. Slime represents a sorption barrier resulting in high accumulation of microelements, as it is supposed for Siberian deposits (Rozanov, Zavarzin, 1999). For example artificial golden microfossils might be obtained in the laboratory imitating gold-bearing Precambrian Witwatersrand stromatolites. However the cementation process is not yet modeled in laboratory. Litiphication of mat into rock makes it mechanically stable. Rifogenic structures change landscape with formation of lagoons as the sites of sedimentation, diagenesis, and microbial life proliferation in the terrestrial-sea ecoton. Cyano-bacterial community represents a materialization of trophic network with sequential decomposition of C_{org} from primary producers. This leads to microzonal distribution of microbial groups ending with complete destruction of organic matter so that cyano-bacterial mats usually do not serve as the sources of buried mortmass. Layered distribution of microbial groups provides optimal exchange of material between subcommunities with the minimal diffusion distance in the range of tenth of millimeter, which provides fast enough exchange and excludes dissipation. Trophic loops are particularly clear for the end of destructive pathway with the catalytical sulfur cycle serving as an electron sink which includes interaction of sulfate reducers, thionic bacteria, purple bacteria like *Thiocapsa* (van Hemerden, 1994). The order of layers is dictated by E_h of catabolic reactions. Perfiliev in late 1920-th demonstrated dependence of microzonation from ox-red reaction with iron minerals as natural indicators visualizing microbial groups (Perfiliev and Gabe, 1961, 1964). For the cyanobacterial community groups of aerobes tolerant to diurnal periods of anoxia and of facultative anaerobes represent dynamic borderline between oxic and reduced zones, which makes and supports geochemical ox-red barrier. Alkalinity barrier is situated within cyanobacterial layer of CO_2-consumption.

3. TOPIC DIVERSITY OF CYANO-BACTERIAL MATS

Multispecies communities of pylogenetically unrelated organisms develop in the same habitats. Desert crusts contain cyano- and algo-bacterial subaerial mats with fungal components for which classification was suggested, indicating disperse, filmy, crustose, slimy, epilitophilic, endolithophilic growth (Gollerbach, Shtina 1969). These designations as well as those for subaqual benthic communities represent microhabitats. One can observe ephemeric oscillatorian mats on humid clay as ubiquitous type of soil microflora.

Bacterial communities were responsible for the formation of the chemical composition of the atmosphere (Zavarzin, 1984). Their action includes production of oxygen as the consequence of incomplete bacterial destruction of C_{org} and sedimentary burial of remnants, making disbalance in CO_2:O_2:C_{org} cycle. Other reactive components of the atmosphere, so called "p.p.m.-gases" were all produced by bacteria. Endogenic origin is ascribed to inert gases and N_2, while CO_2 represents small dynamic reservoir supported by biotic processes.

Important in the biospheric scale are landscapes of larger dimensions. Let us simplify habitats of cyanobacterial mats by three main types: hydrothermal with production of endogenous substances mainly as gases, thalassic on the seashore with carbonates deposition, and athalassic in the intracontinental region (fig.3). These types are permanent during the history of the Earth. Simplifying one can tell that volcanic regions are situated on the active border of continents; terrestrial environments are on intracratonic parts; sedimentation zone and lagoons formation occurs on the passive border. Each zone has its typical cyano-bacterial community. Thermophilic cyano-bacterial mat develops at the temperatures of 65° to about 40°C and is well studied for Yellowstone in USA, Uzon caldera in Russia (Calderic microorganisms, 1989) and few other places around the world. So-called mophetes are most interesting as the sites of endogenic CO_2 emission. Thermophilic cyano-bacterial mats represent complete trophic system with decomposition of organic matter by methanogenic and/or sulfidogenic pathways with remarkable omission of anoxygenic phototrophs, except for *Chloroflexus* in the hottest part.

Usually thermophilic mat is bound to flowing thermal water bringing heat and is crenophilic. Studies of this mat demonstrated its ability to oxidize all reduced gases as H_2, H_2S and produce O_2 from the volcanic gases to the approximate level 20%, thus transforming it into the air (Gerasimenko and Zavarzin, 1982). Volcanic gases are the best possible actualistic models for the transformation of gases from the degasation into the atmosphere during the Earth's history.

4. "SODA CONTINENT"

Not all CO_2 exhalated from the Earth interior is found now in kerogen as the largest biotically produced reservoir. A main sink for CO_2 is represented by carbonates of Ca and Mg, at first dolomite and then calcareous. The main route for CO_2 sedimentation is supposed to be in "carbonate-silicate cycle" that includes weathering reactions of indigenous minerals as spats. The key exchange reaction is simplified as exchange

$CaSiO_3 + CO_2 + H_2O = CaCO_3 + SiO_2$

Alumosilicates are the rest products of weathering. Reaction occurs under acid conditions and is highly accelerated by temperature. Formation of clays in the sites of steam exhaust is observed on thermal springs (Karpov et all, 1984). Concomitant precipitation of carbonate tuffs occurs at lower temperature due to the degazation of $Ca(HCO_3)_2$ bearing waters. CO_2 from the atmosphere is supposed to be eliminated in subaerial reaction of chemical weathering (Kastings, 1993). This process needs acceleration, which is done by concentration of CO_2 into biomass by photoautotrophs and subsequent production of CO_2 by decomposers leading to formation of high CO_2 concentration in soil air in contact with weathering minerals. Calcium is brought out and precipitated outside as it was described yet in 1934 by Polynov in Armenian mountains by the sequence: igneous rocks—calcified weathered rock—sodium solonchak.

Process of rocks transformation leads to liberation not only of Ca and Mg but also of Na. Amount of weathered Na is sufficient to form "soda ocean" (Kempe and Degens, 1985, Kempe and Kazmirzcak, 1998). High pH in «soda ocean» is incompatible with high pCO_2 in the atmosphere and the possibility of runaway greenhouse effect. The hypothesis of Soda Ocean is challenged by most geochemists, which suppose that primitive ocean was neutral to slightly acid as the result of submarine volcanism in mid-oceanic ridges serving as planetary CO_2 source. Another argument for NaCl ocean might be in the similarity of NaCl ocean composition with NaCl fluids produced at the base of meteoric water cell by reaction with superheated rock in volcanic regions. All these considerations do not apply to the subaerial weathering on land in intracratonic regions. Here weathering leads to formation of alkaline waters and soda lakes. Thus I suggest to consider the subaerial CO_2 weathering as leading to the "soda continent". This is what is really observed in the inner parts of continents with alkaline soda lakes. Soda lakes are situated in semiarid climate belt from Big Soda Lake to Mongolia. After Ca^{2+} precipitation in alkaline soil or elsewhere Na^+ remains as a dominant cation.

What might occur within the ecoton, where Ca^{2+} and Mg^{2+} carrying marine waters come in contact with alkaline continental waters? Obviously, precipitation of Ca and Mg carbonates on alkaline geochemical barrier and formation of rifogenic belts and carbonate platforms. This is a type of habitats for thalassic cyano-bacterial communities as predecessors of stromatolites. Lagoons make the best-known place for cyanobacteria, when salinity expels eukaryotes. Mats of this type were studied for Solar Lake, Sivash, Baja California, and Mediterranean. However there occurs no stromatolite formation and lithification is quite limited. The reason is that photosynthesis shifts pH in the interstitial water, where calcium reservoir is limited, while moving water is incompatible with local pH change.

Carbonate deposition in the ecoton between Ca-containing sea and soda continent is illustrated by the simple model: in large cuvette a slope is made from clay, covered for its stabilization by thin layer of hot agar about 0,2 mm thick. Deep part of cuvette is filled with 20mM Ca containing seawater imitating medium of $pH<7$. Above the "sea level" depression is made for "soda lake" with pH10. Precipitation of carbonates occurs in diffusion zone. However this zone is preferred for the development of *Microcoleus* mat, which covers the belt between "lake" and "sea". Trichomes are not calcified and glide between crystals of carbonate of various shape (Zavarzin, unpublished).

Microbiologists disrespected alkaline lakes as unstable shallow bodies of water with varying hydrochemistry reflecting Plinius statement for subaqual weathering: "waters are such as rocks are". We have regarded these waters as some kind of microbial enrichments: the basins accumulating water from surrounding and being more stable habitats than soil. These lakes have varying mineralization up to saturation with trona ($NaHCO_3 \cdot Na_2CO_3 \cdot 2H_2O$) and contain proliferous cyano-bacterial communities by simple reason that phosphates are not bounded by calcium or iron and deliver to the nitrogen-fixing cyanobacterial communities possibilities of high production rate. Cyanobacteria are known as preferring alkaline media. *Spirulina* blooms in African lakes are best known. Decomposition is complete and follows under the strong dominance of sulfur cycle due to evaporative accumulation of salts in waters of endorheic regions. The end steps of destruction are dominated by sulfate reduction followed by H_2S oxidation by anoxygenic sulfur phototrophs giving the water pink to deep red tint. Our idea was that soda lakes with high mineralization might represent refugia for relict microbial communities analogous to those of ancient continents (Zavarzin, 1991).

Main studies were done in equatorial Lake Magadi, Kenya with strong influence of recent volcanogenic processes and weathering of soda containing ashes, and in Central Asia south of Baikal in cryoarid climate and semidesert. Most of lakes studied had no mats because of high productivity and shadowing by plankton. Only few as Lake Khylganta contained true mats of *Microcoleus.* Mats in such lakes are well known in Spain. Alkaliphilic chemotrophic microbial community is now described in its main subsystems (Zavarzin, Zhilina 1999a,b, Jones et all. 1998). Lacustrin origin of stromatolites is by no means novelty. It is known that mats silicified by magadiite are found in Magadi and similar stromatolites with the age of 2 Ga are described for paleo-Magadi in South Africa (Karpeta, 1989). For Baltic shield a landscape of lakes with dolomitic stromatolites of 2.2 Ga is described.

The model of interaction of the "soda continent" with Ca-bearing seawater might be applied to shallow epicontinental seas intruding into the continents and being the main basins for sediments accumulation. This type of seas covered Asian platforms from Spitzbergen through Ural and Siberia to China leaving Proterozoic stromatolites. There is possibility to bring in seawater and mix it with bicarbonate waters of weathered continent with catchments for carbonates. Many of phosphorite deposits with perfectly preserved microfossils resemble these habitats on the margins of continents (Gerasimenko et al, 1998). Now phosphorites throughout the world are regarded as biogenic with cyanobacterial mats as most common component. Formation of the alkaline geochemical barrier on the sea-continent border where cyanobacterial community develops seems to fit to many geochemical microbially mediated processes.

REFERENCES

Biostabilization of sediments. W.E. Krumbein, D.M. Paterson, L.J. Stal (eds) Verlag Rosemeier, Bad Zwischenahn, 1995

Gerasimenko L.M., Zavarzin G.A. (1982) Metabolism of H_2, CO_2, O_2, CH_4 in cyanobacterial communities. *Microbiologia* 51:718 (in Russ.)

Gerasimenko L.M., Zavarzin G.A., Rozanov AYu, Ushatinskaya GT. (1999) Role of cyanobacteria in formation of phosphate minerals. *Zhurn. Obshchei Biologii*, 60(4):415-430 (in Russ)

Gollerbach M.M. and Stina E.A. (1976) Ecology of Soil Algae. Moscow, Nauka (in Russ)

Issatchenko B.L. (1934) Chloride, Sulfate, and Soda Lakes of the Kulundinskaya Steppe and the Biogenic Proccesses therein. (reprinted: Izbrannye trudy [Selected works] Moscow: Acad Nauk SSSR, 1951, 143-162, in Russ.)

Jones B.E., Grant W.D., Duckworth A.W., Oweson G.G. (1998) Microbial diversity of soda lakes. *Extremophiles*, 2:191-200

Jorgensen B.B. (1989) Light penetration, absorbtion, and action spectra in cyanobacterial mats. pp.123-138. In: Microbial mats. Physiological Ecology of Benthic Microbial Communities. Y.Cohen and E.Rosenberg (eds) American Society for Microbiology, Washington , D.C.

Karpeta W.P. (1989) Bedded cherts in the Rietgat formation, Hartbeesfontain, South Africa: a late Archaean to early Proterozoic magadiitic alkaline playa deposit? *South Afr. J. Geol.* 92(1): 29

Karpov G.A., Eroschev-Shakh V.A., Zavarzin G.A. (1984) Role of biogenic factors in the formation of environment for the zone of argillization in the contemporary hydrothermal systems and solfataric fields. *Volcanology and Seismology*, n. **2**:64-74 (in Russ).

Kasting J.F. (1993) Evolution of the Earth's atmosphere and hydrosphere. Hadean to recent. p.611-621 In: Organic geochemistry. Principles and applications M.H. Engel a. S.A. Marko (eds) Plenum Press, N.Y. and London.

Kempe S. and Degens E.T. (1985) An early "soda ocean? Chem. Geol. ***53**:95-108*

Kempe S. and Kazmierczak J. (1994) The role of alkalinity in the evolution of ocean chemistry, organization of living systems and biocalcification processes. Past and present biomineralization processes: considerations about the carbonate cycle. Doumenge F. (ed) *Bull.Inst.Oceanogr.* no.spec 13, 61-117

Novichkova-Ivanova L.N. (1980) Soil algae in phytocoenoses of Sacharo-Gobi desert province. Leningrad, "Nauka". (In Russ)

Perfiliev B.V. and Gabe D.R. (1961) Capillary methods of studying microorganisms. Isd. AN SSSR, Leningrad, (in Russ).

Perfiliev B.V. and Gabe D.R. (1964) Role of microorganisms in formation iron-manganese lacustrine ores. Isd. AN SSSR, Moscow-Leningrad. (in Russ.)

Polynov B.B. (1934) Kora vyvetrivania (Weathered rocks). In: Isbrannie trudi. (Russ)

Rozanov A.Yu., Zavarzin G.A. Bacterial palaeontology. *Vestnik RAS* **67**(3): 241-245 (Engl. translation in *Herald of Russian Ac.Sci*).

Schidlowski M. (1993) The initiation of biological processes on Earth. Summary of empirical evidence. p.639. In: Organic geochemistry. Principles and applications. M.H. Engel a. S.A. Marko (eds) Plenum Press, N.Y. and London.

Schwartzman D., (1999). Life, temperature, and Earth. The self-organizing biosphere. Columbia Univ Press, New York, 190p

Sergeev V.N., Knoll E.H., Zavarzin G.A. 1996 First three billion years of life: from prokaryotes to eukaryotes. *Priroda*, N6:54-67 (in Russ)

Stal L.J. and Caumette P. (eds) (1994) Microbial Mats. NATO ASI Series vol.G 35, Springer Verlag, Berlin.

Stal L.J. (2000) Cyanobacterial mats and stromatolites. In: B.A.Whiton and M.Potts (eds), The Ecology of Cyanobacteria, 61-120

Tyson R.V. (1995) Sedimentary organic matter.Organic facies and palinofacies. Chapman and Hall. London

Zavarzin G.A. (1982) The role of chemosynthesis for the formation of the atmosphere. *Microbiologiya*, 51:

Zavarzin G.A. (1984) Bakterii I sostav atmosphery (Bacteria and the composition of the atmosphere). Moscow, Nauka. (Russ.)

Zavarzin GA (1989) Microbial community in the past and present. *Microbiol. Zhurn.* (Kiev) **51**(6):3-14 (Russ)

Zavarzin G.A. (2000) The non-Darwinian domain of Evolution. *Herald of the Russian Acad. Sci.* **70**(3):252-259

Zavarzin G.A. (1993) Epicontinental soda lakes as probable relict biotopes of terrestrial biota formation. *Microbiology* **62**:789-800

Zavarzina N.B. (1961) Lytic agent in the cultures of *Chlorella pyrenoidosa. Dokladi AN SSSR,* **137**(2):435-437 (Russ)

Zavarzina N.B., Protsenko A.E., (1958) On the lysis in cultures of *Chlorella pyrenoidosa. Dokladi AN SSSR,* **122**(5):936-939. (Russ)

Zavarzin G.A. Gerasimenko L.M., Zhilina T.N. (1993) Cyanobacterial communities of Sivash hypersaline lagoons. *Microbiologiya* **62**(6): 1113-1126 (Russ)

Zavarzin G.A., Stackebrandt E., Murray R.G.E. (1990). A correlation of phylogenetic diversity in the Proteobacteria with the influences of ecological forces. *Canad J. Microbiol.,* **37**:1-6

Zavarzin G.A. (2002) Microbial geochemical calcium cycle. *Microbiology,* **71**(1): 1-17

Zavarzin G.A., Zhilina T.N. (2000) Anaerobic chemotrophic alkaliphiles. In: Journey to Diverse Microbial Worlds (ed. J.Seckbach) Kluver. The Netherlands, 191-208

Zavarzin G.A., Zhilina T.N., Kevbrin V.V. (1999) The Alkaliphilic Microbial Community and Its Functional Diversity. *Microbiology* **68**(5):503-521

Zhilina T.N., Zavarzin G.A. (1991) Anaerobic bacteria-destructors in the halophilic cyanobacterial community. *Zhurn. Obshchei Biologii.* **52**(3):302-318 (Russ).

The Geometry of Ecological Interactions. Simplifying Spatial Complexity. (2000). Eds: U.Dieckman, R.Law, and J.A.J.Metz. Cambridge Studies in Adaptive Dynamics. IIASA Cambridge Univ.Press.

Chapter 9

MICROBIAL SPHERES FROM MICROBIAL MATS

Ulrike BREHM, Wolfgang E. KRUMBEIN and Katarzyna A. PALINSKA
Carl von Ossietzky University of Oldenburg, ICBM,
AG Geomicrobiology, PO Box 2503, Germany
ulibrehm@africa.geomic.uni-oldenburg.de

1. INTRODUCTION

Filamentous cyanobacteria from the LPP-group (*Lyngbya* sp., *Plectonema* sp., *Phormidium* sp.), diatoms and heterotrophic bacteria are the dominant microorganisms of North Sea microbial mats (Gerdes et al., 1986). Cultivated on solid or liquid media the cyanobacteria produce motile trichomes. They employ motility to optimize their positions in their microhabitat using a number of external factors. Light is the major factor in directing their search for a suitable survival and growth niche. Organic substances and their concentrations may be another trigger. However, chemotaxis is essentially very poorly described in cyanobacteria. A large potential area for research has therefore been left untouched (Kangatharalingam et al., 1991).

A description of the coordinated movement and to date unknown three-dimensional assemblage of *Phormidium* sp. filaments with other organisms is reported here. Moreover, the possible strategies for this new type of trichome agglomeration are presented and discussed.

2. MATERIALS AND METHODS

Filamentous cyanobacteria of the LPP-group belonging to the *Phormidium* sp., were isolated from North Sea microbial mat cultures and maintained in liquid as well as ASN III medium solidified with 0.5 % agar (Rippka et al., 1979). The cultures were found to be not axenic, as green algae and diatoms, from the genera *Navicula* and *Nereneis*, were present in the medium, however, filamentous cyanobacteria clearly dominated.

All samples were kept under ambient temperature and light conditions in the laboratory since February 2000. The petri dishes were viewed daily with a help of stereoscope to examine the growth of the organisms. Identification of the cyanobacteria was performed under light microscopy (Axiovert, Zeiss and Axioplan, Zeiss) and by Cryo-Scanning Electron Microscopy (Cryo SEM, Zeiss) described by Sargent (1988).

In further experiments following signal substances in final concentration of 10 mM (cAMP - cyclic Adenosin-Mono-Phosphat, BHL - Butyroyl-Homoserinlacton, OHHL - Oxohexanoyl-Homoserinlacton) were separately added to solid (0.5% agar) and liquid ASNIII media. BHL was additionally used at 1 and 5 mM concentrations.

3. RESULTS AND DISCUSSION

Microbial communities isolated from North Sea microbial mats showed a number of different assemblages in culture.

The most spectacular agglomerations were spheres. Tightly interwined filaments of cyanobacteria formed the surfaces of the spheres. The inner part of the spheres was composed of a varying number of diatoms.

3.1 Principles of movement

Three types of gliding trichomes were observed, displaying a different but very constant velocity as evidenced by repeated microscopic measurements. The majority of the cyanobacteria were sluggish and moved very slowly, packed close together and following the same direction. In contrast, the trichomes, which constructed the spheres, were constantly and actively moving on the surfaces of the spheres. A third group of filamentous cyanobacteria changed the position frequently by forming numerous loops. These, however, never built circles or spheres. Perhaps they were able to form spheres but not all of the needed biochemical parameters were fulfilled.

The velocity of the *Phormidium* sp. trichomes moving through 0.5 % agar medium at room temperature was about 8 μm per second.

From our observations we derive the theory that motility is a significant survival factor for the microorganism that possesses it. Simple gliding movement even without oriented response to the stimuli would allow exposure to a greater number of nutrient ions in a given time.

Motility in some cyanobacteria, e.g. the heterocystous forms, is occasional, and occurs only in some phases of their morphogenetic cycles (hormogonia). Movement and/or growth can serve as a mean of spreading an otherwise attached organism over a wider area.

Although steering by gliding movement of cyanobacteria has not yet been reported, some taxic responses are features, which could add at least the directional orientation to many of their movements.

Gliding is movement in contact with a solid or semi-solid surface without flagella-like organs (Castenholz 1973). This movement can be continuous in one direction for a longer time or can occur in frequent intervals.

Most members of the Oscillatoriaceae rotate when gliding. Trichomes moving on the surface of the agar often move in wide arcs, resulting in flat circles of many trichomes with non-uniform diameters. Castenholz (1973) explained circling of trichomes by a lateral force caused by a rotating trichome. In our studies this phenomenon was observed at the late stage of cultivation.

Such trichome "deformations" or occurrence of unexpected colony forms may occur with collisions and with portions of the same trichome gliding together in opposing directions. However, this explanation does not accommodate for the formation of the three-dimensional structures observed and documented during these studies.

3.2 Assemblages of biofilm communities

The filamentous cyanobacteria studied here are characterized by single segments of 3μm width and 4-4.5μm length. Obviously they belong to the LPP group (*Lyngbya, Plectonema, Phormidium*) (Rippka et al., 1979).

Filamentous cyanobacteria isolated from North Sea microbial mats showed a number of different assemblages in culture.

The formation of these aggregates appeared to be deliberate. Twelve days after incubation cyanobacteria formed circles of microorganisms. During the next days the circles remained stationary but morphed into spheres composed of a small number of individuals which could be clearly identified. Cyanobacterial trichomes seemed to glide inside the spherical envelope, never penetrating into the center of the spheres. The spheres themselves were located within the agar. The number of associated diatoms was found to be considerably higher inside the spheres than outside (Fig. 1).

Other spherical assemblages developed as well, but were considerably smaller in size (0.81 mm, 0.75 mm, 0.46 mm). It was evidenced in several culture sets that the spheres were the center from which a radial arrangement of cyanobacteria emerged (Fig. 2). The surfaces of the spheres were composed of permanently moving trichomes.

The position of the spheres did not change with time, but the number of trichomes involved was variable. After 4-6 weeks the appearance of the three-dimensional figures grew pale, favouring dense concentric circles. Beginning at the surface of the spheres the cyanobacteria moved to form dense rings growing on the centerpart. The number of individuals joining the circles increased with time (Fig. 3). In comparison with the number of cyanobacteria in the surrounding agar, the dark green colour of the rings was conspicious.

Trichomes of the studied "sphere" can move with respect to other trichomes in the colony and, this can result in constant changes in the relative position of the trichomes and shape of the colony. "Spheres" maintain their original shape and place, although the trichomes continuously move. The movement of the aggregated trichomes is clearly coordinated. This means: the trichomes move together as a "sphere". In cases, in which trichomes move away, the "sphere" turns. According to our screening of the literature this pattern of aggregation and locomotion of aggregates has not been recorded before.

The size of the spheres did not change over the whole time of observation. Finally, the rings were reduced in width and number of trichomes.

3.2.1 Formation of spherical aggregates of trichomes

The steps of biofilm sphere formation are as follows:

A marine biofilm community kept in liquid or on solid media excretes various substances. These substances are accumulated in certain places and encircled by microscopically visible envelopes (Fig. 4).

These spherical appearances are recognized by some microorganisms. They approach the spheres and penetrate through the envelopes at specific places. The envelopes seem to be permeable only for specific microorganisms. According to our observations bacteria and diatoms can not penetrate through the flexible but resistant envelopes. As long as the trichomes of *Phormidium* sp. are penetrating the membranes, they still can glide inward and outward. After they enter completly into the sphere, they do not glide back.

The colonization of the spheres leads to recognizable stratifications of:

a) trichomes of *Phormidium* sp. forming a balloon-like layer underneath the surface of the sphere and
b) loosely distributed masses of the diatom *Navicula perminota* rich in EPS, and located in the inner part of the sphere (Fig. 5)
c) chemoorganotrophic bacteria and sometimes
d) a few green algae.

Sometimes cyanobacteria leave the spheres and form hairy outer rims (Figs. 6 and 7). Chemoorganotrophic bacteria and diatoms cannot leave the mature spheres.

During the sphere formation the number of diatoms inside systematically (Figs. 4 and 8) increases.

Figures 9-10 document the penetration of *Phormidium* sp. trichomes through the envelope into the sphere.

The cryo-REM micrograph shows dense layer of thin *Phormidium* sp. trichomes (Figs. 11).

Surface as well as the inside of the sphere was always very rich in EPS. *Phormidium* sp. trichomes in an advance stage of sphere formation created a non-transparent cover. Diatoms inside the sphere were visible only in sections after having cut the sphere (Fig. 12).

Finally, when the growth conditions inside the sphere become less favourable the diatoms abandon this microbial community, however, leaving the cyanobacterial surface network intact in form and shape. Only the framework of the spheres stays visible. This phase remains permanent. Sometimes the number of cyanobacteria increases again, but never reaches the maximum of the initial spherical phase.

3.2.2 Initiation of sphere formation

The activity of the diatoms and the presence of the diatom-cyanobacterial community were conspicuous. To elucidate the role of the bacteria in this community typical signal substances were applied in further investigations. The chosen signal substances favour the metabolism of certain heterotrophic bacteria what results in emphasizing the organisms out of the community.

In order to enhance the sphere building a number of different signal substances were separately added to the media (cAMP, OHHL Oxohexanoyl-Homoserinlacton, BHL Butyroyl-Homoserinlacton). The best results were obtained with BHL. Samples without signal substances had only up to 4 spheres per plate. After adding BHL the number of the spheres raised up to 43, with the average of 25. Temperatures between 16 ° C and 27 °

(degree) C and varied light intensities were tested and revealed not to have influence on the sphere growing. The size of the spheres differed between 0.3 mm and 1.2 mm.

Bacterial chemotaxis is a motile response to chemical gradients and may explain many consortia of cyanobacteria and bacteria. Direct evidence for the role of chemotaxis in forming such consortia is incomplete. Also quorum sensing (Surette et al., 1999) may be an important factor. Formation of spherical aggregates of trichomes could provide evidence for cyanobacterial chemotaxis to bacterial and algal products (e.g. calcite formation, data not shown), implying that chemotactic responses play a role in the formation of certain microbial consortia in nature.

A kind of trichome communication was observed in *Pseudanabaena galeata* (Castenholz, 1982). He observed the formation of swarms of trichomes, and called them "comets". Similar to the phenomenon observed there, an elaborate gliding of aggregates took place in our cultures as well. In case of these comets, light triggered complex movements. In case of sphere formation light has to be excluded as a factor causing movement. The origin of the spheres can not be connected with the existence of air bubbles, as no differences in density inside and outside the sphere within the agar were visible. Similar to our "spheres", the "comets" formed and maintained their integrity during their movement. The manner in which trichome-to-trichome communication is formed with the respect to the movement is, in both cases entirely unknown.

Aggregates of other simpler geometry have been often observed in planktonic *Aphanizomenon flos-aquae* (Janson et al., 1994). The integrity of large flakes in this case is influenced by chemotactic phenomena and is probably maintained by some type of trichome-trichome interaction. Similar reactions could also be involved in the formation of "spheres", in the case of benthic biofilm LPP-forms.

4. CONCLUSIONS

Filamentous microorganisms isolated from North Sea microbial mats showed spherical assemblages in culture. The formation of "spheres" is certainly a result of elaborate gliding of trichome aggregates. The integrity of large "spheres" is most probably influenced by chemotactic phenomena and maintained by some type of trichome-trichome interaction. In the early stage of culture cyanobacteria move radially towards the sphere, later in older culture stages, the direction of the movement turns to the opposite direction. This suggests the presence of methabolic, secondary products, which attract cyanobacteria and influence their movement.

Following hypothesis could be drawn from these studies:

Biofilms can organize themselves in spherical aggregates.
The functions of these aggregates are multiple

- Better organization of light, mineral and other resources
- Better protection from grazing
- Improved propagation
- Mutual symbiotic relationships
- Propagation by transport of spheres in the water column
- Organization of mineral deposits in spherical bodies (ooids).

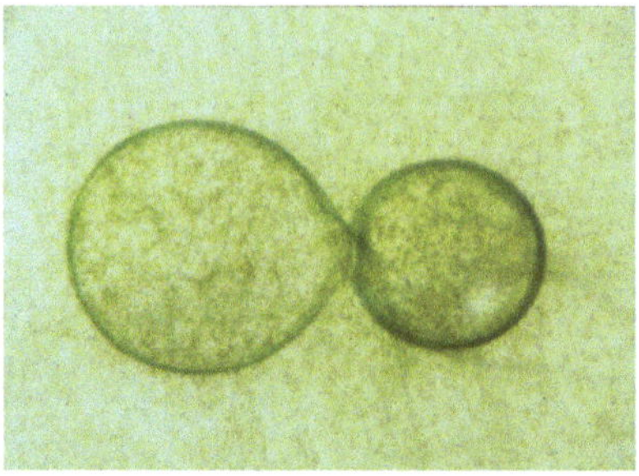

Figure 1. High diatom concentrations inside the spheres. Diameter of both spheres is about 1.2 mm. Light microscopy.

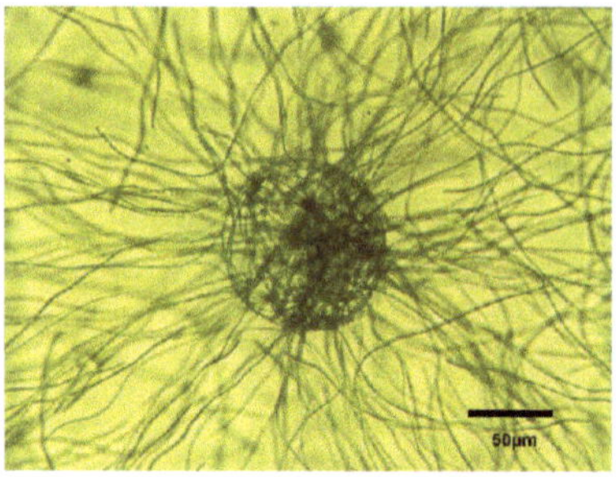

Figure 2. Radial movements of *Phormidium* trichomes towards and against the sphere

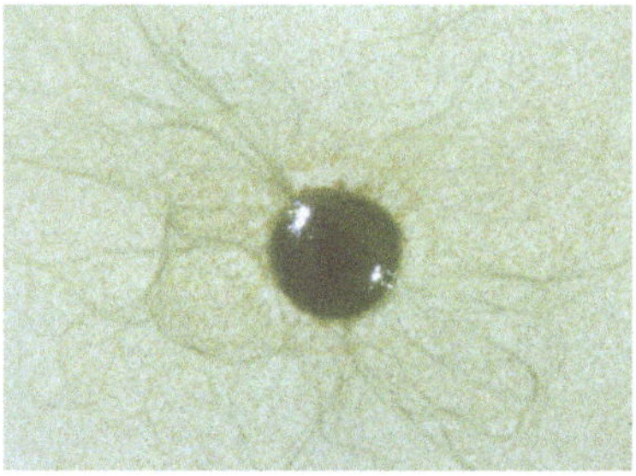

Figure 3. Sphere formation after 1 week. Diameter ca. 700 µm. Light microscopy.

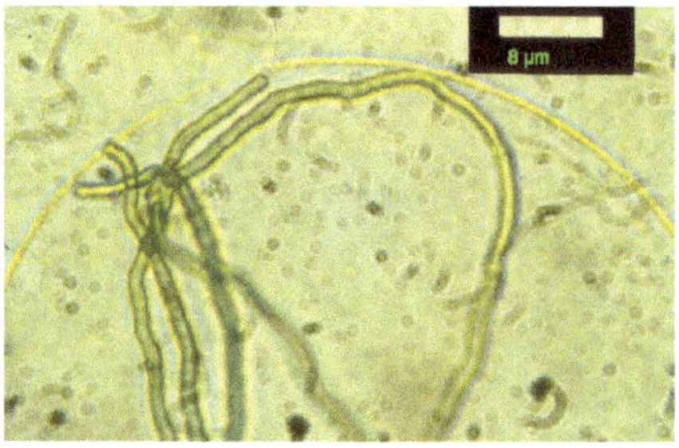

Figure 4. Envelope appearance of a sphere. Light microscopy.

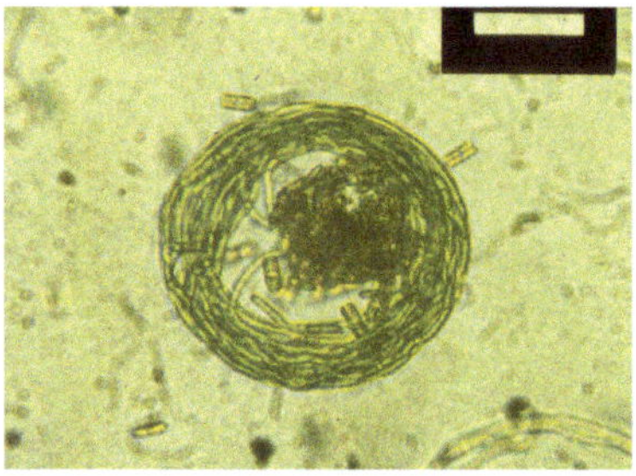

Figure 5. Sphere 8-12 hours after initiation. Diameter ca. 85 µm. Light microscopy

Figure 6. Two types of spheres: A) a smooth and B) a hairy one. Diameter of each ca. 1.0 mm. Light microscopy.

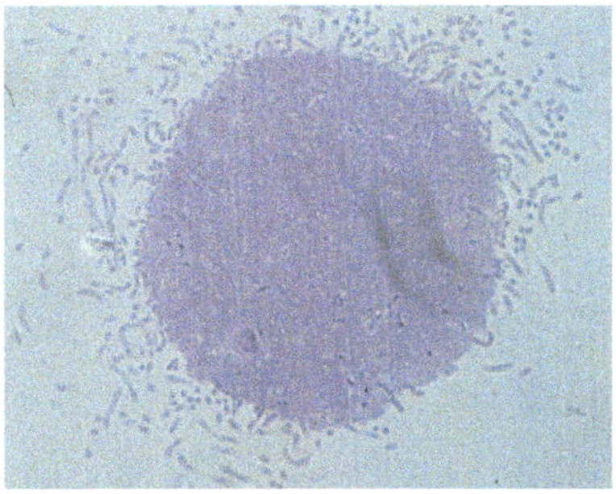

Figure 7. Thin section of a hairy sphere stained with Toluidin. Diameter ca. 1.0 mm. Light microscopy.

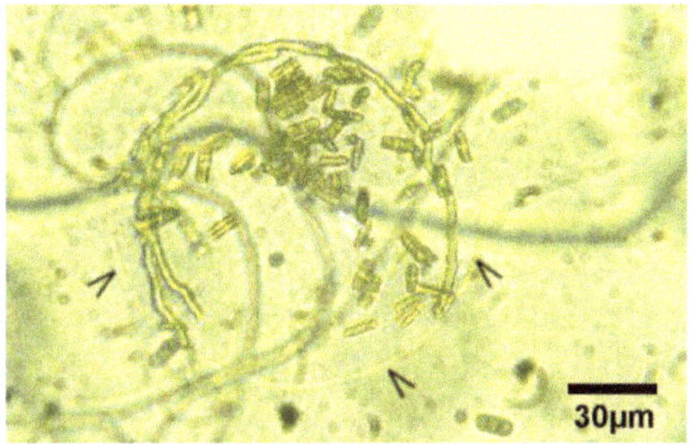

Figure 8. Sphere in an early stage. Diameter 120 µm. Light microscopy.

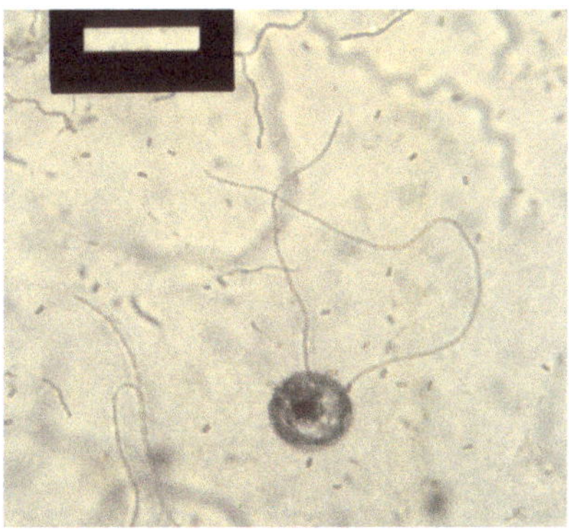

Figure 9. Trichomes of *Phormidium* sp. penetrate a sphere. Diameter ca. 84 µm. Light microscopy.

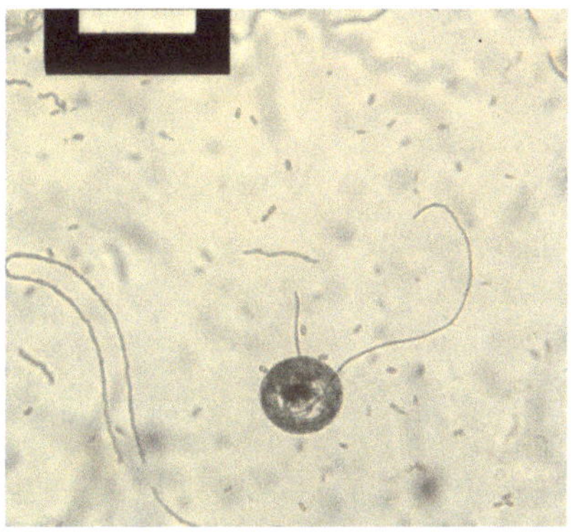

Figure 10. Trichome penetration eight minutes later. Light microscopy.

Figure 11. Surface detail of a sphere. Interwinded *Phormidium* trichomes are clearly visible. Cryo-SEM.

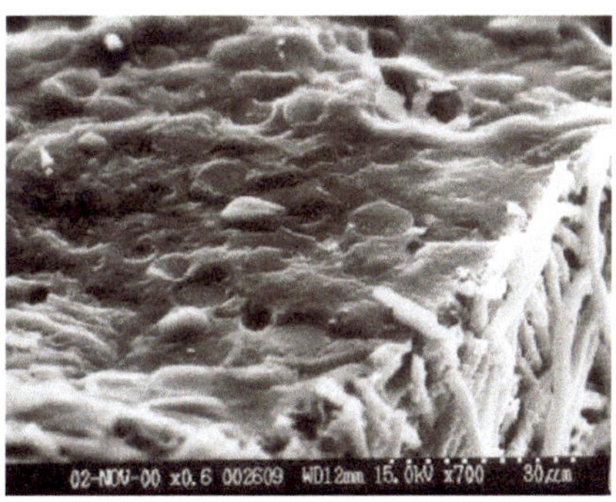

Figure 12. Section detail of a sphere. Cryo-SEM.

REFERENCES

Castenholz, R. W. 1973. Movements. In: (N. G. Carr and B.A. Whitton, eds) The biology of blue-green algae. The biology of blue-green algae. Blackwell Scientific Publications, Oxford pp. 320-339.

Castenholz, R.W. 1982. Motility and taxes. In: (N.G. Carr and B.A. Whitton, eds) The biology of cyanobacteria. Blackwell Scientific Publications, Oxford. pp. 413-438

Famintzin, A. 1867. Die Wirkung des Lichtes auf Algen und einige andere ihnen nahe verwandte Organismem. Bot. Jahrb. 6: 27-44.

Häder, D.-P. 1982. Gated ion fluxes involved in photophobic responses of the blue-green alga, *Phormidium uncinatum.* Arch. Microbiol. 131: 77-80.

Halfen, L.N. and R. W. Castenholz. 1971. Gliding motility in the blue-green alga *Oscillatoria princeps.* J. Phycol. 7: 133-145.

Hoiczyk, E. 2000. Gliding motility in cyanobacteria: observations and possible explanations. Arch. Microbiol. 174: 11-17

Hoiczyk, E. and W. Baumeister. 1999. Oscillin, an extracellular, Ca^{2+}-binding glycoprotein essential for the gliding motility of cyanobacteria. Molec. Microbiol. 26: 699-708.

Janson, S., E. J. Carpenter and B. Bergman. 1999. Fine structure and immunolocalisation of proteins in *Aphanizomenon* sp. from the Baltic Sea. Eur. J. Phycol. 29: 203-211.

Kangatharalingam, N., L. Wang and J. C. Priscu. 1991. Evidence for bacterial chemotaxis to cyanobacteria from a radioassay technique. Appl. Environ. Microbiol. 57: 2395-2398.

Rippka, R., J. Deruelles, J. B. Waterbury, M. Herdman and R. Y. Stanier. 1979. Generic assignments, strain histories and properties of pure cultures of cyanobacteria. J. Gen. Microbiol. 111: 1-61.

Sargent, J.A. 1988. Low temperature scanning electron microscopy: Advantages and applications. Scanning Microscopy 2: 835-849.

Surette, M.G., M. B. Miller and B. L. Bassler. 1999. Quorum sensing in *Escherichia coli, Salmonella typhimurium* and *Vibrio harveyi*: A new family of genes responsible for autoinducer production. Proc. Natl. Acad. Sci. USA. 96: 1639-1644.

Gerdes, G., Krumbein W. E. , Reineck H.-E. (1986) Mellum- Portrait einer Insel. Kramer, Frankfurt/M., (S. 170-187 Das Farbstreifensandwatt).

Chapter 10

PROTOZOANS AND BIOFILMS

Hartmut ARNDT, Kerstin SCHMIDT-DENTER, Brigitte AUER and Markus WEITERE
University of Cologne, Zoological Institute, Dept. General Ecology and Limnology, D-50923 Cologne, Germany
hartmut.arndt@uni-koeln.de

1. INTRODUCTION

The role of protozoans in both pelagic and benthic aquatic food webs has received increasing attention by scientists within the last two decades (e.g. Azam et al. 1983, Alongi 1991, Sherr and Sherr 2000). Protozoans are part of a complex food web, in which they conduct several important functions, such as the control of the bacteria (e.g. Berninger et al. 1991, Arndt et al. 2000) and the enhancement of remineralisation processes (Sherr and Sherr 1984). As main bacterial consumers, protozoans can have a high impact on both the taxonomic and morphologic composition of bacterial communities (e.g. Jürgens and Güde 1994, Jürgens et al. 1997). One defence strategy against protozoan predation is the formation of bacterial colonies and filaments (e.g. Jürgens and Güde 1994, Hahn et al. 2001) possibly also forcing the formation of biofilms.

Parallel to the investigations of the microbial food web, scientists became aware of the importance of interfaces in microbial processes (e.g. Marshall 1976, Krumbein et al. 1994, Costerton et al. 1995). Studies on biofilms generally reflect the point of view of microbiologists, focusing on the prokaryots. In addition, various studies in the limnological field focus on substrate associated algae (e.g. Sommer 1999, Hillebrand et al. 2000). In contrast to the large number of publications on substrate associated bacteria and algae the knowledge on the role of substrate associated protozoans is still low, though protozoans can occur in very high abundances on biofilms. They are found on different interfaces, such as the solid substrate-air interface (e.g. on soil grains or rocks, Cowling 1994), the water-air interface (e.g. Ruttner 1963), or the solid substrate-water interface. On the latter, protozoans are found on various substrates, such as stones (e.g. Foissner et al. 1992), macrophytes (e.g. Baldock et al. 1983), animals (e.g. Fenchel 1965), leave litter (e.g. Franco et al. 1998), various different artificial substrates (e.g. Cairns and Yongue 1968, Jackson and Jones 1991, Foissner et al. 1992) as well as on aggregates in the water column (e.g. Caron 1991, Arndt 1993). Protozoans can also occur in high abundances on microbial mats (e.g. Bernard and Fenchel 1995, Dietrich and Arndt 2000, Garstecki and Arndt 2000). In the water management, biofilm dwelling protozoans are of interest due to their control of bacteria and their impact on the water quality (e.g. Sibille et al. 1998, Fried et al. 2000). In addition, protozoans are used as biological indicators (e.g. Foissner et al. 1992). In the medical disciplines, the interest is on potentially harmful protozoans (or those which host harmful bacteria) on medically relevant substrates, such as contact lenses, teeth, or operation instruments and drinking water pipes (e.g. Gray et al. 1995, Brown and Barker 1999).

In this short article, we will review the current knowledge on the ecology of protozoans on biofilms with a focus on protozoans on solid surfaces in aquatic systems. After giving an overview about the abundance and dynamics of protozoans on biofilms and on the protozoan properties to colonise surfaces, attention will be drawn to the impact of protozoans on the function and structure of biofilms.

2. PROTOZOAN COMMUNITIES ON BIOFILMS

Biofilms can be colonised by various protozoan groups, including naked amoebae (e.g. Jackson and Jones 1991), heterotrophic flagellates (e.g. Zubkov and Sleigh 1999, Strüder-Kypke and Schönborn 1999), testate amoebae (e.g. Jax 1992), foraminiferans (in marine habitats, e.g. Bernhard and Bowser 1992), heliozoans (e.g. Fig. 1, Auer et al. in prep.) and ciliates (e.g. Franco et al. 1998). Most studies on biofilm protozoans concentrate on a single protozoan group. The quantitative and qualitative investigations of protozoans require good experience in the protozoan taxonomy beside the use of different methods for the different protozoan groups. Such methods are for example the direct live counting technique for heterotrophic flagellates (Arndt et al. 2000), quantitative protargol stain for ciliates (Skibbe 1994, Wickham et al. 2000) and the liquid aliquot method for amoebae (Butler and Rogerson 1995, Garstecki and Arndt 2000). The time consuming methods are probably one reason, why comparative studies that considered all of the important protozoan groups are scarce.

Heterotrophic flagellates (HF) are often the most abundant protozoans on biofilms. On artificial substrate (glass slides) exposed in the River Rhine (Germany), abundances ranged between 1.2 and 391 HF cm^{-2} (January until August) with highest abundances in spring (Fig. 1.A, Schmidt-Denter and Arndt subm.). Ciliates occurred here in abundances of 0.2 to 131 ciliates cm^{-2}. However, due to their larger size ciliates occurred in much higher biovolumes than HF. Extremely high abundances of up to 8,600 protozoans cm^{-2} were found on artificial substrate exposed in a lake (Fig. 1.B, Auer et al. in prep.). Again, the most abundant group were HF (6,500 ind. cm^{-2}) followed by amoebae (1,050 ind. cm^{-2}). Ciliates reached densities of up to 640 ind. cm^{-2}. Even much higher ciliate densities (*Euplotes*) on biofilms of 15,000 to 20,000 ind. cm^{-2} have been reported by Glud and Fenchel (1999). Naked amoebae are frequently recorded on biofilms and some studies highlighted their numerical importance (e.g. Jackson and Jones 1991). During our own investigations of biofilms from lakes and rivers naked amoebae were always present, but abundances were distinctly lower than those of HF. Reports on testate amoebae are scarce. However, a detailed study by Jax (1992) on slides exposed in small water bodies characterised testate amoebae as a species rich group with occasionally high abundances

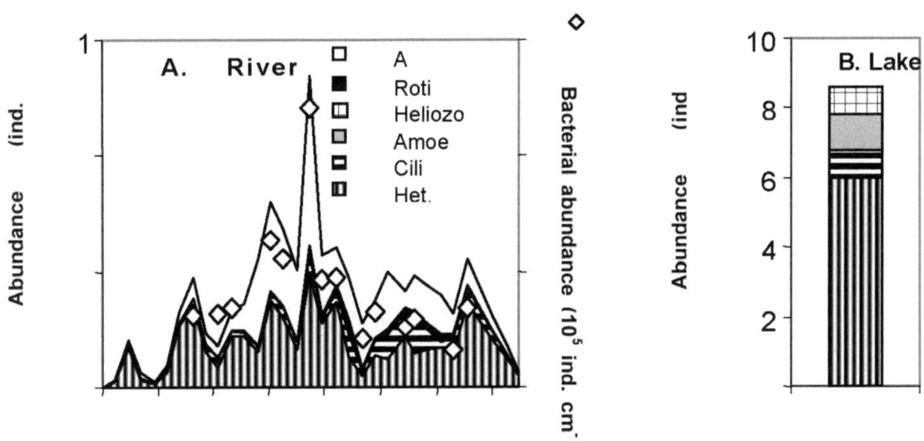

Figure 1. Abundances of biofilm-dwelling protozoans on artificial substrate: A. Seasonal succession in the River Rhine (heliozoans and amoebae occurred only in very low abundances). B. High abundances of protozoans on an algae dominated biofilm exposed for three weeks in the pelagial of Lake Fühlingen (Cologne)(algae, bacteria and rotifers were not considered in this graph).

(up to 1,360 ind. cm^{-2}). In drinking water distribution systems testate amoebae were also reported as being abundant (e.g. Sibille et al. 1998). In marine habitats, foraminiferans can be important grazers on surfaces (Bernhard and Bowser 1992). Heliozoans can change between benthic substrates and the pelagial (Arndt 1993). We observed up to 305 heliozoans cm^{-2} in lake biofilms (Auer et al. in prep.).

Protozoans can not only occur in high abundances, but they may also reach a high diversity. In this respect, ciliates are the most well studied protozoan group. Foissner et al. (1992) found a total number of 209 ciliate taxa on substrates in a river with 122 taxa on foam units, 155 taxa on litter bags and 174 taxa on natural substrates. At one sampling date, 38 ciliate taxa were recorded on average on foam units, 47 on litter bags and 54 on natural substrates, respectively. Franco et al. (1998) could determine a total number of 54 ciliate species on leaves submerged in a river. According to this study, 20 to 35 ciliates species have been found to co-exist within dimensions of only a few cm^2. Studies on flagellates are more scarce. We found 56 heterotrophic flagellate taxa on biofilms (slides) in the River Rhine (Schmidt-Denter and Arndt subm.). Jax (1992) reported up to 63 taxa of testate amoebae from biofilms of small stagnant water bodies. It seems that due to the fractal structure of biofilms the species numbers of different morphotypes (and size classes) is relatively similar.

The types of substrates that are known to be populated by protozoans are very diverse (see introductory remarks). Most microbial biofilms that are wetted for about one day may contain heterotrophic protists. During our investigations on microbial biofilms of the balustrades on the roof of the Cologne Cathedral we found more than 35 protozoan taxa (mostly flagellates). Biofilms can be extremely harsh habitats for protozoans. Temperature, water content and many other parameters can range widely within short periods of time (for reviews see e.g. Flemming 1991, Neu 1994). These extreme conditions might be a reason for the relatively low number of different heterotrophic flagellate species (morphospecies level) that we have found during our analysis of very different types of biofilms such as stones on the roof of the Cologne Cathedral, biofilms of lakes and of the River Rhine, on littoral marine biofilms and even on biofilms in the ultra-oligotrophic deep sea (Angola Basin) at a depth of more than 5000 m. Some species seem to occur on all these different biofilms. Among them were the euglenozoan flagellates *Bodo designis, B. saltans* and *Petalomonas minuta.*

One reason for the general lack of sufficient information on the occurrence of protozoans on biofilms are methodological problems. Taking a biofilm away from its microhabitat may completely change environmental conditions within seconds (e.g. water content, pH, temperature, oxygen content, concentration of exudates released by bacteria and algae, etc.). Protists with their delicate membranes may disintegrate very soon and often leave no remains. Fixatives are only useful for specific species. And in addition, the background of the large bacterial or algal biomass makes it often impossible to identify protozoans by routine staining techniques. Live counting directly after sampling or within the microhabitat is mostly indispensable, mostly cultivation techniques have additionally to be applied. Up to now only a few aquatic biofilms from marine and limnetic sites have been studied regarding their protozoan fauna.

3. PROTOZOANS′ ADAPTATIONS

They can be highly tolerant towards abiotic factors, such as temperature, oxygen supply, salinity and pressure (e.g. Finlay 1990, Arndt et al. 2000). This allows protozoans to inhabit a broad range of different habitats including such extreme habitats as the anoxic environment (e.g. Fenchel and Finlay 1995), groundwaters (e.g. Novarino et al. 1997) the deep sea including hydrothermal vents (e.g. Patterson et al. 1993, Atkins et al. 2000, Hausmann et al. subm.) and sea ice (e.g. Archer et al. 1996). In addition, several species form cysts, which can outlast unfavourable conditions and facilitate airborne distribution (Peterhoff et al. in prep.). Since protozoans can obtain high abundances in the vicinity of biofilms in the plankton (e.g. Berninger et al. 1991), the benthos (e.g. Alongi 1991) and in soils (e.g. Cowling 1994), they may rapidly colonise new substrata (see below). Chemosensory behaviour enables protozoans to congregate at point sources of food within a few minutes (e.g. Fenchel and Blackburn 1999). This ability might also play a role in colonising fresh substrata with newly attached bacteria.

Free-living protozoans can be associated to substrata in different ways. This includes a flexible association, which enables the protozoans to move over the substrate, or to attach by holdfast organelles (Fenchel 1987, Hausmann and Hülsmann 1996). Naked amoebae adhere strongly to the surface due to their large area of contact with the substratum. Many flagellates such as members of the bodonids and euglenids glide over the substrate on their trailing flagellum. Van der Waals forces probably play an important role in holding the protozoans to the surface (Fenchel 1987). Some ciliates (e.g. stichotichs and hypotichs) are able to "walk" over the substrate by ventral bundles of cilia, so called cirri. An attachment to the surface can be attained by specific structures such as flagella (e.g. bicosoecid flagellates, i.e. *Cafeteria,* or *Bodo saltans*), by a thread-like extension of the posterior end of the cell (e.g. in the flagellates *Spumella* and *Paraphysomonas*), by intra- or extracellular stalks (e.g, in peritrich ciliates or the colony forming chrysomonad flagellate *Anthophysa*), by arm-like appendages of the cell (e.g. in the suctorian *Erastophrya*) or even by sucker-like organelles (e.g. in the epizoic ciliate *Trichodina*). In some protozoans, such as the ciliate *Stentor*, the secretion of mucus plays a role in the attachment (see also Fig. 3, below). Various protozoans can be associated to the substrate with the help of loricae (e.g. several choanoflagellate species or folliculinid ciliates). Also an attachment by gelatinous colonies is found (e.g. in the flagellate *Spongomonas*).

The ability of many protozoans to temporarily attach or detach from the biofilm is of great significance to avoid unfavourable conditions. In several species swarming cells can be produced in their sometimes complex life cycles (Hausmann and Hülsmann 1996) which can serve for the dispersal. Many naked amoebae can produce floating forms (e.g. vannellid amoebae) or flagellated stages (e.g. Vahlkamphiidae) (e.g. Page 1988).

Specific feeding types among the protozoans make them well adapted to feed on attached particles. An example is the cytopharynx in kinetoplastid and euglenid flagellates (Nisbet 1974, Fenchel 1986). This structure is formed by microtubular rods, which support the detachment of attached bacteria. Many ciliates are able to browse on loosely attached bacteria or pore water bacteria. On the other hand, several biofilm dwelling protozoans are able to feed on suspended particles (see below).

The small size of the protozoans is of great advantage for the life on and in biofilms. Many protozoans can live in the water boundary layer around soil particles which enables them to live in soils even when the pore water is dried out (Fenchel 1987). In flowing waters, protozoans associated to biofilms live in the water boundary layer where the stream velocity is close to zero. Thus they are sheltered from drift (e.g. Silvester and Sleigh 1985).

4. BIOFILM COLONISATION BY PROTOZOANS

New substrates can be rapidly colonised by protozoans. Own investigation on glass slides exposed to the River Rhine revealed, that heterotrophic flagellates (HF) were already present on the first inspection after 10 minutes of exposition into the river water (Fig. 2.A). After one day, high abundances of about 250 HF cm^{-2} were reached. Numbers did not increase after a longer period of exposition (Fig. 2.B). The maximal HF densities occurred even earlier than the maximal bacterial densities. Algae showed a similar colonisation pattern as HF, whereas the colonisation of the ciliates was slower.

The occurrence of protozoan species on new substrate consists with the general prediction by the MacArthur-Wilson model for the colonisation of islands (e.g. Cairns et al. 1969, Franco et al. 1998). At the early stage of colonisation, the immigration rate of species to the substrate is high, and the

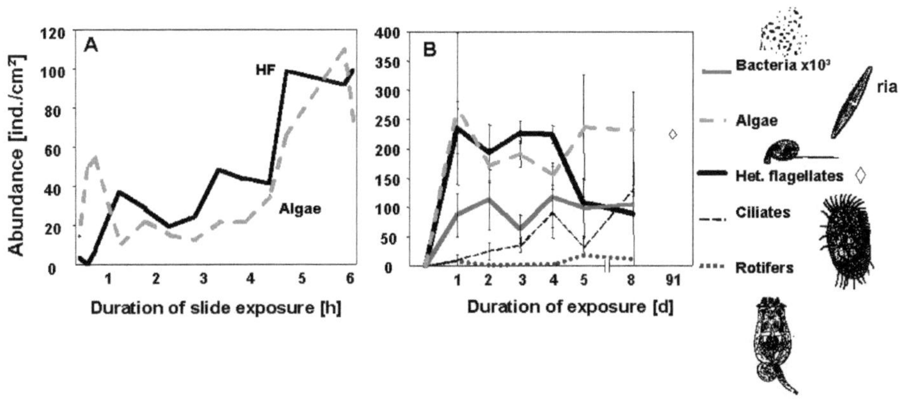

Figure 2. Succession of the colonisation by organisms on glass slide biofilms (River Rhine, summer 1999). A: Colonisation within the first hours. B: Colonisation within the first days of exposure. (after Schmidt-Denter and Arndt subm.).

extinction rate is low. As the colonisation proceeds, the immigration rate declines and the extinction rate rises, reaching an equilibrium, were both rates are equal. With ongoing colonisation, the intensity of biological interactions (e.g. competition, predation) increases and therefore the species composition changes from species that are able to rapidly colonise the substrate to species with high competitive ability. However, the effects of ongoing colonisation might overlap with a seasonal succession. Biofilms are generally strongly influenced by environmental factors such as temperature, light intensity, the activity of grazers and the composition of the planktonic organisms (which colonise the biofilm and can be a food source of biofilm dwelling protozoans). This has been demonstrated in biological reactors, where the biofilm dwelling ciliates (and some metazoans) responded quickly to changes in the plant operation conditions with rapid changes in the community structure (e.g. Fried et al. 2000). In nature, environmental factors show strong seasonal changes. Consequently, the biofilm dwelling protozoans exhibit strong seasonal differences in both species composition and abundance (Fig. 1, see also Jax 1992, Strüder-Kypke and Schönborn 1999).

There may be several advantages for protozoans colonising substrates, e.g.: 1.) food concentrations (bacteria, algae or other protists) can be several orders of magnitude higher than in the surrounding water. 2.) Biofilms often serve as a refuge from predation by suspension feeding predators. In the River Rhine, strong grazing pressure by suspension feeding benthic predators cause low planktonic ciliate abundances. Whereas ciliates find refuges on biofilms and reach high biomasses. 3.) In running waters and in estuaries, an attachment can prevent protozoans from a rapid export. 4.) For some species it has also been suggested, that attachment of suspension feeding protozoans may increase the efficiency of food uptake (Fenchel 1986), for other protozoans no effect was postulated (Boenigk and Arndt subm.). 5.) Rafting biofilms can serve as a vehicle for the distribution of benthic protozoans. Microbial mats disrupted from the sediment due to different reasons often harbour large numbers of protozoans that are distributed via the water column. The large mats serve as a refuge during the travel. In the eutrophic lake Speldrop, cyanobacterial mats transported many ciliates and euglenids into the water column, where they atypically contributed several per cent of the planktonic protozoan biomass (Arndt et al. unpubl.).

5. STRUCTURING OF BIOFILMS BY PROTOZOANS

Due to their occasionally high abundance and their special properties, protozoans may significantly influence the biofilm architecture (Fig. 3, Riedel-Lorje 1980). Martín-Cereceda et al. (2001) highlighted the important role of peritrich ciliates in maintaining the multilayer biofilm organisation in a full-scale treatment plant owing to the intricate entanglement formed by their stalks. Protozoans can completely structure biofilms. Figure 3 (right graphs) shows large extensive colonies of the heterotrich ciliate *Stentor roeseli* completely dominating the structure of the biofilm on sediment from the River Rhine over an area of more than 0.5 m_ (Kureck, Scherwaß & Arndt, unpubl.).

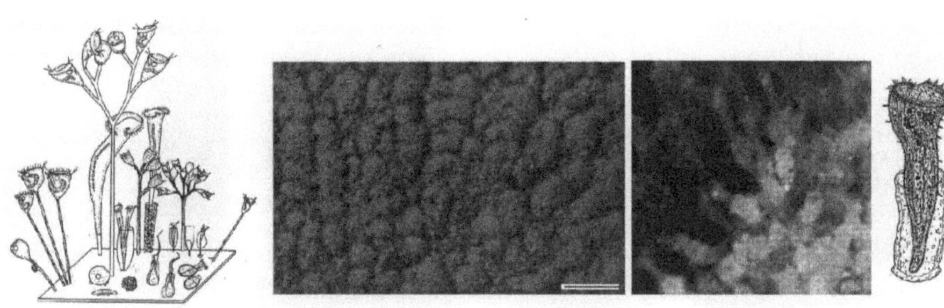

Figure 3. Left graph: Three-dimensional structures of a biofilm created by ciliates grown on a glass slide exposed to a river (modified after Riedel-Lorje 1980). Right graphs: Biofilm structure created by extensive colonies of the ciliate *Stentor roeseli* grown on sediments of the River Rhine (Photos A. Kureck). Scale in the left picture equals 5mm (the right photograph resembles a 10x zoom).

As several protozoans have extracellular structures such as stalks, loricae and shells, which can remain in the biofilms also after cell death, protozoans can persistently influence the three-dimensional structure of biofilms. In addition to this impact on the biofilm architecture, protozoans can enhance the transport of biochemically important solutes into and within the biofilm. It has been demonstrated, that benthic ciliates can enhance the interstitial solute transport in sediment biofilms due to their motion of cilia by a factor of 1.1 to 10 (Glud and Fenchel 1999).

Protozoans influence the architecture of biofilms not only due to their own structures but also due to their feeding activity on biofilm organisms. Protozoans can increase the spatial and temporal heterogenity of the bacterial biofilm by their spatially patched feeding activity, as demonstrated for the ciliate *Euplotes* (Lawrence and Snyder 1998). *Allogromia* sp. (foraminifera) was demonstrated to leave highly denuded regions of an area between 0.3 and 8.6 mm^2 on biofilms (Bernhard and Bowser 1992). The impact of protozoans on the biofilm structure was also highlighted by Jackson and Jones (1991). The authors investigated biofilm protozoans from artificial substrates. Especially the activities of the abundant naked amoebae (feeding on diatoms, movement between substrate and biofilm) were seen as important causes for a disruption and subsequent sloughing of the biofilms.

6. TROPHIC INTERACTIONS WITHIN BIOFILMS

Grazing by protists on biofilms: The grazing of surface bacteria by protozoans has been investigated for naked amoebae (e.g. Rogerson et al. 1996, Butler and Rogerson 1997, Zubkov and Sleigh 1999), foraminiferans (Bernhard and Bowser 1992), HF (e.g. Caron 1987, Zubkov and Sleigh 1999, Boenigk and Arndt 2000) and ciliates (Young et al. 1994, Eisenmann et al. 1998, Lawrence and Snyder 1998). Protozoans can feed efficiently on attached bacteria and obtain high growth rates. However, the grazing efficiency differs strongly between the different protozoan groups. In foraminiferans, two allogromiid foraminifera were demonstrated as effective feeders on biofilm bacteria, whereas the impact of calcareous and agglutinated foraminifera was low (Bernhard and Bowser 1992). Zubkov and Sleigh (1999) investigated the growth and assimilation rates of an amoebae and two HF feeding on attached radioactive labelled *Vibrio* bacteria. The authors demonstrated a high grazing impact of the protozoans, especially of the surface-feeding flagellate *Caecitellus parvulus* and the amoebae *Vannella septemtrionalis*, on biofilm bacteria. Boenigk and Arndt (2000) found the grazing rate of the small kinetoplastid *Rhynchomonas nasuta* to be about 13 biofilm bacteria $flagellate^{-1}\ h^{-1}$. Bacteria that were strongly adhered to the substratum as well as particles that were to big to be ingested were released from the pharynx.

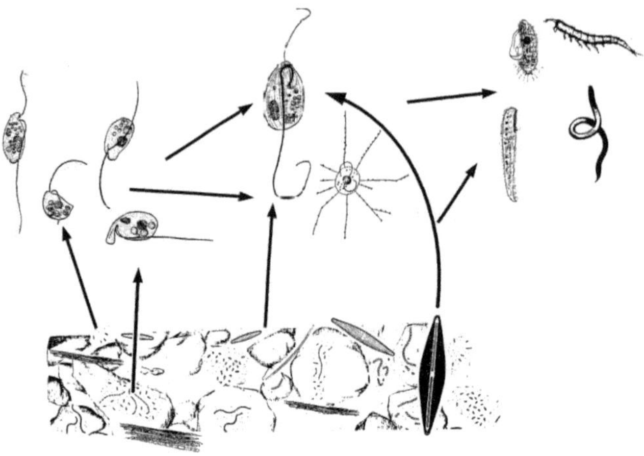

Figure 4. Part of the complex food web on biofilms: Flagellate grazing on attached bacteria/algae and pore water bacteria. Heterotrophic flagellates in turn are grazed by other flagellates, ciliates and meiofauna (modified from Arndt et al. 2000)

A maximal area of about $7 * 10^4$ m^2 was searched by one animal per hour. Fenchel (1986) reported that a *Bodo* species was able to ingest up to 60 bacteria per hour depending on bacteria concentration on the biofilm. Suspension feeding protozoans, such as the flagellate *Pteridomonas danica* and the ciliate *Tetrahymena*, can also significantly graze on loosely attached bacteria, although the grazing rates were lower than when feeding on suspended bacteria (Eisenmann et al. 1998, Zubkov and Sleigh 1999). As protozoans can reduce the biomass accumulation on surfaces, they are seen as a useful tool to control biomass in the waste water treatment, e.g. in biotrickling filters (Cox and Deshusses 1999).

Different trophic levels: The various types of nutrition among both, protozoans and metazoans, suggests biofilms to be highly complex systems with various trophic interactions and several trophic levels also within the protozoans (Fig. 4). However, studies on trophic interactions with biofilm dwelling protozoans are scarce and mostly concentrate on the grazing of bacterivores. Also little is also known about the energy transfer from protozoans to metazoan grazers. Own grazing experiments on biofilms established in the River Rhine with live fluorescently labelled heterotrophic flagellates (Schmidt-Denter and Arndt subm., Premke and Arndt 2000) demonstrated ciliates and rotifers to be important grazers of HF. Both groups were also found to feed frequently on biofilm dwelling algae. The grazing rates on HF exceeded by far their potential growth rates, indicating that a continuous colonisation by HF from the water column is necessary to maintain the HF standing stock on the biofilm (see also Fig. 2). In laboratory experiments a benthic hypotrichous ciliate (*Euplotes*) consumes 2-53 HF per ciliate per hour (Premke and Arndt 2000). Protozoans are a potential food source not only for other protozoans and meiofaunal components, but also for the macrofauna. It is well known, that macrofauna grazing, e.g. by snails can have a high impact on the biofilm structure (e.g. Sommer 1999, Hillebrand et al. 2000). Again, quantitative data are scarce. Schönborn (1998) quantified the grazing interactions on biofilms of the polluted River Saale. He highlighted ciliates as an important link in the food chain bacteria – ciliates – *Chaetogaster diastrophus* (oligochaeta) – *Erpobdella octoculata* (leech).

Nutrient recycling: Due to their feeding activity, protozoans can play an important role in the nutrient recycling on biofilms. Zubkov and Sleigh (1999) estimated, that 70-85% of the nutrients present in biofilm bacteria ingested by the protozoans were regenerated and released into the immediate vicinity of the bacterial biofilm. In addition, protozoans can stimulate the bacterial production and therefore enhance the carbon and energy flow (Sherr and Sherr 1984). Verhagen and Laanbroek (1992) reported an increased nitrification by protozoan activity. On the other hand, also the opposite, a reduction of the nitrification rate by the grazing of eukaryotic grazers, was demonstrated for biofilms (Lee and Welander 1994). The authors concluded, that the slow growing nitrifiers were more effected by grazing than other bacteria. We found that protozoans on lake biofilms seem to contribute significantly to the change of size classes of biofilm algae from small to large species by selective feeding on small algae. Large algae could make use of recycled nutrients (Auer et al. in prep.).

7. TROPHIC INTERACTIONS WITH THE SURROUNDING MEDIUM

Biofilms are open systems which interact with the surrounding media. This includes attachment and detachment of organisms as well as the biological matter turnover, which can be influenced directly or indirectly by the protozoans (see above). In addition, biofilm dwelling protozoans can have a direct grazing impact on particulate organic matter in the water column. Looking at the functional groups of biofilm dwelling protozoans, a large part of suspension feeding protozoans is often present. During own studies on biofilms on artificial substrate in the River Rhine (Weitere and Arndt subm.), between 8 and 54 % of the HF abundance were composed of suspension feeding chrysomonads and choanoflagellates and as much as 63 to 95 % of the biofilm dwelling ciliates were composed of suspension feeding heterotrich and peritrich ciliates (see also Fig. 3). On leaves in a stream, the majority of the cilitates have been shown to be filter feeders (Franco et al. 1998). Suspension feeding protozoans are also able to consume loosely attached bacteria (Eisenmann et al. 1998, Zubkov and Sleigh 1999), but the majority of their food is presumably composed of planktonic organisms (Caron 1987). Sibille et al. (1998) demonstrated a significant decrease of *Esherichia coli* bacteria added to a drinking water distribution system, when both suspended and attached protozoans were present. Own laboratory experiments demonstrate, that small plankton organisms (bacteria, algae, HF, ciliates) can be strongly reduced by biofilm dwelling ciliates (especially *Stentor* sp. and peritrich ciliates) and rotifers (Weitere and Arndt subm.). Biofilms acted as a significant sink for plankton organisms of the River Rhine (Germany) when biofilms were densely populated by ciliates and rotifers in June while much lower effects were recorded in winter at low predator density (Fig. 5). Drastic changes in the water level regulate the ratio of water volume to the colonised river bottom area which influences the losses of plankton organisms in the river (Weitere and Arndt in press).

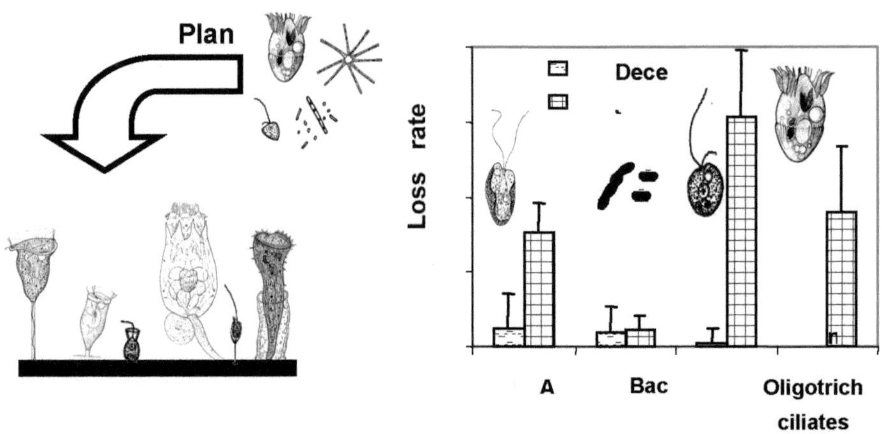

Figure 5. Biofilms as a sink for plankton communities: Loss rates of different plankton groups to biofilm dwelling organisms on two occasions (December and June). The experiments were conducted with natural plankton assemblages in flow vessels (Weitere and Arndt subm.), which were pre-exposed for 14 days in the River Rhine (establishment of natural biofilms on the inner vessel walls). Loss rates were calculated as the difference between growth rates in control vessels (without biofilms) and in experimental vessels (with biofilms).

8. CONCLUSIONS

Protozoans have been demonstrated as important components of biofilms, which are able to rapidly colonise new substrata and which generally occur in high abundances on biofilms. They may structure the biofilms by their activity (feeding, movement) as well as due to loricae, stalks and excretions. They can enhance the transport of solutes onto and within the biofilm and stabilise the matter flux as part of a complex food web. By their feeding on plankton organisms, protozoans can be important in connecting the biofilms with other parts of the ecosystem. A very important point is that protozoans should be considered as important members of biofilms even if they are not visible using traditional methods of bacteriology and algology. There are many open questions for further research such as the quantification of grazing impacts and recycling effects of protozoans, their impact on the coexistence of bacteria on biofilms, the specification of protozoan niches on biofilms, their three-dimensional activity patterns, the role of extracellular polymers for the interaction between bacteria/algae and protozoans, etc.

ACKNOWLEDGEMENT

The studies were supported by grants from the Deutsche Forschungsgemeinschaft to H.A. and by the Deutsche Bundesstiftung Umwelt to M.W. The authors wish to express their sincere thanks to Wolfgang Krumbein for organising a very stimulating and fruitful community working on Microbial Mats.

REFERENCES

Alongi, D.M. (1991) Flagellates of benthic communities: characteristics and methods of study, in D.J. Patterson and J. Larsen (eds.), *The biology of free-living heterotrophic flagellates*, Clarendon Press, Oxford, pp. 57-75.

Archer, S.D., Leakey, R.J.G., Burkill, P.H., Sleigh, M.A. and Appleby, C.J. (1996) Microbial ecology of sea ice at a costal Antarctic site: community composition, biomass and temporal changes, *Mar. Ecol. Progr. Ser.* 135, 179-195.

Arndt, H. (1993) A critical review of the importance of rhizopods (naked and testate amoebae) and actinopods (heliozoa) in lake plankton, *Mar. Microb. Food Webs,* 7, 3-29.

Arndt, H., Dietrich, D., Auer, B., Cleven, E.J. Gräfenhan, T. Weitere, M. and Mylnikov, A. P. (2000) Functional diversity of heterotrophic flagellates in aquatic ecosystems, in B.S.C. Leadbeater and J.C. Green (eds.), *TheFflagellates,* Taylor & Francis, London, New York, pp. 240-268.

Atkins, M.S., Teske, A.P. and Anderson, O.R. (2000) A survey of flagellate diversity at four deep-sea hydrothermal vents in the eastern Pacific ocean using structural and molecular approaches, *J. Euk. Microbiol.* 47, 400-411.

Azam, F., Fenchel, T., Field, J. G., Gray, J. S., Meyer-Reil, L.A. and Thingstad, F. (1983) The ecological role of water column microbes in the sea, *Mar. Ecol. Progr. Ser.* 10, 257-263.

Baldock, B., Baker, J., and Sleigh, M.A. (1983) Abundance and productivity of protozoa in chalk streams, *Holarctic Ecology* 6, 238-246.

Bernard, C. and Fenchel, T. (1995) Mats of colourless sulphur bacteria. II Structure, composition of biota and successional patterns, *Mar. Ecol. Prog. Ser.* 128, 171-179.

Bernhard, J. and Bowser, S. (1992) Bacterial biofilms as a trophic resource for certain benthic foraminifera. Mar. Ecol. Prog. Ser. 83, 263-272.

Berninger, U.G., Caron, D.A., Sanders, R.W. and Finlay, B.J. (1991) Heterotrophic flagellates of planktonic communities: their characteristics and methods of study, in D.J. Patterson and J. Larsen (eds.), *The biology of free-living heterotrophic flagellates*, Clarendon Press, Oxford, pp. 39-56.

Boenigk, J. and Arndt, H. (subm.) Bacterivory by heterotrophic flagellates: community structure and feeding strategies,

Boenigk, J. and Arndt, H. (2000) Comparative studies on the feeding behavior of two heterotrophic nanoflagellates: the filter-feeding choanoflagellate *Monosiga ovata* and the raptorial-feeding kinetoplastid *Rhynchomonas nasuta, Aquat. Microb. Ecol.* 22, 243-249

Brown, M. and Barker, J. (1999) Unexplored reservoirs of pathogenic bacteria: protozoa and biofilms, *Trends in Microbiology* 7, 46-50.

Butler, H. and Rogerson, A. (1995) Temporal and spatial abundance of naked amoebae (Gymnamoebae) in marine sediments of the Clyde Sea area, Scotland, *J. Euk. Microbiol.* 42, 724-730.

Butler, H. and Rogerson, A. (1997) Consumption rates of six species of marine benthic naked amoebae (Gymnamoebia) from sediments in the Clyde Sea area, *J. mar. biol. Ass. UK* 77, 989-997.

Cairns, J. and Yongue, W. (1968) The distribution of fresh-water protozoa on a relatively homogeneous substrate, *Hydrobiologia* 31, 65-72.

Cairns, J., Dahlberg, M., Dickson, K., Smith, N., and Waller, W. (1969) The relationship of freshwater protozoan communities to the MacArthur-Wilson Equilibrium model, *Am. Nat.* 103, 439-454.

Caron, D.A. (1987) Grazing of attached bacteria by heterotrophic microflagellates, *Microb. Ecol.* 13, 203-218.

Caron, D.A. (1991) Heterotrophic flagellates associated with sedimenting detritus, in D.J. Patterson and J. Larsen (eds.), *The biology of free-living heterotrophic flagellates*, Clarendon Press, Oxford, pp. 77-92.

Costerton, J.W., Lewandowski, Z., Caldwell, D.E., Korber, D.R., Lappin-Scott, H.M. (1995) Microbial biofilms, *Ann. Revs. Microbiol.* 49, 711-745.

Cowling, A.J. (1994) Protozoan distribution and adaptation, in: J.F. Darbyshire (ed.) *Soil Protozoa*, CAB International, Wallingford, pp. 5-32.

Cox, H. and Deshusses, M. (1999) Biomass control in waste air biotrickling filters by protozoan predation, *Biotechnology and Bioengineering* 62, 216-224.

Dietrich, D. and Arndt, H. (2000) Biomass partitioning of benthic microbes in a Baltic inlet: Relationships between bacteria, algae, heterotrophic flagellates and ciliates, *Marine Biology* 136, 309-322.

Eisenmann, H., Harms, H., Meckenstock, R., Meyer, E., and Zehnder, A. (1998) Grazing of a *Tetrahymena* sp. on adhered bacteria in percolated column monitored by in situ hybridization with fluorescent oligonucleotide probes, *Appl. Environ. Microbiol.* 64, 1264-1269.

Fenchel, T. (1965) On the ciliate fauna associated with the marine species of the amphipod *Gammarus* J. G. Fabricius, *Ophelia* 2, 281-303.

Fenchel, T. (1986) The ecology of heterotrophic flagellates, in K. C. Marshall (ed.) *Advances in Microbial Ecology*, Plenum Press, New York , pp. 57-97.

Fenchel, T. (1987) *Ecology of Protozoa,* Springer-Verlag, Berlin.

Fenchel, T. and Blackburn, N. (1999) Motile chemosensory behaviour of phagotrophic protists: mechanisms for and efficiency in congregating at food patches, *Protist* 150, 325-336.

Fenchel, T. and Finlay, B.J. (1995) *Ecology and Evolution in Anoxic Worlds*. Oxford University Press, Oxford.

Finlay, B.J. (1990) Physiological ecology of free-living Protozoa, in K. C. Marshall (ed.) *Advances in Microbial Ecology*, Plenum Publishing Corp., New York, pp. 1-35.

Flemming, H.-C. (1991) Biofilms as a particular form of microbial life, in H.-C. Flemming and G.G. Geesey (eds.), *Biofouling and biocorrosion in industrial water systems*, Springer-Verlag, Heidelberg, pp. 3-9.

Foissner, W., Unterweger, A., and Henschel, T. (1992) Comparison of direct stream bed and artificial substrate sampling of ciliates (Protozoa, Ciliophora) in a mesosaprobic river, *Limnologica* 22, 97-104.

Franco, C., Esteban, G., and Téllez, C. (1998) Colonization and succession of ciliated protozoa associated with submerged leaves in a river, *Limnologica* 28, 275-283.

Fried, J., Mayr, G., Berger, H., Traunspurger, W., Psenner, R., and Lemmer, H. (2000) Monitoring protozoa and metazoa biofilm communities for assessing wastewater quality impact and reactor up-scaling effects, *Wat. Sci. Tech.* 41, 309-316.

Garstecki, T. and Arndt, H. (2000) Seasonal abundances and community structure of benthic rhizopods in shallow lagoons of the southern Baltic Sea, *Europ. J. Protistol.* 36, 103-115.

Glud, R.N. and Fenchel, T. (1999) The importance of ciliates for interstitial solute transport in benthic communities, *Mar. Ecol. Prog. Ser.* 186, 87-93.

Gray, T., Cursons, T., Sherwan, J., and Rose, P. (1995) *Acanthamoeba,* bacterial, and fungal contamination of contact lens storage cases, *British Journal of Ophthalmology* 79, 601-605.

Hahn, M.W., Moore, E.R.B. and Höfle, M.G. (2001) Role of microcolony formation in the protistan grazing defence of the aquatic bacterium *Pseudomonas* sp. MWH1, *Microb. Ecol.* 39, 175-185.

Hausmann, K. and Hülsmann, N. (1996) *Protozoology.* 2nd ed., Thieme, Stuttgart.

Hillebrand, H., Worm, B., and Lotze, H. (2000) Marine microbenthic community structure regulated by nitrogen loading and grazing pressure, *Mar. Ecol. Prog. Ser.* 204, 27-38.

Jackson, S.M. and Jones, E.B.G. (1991) Interactions within biofilms: the disruption of biofilm structure by protozoa, *Kieler Meeresforsch. Sonderh.* 8, 264-268.

Jax, K. (1992): Investigations on succession and long-term dynamics of testacea assemblages (Protozoa: Rhizopoda) in the Aufwuchs of small bodies of water, *Limnologica* 22, 299-328.

Jürgens, K., Arndt, H. and Zimmermann,H. (1997) Impact of metazoan and protozoan grazers on bacterial biomass distribution in microcosm experiments, *Aquat. Microb. Ecol.* 12, 131-138.

Jürgens, K. and Güde, H. (1994) The potential importance of grazing-resistant bacteria in planktonic systems. *Mar. Ecol. Prog. Ser.* 112, 169-188.

Krumbein, W.E., Paterson, D.M. and Stal, L.J.(eds.) (1994) *Biostabilization of Sediments.* Bibliotheks- und Informationssystem der Carl von Ossietzky Universität Oldenburg.

Lawrence, J. and Snyder, R.A. (1998) Feeding behaviour and grazing impacts of a *Euplotes* sp. on attached bacteria, *Can. J. Microbiol.* 44, 623-629.

Lee, N. and Welander, T. (1994) Influence of predators on nitrification in aerobic biofilm processes. *Wat. Sci. Tech.* 29, 355-363.

Marshall, K.C. (1976) *Interfaces in Microbial Ecology*, Harvard University Press, Cambridge.

Martín-Cereceda, M., Álvarez, A., Serrano, S., and Guinea, A. (2001) Confocal and light microscope examination of protozoa and other microorganisms in the biofilms from a rotating biological contactor wastewater treatment plant, *Acta Protozool.* 40, 263-272.

Neu, T.R. (1994) Biofilms and microbial mats, in W.E. Krumbein, D.M. Paterson and L.J. Stal.(eds.) *Biostabilization of Sediments.* Bibliotheks- und Informationssystem der Carl von Ossietzky Universität Oldenburg. pp. 9-15.

Nisbet, B. (1974) An ultrastructural study of the feeding apparatus of *Peranema trichophorum*. *J. Protozool.*, 21, 39-48.

Novarino, G., Warren, A., Butler, H., Lambourne, G., Boxshall, A., Bateman, J., Kinner, N.E., Harvey, R.W., Mosse, R.A. and Teltsch, B. (1997) Protistan communities in aquifers: a review, *FEMS Microbiology Reviews* 20, 261-275.

Page F.C. (1988) *A New Key to Freshwater and Soil Gymnamoebae*. The Freshwater Biological Association, Cumbria.

Patterson, D.J., Nygaard, K., Steinberg, G. and Turley, C.M. (1993) Heterotrophic flagellates and other protists associated with oceanic detritus throughout the water column in the mid North Atlantic, *J. Mar. Biol. Ass. UK* 73, 67-95.

Premke, K. and Arndt, H. (2000) Predation on heterotrophic flagellates by protists: Food selectivity determined using a live-staining technique, *Arch. Hydrobiol.* 150, 17-28.

Riedel-Lorje, J.C. (1980) Untersuchungen über den Indikationswert von Aufwuchs in Süss- und Brackwasserzonen des Elbe-Aestuars unter Berücksichtigung industrieller Einleitungen, Doctoral thesis, Universität Hamburg.

Rogerson, A., Hannah, F. and Gothe, G. (1996) The grazing potential of some unusual marine benthic amoebae feeding on bacteria, *Europ. J. Protistol.* 32, 271-279.

Ruttner, F. (1963) *Fundamentals of Limnology*, 3rd ed., University Press Toronto, Toronto.

Schmidt-Denter, K. and Arndt, H. (subm.) Heterotrophic flagellates as a major component of biofilms of the River Rhine: Community structure and trophic interactions. Submitted to *Arch. Hydrobiol.*

Schönborn, W. (1998) Changes of biological time patterns and of the energy transfer on exposed slides and stone surfaces (boundary layers with fractal structures) by organic pollution in an oligotrophic stream (Ilm, Thuringia, Germany), *Limnologica* 28, 347-361.

Sherr, B.F. and Sherr, E.B. (1984) Role of heterotrophic protozoa in carbon and energy flow in aquatic ecosystems, in M.J. Klug and C.A. Reddy (eds.) *Current Perspectives in Microbial Ecology*, American Society of Microbiology, Washington D.C., pp. 412-423.

Sherr, E.B. and Sherr, B.F. (2000) Marine microbes: an overview, in D.L. Kirchman (ed.) *Microbial Ecology of the Oceans,* Wiley-Liss, New York.

Sibille, I., Sime-Ngando, T., Mathieu, L., and Block, J. (1998) Protozoan bacterivory and *Esherichia coli* survival in drinking water distribution systems, *Appl. Environ. Microbiol.* 64,197-202.

Silvester, N. and Sleigh, M.A. (1985) The forces on microorganisms at surfaces in flowing water. *Freshw. Biol.* 15, 433-448.

Skibbe, O. (1994) An improved quantitative protagol stain for ciliates and other planktonic protists. *Arch. Hydrobiol.* 130, 339-347.

Sommer, U. (1999) Periphyton architecture and susceptibility to grazing by periwinkles (*Littorina littorea*, Gastropoda), *Int. Revue Hydrobiol.* 84, 197-204.

Strüder-Kypke, M.C. and Schönborn, W (1999) Periphyton and Sphagnicolous Protists of Dystrophic Bog Lakes (Brandenburg,Germany).II.Characteristic species and trophy of the lakes, *Limnologica* 29, 407-424.

Verhagen, F.J.M. and Laanbroek, H.J. (1992) Effects of grazing by flagellates on competition for ammonium between nitrifying and heterotrophic bacteria in chemostats, *Appl. Environ. Microbiol.* 58, 1962-1969.

Weitere, M. and Arndt, H. (in press) Top-down effects on pelagic heterotrophic nanoflagellates (HNF) in a large river (River Rhine): Do losses to the benthos play a role? *Freshw. Biol.*

Weitere, M. and Arndt, H. (subm.) Periphyton/biofilm communities of the River Rhine: Its function as a significant loss factor of potamoplankton, submitted to *Freshw. Biol.*

Wickham, S.A., Gieseke, A, and Berninger, U. G. (2000) Benthic ciliate identification and enumeration: an improved methodology and its application, *Aquat. Microb. Ecol.* 22, 79-91.

Young, I.M., Roberts A., Griffiths, B.S., and Caul, S. (1994) Growth of a ciliates protozoan in model ballotini systems of different particle sizes, *Soil Biol. Biochem.* 26, 1173-1178.

Zubkov, M.V. and Sleigh, M.A. (1999) Growth of amoebae and flagellates on bacteria deposited on filters, *Microb. Ecol.* 37, 107-115.

Chapter 11

BRYOZOAN MORPHOPROCESSES

Joachim SCHOLZ
Forschungsinstitut und Naturmuseum Senckenberg, Sektion Marine Evertebraten III, Senckenberganlage 25, D-60325 Frankfurt, Germany
jscholz@sngkw.uni-frankfurt.de

George S. LEVIT
Friedrich-Schiller-University of Jena
Institut for the History of Medicine, Natural Science and Technology (Ernst-Haeckel-Haus), D-07745 Jena, Germany

1. INTRODUCTION

Bryozoans represent a phylum of colonial aquatic suspension feeding animals. They are abundantly found in both, cool water and tropical reefs, silicoclastic shelfs, and even the soft bottoms of deep sea waters. Colonies are formed by repeated budding of genetically but not always morphologically identical, physically connected animals (Fig. 1). Since the modular construction of mostly calcifying colonies makes it possible to reconstruct interactive relationships to the environment, and timing of environmental factors, bryozoans have been traditionally used a facies fossils and bioindicators (McKinney & Jackson 1989, with review of literature).

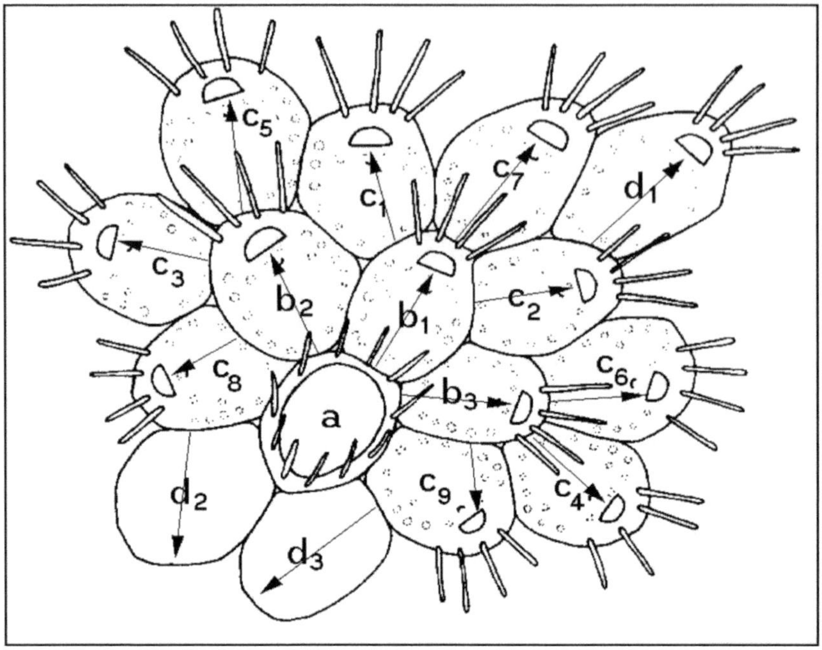

Figure 1. Budding sequence in a cheilostome bryozoan of the genus *Fenestrulina* belonging to the zooidal (z) lamina type (see text). Arrows indicate polar orientation such as determined by geometrical arrangement of pore chambers which in turn give rise to a daughter zooid. Redrawn from Gordon 1971.

Microbial mats and biofilms represent one of the most important factors influencing bryozoan settlement and distribution. Since the majority of post-paleozoic bryozoan growth types occur laminar within the boundary layer, bryozoans are also influencing microbial mat coverage on the sediment water interface. They are able to cover considerably large substrate areas with small investment of biomass. In contrast to occasional statements in marine biology textbooks, there are many bryozoan species present in coastal seas around the globe that compete successfully with weedy "algal mats" (Scholz 2000; see fig. 5). Judging from data collected in reefs of the Philippines, and the Gulf of Aqaba region, the bathymetric occurrence of various bryozoan growth types is mainly linked to the competitive pressure of microbial mats in shallow waters, and mutalistic interaction with epizoic mats in shaded microhabitats (Scholz 1993, 2000). Accordingly, benthic micro-organisms are one of the key factors to understand bryozoan diversity patterns. Yet, it is our basic assumption that any conclusion of "microbial control" of bryozoans or other benthic marine invertebrates is a too cybernetic simplification.

Therefore, this essay addresses the following two questions in the interpretation of bryozoan occurrence and bryozoan exology:

1. Does the spatial occurrence of bryozoans depend on species-specific biological potential (such as growth rates, considered to be crucial in competition for substrate space), or rather depend on the environmental control, and the availability of suitable niches?
2. Is a distinction between "biofactors" and "ecofactors" (such as suggested by Lapo, 1987) always applicable in the interpretation of bryozoan settlement and settlement success?

The scientific idea that certain features of an organism are genetically determined, while others are acquired by interaction with the environment, is well embedded in popular culture. Nowadays, such a dichotomous account on development does not remain unchallenged (Griffiths & Gray 1994). In communications utilising the English language, some of the critical approaches of fitness as a matter of "fittedness" to an independent environment are known as "theory of developmental systems". There were also earlier attempts (Beklemishev, 1964) to overcome two implicit dichotomies dominating in the scientific literature: (1) a dichotomy between genes and the remaining living system, (2) a dichotomy between organism and environment.

The concept of a living organism itself is rather obscure. Already in the first quarter of the century it was supposed (Kozo-Poljanskij, 1921) and in the recent decades proven (e.g., Margulis, 1993) that an eukaryotic cell can be treated as a complex symbiotic system, because mitochondria and chloroplasts show many properties of independent living entities. The precise analysis of the relationships between parts of a living system, as their separate and common interactions with the environment makes the traditional concept of organism inapplicable to the real biological systems. Nevertheless even in very recent publications (e.g., Mahner & Bunge, 1997) one can find attempts to give an ultimate definition of this last metaphor of the romantic biology. In accordance with the contemporary scientific data, only a prokaryotic cell can be seen as an "elementary biosystem", a unite «organism», although also here horizontal gene transfer and other means of mixing information and morphological pattern allows to talk of a chimeric origin of many prokaryotes. All biosystems of any higher level are complex entities composed of biosystems of the lower order. There is no impassible border between the «organism» and its «environment». Likewise there is no impassible border between «organism» and other higher level organizations of living matter such as, for example, biogeocoenoses. The concept of organism is not only contestable (Sterelny & Griffiths 1999), there is no real biological object that unequivocally corresponds to this concept. Roughly speaking there are no organisms. Instead, there are biosystems that more or less correspond to the traditional concept of an organism. In some areas of biology the concept of organism is still applicable, while many species (e.g., fungi, bryozoa) do not fit to this traditional concept.

In the following remarks, bryozoan regular growth and change is described in terms of a morphoprocess as proposed by the Russian zoologist V. N. Beklemishev (1890-1962). Consistent with the ideas of Beklemishev we define morphoprocess as a concept describing the dynamic form of organisation of living matter. Morphoprocess is not necessarily interrupted when an individual organism disintegrates. The whole biosphere can be described as a global morphoprocess consisting of lower order morphoprocesses. According to this dynamic vision, morphoprocess is the ultimate object of the biological systematics. The concept of a morphoprocess involves regular growth and change of the identifiable lasting form. At the same time, morphoprocess is defined as a 'lasting form', which is in the regular growth and change and is seen as a self-organising process. The concept of morphoprocess allows to overcome the deficiencies of thinking in the traditional terms such as 'organism' and 'community' and is especially appropriate for describing moderately individualised biosystems with much more complex structural relationships than can be represented in the traditional schemes. Thereby, morphoprocess is a concept, which may - from a certain viewpoint- replace the present day concept of an organism.

2. BIOLOGICAL POTENTIAL OF BRYOZOANS TO REFLECT ENVIRONMENTAL INTERACTION

Before dwelling upon some detail we will provide some preliminaries on bryozoan morphology.

Bryozoan polypides consist of a U-shaped gut and a ring of tentacles. In feeding position, they protrude from their rigid zooidal skeleton (zooecium). The zooecial chamber is very small (< 0.5mm in most species). The protrusion of the tentacle sheet depends on the muscular contraction of membranes and the increase of hydrostatic pressure of the coelomic fluid. Spatial and functional arrangement of membranes, zooecial walls and annual or parietal muscles is a criteria for classification of bryozoans up to the level of orders and infraorders (see Ryland 1970; McKinney & Jackson 1989; Gordon 1993).

Furthermore, bryozoan zooids and colonies are distinctively polar structures. The structural polarity is e.g. evident by the arrangement of pore chambers, which shall give rise to a daughter zooid (fig. 1). In planar growth, the sequences of mother-daughter zooids result in a geometrical void-filling arrangement. Environmental factors such as competition for space may cause a change in the arrangement of autozooids. As an induced (reactive) polarisation of colonial (zoarial) growth, they may superimpose the structural polarity of zooids and thus result in "equipotential", aclonal areas of the zoarium. The presence of highly reactive equipotential areas characterises the zoarial margins and/or frontal surfaces of many but not all planar-encrusting bryozoans.

There are various species-specific degrees of growth determined by structural polarity, or biological potential of growth on the grounds of environment induced polarity. It has thus been possible to group encrusting multilinear zoarial types into four basic classes. Increasing degrees in the formation of an equi-potential marginal area are achieved in the following order: zooidal (z) laminae, multizooidal (m) laminae, celleporiform (c) laminae, and sheet (s) laminae.

In zooidal laminae, growth is mainly determined by structural polarity of a single zooid. New buds derive in a geometrically predictable way from pore chambers. During interaction with microbial mats, stagnation of growth is the usual outcome; or changing direction of growth away from the contact zone if that is possible.

Sheet laminae show a much more environment induced way of growth, with various types or sizes of giant buds, stolonal outgrowths, or weakly calcified zoarial margins overruling the structural polarity of the single colonial module (fig. 5).

Multizooidal laminae are a common intermediate type, producing comparatively short giant buds (see Fig. 4), whereas celleporifirm laminae perform increased rapid frontal budding and self-cleaning of zoarial surface through frequent replacement of frontal zooids (see Scholz 2000).

Similar to s-laminae, c-laminae (celleporiform laminae) are comparatively strong in overgrowth interaction. While contacting a competing biosystem on a substrate surface, vertical growth is achieved by rapid frontal budding, and the colony is thereby able to overtop for example a sponge mat, or another bryozoan. Typical species showing this reaction can be found in the genera *Celleporaria* and *Rhynchozoon*.

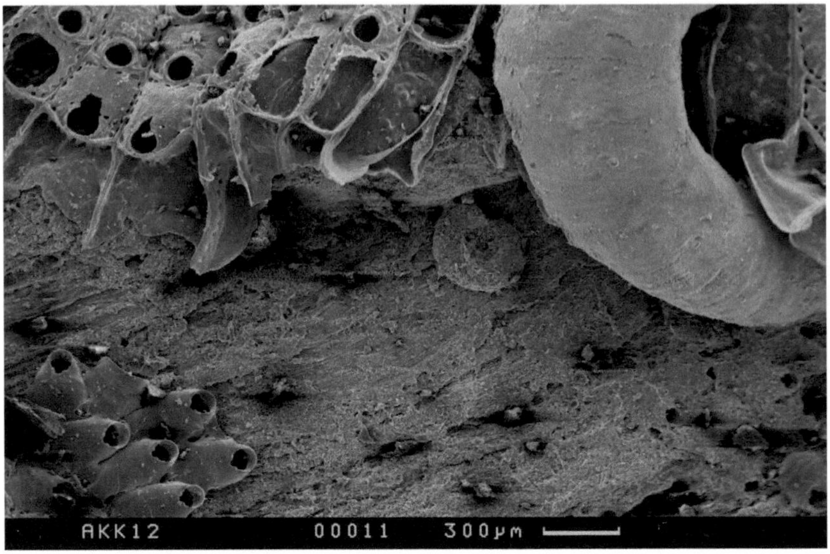

Figure 2. A multizooid (m) lamina of the ascophoran bryozoan *Schizomavella auriculata* (HassaL) (top) produces a weakly calcified colonial margin composed of short giant buds. In the interaction zone with a serpulid worm (right), the giant buds grow significantly longer. Encountering a microbial mat will eventually result in a similar morphological signal. At the lower margin of the SEM micrograph, a zooid (z) lamina of the genus *Celleporella hyalina* (Linnaeus) is illustrated. Zooid lamina do not produce or produce very rarely, under exceptional circumstances giant buds. Growth is thus determined by the structural polarity of void filling zooids that are budded from a single mother zooid, and not from the colonial margin.
Bleached sample, collected off Akkeshi Marine Station, Hokkaido, Japan.

3. BRYOZOANS AS A MICROBIAL COLONY: MUTUALISTIC INTERACTIONS

Bryozoans are only one out of many examples that have been documented as complex entities composed of many biosystems such as associated microbial cells living on and within bryozoan tissues (Scholz & Krumbein 1994, 1996). To see Bryozoans as a genetically defined species is thus of rather relative nature when it comes to environmental interaction. Some bryozoan species are actually a microbial mat with underlying bryozoan zooids. Bryozoans supporting an epizoic microbial mat on a species-specific rather than colony-specific base have been named "bryozoan micro-reefs" (Scholz 1995, 2000; Scholz & Hillmer 1995; see Fig. 2).

Figure 3. A bryozoan micro-reef: *Chaperiopsis cervicornis* (Busk), collected from the intertidal zone of Evans Bay, Wellington, New Zealand. The frontal surface is covered by a fungal mat with associated diatoms and bacteria. The presence of an epizoic fungal mat is a typical feature of colonies of *C. cervicornis* studied from New Zealand sites.

In bryozoan micro-reefs, a calcifying bryozoan provides the principal "frame builder", whereas epizoic micro-algae, (cyano-)bacteria, fungi, ciliates and/or small metazoans are treated as "reef dwellers".

However, is it justified to consider a structure that is smaller than a centimeter, as a "reef"? Flügel & Flügel-Kahler (1992:170) pointed out two important criteria to define any organogenic build-up as reef;

a) Biological control during the formation of the structure (especially by sessile organisms);

b) rigidity of the structure and laterally restricted topographic relief.

Since this definition does not include any limitations of size (including the postulate that a reef structure reaches sea-level), the criteria may be applied as well to the micro-scale: (Bryozoan-) micro-reefs are reefs which build a micro-topographic relief at the substrate-water interface, thus remaining largely within the boundary layer.

Although the recognition of bryozoan micro-reefs has been based on recognisable morphological features of bryozoan skeletons, bryozoans are not solely a supporting system for microbial life. Instead, there is no clear separation possible between bryozoan and microbial environment as between the bryozoa part and the bacterial/fungal part of such micro-reef structures.

Bryozoan micro-reefs are at least partly self-controlling (-organizing), judging from examinations of the anascan cheilostome bryozoan *Chaperiopsis cervicornis* (Busk), a species common in intertidal zones around New Zealand. The same is true for several species of the cribriomorph genus *Puellina* spp, collected from various sites in thze Philippines, the Red Sea, and waters around Japan.

The spatial distribution of the epizoic community clearly indicates that bryozoan generated feeding currents favour the development of an epizoic micro-flora. The maximal velocity of feeding currents outlines microbial nano-habitats on bryozoan surfaces. On the other hand, the epizoic biomat protects the bryozoan surface against predatory cnidarians, and helps to delay or halt overgrowth of the colonial surface by faster growing bryozoan species and other competitors for space.

The mutual benefit of bryozoans and epizoic communities is also indicated by the spatial distribution of benthic micro-organisms in the surrounding surfaces. In the New Zealand collection sites of *Chaperiopsis,* only biofilms are seen but hardly any microbial mats are found outside the area encrusted by the bryozoan. In this way, a highly diverse (micro-)biosystem has distinguished itself from the environment.

Not all inhabitants of a bryozoan micro-reef are beneficial for the "host". It has been possible to isolate several species of zoo-pathogenic fungi from bryozoans including "bryozoan micro-reefs" (Sterflinger & Scholz 1997, Scholz et al. 2000, Sterflinger et al. 2001). Therefore, the statement of "mutualistic" interactions is again a rather relative one. However this relativity does not yet differ micro-reef from the «trivial organism», because every organism is in the nature of a semi-parasitic, semi-mutualistic community (Beklemishev, 1964).

What appears to be typical for bryozoan micro-reefs is the observation that

1. bryozoan exoskeletons ("gymnocysts") favour the growth of an epizoic microbial community
2. bryozoans providing gymnocystal settlement space are usually weak in overgrowth competition, since the ability to perform frontal budding and lateral outgrowths depends on the presence of hypostegal coeloms, and thus on the development of interior walls (e.g. cryptocysts).

The slow growing zooidal (z) laminae are therefore the most likely candidate for giving rise to a bryozoan microreef. Z-laminae are characterized by a low degree of morphological variation.

Most biosystems consist of parts which, in their turn, can possess a certain degree of individuality, but can be also very simply organized. The general principle claims that the better every living part supports the development of the whole system, and the better the whole supports the functions of individual morphoprocesses, the higher is the degree of mutuality. In other words, the individuality of parts and the individuality of the whole biosystem are two pans of the scales.

Z-laminae are highly individualized on the level of the single colonial module (zooid), and due to their high degree of zooidal individualization (in the sense of autonomy of the colonial module from the colony), a z-lamina colony that is composed of highly individualized modules is only poorly individualized. The low degree of individuality of the colony thus limits the ability to change colonial growth directions through reactive budding, in situations of interaction with other biomats.

Usually, a bryozoan microreef can be readily recognized by extensive microbial mat growth on the surface. On the other hand, colonies of other bryozoans may look clean, but support extensive fungal mycelia growth inside their coelomic cavities (Hain et al. 1999, Kühner 2000; see Fig. 5). Little is known on the true mutualistic or antagonistic nature of those associations that are hidden in the coelomic cavities of bryozoans.

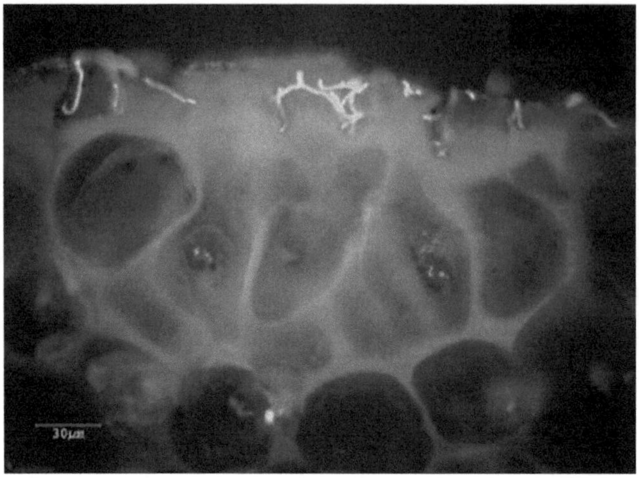

Figure 4. Thin section of the ctenostome bryozoan *Alcyonidium gelatinosum* (Linnaeus). Near the frontal surface, some endozoic fungal hyphae occur. Due to calcofluorwhite staining, the fungi are clearly visible as light blue structures. Fotograph courtesy by Mr. Simon Kühner (Lübeck). From Kühner (2000: 29).

4. DOCUMENTATION OF ANTAGONISTIC INTERACTIONS IN BRYOZOAN SKELETONS

The higher the degree of mutuality and, hence, the level of functional harmony of the bryozoan biosystem, the more morphogenetically secluded it becomes (e.g. bryozoan micro-reefs). On the other hand, competitive interactions of bryozoans with biomats result in a high degree of variation of zoarial forms (e.g. bryostromatolites).

In such cases where microbial mats expand over bryozoan frontal surfaces, certain species of bryozoan have the biological potential of a dynamical self-overgrowth. If microbial mat expansion and tactical bryozoan self-overgrowth is balanced for a certain period of time, a laminated carbonate rock is generated under microbial control. Such bryozoan laminations have been named "bryostromatolites" (Scholz 2000, Scholz et al. 2000). Bryostromatolites are extremely variable in morphology, and sometimes produced by species that usually grow as a sheet like crust not unlike microbial microstromatolites

It is important to note that the overgrowth interactions between bryozoans and (epizoic) microbial mats show a reversible pattern. It is a cyclic competitive intergrowth that results in the formation of bryostromatolites. Being seen as a quasi-infinite morphorpocess the bryozoan biosystems can be described in terms of periodic and aperiodic morphoprocesses. The general principle of a morphoprocess declares in this case that the more periodic a morphoprocess is, the higher will be the degree of individuality it possesses. The cyclic-periodical morphoprocess of bryostromatolites can be in this respect contrasted to the aperiodic rhythm of a normal overgrowth, which results in decease of a defeated competitor.

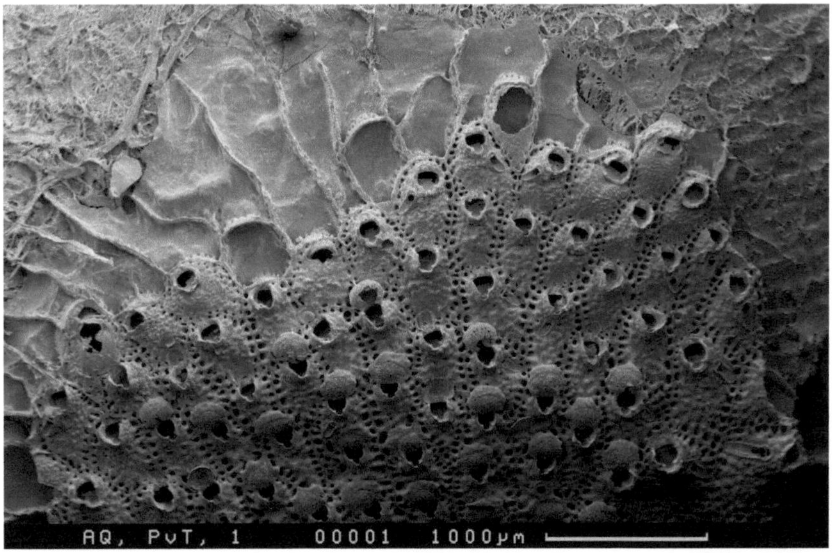

Figure 5. Biomat contra biomat: *Parasmittina tropica* (Waters) shown below, is belonging to the bryozoans of s-lamina type. The weakly calcified colonial margin is composed of giant buds that are considerably larger than a single feeding zooid. The giant bud zone does most probably not contribute to the nutrient supply of the colony. The functional relevance is to facilitate overgrowth of competitors for space such as the microbial mat illustrated at the top margin. Accordingly, the size giant buds may achieve depends on confrontational situations - they are generated when they are needed, since they are "expensive". Henceforth, the shape of the colony is subject to considerable variation.
Locality: Gulf of Aqaba, Red Sea. Sample courtesy of Helmut Schuhmacher (Essen, Germany).

Being the ultimate cadence of what the bryozoan constructional morphology is able to perform in a situation of stiff competition for space, bryostromatolites reconfirm some predictions made by Beklemishev:

Antagonistic interaction results in variation. Mutualistic interactions (such as present in bryozoan microreef), stabilise the shape of the colony due to an increase of independence from the environment. In bryostromatolites and in s-lamina, an epizoic microbial community is either lacking or harmful. However it must be stressed that mutualistic interactions which are not counter-balanced by the antagonistic relations of a certain kind do not lead to long-time stabilisation.

Generally both micro-reefs and bryostromatolites can be seen as relatively highly individualised biosystems. The degree of individuality in both cases is comparable to some low organized «trivial organisms».

We may feel tempted to argue, that due to presence of certain "adaptations" such as giant buds, the outcome of competitive interactions between bryozoans and microbial mats may be determined: if giant buds are present, the respective bryozoan species is probably able to overgrow another biomat. If giant buds are lacking such as being the case in z-laminae, interactive cooperatio is favoured. Mutual adaptations in different stages of bryozoan morphoprocesses can produce either a bryozoan micro-reef or alternatively a bryostromatolite. Microbial environment is one of the most decisive factors determining formation of a bryostromatolite. We do not know a priori what kind of biosystem we will have afterwards. The process is not «genetic programmed». That is why, there is also no clear distinction possible between genetically transmitted and environmentally determined traits.

Figure 6. A bryostromatolite collected from Akkeshi, Hokkaido, Japan: *Schizomavella auriculata* (Hassal) The same species has been illustrated growing as a m-lamina in Fig. 3. The laminated structure is composed of bryozoan sheets, and agglutinated sediments as a trace of microbial mat intergrowth. (Explanation on bryostromatolites: see text). Critical point dried sample. Photograph by Mr. Juergen Kaselowsky, Frankfurt.

5. CONCLUSIONS

The attempts to describe bryozoan growth and change and their interactions with the microbial and inanimate environment in traditional terms of a genetically driven «organisms» and an «environment» are faced with fundamental difficulties. The proper description and analysis implies a dynamic approach, which is possible in terms of the morphoprocess. Diverse elements of a developing bryozoan biosystem must be assessed in respect to their contribution to the whole picture. The totality of different factors determines different degrees of individuality of the same morphoprocess, which in its turn indicate and influence structural variations. The degree of individuality can be determined by studying the balance between mutualistic and antagonistic tendencies, the correspondence between rhythm of the whole morphoprocess and the rhythms of its parts and other factors. Every stage (e.g. a terminal process) and every element of the developing system are seen in the context of a theoretically infinite process.

ACKNOWLEDGEMENTS

Financial support by the German Research Agency (DFG-grants Sch581/2, Ge64/8, HO 2143/5) is gratefully acknowledged.

REFERENCES

Beklemishev, V.N. On General Principals of Organisation of Life. Bulletin of MOIP, section Biology 1964; 69 (2): 22-38. [in Russ]

Beklemishev, V. N. *Methodology of Systematics*. Moscow: KMK Scientific Press Ltd., 1994. [written in 1928] [in Russ]

Flügel, E., Flügel-Kahler, E. Phanerozoic Reef Evolution: Basic Questions and Data Base. Facies 1992; 26: 167-278.

Gorbushina A.A., Krumbein W.E. «Rock Dwelling Fungal Communities: Diversity of Life Styles and Colony Structures.» In: *Journey to Diverse Microbial Words,* Seckbach J. ed.: Kluwer, 2000.

Gordon, D. P. Colony Formation in the Cheilostomatous Bryozoan Fenestrulina malusii var. thyreophora. New Zealand Journal of Marine and Freshwater Research 1971; 5/2: 442-351

Gordon, D. P. Bryozoan frontal shields: studies on umbonulomorphs and impacts on classification. Zoologica Scripta 1993; 22/2: 203-221

Griffiths, P., Gray, R. "Developmental Systems and Evolutionary Explanation." In: *Philosophy of Biology*, Hull & Ruse, eds. Oxford University Press, 1998.

Hain, M., Scholz, J., Sterflinger, K. Marine fungi in epizoic biofilms on bryozoans. Poster. Vereinigung für Allgemeine und Angewandte Mikrobiologie. Frühjahrstagung in Göttingen; März 7-10; 1999.

Kozo-Poljanskij, B.D. *Novyj prinzyp biologii* (A new principle of biology). Leningrad-Moscow: Putchina, 1924.

Kühner, S. Mykologische Untersuchungen an epi- und endozoischen Biofilmen von Bryozoen. Diplomarbeit, Studiengang Marine Umweltwissenschaften. Carl v. Ossietzky Universität Oldenburg, 2000.

Lapo, A. V. *Traces of Bygone Biospheres*. 2 Auflage. Moskau: Mir Publishers, 1987.

Lovelock J. E. *Gaia: a New Look at life on Earth*. Oxford University Press, 1989.

Lovelock J. E. "The Gaia Hypothesis." In: *Gaia in Action,* ed. by P. Bunyard, Floris Books: Edinburgh, 1996.

Mahner M., Bunge M. *Foundations of Biophilosophy*. Springer, 1997.

Margulis, Lynn. *Symbiosis in cell evolution : microbial communities in the Archean and Proterozoic eons*. New York: 2. ed. Freeman, 1993.

Margulis L., Guerrero R., Bunyard P. "We are all Symbionts." In: *Gaia in Action,* ed. by P. Bunyard, Floris Books: Edinburgh, 1996.

McKinney, F. K., Jackson, J. B. C. Bryozoan Evolution. Special Topics in Paleontology 1989; 2: 1-238

Oyama, S. *The Ontogeny of Information*. Cambridge University Press, 1985.

Ruiz-Mirazo K, Etxeberria A., Moreno A. and Ibánez J. Organisms and their place in biology. Theory Biosci, 2000; 119: 276-317.

Ryland, J. S. *Bryozoans*. London: Hutchinson, 1970.

Scholz, J. Indications for Microbial Clues for Bryozoans when settling. Facies 1993; 29: 107-118.

Scholz, J., Krumbein, W. E. Entwicklung tropischer Korallenriffe. Biologie in unserer Zeit 1994; 2: 96-102.

Scholz, J. Bryozoan "Micro-Reefs". Epibiontic microorganisms as local control factor of bryozoan distribution. Beiträge zur Paläontologie, 1995; 20: 75 - 87

Scholz, J., Hillmer, G. Reef-Bryozoans and Bryozoan-Microreefs - Control Factor Evidence from the Philippines and other Regions. Facies 1995; 32: 109-144.

Scholz, J., Krumbein, W. E. «Microbial mats and biofilms associated with bryozoans.» In: *Bryozoans in space and time*. Gordon, D. P., Smith, A. M.,Grant-Mackie, J. A. National Institute of Water & Atmospheric Research. Frankfurt a.M.: Wellington.haft, 1996.

Scholz, J. *Eine Feldtheorie der Bryozoen, Mikrobenmatten und Sedimentoberflächen. Abhandlungen der Senckenbergischen Naturforschenden Gesellschaft*. Frankfurt am Main: Waldemar Kramer, 2000.

Scholz, J., Sterflinger, K., Junge, C., Hillmer G. A Preliminary Report on Bryostromatolites. Proceedings of the 11th International Bryozoology Association Conference; 2000 376-384 (Allen Press).

Sterelny K., Griffiths P. *Sex and Death. An introduction to philosophy of biology*. Chicago: The University of Chicago Press, 1999.

Sterflinger, K., Scholz, J. Fungal infection and bryozoan morphology., Courier Forschungsinstitut Senckenberg, Vogel-Festschrift, 1997; 201: 433-447.

Sterflinger, K., Hain, M., Scholz, J., K. Wasson. Fungal infections of a Colonial Marine Invertebrate: Diversity and Morphological Consequences. Facies 2001; 45: 31-38

Vernadsky, V.I. The Chemical Structure of the Biosphere of the Earth
and of its Environment. Moscow, 1965. [in Russ].

Vernadsky, V.I. Scientific Thought as a Planetary Phenomenon, Nongovernmental Ecological V.I. Vernadsky Foundation. Moscow, 1997.

Zavarzin, G.A. Individualistic and systematic approaches to biology. Vop. Filos., 1999; 4: 89-106.

Zavarzin, G.A., Zhilina, T.N., Kevbrin, V.V. The alkaliphilic microbial community and its functional diversity, Microbiology, 1999, 68/5: 503-521.

Chapter 12

BIOFILMS AND MACROORGANISMS

Gisela GERDES
Carl von Ossietzky University of Oldenburg,
Institute for Chemistry and Biology of the Marine Environment (ICBM) - Marine Laboratory
Schleusenstrasse 1, D-26382 Wilhelmshaven, Germany
g.gerdes@icbm.terramare.de

1. INTRODUCTION

The heading frames a wide range of possible meeting points between biofilms and macroorganisms. To begin this chapter, it seems thus neccessary to select. In addition to the chapter in this volume on epizoic microbial growth on sessile encrusting metazoans (bryozoans), the main emphasis will be on non-sessile and non-encrusting metazoans co-existing with laminated microbial mats of shallow marine and coastal settings.

Laminated microbial mats are a specific sedimentary aspect of biofilms. The classical definition of a biofilm frames microbe populations embedded in extracellular polymeric substances (EPS) that adhere to living or non-living surfaces and are able to immobilise water. While adhesion usually leads to cell immobilisation, the laminated microbial mats are made of microorganisms that at least in part remain motile and are able to glide up and down in order to position themselves as optimal as possible into environmental gradients. Triggers are local environmental changes, for example burial by sediment overlays. Such repositioning establishes well-laminated buildups (Gerdes et al., 1991, 2000). These can change the sedimentary environment completely in biogeochemical aspects. Fossil analogues are also termed stromatolites.

Macroorganisms per definition are large enough to be seen with the naked eye. Syntopic meetings between microbial mats and macrozoobenthic biota are strained since Vendian time (Seilacher, 1999). Recent studies of

Proterozoic-Cambrian marine strata have emphasized the restriction of microbial mat-related sedimentary structures in subtidal paleoenvironments due to the increasing vertically-oriented metazoan bioturbation (Gehling, 1999, Hagadorn and Bottjer, 1999, Pflüger, 1999, Seilacher, 1999, Seilacher & Pflüger, 1994). Modern refuges of biolaminated deposits are said to be intertidal, supratidal, hypersaline, hyperthermal and anoxic environments (Neu, 1994, Neumann et al., 1970, Pflüger, 1999, Pierson et al., 1992, Seilacher & Pflüger, 1994, Walter et al. 1972).

It may be evident that microbially laminated sediments disappear from shallow marine environments when grazing and burrowing occurs (Seilacher & Pflüger, 1994). This does not mean, however, that benthic animals and mat-forming microbiota never would meet themselves. Consequences of these meetings can be studied in various settings. This chapter will review tracemakers in modern and ancient biofilm-laminated deposits to elucidate the degree of ichnodiversity even in settings considered as microbial mat refuges, i.e., environments under ecological stress. Further the question of survival of microbial mats in the presence of faunal activity, but also the survival of metazoa in the presence of microbial mats will be discussed. The final aim is to contribute to an ichnology of microbially laminated sediments.

2. TRACEMAKERS IN RELATION TO MICROBIAL BIOFILMS – TAXONOMIC/ETHOLOGICAL SURVEY

Tracemakers of various lifestyles and systematic groups have been observed in modern and ancient sediments coated by microbial mats (Table 1).

Vertebrates: Sea gulls and limicoles leave traces when resting and/or feeding on subaerially exposed microbial mats in the supratidal zone (Fig. 1a; Gerdes et al. 1985a). Locomotion tracks of reptiles (dinosaurs, chelonids) on cyanobacterial mat-stabilized bedding planes were reported from Cerin lithographic limestones (Kimmeridgian, south Jura Mountains; Bernier et al., 1982, Gall, 1989). Fishes (Moguliidae) during higher tides typically browse on supratidal cyanobacterial mats and leave parallel-lining scratches known from the coasts of Mauritania (Schwarz et al., 1975) and the southern North Sea (Fig. 1b; Gerdes et al., 1985a).

Echinoderms: Echinoids making scratching traces on microbial films with their characteristic mouth parts (lantern of Aristoteles) are reviewed by Boucot (1990, Fig. 154d). White and Findley (1988) reported bioturbation structures of sand dollars in microbial sediments, and Carpenter (1981) studied grazing effects of *Diadema entillarum* on benthic microbial communities.

Gastropods: Cameron et al. (1985) reported snails of the genus *Littorina* grazing on supratidal microbial mats. Moore & Burne (1994) found hydrobiid snails in thrombolitic buildups. On Mellum Island (North Sea), hydrobiid snails leave resting traces in the mats during low tides (Fig. 1c). Particularly famous are cerithic snails because they are able to completely disrupt even tough filamentous microbial mats with their radula. At metahaline fringes, they often are extraordinarily abundant (Fig. 2a-c). The re-establishment of mat laminae in cerithid snail habitats is documented in Fig. 2d. It seems, however, that such criteria first are realistic if grazing stress is mildered at increasing hypersalinity (Gerdes et al., 1985b).

Chelicerata: Modern examples are spiders (Clubionidae) which form their webs below domed, evaporite-encrusted microbial mats (petees) at hypersaline lagoonal margins. Pseudoscorpions also occur in wetland mats (Gerdes et al., 1985b).

Crustaceans: Moore & Burne (1994) report fenestrae of modern thrombolites colonized by a wide range of crustaceans (Anostraka, Ostracoda, Amphipoda, Isopoda and Decapoda). While the cavities of the thrombolites are filled with fecal pellets and carapaces, real trace records are not mentioned. The authors assume that the thrombolitic cavities and tubes are intrinsic due to the microbial communities rather than produced by the metazoans living in between them. Burrowing amphipods of the genus *Corophium* occur in modern sandy tidal flats of the North Sea. These species are immediately excluded if the mats completely seal the surface and change intrasedimentary redox conditions towards anoxia. The burrows of *C. arenarium* are vertically straight, while the sympatric species *C. volutator* forms U-shaped burrows. Using their second pair of antenna, the animals scratch biofilm-coated mineral grains radially off around their burrows and feed on the biofilms within the protective burrows (Fig. 3a). After doing this, they deposit the „cleaned" sand grains outside the burrow. Such spezialised feeding habit leeds to characteristic scrap traces and little sand grain hillocks around the burrows (Fig. 3b). Amphipod burrows are also reported from Cerin lithographic limestones (Gall, 1989).

Insects: Various taxonomic groups live in co-existence with microbial mats. Hypersaline lagoons of the Sinai coast harbour various aquatic and subaquatic beetles and diptera (adults, larvae and pupae). The animals particularly are abundant in the flocculous microbial scum drifting in the water (Gerdes et al., 1985b). Moore & Burne (1994) reported the occurrence of Hydrophiliidae, tabanids and trichopteran larvae in the fenestrae of thrombolites.

Table 1: Selection of trace making animals on or in microbial mats, modern and ancient

Systematic group	Trace makers	Traces	Environment, age	Source
Birds	Oyster catchers	Picking traces	cmat-covered sandy tidal flats, supratidal zone, modern	Gerdes et al. 1985a
	Seagulls	Spit balls consisting of s keletal hard parts of marine invertebrates, regularly broken by crop muscles	mat-covered sandy tidal flats, supratidal zone, modern	Gerdes et al. 1985a
Reptiles	Chelonien	Locomotion tracks	Tropical lagoon above ancient coral barrier reef, lithographic limestone, Cerin, Kimmeridgian, southern Jura Mountains France	Bernier et al. 1982
Fishes	Mullets	Feeding traces	mat-covered sandy tidal flats, supratidal zone, modern	Gerdes et al. 1985a; Schwarz et al. 1975
	Gobiidae	Suspect-trace: knabbern? (no real trace described by the authors)	Thrombolitic buildup in a c oastal lake, Lake Clifton, Western Australia, modern	Moore & Burne 1994
Echinoderms	Sea urchins	Scratching trails	Helgoland rocky sublittoral	Krumbein & v an der Pers 1974; Boucot 1990 Fig. 154d
Molluscs	Gastropods, cerithic snails	Disruption of mats	Lagoonal margins, beaches, spits, metahaline	Gerdes et al. 1985b; Davaud et al. 1994; Friedman et al. 1992
Insects	Staphilinid beetles Bledius capra B. spectabilis B. subniger B. angustus	Burrows, excavation pellets, aeration chimneys, undermat mining	Supratidal zone, siliciclastic tidal flats, hypersaline lagoonal margins	Gerdes et al. 18985a,,b; Seilacher 1999
	Beetles Heterocerus sp.	Undermat mining	Supratidal zone, siliciclastic tidal flats	Gerdes et al. 1985a
	Bark beetles	Undermat mining	Shallow marine, supratidal zone	Seilacher 1999
Arachnoids	Limulus	Locomotion trace (maker is still present)	Solnhofen formation, Malm, Jura	Boucot 1990 p. 314/315 Fig. 255
	Clubionidae	Spider webs	Below salt encrusted mats, petees, modern	Gerdes pers. observation
Decapods		Locomotion traces, maker still present	Solnhofen formation, Malm, Jura	Boucot 1990 p. 319, Fig. 260; Gerdes pers. observation (Bretagne abandoned salinas)
Amphipods	Corophionidae	Burrows, straight and U-shaped		Gerdes & Reineck 1987

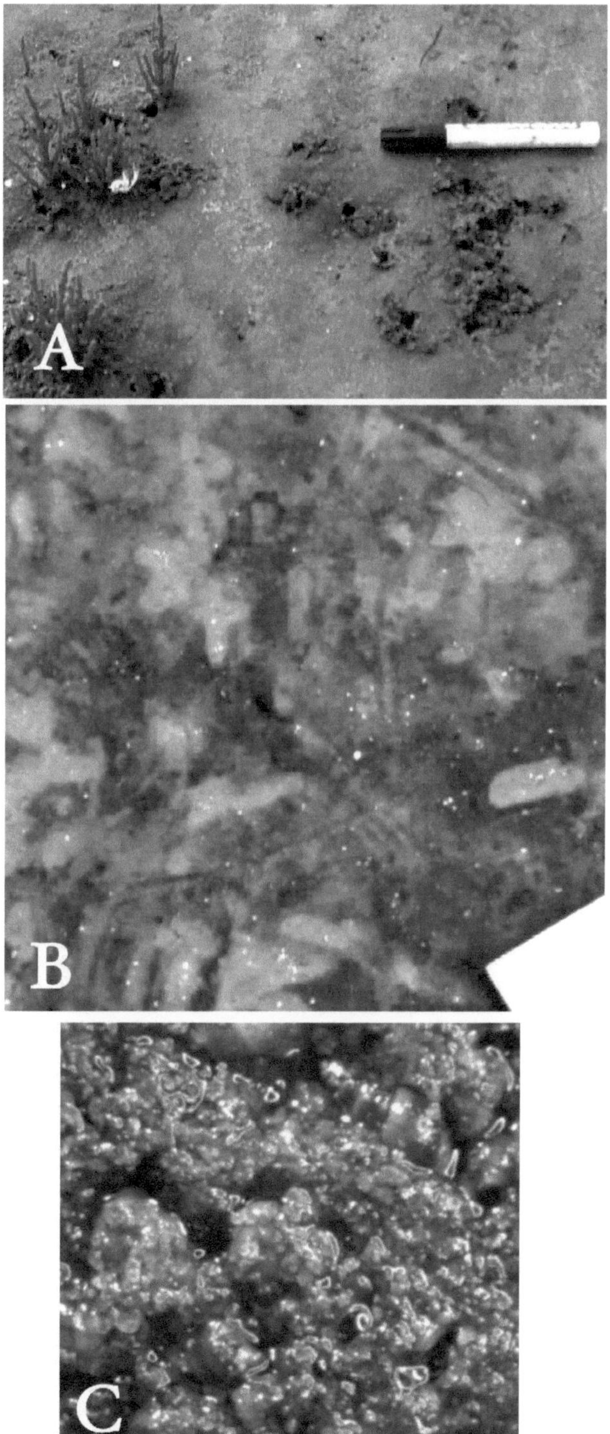

Figure 1. Modern trace records on Mellum Island, North Sea. a) Picking traces of oystercatchers resting on supratidal microbial mats, length of pencil 12 cm; b) Parallel-lining scratches of mullets (Mugilidae) on supratidal cyanobacterial mats, scale 5 cm; c) Resting traces of *Hydrobia ulvae* which hide in microbial mats to escape desiccation during low tide, scale 3 mm.

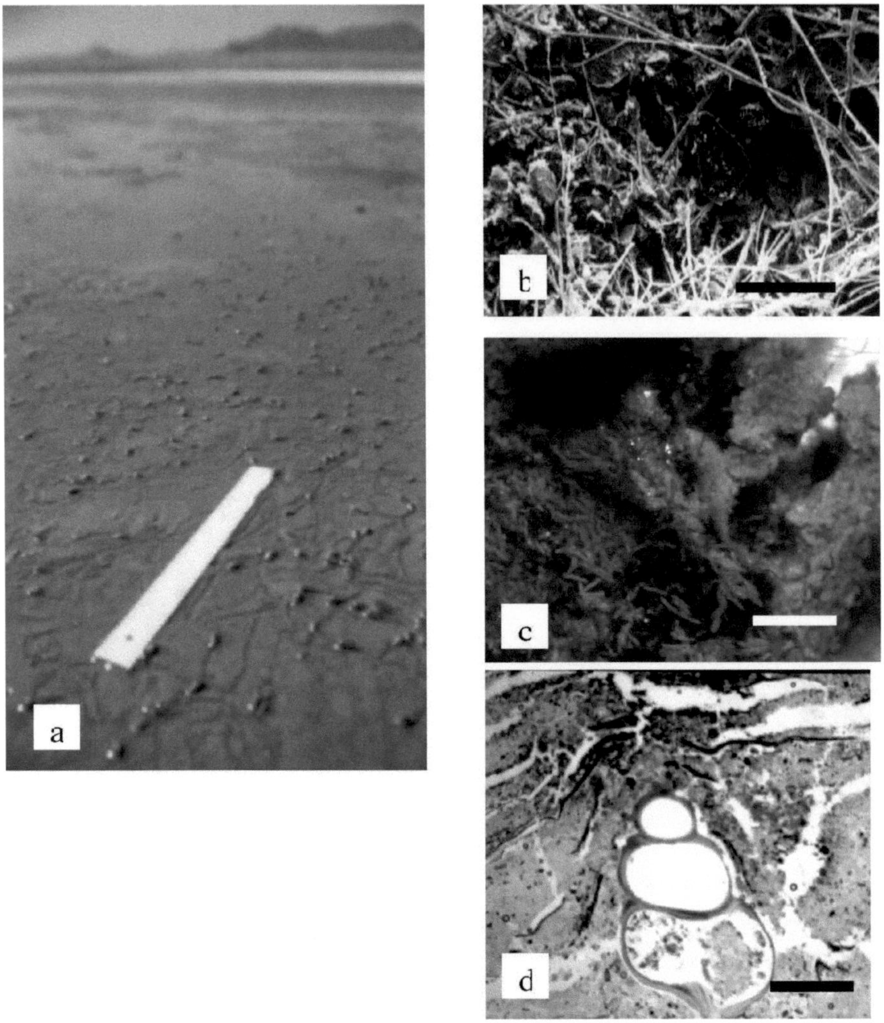

Figure 2. Cerithic snails (*Pirenella conica*) at the metahaline fringe of a hypersaline lagoon (Gulf of Aqaba, Sinai). a) View of the environment; b) Surface layer of a filamentous microbial mat disrupted by the radula of the snails, scale 1 cm; c) As a consequence of grazing, the laminated mat structure becomes reworked and changed into a thicket of microbial scum interwoven with fecal pellets, scale 5 mm; d) Re-establishment of mat laminites on top of a gastropod shell embedded in microbial sediments (thin section microphotography), scale 5 mm.

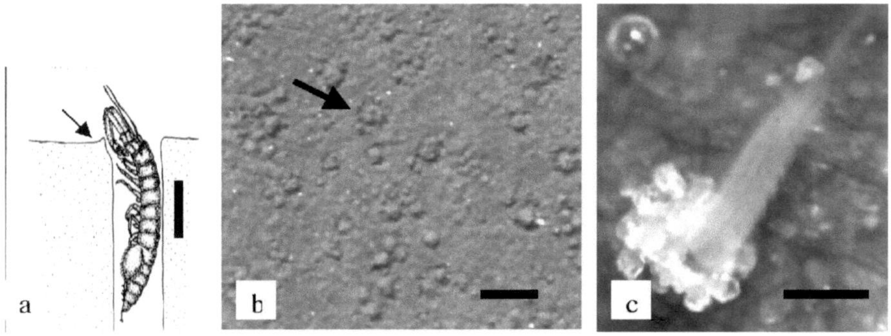

Figure 3. a) Schematic drawing of the amphipod crustacean *Corophium arenarium* in its burrow from where the animals scratch off biofilm-coated mineral grains from the surface around their burrows (arrow; modified after Theede, 1981), scale 5 mm; b) Radial scratch traces around the burrow entrances (arrow) and little hillocks of cleaned sand grains deposited by the animal at the surface after feeding on the biofilms (modified after Gerdes et al., 1985a) , scale 1 cm; c) *Pygospio elegans*, a small hemisessile polychaete; the animal comes out of its tube with its front part to filter for food at the surface of a filamentous cyanobacterial mat (intertidal flats, Mellum Island), scale 2 mm.

Figure 4. Traces of mat-undermining salt beetles of the genus *Bledius*. a) gypsum-encrusted aeration chimneys of *B. angustus* (Gulf of Aqaba coast, Sinai), scale 2 mm; b) feeding traces of *B. angustus* undermining the gypsum-encrusted microbial mat, scale 2 mm; c) elongated piles of excavation pellets of the salt beetle *B. subniger* removed from its subsurficial browsing traces (Mellum Island, North Sea), scale 2 mm.

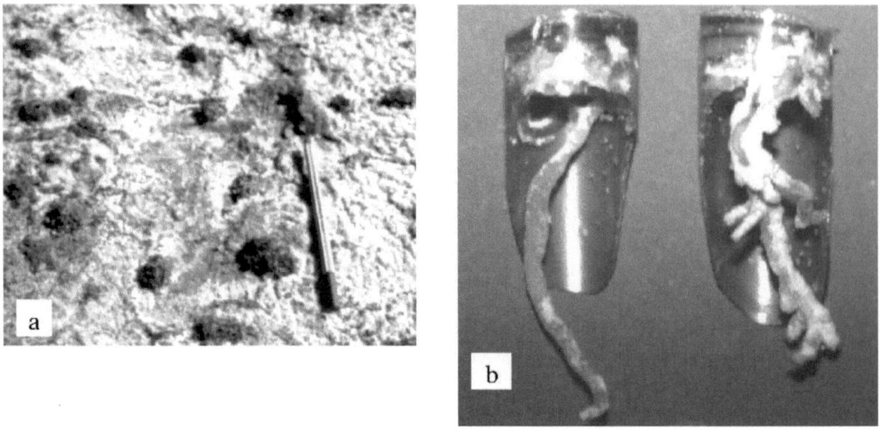

Figure 5. Traces of vertical burrowing salt beetles of the genus *Bledius*. a) round piles of excavation pellets of *B. capra* on the gypsum-encrusted surface of microbial mats (southern Tunisia coast), piles in nature 2 cm in diameter; b) resin casts of burrows of *B. capra* (left: male, viz. burrow entrance formed like a „bottle neck“, right: female, viz. breeding chambers branching sidewards off the main shaft; appendices filled with feces), length of burrows 5 cm.

Prominent tracemakers among insects are staphilinid beetles of the genus *Bledius*. Their excavation pellets line the fringes of subaerially exposed microbial mats along arid as well as temperate shorelines. Two different lifestyles can be distinguished: (i) undermat mining (Seilacher, 1999), and (ii) vertical burrowing. On each of the shores of hypersaline lagoons of the Sinai and the temperate humid North Sea coast, we found both lifestyles.

Mat underminers are *B. angustus* (Sinai coast) and *B. subniger* (North Sea coast). The arid shoreline beetle *B. angustus* undermines salt-encrusted microbial mats and forms vertically directed aeration chimneys through the salt crusts (Fig. 4a, b). On the shores of the North Sea, the allopatric species *B. subniger* forms horizontal feeding traces below and elongated excavation pellets above the mat-stabilized sandy surface (Fig. 4c). Another species making horizontal undermat traces is the carabiniid beetle *Heterocerus flexuosus* on Mellum Island (Gerdes and Krumbein, 1987).

The two allopatric salt beetles *Bledius capra* (arid) and *B. spectabilis* (temperate humid coast) form vertical burrows. Both deposit characteristic rounded hillocks of excavation pellets on the mat-covered surface of supratidal flats (Fig.5a). Below are very conspiuous burrows with fecal chambers and bottle-necked entrances which have the function to trap air bubbles. These block the entrance and protect the animals against inflowing seawater (Fig. 5b). The beetles store lumps of microbial mats in chambers just below the bottle-necked entrance. At this position the mat fragments still have sufficient light and moisture and remain alive. This store serves for alimentation of the beetles during high floods. Burrows of females also show breeding chambers branching off the main tunnel (Fig. 5b). According to Seilacher (1999), these specialized burrows represent the antithesis to real bioturbation of mats.

Polychaetes. Real bioturbation of mats may occur among polychaetes. Lab experiments with the hemisessile endobiontic polychaete *Nereis diversicolor* which lives on modern tidal flats have shown that this worm is able to rip with its strong jaws (Fig. 6a) at tough surface mats to re-establish its burrow. These burrows resemble fossil *Chondrites* (Savrda et al., 1991).

An unique association between biolaminated columnar buildups and burrowing polychaetes on modern Persian Gulf tidal flats was described by Shinn (1972). The author traces these biolaminated, about 10 cm high mounds back to specific high accretion rates of sediments at the slope of a cyanobacteria-coated, wide shallow channel which migrates laterally. The worms continue to build their mounds upward when the accreting bank moves over them. Once the prograding channel bank begins to cover established mounds, alternating growth and excreted pellets produce upward growth of laminated columns.

Finally, also associations between sessile polychaetes and stromatolitic mats may be mentioned. Toomey and Cys (1977) observed well-developed stromatolites containing abundant remains of spirorbid worms "enmeshed within the laminae" (Lower Permian of New-Mexico). Wright and Mayall (1981) described Upper Triassic stromatolites showing a conspicuous alternation of flat or domal laminated units with generally thicker layers characterised by the development of dendritic, arborescent structures. These, reaching a few centimetres in height, were composed of micritic laminations coating tubes of serpulid worms. Davaud et al. (1994) reported stromatolite and serpulid bioherms from a Holocene lagoonary environment. The authors assumed that the stromatolites progressively replaced the serpulid reefs due to changes from euryhaline to hypersaline environmental conditions in the course of regression of the Flandrian sea-level highstand.

Oligochaetes also are common in laminated microbial sediments. On Mellum Island, dense populations of the genus *Lumbrificoides* occur in thick surface scums of microbial mats which look like pincushins. The worms settle vertically straight and feed on bacteria under the top mat.

Helminthoids: Traces are reported on top of wrinkle structures, associated with suspect-microbial structures, from Vendian and Lower Cambrian siliciclastic deposits (Hagadorn & Bottjer, 1999). An analogous modern trace is made by *Lineus ruber,* an up to 10 cm long nemertine which preys on evertebrates on biofilm-stabilised intertidal flats. Kruse and Buhs (2000) highlight the strong impact of a nemertine species on *Corophium arenarium* on high intertidal sandflats.

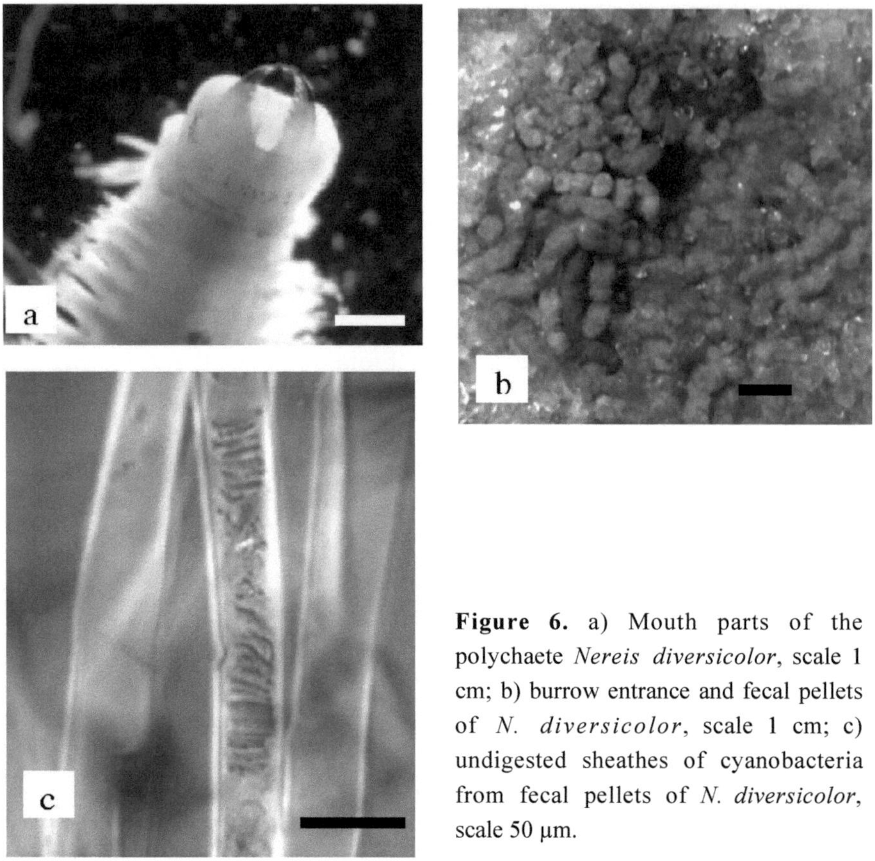

Figure 6. a) Mouth parts of the polychaete *Nereis diversicolor*, scale 1 cm; b) burrow entrance and fecal pellets of *N. diversicolor*, scale 1 cm; c) undigested sheathes of cyanobacteria from fecal pellets of *N. diversicolor*, scale 50 µm.

3. MUTUAL RELATIONSHIPS BETWEEN TRACEMAKING ANIMALS AND MICROBIAL MATS

3.1 Foraging strategies of animals coexisting with microbial mats

The question whether and how animals may use mat-forming primary producers as a source of food may be of partic ular interest. Of the animal groups selected for the trace records in this chapter, feeding on microbial mats was observed for mullets, gastropods (cerithids, hydrobiids), echinoids, insects (diptera, beetles), amphipods, polychaetes and oligochaetes. However, it seems that cyanobacteria often make only little contribution to the diets of macrobenthic grazers. Nicotri (1977) experimentally studied grazing of limpets and periwinkles on a microbial mat community dominated by coccoid and filamentous cyanobacteria.The studies resulted in the observation that non of the cyanobacterial taxa were strongly affected by the presence of the grazers. Fecal analysis showed that cyanobacteria did not form an important portion of the diet such as the diatoms associated with the bluegreens. At least when the diet was mixed of bluegreens and diatoms, the limpets selectively preferred diatoms. Also *Hydrobia ulvae*, a common tidal flat species of the North Sea, browses mainly on diatoms associated to microbial mats (Gerdes and Krumbein, 1987).

Aquatic and subaquatic beetles and diptera occurring in the flocculous microbial scum of shallow hypersaline lagoons feed, together with various meiofauna, on drifting diatoms and coccoid cyanobacteria rather than on the tough filamentous aggregates (Gerdes et al., 1985b). Also oligochaetes prefer bacterial and diatom coatings of mineral grains (Giere, 1975).

The polychaete *Nereis diversicolor* (Fig. 6a) has a wide range of foraging strategies. In microbial mat habitats, it is able to feed also on filamentous cyanobacteria as indicated by undigestible sheathes found in feces which the worm deposit outside of its burrow (Fig. 6b, c). The animal is not specialized in this behaviour, but is omnivorous. Also chemosymbiosis and farming is possible (Savrda et al., 1991). The species forms slime nets close to the branched burrow entrances and attracts bacterial or fungal films by the respiration currents. It utilises these films as a source of food. Although this species seems to be highly specialised for the life in microbial matgrounds, it is able to flourish in a variety of environments.

3.2 Migratory animals and syntopic meetings between mats and evertebrates

Broadly, active animal migration and passive dispersal may contribute to traces on or in microbial mats. Some migratory animals are casual, and others specific visitors. Traces of oystercatchers (Fig. 1a) occur casual. They utilize the highshore mats for resting during high tide and leave their picking traces at random (Fig. 1a). Specific visitors utilizing the primary production are mullets (Fig. 1b) and hydrobiid snails. *Hydrobia ulvae* forms slime rafts which support their transport by the incoming tide into higher tidal flat areas. To survive desiccation during low tide, the snails hide and rest in the softground mats (Fig. 1c).

Figure 7. Resin-hardened relief casts, Mellum Island, North Sea. a) L-shaped trace made by a juvenile lugworm (*Arenicola marina*) in evenly laminated sand, topped by a laminated microbial mat which sealed the surface and interrupted the neccessary connection of the animal with its external environment (arrow), („killermat" phenomenon"), scale 5 cm; b) Vertical section showing tiering due to sharply projecting buried microbial mats which grew on top of the sandy interlayers, figurative dwelling traces of hemisessile infauna community restricted to the sandy interlayers; arrow 1: Tubes and burrows piercing through mats; arrow 2: compact and multilaminated microbial mats without penetration of burrows or tubes ("killermat" phenomenon), scale 5 cm.

Although real stenotopic fauna restricted to microbial mats seems to be lacking or is rare (not considered here the great variety of meio- and microbenthos which perhaps is linked to matground environments by physiological adaptation; Fenchel, 1998), the passive dispersal of benthos from other sites of the environmental hyperspace is rather important. Examples are marine epibenthic and endobenthic species which spread into highshore mat habitats. Some of them only form scattered peripheral populations such as the lugworm (Fig. 7a), while others, having a wide range of environmental tolerance, also reproduce in these areas. Their density can be explosive in number. One example are the cerithic snails which concentrate in herds at arid shorelines (Fig. 2a). Their grazing and locomotion destroys the laminated structure, but not the whole microbial community. The mat changes into a thicket of microbial scum interwoven with fecal pellets (Fig. 2c). Bacterial growth rates even can be enhanced due to such intensive bioturbation (Yingst, 1980). Seilacher and Pflüger (1994) when collecting data from Vendian shallow water deposits stated that grazing and bioturbation elevate the microbial system to a new mode with higher levels of productivity (benthic agriculture; see also McIlroy and Logan, 1999).

On North Sea tidal flats, an infauna community of small hemisessile amphipods and polychaetes (*Corophium arenarium, Pygospio elegans* and *Nereis diversicolor*) provides another example of increasing concentrations of eurytopic animals in highshore, mat-covered habitats (Figs. 3 and 6). They all live also in other parts of the tidal flat environment. Apparently, mats and fauna both prefer the same high level since physical turbation by currents and waves is lowered (syntopic meeting). At these levels, burrows and tubes of the infauna are very abundant (Fig. 7b). The term bioturbation mainly is concerned to undistinct, mottled textures left behind by faunal activity (Bromley, 1990). This faunal association, however, leaves behind figurative traces (Gerdes and Reineck, 1987). Mainly *P. elegans* and *Corophium* sp. are not capable to markedly restrict microbial mat development (viz. the surface mat in. Fig. 3c which formed in the presence of populations of the polychaete *Pygospio elegans*). Vertical sections through sediments of the upper intertidal zone show that the mats are able to succeed, in spite of burrowing, with any new deposited sandy layer. Microbial bedding plane stabilisation and biomass production makes tiering visible (Fig. 7b). The tubes and burrows of the infauna mainly dominate the sandy interlayers, but also penetrate thinner buried mats (Fig. 7b arrow-1; compare also Fig. 3c). In thicker organic horizons, however, burrows fail (Fig. 7b, arrow-2). These patterns reveal that the fauna coexists in a very delicate equilibrium with the mats which is mainly controlled by the sedimentation rate. During longer-lasting periods of low or zero sedimentation, the mats are able to produce thick fibrillar meshworks on top of the sediments. More often the cyanobacterium *Microcoleus chthonoplastes* is involved which is known for its effective surface sealing. During such periods, burrows of the figurative infauna disappear (Fig. 7b). There are two reasons for this „killermat“ phenomenon (Gerdes and Reineck, 1987): (1) The burrowing animals are surface feeders and require a connection with the surface. They must expend a great deal of energy if they have to burrow upwards through thick, filamentous cyanobacterial mats. (2) The large amounts of biomass in the mats cause changes unfortunate for these animals. Oxygen becomes limited, and H_2S concentration increases. These conditions exclude the tube and burrow-building polychaetes and amphipods. First after burial of the killer mats, e.g., by tempestites, the fauna has a chance to re-colonize the area (Fig. 7b).

A similar syntopic meeting associated with the killermat phenomenon occurs at upper intertidal and lower supratidal level. In this zone, juveniles of the lugworm *Arenicola marina* are able to survive in lower-lying wet puddles. Vertical sections of sediments indicate that the animals are killed or driven out from this area due to extensively growing microbial mats which overgrow the sandy surface and seal burrow entrances (Fig. 7a).

The only polychaete in this sandy tidal flat community which is able to withstand extended mat growth is *Nereis diversicolor* (Fig. 6a). This species has a wide range of tolerance to H_2S and oxygen deficiency (Theede et al., 1969) and is less sensitive against surface sealing by the mats.

3.3 Sublethal or lethal effects of toxic cyanobacterial blooms

Boucot (1990, Fig. 255) visualised a locomotion track on the lithographic limestones of Solnhofen (Jura Mountains of Southern Germany) of which the tracemaker, a limuloid species, is still present. As suggested by Keupp (1977), the Solnhofen paleoenvironment was a deeper water lagoon. It is assumed that the bottom became periodically covered by a scum of coccoid cyanobacteria which also made the excellent fossil preservation possible. Gehling (cit. in Seilacher et al., 1985, pl. 1 fig. 4-9) interpreted fine, reticulate ridge patterns on bedding surfaces of the Solnhofen lithographic limestone as the macroscopic imprint of prokaryotic scum. They also argued that cyanobacterial films were instrumental in the preservation of animal tracks and drag marks, and important ingredients in the chemistry of diagenesis.

Various studies of modern freshwater, brackish and shizohaline lagoons suggest that planktonic as well as benthic cyanobacteria periodically develop blooms which release poisonous metabolic substances into the surrounding water. Even mammals have been killed drinking from this water (Chorus & Bartram, 1999).

The characteristic limuloid trail on the Solnhofen lithographic limestone with the tracemaker at the end of its track, and also similar traces of decapods (Keupp, 1977, Boucot, 1990, Fig. 260) thus may represent sublethal traces produced by poisoned organisms just before death.

4. ICHNOLOGY OF MICROBIALLY LAMINATED SEDIMENTS

Ichnology is concerned with effects of biological activity on deposits (Ekdale et al. 1984). Following this definition, effects of microbial biofilms on sediments are part of ichnology. On the other hand, the many records of animal traces on or in microbially laminated sediments of modern and ancient depositional environments make it valuable to discuss lifestyle (behavioural) significance of animals making traces on or in microbial mats (Seilacher, 1999). To classify traces associated mainly to Phanerozoic and modern microbial mat environments, the scheme of Ekdale et al. (1984) was modified (Fig. 8). The following examples assign to these items:

Traces of *Hydrobia ulvae*, a common tidal flat species of the North Sea which hides and rests in the softground mats to survive desiccation during low tide (Fig. 1c) account to resting traces (Cubichnia).

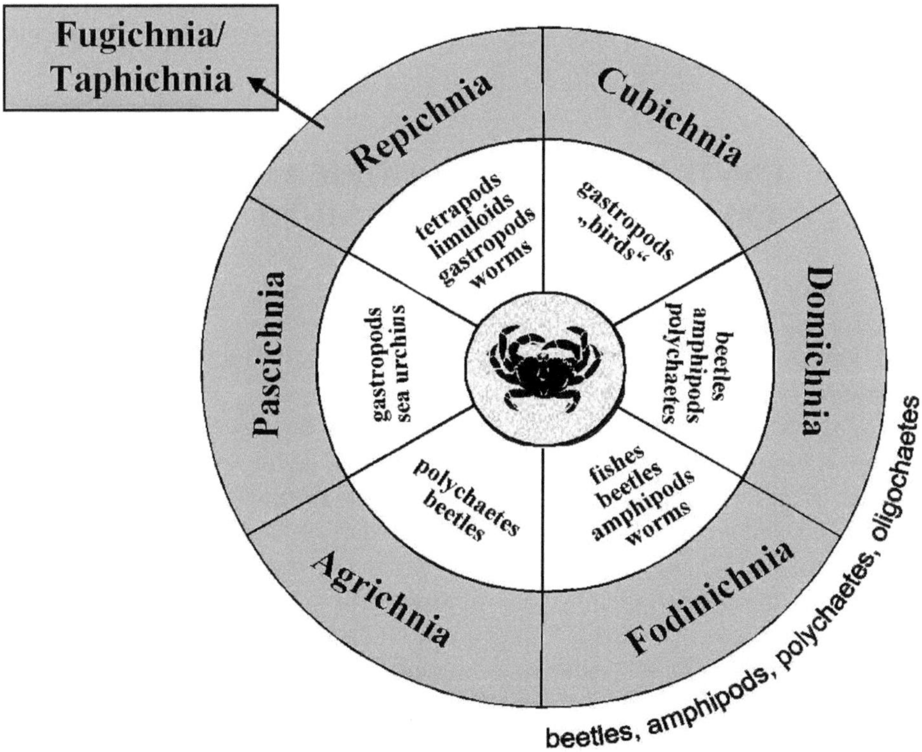

Figure 8. Trace categories on or in microbial mats grouped according to their facies-indicative behavioural criteria

The tubes of the small hemisessile polychaete *Pygospio elegans* (Fig. 3c), and the burrows of amphipods (Corophiidae, Fig. 3a) account to dwelling traces (Domichnia).

Examples of Fodinichnia (feeding burrows) as well as of Domichnia are the burrows of the vertically burrowing salt beetles (Fig. 5) and of the polychaete *Nereis diversicolor* (Fig. 6). The same species (salt beetles and polychaetes) also account for the trace category of Agrichnia (see stores of microbial colonies established by the beetles, and mucus traps for bacterial films formed by *Nereis* sp.)

Examples of Pascichnia (grazing tracks) are the scratching traces of cerithic snails (Fig. 2a), mullets (Fig. 1b), echninoids, hydrobiid snails and dipterans. Also the mat-undermining salt beetles (Fig. 4) and other beetles (carabiniidae) contribute to this trace category.

Repichnia (locomotion traces): examples are traces of crabs, reptiles, snails and helminthoids.

Finally Fugichnia (Taphichnia; Pemberton et al., 1992) are made by animals which try to escape from any touch to microbial mats or cyanobacterial scums. The killermat phenomenon shown in Fig. 7 also may account to this category.

5. ENVIRONMENTAL SIGNIFICANCE OF ANIMAL/MAT RELATIONSHIPS

Behavioural patterns mirrored in a trace are valuable for facies interpretations (Pemberton et al., 1992, Seilacher, 1999).

Traces of mullets (Fig. 1b) and insects (Fig. 4) indicate systematic foraging in relation to microbial mats at supratidal level.

The specific dwelling/feeding burrows of salt beetles (Fig. 5) may testify a microbially laminated sediment at supratidal level which still is in the range of spring tides or slightly above. Both tropical arid and temperate humid coastal wetlands yield such trace records.

Tiered sediments with figurative burrows in sandy interlayers between microbial mat horizons (Fig. 7b) may point to a delicate interrelationsship between hemisessile marine infauna and intertidal microbial mats which is mainly controlled by the sedimentation rate.

Sublethal traces of animals associated with reticulate ridge patterns on bedding surfaces may be characteristic of poisoning of the tracemakers by toxic cyanobacterial blooms.

Stromatolitic biolaminites overgrowing vermetid, spirorbid or serpulid reefs (Burchette and Riding, 1977, Davaud et al., 1994, Tomey & Cys, 1977) may attest environmental changes facilitating overgrowth and replacement of animal colonies by microbial mats.

Recent studies of Proterozoic-Cambrian marine strata have emphasized the restriction of microbial mat-related sedimentary structures from subtidal paleoenvironments due to the increasing vertically-oriented metazoan bioturbation (Seilacher & Pflüger 1994, Gehling 1999, Pflüger 1999, Hagadorn & Bottjer 1999, Seilacher 1999).

The noncalcifying softground microbial mats emphasized in this paper need calm conditions for their initial development, and particular for their subsequent growth and expanse. In many shallow marine environments, however, including subtidal and lower intertidal areas, hydrodynamic energy would not allow for a longer period of non-erosion and low sedimentation which is neccessary for the establishment of multilaminated, preservable mats. In facies analyses, one should thus carefully proof the conditions of the physical environment. Perhaps "wave base" criteria important for shallow marine deposits (Flemming, 2001) have been the main factor which excluded the development of microbial mats from a subtidal depositional environment.

6. SUMMARY AND CONCLUSIONS

The examples given in this paper are borrowed from such habitats described by the word „extreme": Mellum Island intertidal regions and intertidal to supratidal transitions, shizohaline lagoons (Solnhofen example), metahaline and hypersaline lagoons (modern examples at the Sinai coast), supratidal sabkhas (modern examples at the southern Tunisia coast). It was one of the aims of this paper to emphasize that a great diversity of faunal traces even is possible in microbial mats growing in environments under stress.

Care should be taken to state that microbial mats and macrofauna have to be mutually exclusive. The examples show that competition or coexistence would rather be a matter of the individual lifestyles of animal species meeting the mats:

Some are able to live in coexistence with mats (see for example salt beetles), others are capable to completely destruct the mats (see for example cerithic snails), others demonstrate an apparent inability to markedly restrict microbial mat development, and again others are incapable to survive in the presence of mature, multilaminated microbial mats (see for example burrowing and tube-forming surface feeders).

REFERENCES

Bernier, P., Barale, G., Bourseau, J.-P., Buffetaut, E., Demathieu, G., Gaillard, C. and Gall, J.-C. (1982) Trace nouvelle de locomotion de chelonien et figures d'emersion associees dans les calcaires lithographiques de Cerin (Kimmeridgian Superieur, Ain, France), Geobios 15, 447-467.

Boucot, A.J. (1990) Evolutionary Paleobiology of Behavior and Coevolution, Elsevier, Amsterdam.

Bromley, R.G. (1990) Trace fossils. Biology and taphonomy, Unwin Hyman, London.

Burchette, T.P. and Riding, R. (1977) Attached vermiform gastropods in Carboniferous marginal marine stromatolites and biostromes, Lethaia 10, 17-28.

Cameron, B., Cameron, D. and Jones, J.R. (1985) Modern algal mats in intertidal and supratidal quartz sands, Northeastern Massachusetts, U.S.A.. in H.A. Curran (ed.), Biogenic structures: Their use in interpreting depositional environments, SEPM Special Publ. 35, Tulsa, Oklahoma, pp. 211-223.

Carpenter, R.C. (1981) Grazing by *Diadema entillarum* (Phillippi) and its effects on the benthic algal community, J. Mar. Res. 39, 749-765.

Chorus, I. and Bartram, J.(eds.) (1999) Toxic Cyanobacteria in Water, WHO, E and FN Spon, London.

Davaud, E., Strasser, A. and Jedoui, Y. (1994) Stromatolite and serpulid bioherms in a Holocene restricted lagoon (Sabkha El Melah, southeastern Tunisia), in J. Bertrand-Sarfati and C. Monty (eds.), Phanerozoic stromatolites II, Kluwer Academic Publishers, Dordrecht, pp. 131-151.

Ekdal, A.A., Bromley, R.G. and Pemberton, S.G. (1984) Ichnology. The use of trace fossils in sedimentology and stratigraphy, SEPM, Tulsa, Oklahoma.

Fenchel, T. (1998) Formation of laminated cyanobacterial mats in the absence of benthic fauna, Aquatic Microbial Ecology 14, 235-240.

Flemming, B.W. (2001) Was ist die "Wellenbasis"?: Eine kritische Analyse, in R. Gaupp and S. van der Klauw (eds.), Sediment 2001, Kurzfassungen, Schriftenreihe der Deutschen Geologischen Gesellschaft, 13, 36.

Friedman, G.M., Sanders, J.E. and Kopaska-Merkel, D.C. (1992) Principles of Sedimentary Deposits, Macmillan Publishing Company, New York.

Gall, J.-C. (1989) Die Grabungsstelle der oberjurassischen Plattenkalke von Cerin (Südjura, Frankreich), Archaeopteryx 7, 1-11.

Gehling, J.G. (1999) Microbial mats in terminal Proterozoic siliciclastics: Ediacaran death masks, Palaios 14, 40-57.

Gerdes, G. and Krumbein, W.E. (1987) Biolaminated deposits, in S. Bhattacharji, G.M. Friedman, H.J. Neugebauer and A. Seilacher (eds.), Lecture Notes in Earth Sciences 9, Springer, Berlin, pp. 1-183.

Gerdes, G. und Reineck, H.-E. (1987) Aktuogeologie und Aktuopaläontologie des Farbstreifen-Sandwattes von Mellum, in G. Gerdes, W.E. Krumbein und H.-E. Reineck (eds.), Mellum, Portrait einer Insel, Kramer, Frankfurt/M., pp. 219-233.

Gerdes, G., Klenke, T. and Noffke, N. (2000) Microbial signatures in peritidal siliciclastic sediments: a catalogue, Sedimentology 47, 279-308.

Gerdes, G., Krumbein, W.E. and Reineck, H.-E. (1985a) The depositional record of sandy, versicolored tidal flats (Mellum Island, Southern North Sea), Journal of Sedimentary Petrology 55, 265-278.

Gerdes, G., Krumbein, W.E. and Reineck, H.-E. (1991) Biolaminations – ecological versus depositional dynamics, in G. Einsele, W. Ricken and A. Seilacher (eds.), Cycles and Events in Stratigraphy, Springer, Berlin, pp. 592-607.

Gerdes, G., Spira, J. and Dimentman, C (1985b) The fauna of the Gavish Sabkha and the Solar Lake – A comparative study, in G.M. Friedman and W.E. Krumbein (eds.), Hypersaline Ecosystems, Ecological Studies 53, Springer, Berlin, pp. 322-345.

Giere, O. (1975) Population structure, food relations and ecological role of marine oligochaetes, with special reference to meiobenthic species, Marine Biology 31, 139-156.

Hagadorn, J.W. and Bottjer, D.J. (1999) Restriction of a Late Neoproterozoic biotope: Suspect-microbial structures and trace fossils at the Vendian-Cambrian transition, Palaios 14, 72-85.

Keupp, H. (1977) Ultrafazies und Genese der Solnhofener Plattenkalke (Oberer Malm, Südliche Frankenalb), Abhandlung der Naturhistorischen Gesellschaft Nürnberg e.V. 37, 1-128.

Kruse, I. and Buhs, F. (2000) Preying at the edge of the sea: the nemertine *Tetrastemma melanocephalum* and its amphipod prey on high intertidal sandflats, in G. Liebezeit, S. Dittmann and I. Kröncke (eds.), Life at Interfaces and Under Extreme Conditions, Developments in Hydrobiology, Kluwer Academic Publishers, Dordrecht, 43-55.

McIlroy, D. and Logan, G.A. (1999) The impact of bioturbation on infaunal ecology and evolution during the Proterozoic-Cambrian transition, Palaios 14, 58-72.

Moore, L.S. and Burne, R.V. (1994) The modern thrombolites of Lake Clifton, Western Australia, in J. Bertrand-Sarfati and C. Monty (eds.), Phanerozoic Stromatolites II, Kluwer Academic Publishers, Dordrecht, pp. 3-29.

Neu, T.R. (1994) Biofilms and microbial mats, in W.E. Krumbein, D.M. Paterson and L.J. Stal (eds.), Biostabilization of Sediments, Bibliotheks- und Informationssystem der Universität Oldenburg, pp. 9-15.

Neumann, A.C., Gebelein, C.D. and Scoffin, T.P. (1970) The composition, structure, and erodibility of subtidal mats, Abaco, Bahamas, Journal of Sedimentary Petrology 40, 274-297.

Nicotri, M.E. (1977) Grazing effects of four marine intertidal herbivores on the microflora, Geology 58, 1020-1032.

Pemberton, S.G., MacEachern, J.A. and Frey, R.W. (1992) Trace fossil facies models: environmental and allostratigraphic significance, in R.G. Walker and N.P. James (eds.), Facies Models, Response to Sea-level change, Geol. Assoc. of Canada, St. Johns, pp. 47-72.

Pflüger, F. (1999) Matground structures and redox facies, Palaios 14, 25-39.

Pierson, B.K., Bauld, J., Castenholz, R.W., D'Amelia, E., DesMarais, D.J., Farmer, J.D., Grotzinger, J.P., Jorgensen, B.B., Nelson, D.C., Palmisano, A.C., Schopf, J.W., Summons, R.E., Walter, M.R. and Ward, D.M. (1992) Modern mat-building microbial communities: a key to interpretation of Proterozoic stromatolitic communities, in J.W. Schopf and C. Klein (eds.), The Proterozoic biosphere, a multidisciplinary study, Cambridge University Press, Cambridge, pp. 245-342.

Por, F.D. (1980) A classification of hypersaline waters based on trophic criteria, Mar. Ecol. 1, 121-132.

Savrda, C.E., Bottjer, D.J. and Seilacher, A. (1991) Redox-related benthic events, in G. Einsele, W. Ricken and A. Seilacher (eds.), Cycles and Events in Stratigraphy, Springer, Berlin, pp. 525-541.

Schwarz, H.-U., Einsele, G. and Herm, D. (1975) Quartz-sandy, grazing-contoured stromatolites from coastal embayments of Mauritania, West Africa, Sedimentology 22, 539-561.

Seilacher, A. (1999) Biomat-related lifestyles in the Precambrian, Palaios 14, 86-93.

Seilacher, A. and Pflüger, F. (1994) From biomats to benthic agriculture: a biohistoric revolution, in W.E. Krumbein, D.M. Paterson and L.J. Stal (eds.), Biostabilization of sediments, Bibliotheks- und Informationssystem der Universität Oldenburg, pp. 97-105.

Seilacher, A., Reif, W.-E. and Westphal, F. (1985) Sedimentological, ecological and temporal patterns of fossil Lagerstätten, Phil. Trans. R. Soc. London B. 311, 5-23.

Shinn, E.A. (1972) Worm and algal-built columnar stromatolites in the Persian Gulf, Journal of Sedimentary Petrology 42, 837-840.

Theede, H. (1981) Kleintiere im Watt, in F. Gerhard (ed.) Naturraum Wattenmeer, Meyster Verlag, München, pp. 113-128.

Theede, H., Ponat, A., Hiroki, K. and Schlieper, C. (1969) Studies on the resistance of marine bottom invertebrates to oxygen-deficiency and hydrogen sulphide, Marine Biology 2, 325-337.

Toomey, D.F. and Cys, J.M. (1977) Spirorbid/algal stromatolites, a probable marginal marine occurence from the Lower Permian of New Mexico, USA, Neues Jahrbuch Geolog. Paläontol. Abh., 331-342.

Walter, M.R., Bauld, J. and Brock, T.D. (1972) Siliceous algal and bacterial stromatolites in hot springs and geyser effluents of Yellowstone National Park, Science 178, 402-405.

White, D.C. and Findlay, R.H. (1988) Biochemical markers for measurement of predation effects on the biomass, community structure, nutritional status, and metabolic activity of microbial biofilms, Hydrobiologia 159, 119-132.

Wright, V.P. and Mayall, M. (1981) Organism-sediment interactions in stromatolites: an example from the Upper Triassic of South West Britain, in C. Monty (ed.) Phanerozoic stromatolites, Springer, Berlin, pp. 74-84.

Yingst, J.Y. (1980) Role of bioturbation in the enhancement of bacterial growth rates in marine sediments, in D.C. Rhoads (ed.), Marine Benthic Dynamics, Univ. South Carolina Press, Columbia, pp. 407-422.

Chapter 13

"FARMING" OF MICROBIAL MATS IN THE HINDGUT OF ECHINOIDS.

Chantal De RIDDER [1] & Robin L. BRIGMON [2]
(1) Université Libre de Bruxelles,
Marine Biology Laboratory (C.P. 160/15),
50 av. FD Roosevelt, 1050 Brussels, Belgium
(2) Westinghouse Savannah River Company,
Environmental Biotechnology Section, Savannah River Technology Center, Aiken, SC 29808, USA
cridder@ulb.ac.be

1. INTRODUCTION

The hindgut of several species of spatangoids (burrowing echinoids) harbours a highly contractile and extensile caecum filled up with spherical nodules (De Ridder *et al.*, 1985, De Ridder 1994). A nodule reaches up to 7 mm in diameter and the caecum usually contains numerous nodules of varied size (Figs 1-2). Nodules are kept for days before to be expelled with the feces. Each nodule is made of a central detrital particle, the nucleus, which is coated with a laminated microbial mat (Fig. 1). The nature of the nucleus consists of skeletal remains of animals (e.g., shells, fragments of hydrozoans or bryozoans colonies, pieces of bone), of wood or algae pieces, or of organic aggregates. Sheathed filamentous bacteria form the "frame" of the mat (De Ridder *et al.*, 1985). Previous works have indicated that these bacteria can display a sulfur-oxidative metabolism (Temara *et al.* 1993). Enzyme-linked immunosorbent assay (ELISA) and indirect imunofluorescence (IIF), have suggested that these bacteria could belong to the genus *Thiothrix* (Brigmon and De Ridder 1998).

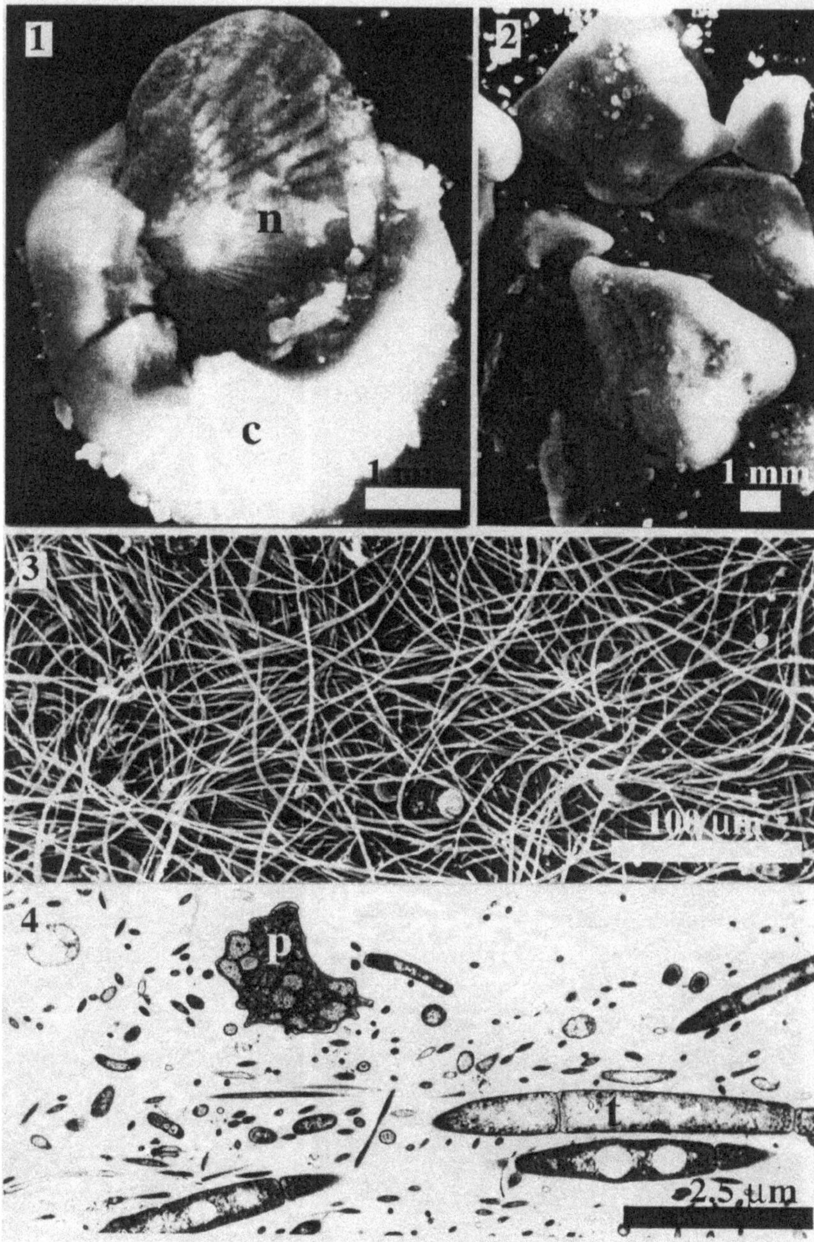

Figure 1. Dissected nodule showing the nucleus (n) and the microbial coat (c). Figure 2. A set of nodules (SEM). **Figure 3-4.** Detailed SEM (3) and TEM (4) views of the nodule surface. P: protozoan, t: *Thiothrix* spp bacteria.

2. STRUCTURE OF THE MICROBIAL MAT

The microbial mat that coats fully developed nodules consists of three successive layers. Living filamentous bacteria occur in surface and subsurface layers of the mat, respectively named outer and intermediate layers. "Young" bacterial filaments, i.e., with still an inconspicuous sheath, occur in the outer layer and form a network of interlacing filaments (Figs 3-6) while "older" filaments, i.e., with a well-developed sheath, occur in the intermediate layer (Fig. 7). The deep layer of the mat consists of left bacterial sheaths that are piled and tightly packed around the nucleus, forming an internal layer of exopolymeric substances (EPS) (Figs 8-9). Diverse microorganisms coexist in the mat layers, e.g., filamentous, coccoid, rod- and spiral-shaped bacteria, and protozoans (Fig. 4).

3. FUNCTIONAL ASPECTS OF THE MICROBIAL MAT

Sulfide and oxygen are available in the intestinal caecum of the echinoid (Temara *et al.* 1993, Brigmon and De Ridder 1998, Thorsen 1998, 1999). Microbial production of sulfide can be related to the activities of heterotrophic bacteria that create anoxic conditions while oxidizing organic matter, and then to the activities of sulfur-reducing bacteria. In the nodule, hydrogen sulfide forms in the nucleus and in the layers of the coating (i.e., in regions rich in organic matter), it could diffuses then centrifugally and reach the nodule superficial layer (Fig. 10 A). On its way towards the nodule surface, sulfide may react with iron –presumably immobilized in the EPS- to form iron sulfide. Oxygen originates from the coelomic fluid that surrounds the caecum. It reaches the nodules by diffusing through the caecal wall. The diffusion must be enhanced by the close vicinity of the ampullae of the respiratory tube feet (located just above the caecum). Such conditions could support the growth of sulfur-oxidizers like *Thiothrix* bacteria that are aerobic and have to grow in chemical gradients where both sulfide and oxygen are accessible to them.

The intestinal contents can provide additional sources of sulfide and oxygen (Fig. 10 B). Indeed, most of the time, the intestine is obstructed by stagnating sediments where anoxic conditions allow the activities of sulfur-reducing bacteria, and the subsequent production of sulfide. Such anoxic conditions are easily reached when low permeable sediments are kept in the digestive tube. Ingested water -transfered to the intestine via the siphon- can sporadically bring oxygen to the intestinal lumen, every time the sediments are eliminated through the anus.

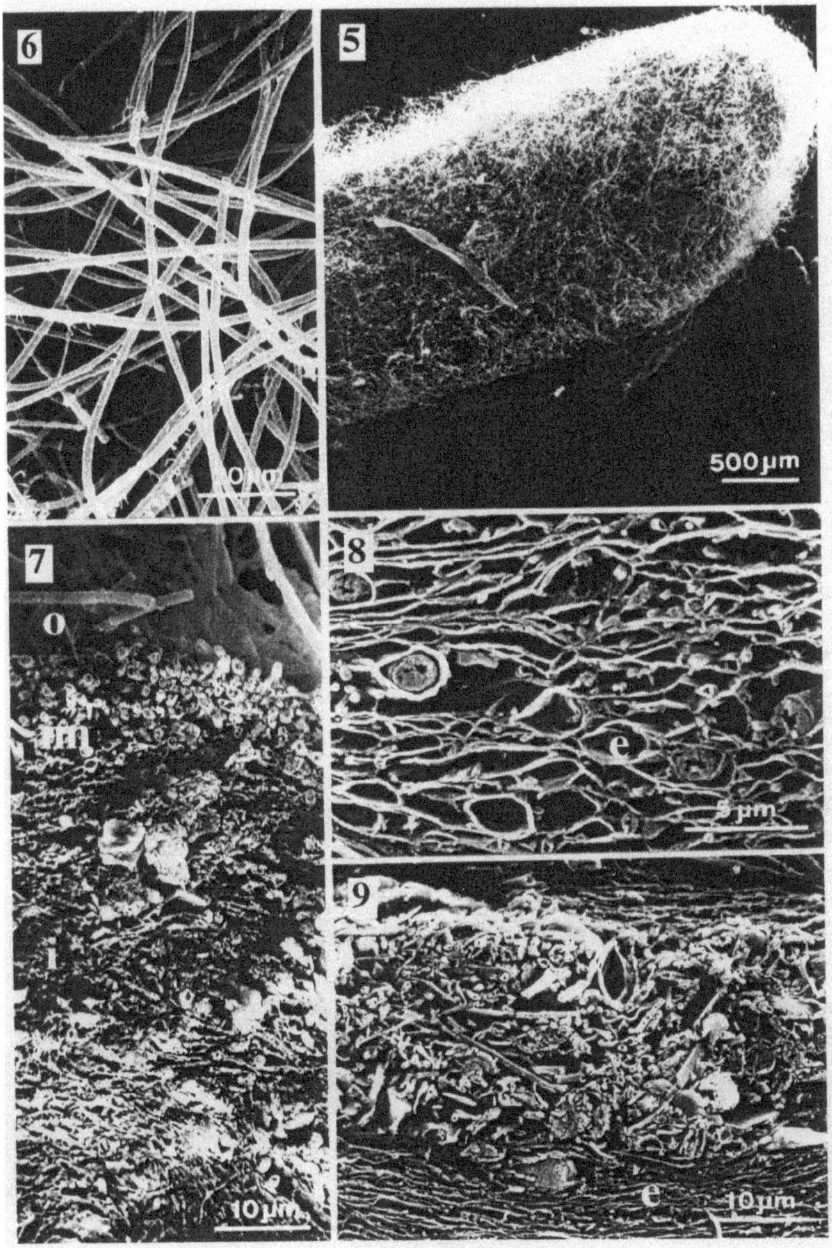

Figure 5. Nodule surface (SEM). **Figure 6.** *Thiothrix* spp bacteria in the outer layer (SEM). **Figure 7.** Cross-section through the microbial mat (SEM). **Figure 8.** *Thiothrix* spp bacteria in the intermediate layer (SEM). **Figure 9.** Empty bacterial sheats and detrital particles in the inner layer (SEM). I: inner layer, e: empty bacterial sheath, im: intermediate layer, o: outer layer.

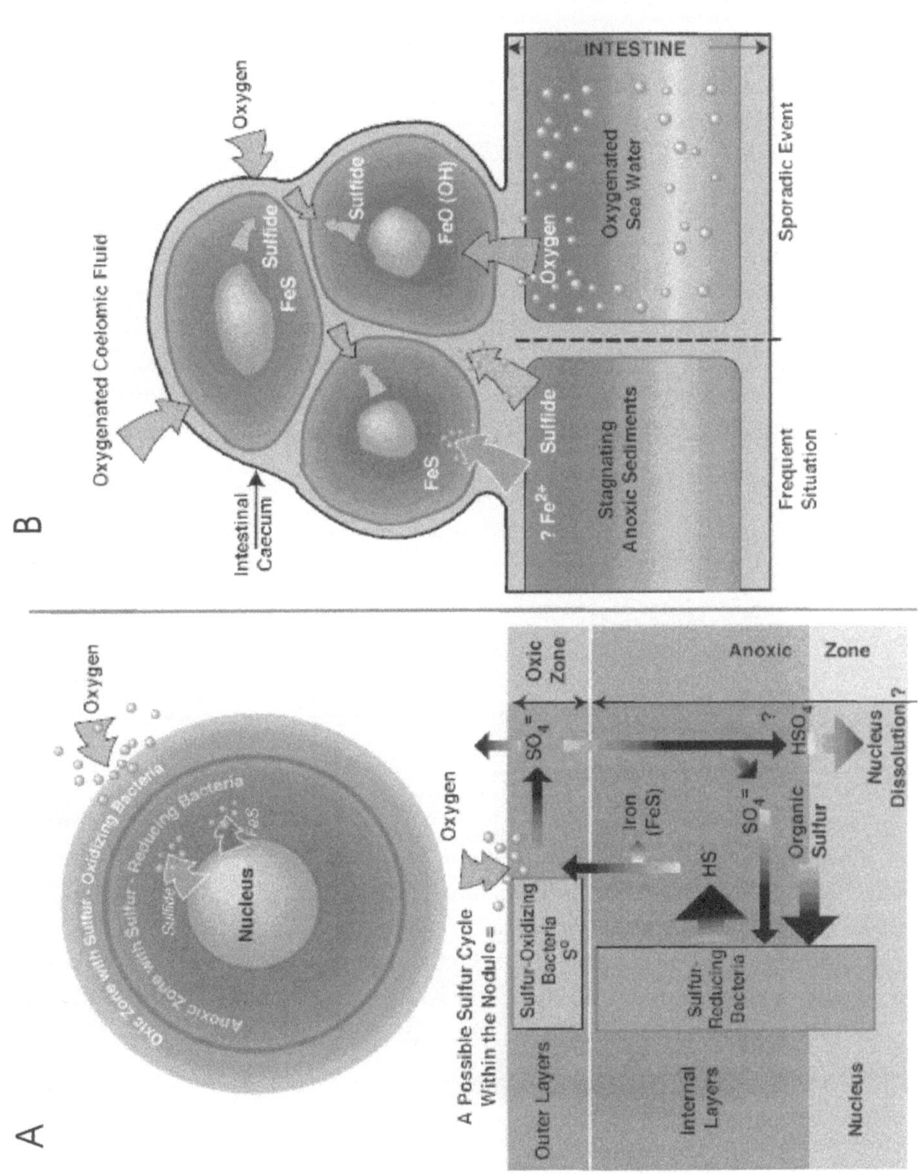

Figure 10. Hypothetical function of the nodules. A. Sulfur cycle within the nodule. B. Sources of sulfur and oxygen for the symbiotic bacteria.

4. NODULE FORMATION

The nodule is built within the intestinal caecum of the echinoid. Detrital particles from the digestive contents (i.e., sediments) regularly enter the caecum where they are colonized by the filamentous bacteria. These bacteria seem to be permanent members of the caecum microflora (De Ridder 1994). A mat progressively forms around each particle (Fig. 11). It becomes three-layered with time as a result of the growth of the filamentous bacteria. Indeed, these bacteria keep themselve positioned in the peripheral layers of the nodule where optimal growth conditions are met, young filaments being most superficial. With time, old filaments die and decay but their sheaths persist and accumulate below the living population of fimentous bacteria. An internal layer of EPS progressively forms that will characterize the "old" nodules.

5. THE SPATANGOID-"BACTERIAL MAT" SYMBIOSIS

One could hypothesize that sulfur-oxidizing bacteria utilize sulfide that regularly forms both in the nodule and in the intestinal content (Temara *et al.* 1993, Brigmon and De Ridder 1998, Thorsen 1998, 1999). They could prevent critical accumulation of sulfide within their host. This detoxifying effect could have nutritional consequences for the animal by allowing the maintenance of an active microflora known to produce nutrients (Temara *et al.* 1991, Thorsen 1999) and by facilitating a more advanced hydrolysis of the organic matter (De Ridder 1994, Brigmon and De Ridder 1998). In this hypothetical functional pattern , the symbiosis is mutually beneficial. Molecular analyses are undertaken in order to identify the members of the bacterial consortium and to infer their metabolic abilities.

Interestingly, several species of spatangoids from various geographic and bathymetric origins possess intestinal bacterial nodules built by filamentous bacteria (De Ridder 1994). This symbiosis appears unique to the order Spatangoida in the class Echinoidea and occurs as synapomorphies in three phylogenetically dispersed families (Loveniidae, Schizasteridae and Asterostomatidae). All the symbiotic spatangoids feed on fine-grained sediments, i.e. few permeable sediments. It is suggested that the symbiosis has allowed the spatangoids to acquire a new digestive mechanics (intermittent feeding and long digestive transit time).

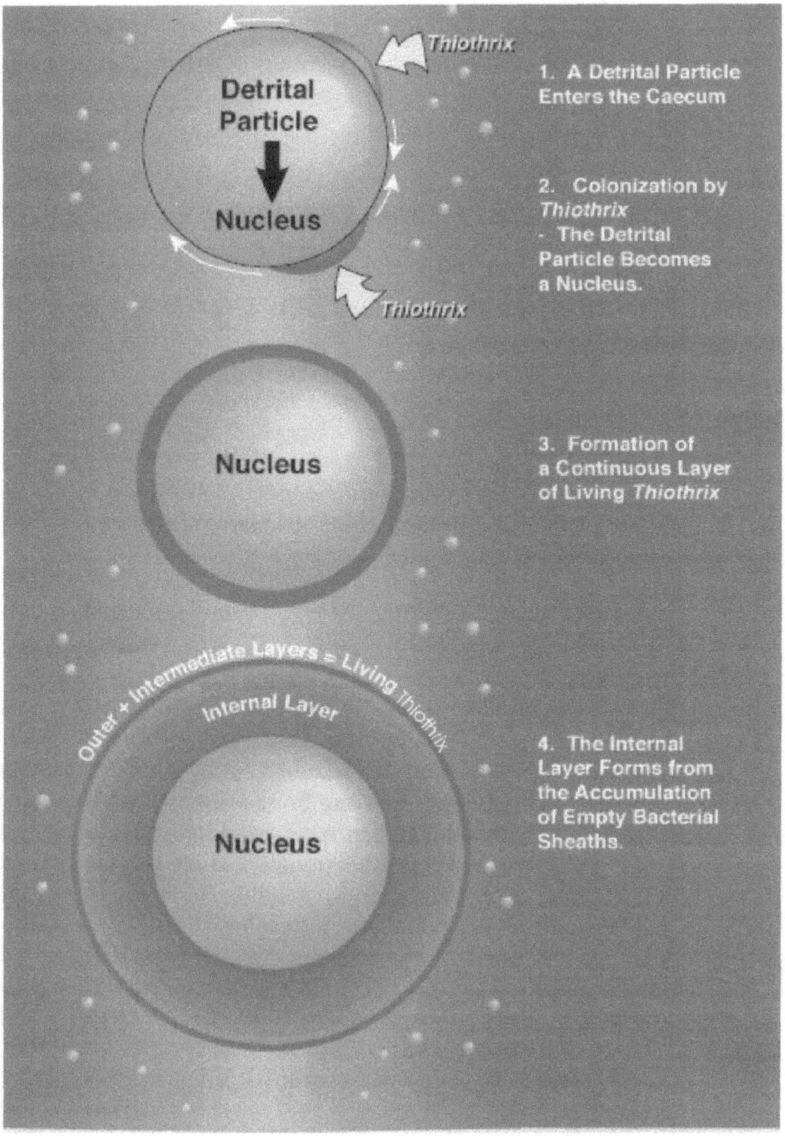

Figure 11. Stages of formation of a bacterial nodule.

6. CONCLUSION

Symbiotic filamentous bacteria form laminated mats around detrital particles and build nodules in the hindgut caecum of spatangoids. Anaerobic conditions that prevail in the nodule and in the hindgut enhance microbial production of sulfide. This seemingly does not affect the animal and hence the mutualism presumably because the bacterial symbionts could remove sulfide. It is thought that the symbiosis provides a detoxifying effect with nutritional consequences to the echinoid.

The nodule, through its structure (i.e., a nucleus wrapped by a layered microbial mat) reminds oncoids, i.e., a type of biolaminated particle in sedimentary structures (Dahanayake and Krumbein 1986). It could constitute a living model of ancient microbial ecosystems and provide informations on paleoenvironmental conditions.

ACKNOWLEDGEMENTS.

We thank E. Bricourt, M. Klinkert and D. Gillan for their help with the preparation of the illustrations. Research was supported by an FRFC grant to C. De Ridder (convention n°2-4515-99). Contribution to the CIBIM.

REFERENCES

Brigmon, R.L. and C., De Ridder (1998) Symbiotic relationship of *Thiothrix* spp. with an echinoderm, *Appl. Environm. Microbiol.* 64 (9), 3491-3495.

Dahanayake, K. and W.E., Krumbein (1986) Microbial structures in oolitic iron formations, *Miner Deposita* 21, 85-94.

De Ridder, C. (1994) Symbioses between spatangoids (Echinoidea) and *Thiothrix*-like bacteria (Beggiatoales), in David B., Guille A., Féral J.-P. & M. Roux (eds), *Echinoderms through time*, Balkeman Publisher, Rotterdam, pp. 619-625.

DeRidder, C., Jangoux, M., and De Vos, L.(1985) Description and significance of a peculiar intradigestive symbiosis between bacteria and a deposit-feeding echinoid, *J. exp. mar. Biol. Ecol.* 91, 65-76.

Temara, A., De Ridder, C., and Kaisin, M. (1991) Presence of an essential polyunsaturated fatty acid in intradigestive bacterial symbionts of a deposit-feeder echinoid (Echinodermata), *Comp. Biochem. Physiol.* 100 B (3), 503-505.

Temara, A., De Ridder, C., Kuenen, J.G., and Robertson, L.A. (1993) Sulfide-oxidizing bacteria in the burrowing echinoid, *Echinocardium cordatum* (Echinodermata), *Mar. Biol.* 115, 179-185.

Thorsen, M.S. (1998) Microbial activity, oxygen status and fermentation in the gut of the irregular sea urchin *Echinocardium cordatum* (Spatangoida: Echinodermata), *Mar. Biol.* 132, 423-433.

Thorsen, M.S. (1999) Abundance and biomass of the gut-living microorganisms (Bacteria, Protozoa and Fungi) in the irregular sea urchin *Echinocardium cordatum* (Spatangoida: Echinodermata), *Mar. Biol.* 133, 353-360.

Chapter 14

EXTRACELLULAR POLYMERS (EPS) AND CALCIFICATION WITHIN MODERN MARINE STROMATOLITES

Alan W. DECHO, Tomohiro KAWAGUCHI
University of South Carolina,
Department of Environmental Health Sciences,
School of Public Health,
Columbia, SC. 29208. USA
adecho@sph.sc.edu

1. INTRODUCTION

Biologically-induced calcification within marine systems is by no means a singular nor unique process. The wide range of organisms and systems in which it occurs suggest a diversity of mechanisms. These have been summarized in reviews (Pentecost 1991; Kempe and Kazmierszak 1994; McConnaughey and Whelan 1997; Simkiss and Wilbur 1989). This work will focus on the role of extracellular polymers (EPS) in calcification within the bacterial biofilms of modern marine stromatolites.

Stromatolites are lithified, laminated sedimentary structures formed by complex interactions of microbes, sediments and the environment. Marine stromatolites provide a potentially interesting example to examine the calcification process, since they represent one of the earliest-known biofilms and perhaps the most (simplistic) rudimentary system of calcification. Stromatolites represent the oldest macroscopic evidence of life on earth (Walter 1993; Riding 1994), and one of the oldest (3.5 by) and most persistent biological communities in earth's history.

2. STROMATOLITES OF HIGHBORNE CAY, BAHAMAS

Modern subtidal and intertidal marine stromatolites are found on the margin of Exuma Sound at Highborne Cay, Bahamas (76° 49' W, 24° 43' N) (Reid and Browne 1991; Reid et al., 1995). These stromatolites are considered potential analogues of ancient stromatolites extending back to the Precambrian Era (Reid et al., 2000). The surface of these stromatolites consist of alternating lithified layers and unlithified layers. The subsurface positions of these lithified layers are millimeter-scale spaced laminations that indicate periodic episodes of lithification, i.e. $CaCO_3$ precipitation (Reid *et al.*, 2000). Unlithified layers are composed mainly of well-sorted carbonate ooids (i.e. approximately 125 – 250 um), and near the surfaces of stromatolites are often laden with copious extracellular polymeric secretions (EPS) produced by cyanobacteria (Visscher *et al.* 1998; Decho *et al.* submitted).

Three major bacterial communities, forming distinctly different mat types, are present on the surfaces of Exuma stromatolites. These mats are responsible for formation of the subsurface layers (Reid *et al.*, 2000). The Stage-I mats are composed of sparse populations of the filamentous cyanobacterium *Schizothrix* sp., with dispersed populations of heterotrophic bacteria. These are considered the "Pioneer" community, since these mats are an initial stage of growth in stromatolite formation, but do not lithify. In this system, *Schizothrix* sp. is a dominant cyanobacterium that secretes EPS. The sticky EPS matrix allows relatively slow accretion rates (one grain-layer per d) of sediment, and fosters the upward growth of the mat. Stage-II mats are composed of organized "biofilm" communities exhibiting a distinct microspatial arrangement of cyanobacteria and sulfate-reducer bacteria (SRB), and are associated with thin layers of precipitated calcium carbonate. Finally, Stage-III mats may develop from Stage-II mats, and consist of abundant populations of the coccoid microboring cyanobacterium *Solentia* sp. These are associated with mm-thick lithified layers.

The following mechanisms have been proposed for calcification processes within stromatolites; (1) abiotic processes of submarine cementation (Logan, 1961; Dill *et al.*, 1986); (2) calcification of cyanobacterial sheaths related to photosynthetic activities (Monty, 1976); (3) heterotrophic bacterial decomposition of cyanobacterial sheaths in subsurface layers (Chafetz and Buczynski, 1992; Bartley, 1996); (4) involvement of endolithic cyanobacteria for micritized grains (MacIntyre *et al.*, 2000); and (5) a close coupling of cyanobacterial photosynthesis, sulfate reduction and anaerobic sulfide oxidation activities within lithified layers (Visscher *et al.*, 1998; Reid *et al.*, 2000). Among those mechanisms, endolithic cyanobacteria (4) and close coupling of bacterial metabolic activities (5) have been attributed to lithification of surface mats in Exuma stromatolites. Heterotrophic bacteria may facilitate $CaCO_3$ precipitation by increasing the pCO_2 and by active excretion of Ca^{2+} ions through bacterial degradation of the EPS (Visscher *et al.*, 2000). Also, Visscher and colleagues (2000) showed a micrometer-scale correlation of sulfate reducing bacterial (SRB) activities and micritic laminae formation. However, there are several questions have not been answered in the formation of lithified layer. Reid *et al.* (2000) concluded that the lithification process of surface mat layers occurs by decomposition of an amorphorus bacterial EPS within the photic zone across the stromatolite surface.

3. EPS MATRIX OF STROMATOLITE BIOFILMS

The EPS matrix is a key structuring component of microbial biofilms (Decho 1990), and as such, EPS may influence a number of biological and geochemical processes that are critical to stromatolite formation. These include: (1) physically stabilizing microbial cells and calcium carbonate ooids against the high-energy environments, *e.g.*, waves, tidal currents, in which these structures commonly occur (Reid *et al* 1995); (2) acting as a Ca^{++} buffer, perhaps preventing precipitation of $CaCO_3$ (Arp *et al.* 1999; Reitner *et al.*, 2000); (3) inducing precipitation of $CaCO_3$ (Pentecost and Riding 1986; Defarge *et al.* 1994; Reitner *et al.*, 2000); and (4) may provide a chemically protective microenvironment for cells (Decho 1990; Arp *et al.* 1999).

Although the creation of highly alkaline conditions by photosynthetic microbial activities may create geochemical conditions for calcification in marine stromatolites, $CaCO_3$ precipitation may not occur due to the production of freshly-secreted EPS by cells. Decho *et al* (submitted) showed that new EPS is secreted by cyanobacteria on a daily basis. However, a relatively large portion (ca. 50%) of this newly-secreted EPS is rapidly, but only partially, degraded by heterotrophic bacteria within 48h. Therefore, at any given time, both newly-secreted EPS and partially-degraded EPS exist within the stromatolite biofilm.

EPS is generally composed of carbohydrate, protein and acidic sugar components (Sutherland 1990). These EPS components, however, may act very differently with regard to calcification. Many EPS contain acidic polysaccharides, mainly uronic acids, and proteins (Sutherland 1995; Decho 1990; Trichet and Defarge 1994; Fortin *et al.* 1997). The negative charges of uronic acids may play important roles in the chelation of Ca^{++} ions, and hence inhibit precipitation.

Previous workers have speculated that EPS inhibit $CaCO_3$ precipitation in stromatolites (Arp *et al.* 1999a; 1999b). Recently, Kawaguchi and Decho (submitted) utilized various experimental approaches used in biomineralization research, e.g. pH drift assays, SDS-PAGE, immunological techniques and $CaCO_3$ induction assays, to understand the role of EPS in $CaCO_3$ precipitation. Their results suggest that EPS, when freshly secreted by cyanobacteria, binds Ca^{++} ions and thus inhibits $CaCO_3$ precipitation. EPS isolated from unlithified layers of stromatolites also inhibited $CaCO_3$ precipitation *in vitro*. Results from pH drift assays showed specifically that both the protein and carboxylated polysaccharide fractions of EPS from unlithified layers, were involved in the inhibition of $CaCO_3$ precipitation.

3.1 Role of EPS proteins

The proteinaceous components of EPS have received comparatively less study (compared to carbohydrates), although they have been attributed to the enhancement of the mechanical stability and adhesive capacity of microbial cells to surfaces (Neu and Marshall 1991; Flemming 1996; Mayer *et al.* 1999). Although protein contents of cyanobacterial EPS are generally small, protein has been found in almost every cyanobacteria investigated thus far (see De Philippis and Vincenzini 1998 for review). Kawaguchi and Decho (2000) found a range of small (< 40 kDa) proteins in EPS from two cyanobacterial isolates from Bahamian stromatolites, and fractionated them using two-dimensional gel electrophoresis. Acidic glycoproteins also may be involved in calcification, as has been previously demonstrated in eukaryote organisms (see Simkiss and Wilbur 1989; Lowenstam and Weiner 1989, for reviews).

Studies of biomineral formation in other systems have emphasized the role of proteins in the calcification process (Wheeler and Sikes, 1984; 1989). Many proteins isolated from biominerals were found to be highly acidic (Lowenstam and Weiner 1989). These proteins are typically enriched in glutamic and/or aspartic acids, and often with some phosphorylation.

3.2 Steric conformation of EPS

Kawaguchi and Decho (submitted) showed, using immunological techniques and FT-IR profiling, that the natural EPS isolated from both the lithified surface mat layer and adjacent unlithified layers were specifically produced by cyanobacteria. The two divergent patterns of how EPS may either inhibit or enhance precipitation, may seem contradictory until one additionally considers the role of EPS steric conformation. The steric conformation of the EPS component molecules may influence their ultimate role in regulating the calcification process. Given the geochemical conditions that favor precipitation, the soluble components of EPS act to inhibit calcification. Laboratory studies have shown they do so in a concentration-dependent manner (Kawaguchi and Decho, submitted). This likely occurs because there is a greater 'availability' of Ca^{++}-chelating functional groups when EPS is in a soluble form. When the same EPS is attached to a substrate, precipitation rapidly occurs. This may result from steric 'blocking' of these functional groups. Their data further suggested that, specifically, the small molecular weight (MW) protein fraction (i.e. < 30 kDa) of EPS may be the critical (i.e. functional group containing) component in regulating this calcification process. Amino acid compositions of EPS from marine stromatolites (Kawaguchi and Decho, submitted) showed a distinct difference, being specifically enriched in aspartic acid and glutamic acid. An enrichment of these key amino acids have been found in the organic matrix associated with other types of EPS (Reitner et al., 2000) and other biominerals (Wheeler *et al.*, 1988; Wheeler and Sikes 1989).

Wheeler *et al.* (1981) showed that the EDTA-soluble organic matrix from oyster shell inhibited $CaCO_3$ precipitation *in vitro*. Subsequently, many researchers have posited that a function in biomineralization of the EDTA-soluble protein fraction was an inhibition and/or nucleation site, owing to its negative charges.

When the EDTA-soluble fraction of EPS was isolated from the lithified layers of stromatolites, this fraction inhibited $CaCO_3$ precipitation *in vitro* when in solution, but induced $CaCO_3$ precipitation when they attached to a solid surface. EPS isolated from lithified layers induced $CaCO_3$ crystal formation *in vitro*. However, when the EPS was chemically modified, it was found that the protein fractions were more important in CaCO3 precipitation than carbohydrates. FT-IR studies also showed that the EPS-derived crystals which formed on agarose beads were aragonite, not calcite (Kawaguchi and Decho, submitted).

When crystals form within a matrix of EPS, the presence of EPS may influence the growth and/or shape of the crystals. Aragonite needles are often found within the EPS matrix in the surface layers of stromatolites (Reid XX). These may form within hydrated pockets (sensu Reitner *et al.*, 2000) of the EPS matrix (where partial EPS degradation may have occurred), and crystals may grow unimpeded. Wada *et al.*, (1993) also reported that acidic polysaccharides isolated from calcareous algae induced magnesium calcite precipitation in a double-diffusion experiment. However, the acid polysaccharides inhibited the continued growth of aragonite crystals by sorption on to the surfaces of the crystals (Wada *et al.* 1993). Therefore, EPS from the lithified layer may play an important role in the determination of crystal mineralogy and morphology in stromatolites.

The immobilized-matrix/crystal nucleation (IC) hypothesis has been previously proposed for the mechanism of $CaCO_3$ precipitation in other biominerals (Crenshaw 1989; Sikes and Wheeler 1988a; Wheeler and Sikes 1989). The IC hypothesis asserts that the EDTA-soluble matrix isolated from biominerals when free in solution inhibits *in vitro* $CaCO_3$ precipitation, but when attached to an insoluble matrix framework, may promote precipitation (Crenshaw *et al.* 1988; Sikes and Wheeler 1988a; Wheeler *et al.* 1988b: Wheeler and Sikes 1984, 1989; Campbell *et al.* 1989; Linde *et al.* 1989). In stromatolites, as heterotrophic microbial degradation of EPS proceeds, the protein fractions of EPS may become more accessible to the outside environment. Kawaguchi and Decho (submitted) showed that when protein fractions (of EPS) were attached to a solid surface, they could provide a nucleation site for $CaCO_3$ precipitation. Also, while the contents of uronic acids and carbohydrates in EPS from lithified layers decreased compared to unlithified layer EPS, the protein content remained relatively constant. This result suggested that as microbial degradation of EPS proceeds, protein fractions in EPS remain and may become increasingly exposed.

4. MICROBIAL COMMUNITY DEVELOPMENT AND CALCIFICATION

The three major microbial communities of Highborne stromatolites, through their activities and relative microspatial arrangements, may influence the type(s) and extent of calcification occurring with stromatolites. Results of confocal imaging showed that the EPS of cyanobacteria appeared to be closely associated with $CaCO_3$ precipitation. Immuno-localization of *Schizothrix* sp. EPS confirmed the presence of this EPS among $CaCO_3$ minerals in lithified layers. Therefore, when heterotrophic bacteria such as sulfate reducing bacteria degrade carbohydrate in EPS, labile acidic proteins may remain after bacterial degradation, attached to a solid surface such as polymerized insoluble EPS. This EPS remnant may act as a substrate to induce $CaCO_3$ precipitation. Previous researchers reported that high sulfate reduction activities could be correlated with bands of $CaCO_3$ precipitation in stromatolites (Visscher *et al.* 1998; 2000). It has been suggested that cyanobacterial photosynthesis, sulfate reduction and anaerobic sulfide oxidation, together, are responsible for $CaCO_3$ precipitation by creating highly-alkaline conditions (Visscher *et al.* 1998; 2000).

In the Stage I "pioneer" communities, there are typically fast-growing mats of cyanobacteria that are producing copious amounts of EPS (Decho *et al.*, in review). While there are no apparent spatial arrangements of cyanobacteria and SRB, there is rapid secretion of EPS by cyanobacteria, and relatively rapid (albeit partial) degradation of EPS by heterotrophic bacteria. This may result in random precipitation within localized "hydrated pockets" within the EPS matrix. In the Stage II' biofilm" communities, there are organized microspatial arrangements of cyanobacteria and heterotrophic bacteria. There is a close diel coupling of autotrophic production by cyanobacteria of dissolved organics, and their consumption by heterotrophs (Decho *et al.*, in review). The closely linked microspatial arrangements and diel activities of cyanobacteria and sulfate reducers mediates an organized horizontal precipitation of $CaCO_3$ micritic laminae, as evidenced by S-35 autoradiography (Visscher *et al.*, 2000). Therefore precipitation is largely localized to thin horizontal regions of the stromatolite; a process that physically strengthens the macrostructure. In Stage III communities, there are abundant endolithic coccoid cyanobacteria (*Solentia* sp.), that colonize and microbore ooid grains near the laminae (see Reid *et al.*, this volume). Extensive reworking of ooids by endoliths results in the rapid dissolution/reprecipitation of grains, and the "cementing together" of neighboring ooids. This results in a thickened laminae that contributes to the longer-term preservation of the macrostructure. A repeated cycling of these three communities leads to observable laminations every 1- 2 mm.

5. SUMMARY

The marine stromatolite, owing to its long and persistent history, may represent evolution's first successful effort at 'organized calcification'. The coordination of microbial autotrophic and heterotrophic processes, in the formation of micritic lithified layers, results in a macrostructure (*i.e.* a stromatolite) that lends longer-term ecological stability to the overall system.

Bahamian stromatolites produce a variety of forms of calcium carbonate precipitates that occur in close association with cyanobacterial biofilms and their EPS. Cyanobacterial EPS have divergent roles in the regulation of calcification in stromatolites. A major role of freshly-secreted cyanobacterial EPS is to "inhibit" calcification via binding of Ca^{++} ions by specific functional groups. Geochemical conditions generated by photosynthetic activities are consistent with precipitation in these supersaturated environments, therefore, cyanobacteria must actively maintain fresh EPS in proximity to cells in order to resist precipitation. This is observed using CSLM. A second role of EPS occurs in facilitating concentrated horizontal precipitation in specific regions (e.g. laminae) of the stromatolite. This occurs in the stage-II "biofilm" community- a process that is facilitated by the unique microspatial positioning of autotrophic and heterotrophic bacteria. Steric conformation of EPS appears to be critical in whether there is inhibition or facilitation of precipitation. Compositional analyses and pH drift assays show that environmentally-altered EPS (i.e., partially degraded or cross-linked), when attached to a surface, may act as a nucleation site and facilitate precipitation. However, soluble forms of this same EPS inhibit $CaCO_3$ precipitation. Both uronic acids and certain protein fractions are involved in inhibition of precipitation; through chelation of Ca^{++} ions. However, only specific protein fractions of EPS may be involved in $CaCO_3$ precipitation. Once attached to an insoluble substrate, EPS may provide the nucleation site of $CaCO_3$ precipitation. Therefore, EPS may accomplish several roles in $CaCO_3$ precipitation by providing a nucleation site in the lithified layer of marine stromatolites, and in inhibition. The conspicuous lack of bacterial microfossils, in association with the laminae of many Precambrian stromatolites may be due to precipitation events which occur within the amorphous EPS matrix, rather than in close association with cells (Reid *et al.* 2000). It is therefore, important to view EPS in natural systems as a potential structuring matrix that occurs, not as a single type of molecule, but rather as a complex of large molecules in various stages of degradation or steric-modifications.

While microbial activities may create appropriate geochemical microenvironments for $CaCO_3$ precipitation, EPS associated with these communities may posit divergent roles in regulating calcification within stromatolites. The types of precipitates are closely aligned with the three distinct microbial communities present in stromatolites, as outlined by Reid et al (2000). (1) Within the pioneer (Stage-I) community, small clusters of needle precipitates typically form within EPS. These occur as random CaCO3 needles in pockets within a diffuse EPS matrix of rapidly growing *Schizothrix* sp. mats. These mats do not typically exhibit calcified laminae. (2) The more stabilized biofilm that develops in the Stage-II community, enables an organized spatial structuring of cyanobacteria (*Schizothrix*) and heterotrophic bacteria (e.g. sulfate reducers). Partial degradation of EPS within this biofilm coupled with heterotrophic activities, results in the formation of thin horizontal calcified laminae (10 - 30 um thickness). (3) Finally, a prolonged stabilization by the stage II community enables colonization of ooid grains by the microboring cyanobacterium *Solentia* sp. Activities of *Solentia* sp, result in both dissolution and rapid re-precipitation of ooid grains into a thick solid micritized layer of CaCO3. This final level of calcification contributes to the longer-term preservation of the stromatolite.

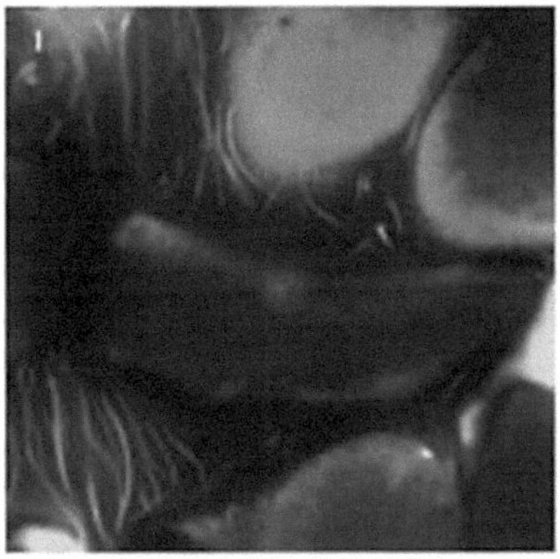
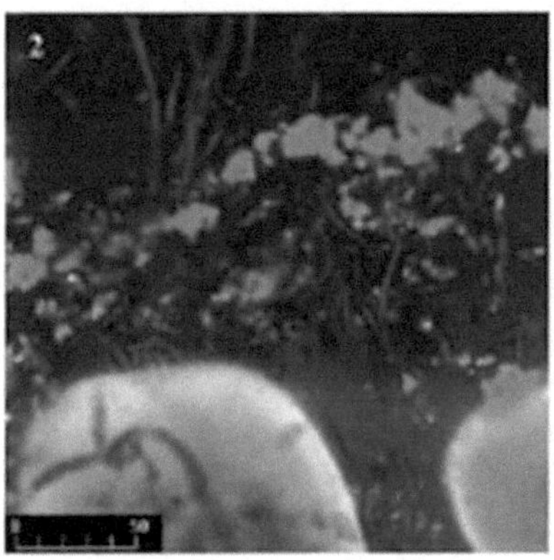
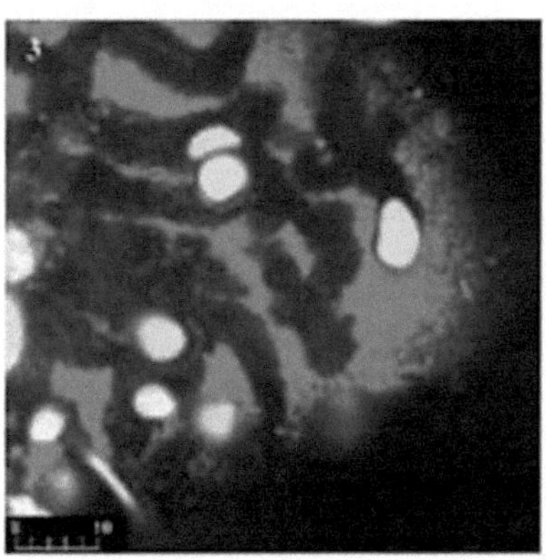

Figure 1. (1) Confocal scanning laser micrograph (CSLM) showing the Stage-I "Pioneer" community that typically occurs near the surface of a stromatolite, and is composed largely of the cyanobacteria *Schizothrix* sp.(DAPI-stained blue) embedded within a gel matrix of EPS (green). The gel-like EPS matrix surrounds $CaCO_3$ sediment grains called 'ooids' (red), and provides initial cohesive stability that allows the mat to grow higher on the stromatolite. Mats that grow too rapidly will be washed away by high-energy wave action. This *Schizothrix* community is rapidly growing and generally resists the formation of organized precipitates. Precipitates occur as small random aragonite "needles" in pockets throughout the EPS. (2) During the Stage-II, "biofilm" community, a spatially-organized community of the cyanobacterium *Schizothrix* (producers) and sulfate reducer bacteria (SRB) develop, and effect the precipitation of a thin horizontal CaCO3 layer (red). While cyanobacterial EPS generally inhibits precipitation in proximity to their own cells, massive precipitation (red) occurs close by, perhaps through partial EPS degradation and geochemical gradients resulting from activities of SRB in proximity to the cyanobacteria. SRB rapidly utilize small-MW metabolites released by cyanobacteria during photosynthesis. The spatially organized arrangement of these two groups results in a thin (10 - 50 m) horizontal precipitate layer near the interface.

(3) During Stage III, ooid grains (red) acquire abundant populations of "endolithic" cyanobacteria, *Solentia* sp (green fluorescence). Microboring by these cyanobacteria in a single ooid (shown here) dissolves the $CaCO_3$ matrix of ooid grains as it bores forward, creating "tunnels" as it proceeds. However, rapid reprecipitation occurs just behind the organism, perhaps to prevent disintegration of the ooid. It is speculated that this rapid precipitation may be mediated by a special EPS secreted by *Solentia*. The microboring dissolution/precipitation process contributes to micritization of grains, and the formation of thickened micritic layers that are characteristic of stromatolites.

ACKNOWLEDGEMENTS

This work was supported by grants from the U.S. National Science Foundation (OCE 96-17738) and the Office of Naval Research (NOOO14-97-10024). This represents RIBS (Research Initiative on Bahamian Stromatolites) Contribution # 16.

REFERENCES

Arp, G., Reimer, A, and Reitner, J. (1999a) Calcification in cyanobacterial biofilms of alkaline salt lakes. *Europ J Phycol* 34, 393-403.

Arp, G., Thiel, V., Reimer, A., Michaelis, W., and Reitner, J. (1999b) Biofilm exopolymers control microbialite formation at thermal springs discharging into the alkaline Pyramid Lake, Nevada, USA. *Sed Geol* 126,159-176.

Awramik, S.M. (1992) The history and significance of stromatolites, in M.Schidlowski et al., (eds.), *Early Organic Evolution: Implications for Mineral and Energy Resources* Springer-Verlag, Berlin.

Bathurst, R.G.C. (1975) *Carbonate sediments and their diagenesis. Developments in Sedimentology* Vol. 12. Elsevier, Amsterdam. Borowitzka, M.A. (1982) Morphological and cytological aspects of algal calcification. *Int Revs Cytol* 74, 127-162.

Borman, A.H., de Jong, E.W., Thierry, R., Westbroek, P., Bosch, L., Gruter, M., Kamerling, J.P. (1987) Coccolith-associated polysaccharides from cells of *Emiliana huxleyi* (Haptophyceae). *J Phycol* 23, 118-123

Campbell, A.A., Ebrahimpour, A., Perez, L., Smesko, S.A., Nancollas, G.H. (1989) The dual role of polyelectrolytes and proteins as mineralization promoters and inhibitors of calcium oxalate monohydrate. *Calcif Tiss Internat* 45, 122-128

Castanier, S., Le Metayer-Levrel, G., Perthuisot, J.-P. (2000) Bacterial roles in the precipitation of carbonate minerals, in R.E. Riding and S.M. Awramik (eds.), *Microbial Sediments*, Springer-Verlag, Berlin, pp. 32-39.

Chavetz, H.S., and Buczynski, C. (1992) Bacterially induced lithification of microbial mats. *Palaios* 7, 277-293

Decho, A.W. (1990) Microbial exopolymer secretions in ocean environments: their role(s) in food webs and marine processes. *Oceanogr Mar Biol Annu Revs* 28, 73-154.

Decho, A.W. and Kawaguchi, T. (1999) Confocal imaging of *in-situ* natural microbial communities and their extracellular polymeric secretions using nanoplast resin. *BioTechniques* 27, 1246-1252.

Decho, A.W., Visscher, P.T. , and Reid, R.P. (submitted) Cycling and production of natural microbial exopolymers (EPS) within a marine stromatolite. *Aquat Microb Ecol*

Defarge, C., Trichet, J., and Coute, A. (1994) On the appearance of cyanobacterial calcification in modern stromatolites. *Sed Geol* 94, 11-19.

Defarge, C., Trichet, J., Jaunet, A.M., Robert, M., Tribble, J., and Sansome, F.J. (1996) Texture of microbial sediments revealed by cryoscanning electron microsopy. *J Sed Res* 66, 935-947.

De Jong, E.W., Bosch, L., and Westbroek, P. (1976) Isolation and characterization of a Ca2+ -Binding polysaccharide associated with coccoliths of *Emiliana huxleyi* (Lohmann) Kamptner. *Europ J Biochem* 70, 611-621.

De Philippis, R., and Vincenzini, M. (1988) Exocellular polysaccharides from cyanobacteria and their possible applications. *FEMS Microbiol Revs* 22,151-175.

Dignac, M-F, Urbain, V., Ryacki, D., Bruchet, A., Snidaro, D., and Scribe, P. (1998) Chemical description of extracellular polymers: implication on activated sludge floc structure. *Water Sci Technol* 38, 45-53.

Dill, R.F., Shinn, E.A., Jones, A.T., Kelly, K., Steinen, R.P. (1986) Giant subtidal stromatolites forming in normal salinity waters. *Nature (London)* 324, 55-58

Dubois, M., Gilles, K.A., Hamilton, J.K., Rebers, P.A., and Smith, F. (1956) Colorimetric methods for determination of sugars and related substances. *Anal Chem* 28, 350-356.

Fichtinger-Schepman, A.M., Kamerling, J.P., Versluis, C., and Vliegenthart, J.F.G. (1981) Structural studies of the methylated, acidic polysaccharide associated with coccoliths of *Emiliania huxeyi* (Lohmann) Kamptner. *Carbohydr Res* 93, 105-123.

Flaibani, A., Olsen, Y., and Painter, T.J. (1989) Polysaccharides in desert reclamation: composition of exocellular proteoglycans complexes produced by filamentous blue-green and unicellular green edaphic algae. *Carbohydr Res* 190, 235-248

Flemming, H-C. (1996) The forces that keep biofilms together, in W. Sand (ed.), *Biodeterioration and biodegradation*. Dechema Monogr 133. VCH, Weinheim.

Folk, R.L. (1993) SEM imaging of bacteria and nannobacteria in carbonate sediments and rocks. *J Sed Petrol* 63, 990-999.

Folk, R.L., and Chafetz, H.S. (2000) Bacterially induced microscale and nanoscale carbonate precipitates, in R.E. Riding, and S.M. Awramik (eds.), *Microbial Sediments* Springer-Verlag, Berlin, pp. 40-49.

Grotzinger, J.P., and Knoll, A.H. (1999) Stromatolites in Precambrian carbonates: evolutionary mileposts or environmental dipsticks?. *Ann Rev Earth Planet Sci* 27, 313-358.

Gunthorpe, M.E., Sikes, C.S., and Wheeler, A.P. (1990) Promotion and inhibition of calcium carbonate crystallization *in vitro* by matrix protein from blue crab exoskeleton. *Biol Bull* 179, 191-200.

Kawaguchi, T., and Watabe, N. (1993) The organic matrices of the shell of the american oyster *Crassostrea virginica* Gmelin. *J Exper Mar Biol Ecol* 170, 11-28.

Kawaguchi, T., and Decho A.W. (2000) Biochemical characterization of cyanobacterial extracellular polymers (EPS) from modern marine stromatolites. *Prep Biochem BioTechnol* 30, 321-330.

Kawaguchi, T., and Decho A.W. (submitted) How do extracellular polymers (EPS) of cyanobacteria regulate calcification?: A study of marine stromatolites, Bahamas. *Amer Mineral.*

Kempe, S., and Kazmierszak, J. (1994) The role of alkalinity in the evolution of ocean chemistry,organization of living systems and biocalcification processes. *Bull L'Inst Oceanogr Monaco* 13, 61-117.

Knorre, H.V., and Krumbien, W.E. (2000) Bacterial Calcification, in R.E. Riding, and S.M. Awramik (eds.), *Microbial Sediments*. Springer-Verlag, Berlin pp. 25-31.

Linde, A., Lussi, A., and Crenshaw, M.A. (1989) Mineral induction by immobilized polyanionic proteins. *Calcif Tiss Internat* 44, 286-295

Logan, B.W. (1961) Cryptozoon and associate stromatolites from the Recent, Shark Bay, Western Australia. *J Geol* 69, 517-533.

Lowenstam, H., and Weiner, S. (1989) *On Biomineralization*, Oxford Univ. Press, N.Y.

Lyons, W.B., Long, D.T., Hines, M.E., Gaudette, H.E., and Armstrong, P.B. (1984) Calcification of cyanobacterial mats in Solar Lake, Sinai. *Geology* 12, 623-626.

Macintyre, I.G., Prufert-Bebout, L., and Reid, R.P. (2000) The role of endolithic cyanobacteria in the formation of lithified laminae in Bahamian stromatolites. *Sedimentology* 47, 915-921.

Marra, M., Palmeri, A., Ballio, A., Segre, A., and Slodki, M.E. (1990) Structural characterization of the exocellular polysaccharide from *Cyanospira capsulata*. *Carbohydr Res*, 197, 338-344.

Marsh, M.E. (1994) Polyanion-mediated mineralization - assembly and reorganization of acidic polysaccharides in the Golgi system of a coccolithophorid alga during mineral deposition. *Protoplasma* 177, 108-122.

Marsh, M.E., Chang, D.-K., and King, G.C. (1992) Isolation and characterization of a novel acidic polysaccharide containing tartrate and glyoxylate residues from the mineralized scales of a unicellular coccolithophorid alga *Pleurochrysis carterae*. *J Biol Chem* 267, 20507-20512

McConnaughey, T.A., and Whelan J.F. (1997) Calcification generates protons for nutrient and bicarbonate uptake. *Earth Sci Revs* 42, 95-117.

Meril, C.R., Goldman, D., Sedman, S.A., and Ebert, M.H. (1981) Ultrasensitive stain for proteins in polyacrylamide gels shows regional variation in cerebrospinal fluid proteins. *Science* 211, 1437-1438.

Merz, M.U.E. (1992) The biology of carbonate precipitation by cyanobacteria. *Facies* 26, 81-102.Merz-PreiB, M. (2000) Calcification in cyanobacteria, in R.E. Riding, and S.M. Awramik (eds.), *Microbial Sediments* Springer-Verlag, Berlin, pp. 50-56.

Monty, C.L.V. (1976) The origin and development of cryptalgal fabric. *Devel Sedimentol* 20,193-249.

Myklestad, S., and Haug A. (1972) Production of carbohydrates by the marine diatom *Chaetoceros affinis* (var. willei Gran) Hustedt. 1. Effect of the concentration of nutrients in the culture medium. *J Exp Mar Biol Ecol* 9, 125-136.

Neu, T.R., and Marshall, K.C. (1991) Microbial "footprints" – a new approach to adhesive polymers. *Biofouling* 3, 101-112.

Pedley, M. (1992) Freshwater (phytoherm) reefs: the role of biofilms and their bearing on marine reef cementation. *Sediment Geol* 79, 255-274.

Pentecost, A. (1991) Calcification processes in algae and cyanobacteria, in R.Riding (ed.), *Calcareous Algae and Stromatolites*, Springer-Verlag, Berlin, pp. 3-20.

Pentecost, A., and Riding R. (1986) Calcification in cyanobacteria, in B.S.C. Leadbeater, and R. Riding (eds.), *Biomineralization in lower plants and animals*. Clarendon Press, Oxford, pp. 73-90.

Ramsing, N.B., Kuhl, M., and Jorgensen, B.B. (1993) Distribution of sulfate-reducing bacteria, O2, and H2S in photosynthetic biofilm determined by oligonucleotide probes and microelectrodes. *Appl Environ Microbiol* 59, 3840-3849.

Reid, R.P., and Browne K.M. (1991) Intertidal stromatolites in a fringing Holocene reef complex, Bahamas. *Geology* 19, 15-18

Reid, R.P., Macintyre, I.G., Steneck, R.S., Browne, K.M., and Miller, T.E. (1995) Stromatolites in the Exuma Cays, Bahamas: Uncommonly common. *Facies* 33, 1-18.

Reid, R.P., Visscher, P.T., Decho, A.W., Stolz, J.F., Bebout, B.M., Dupraz, C., Macintyre, I.G., Pinckney, J., Paerl, H.W., Prufert-Bebout, L., Steppe, T.F., and Des Marais, D.J. (2000) The role of microbes in accretion, lamination, and lithification in modern marine stromatolites. *Nature (London)* 406, 989-992.

Reitner, J., Thiel, V., Zankl, H., Michaelis, W., Worheide, G., and Gautret, P. (2000) Organic and biogeochemical patterns in cryptic microbialites, in R.E. Riding, and S.M. Awramik (eds.), *Microbial Sediments*. Springer-Verlag, Berlin, pp.149-160.

Riding, R. (1991) Calcified cyanobacteria, in R. Riding (ed.), *Calcareous algae and stromatolites*. Springer-Verlag, Berlin.

Riding, R. (1994) Evolution of algal and cyanobacterial calcification, in S. Bengston (ed.), *Early evolution on Earth*. Nobel Symposium No. 84, Columbia UP, N.Y, pp. 426-438.

Rippka, R., Deruelles, J., Waterbury, J.B., Herdman, J., and Stanier, R.Y. (1979) Generic assignments, strain histories and properties of pure cultures of cyanobacteria. *J Gen Microbiol* 111,1-61.

Seong-Joo, L., Browne, K.M., and Golubic, S. (2000) On stromatolite lamination, in R.E. Riding, and S.M. Awramik (eds.) *Microbial Sediments*. Springer-Verlag, Berlin, pp. 16-24.

Simkiss, K., and Wilbur, K.M. (1989) *Biomineralization: Cell Biology and Mineral Deposition*. Academic press, San Diego, CA.

Sutherland , I.W. (1990) *Biotechnology of microbial exopolysaccharides*. Cambridge Univ. Press, Cambridge.

Tazaki, K., and Ishida, H. (1996) Bacteria as nucleation sites for autigenic minerals. *Geol Soc Japan* 102, 866-878.

Towbin, H., Staehelin, T., and Gordin, T. (1979) Electrophoretic transfer of proteins from Polyacrylamide gels to nitrocellulose sheets; procedure and some applications. *Proc Natl Acad Sci* 76, 4350-4354 Vasconcelos, C., and McKenzie, J.A. (1997) Microbial mediation of modern dolomite precipitation and diagenesis under anoxic conditions (Lagoa Vermelha, Rio de Janeiro, Brazil). *J Sed Res* 67, 378-390

Visscher, P.T., Reid, R.P., Bebout, B.M., Hoeft, S.E., Macintyre, I.G., Thompson, J.A. (1998) Formation of lithified micritic laminae in modern marine stromatolites (Bahamas): The role of sulfur cycling. *Amer Mineral* 83, 1482-1493.

Visscher, P.T., Reid, R.P., and Bebout, B.M. (2000) Microscale observations of sulfate reduction: correlation of microbial activity with lithified micritic laminae in modern marine stromatolites. *Geology* 28, 919-922.

Wadda, N., Okazaki, M., and Tachibana, S. (1993) Effects of calcium-binding polysaccharrides from calcareous algae on calcium carbonate polymorphs under conditions of double diffusion. *J Crystal Growth* 132, 115-121.

Walter, M.R. (1977) Interpreting stromatolites. *Amer Sci.* 65, 563-571.

Walter, M.R. (1983) Archean stromatolites: evidence of the earth's earliest benthos, in J.W. Schopf (ed.),*Earth's earliest biosphere, its origin and evolution.* Princeton University Press, N.J, pp.187-213.

Westfall, F., and Rince, Y. (1994) Biofilms, microbial mats and microbe-particle interactions: electron microscope observations from diatomaceous sediments. *Sedimentology* 41, 147-162.

Wheeler, A.P., and Sikes, C.S. (1984) Regulation of carbonate calcification by organic matrix. *Amer Zool* 24, 933-944.

Wheeler, A.P., George, J.W., and Evans, C.A. (1981) Control of calcium carbonate nucleation and crystal growth by soluble matrix of oyster shell. *Science* 212, 1397-1398.

Wheeler, A.P., Rusenko, K.W., and Sikes, C.S. (1988) Organic matrix from carbonate biomineral as a regulator of mineralization, in C.S. Sikes, and A.P. Wheeler (eds.), *Chemical aspects of regulation of mineralization.* Univ. S. Alabama Publ. Serv, pp. 9-13.

Chapter 15

THE STUDY OF A RECENT IRON-ENCRUSTED BIOFILM IN THE MARINE ENVIRONMENT

David C. GILLAN
Université Libre de Bruxelles
Laboratoire de Biologie marine, CP 160/15,
av. Roosevelt 50, 1050 Bruxelles, Belgium.
dgillan@ulb.ac.be

1. INTRODUCTION

Intimate associations of microorganisms with ferric iron deposits are widespread in nature. Such microbial communities occur in fresh and saltwater environments, in soils and in desert rock varnishes (for reviews see Ghiorse 1984 and Konhauser 1998). In saltwater environments, these communities are often remote and therefore difficult to study. In the pelagic zone they have mostly been reported in sinking particles below 100 meters (Cowen & Silver 1984), and in the deep-sea they have been found in ferromanganese concretions (Burnett & Nealson 1981) and in some microbial mats (Karl et al. 1989, Juniper & Tebo 1995). In a few cases, epibiotic ferric iron-encrusted microbial communities have been studied on deep-sea mussel and limpet shells as well as worm tubes (Jannasch & Wirsen 1981, Baross & Deming 1985). Iron-encrusted microbial communities have also been reported in reef caves (Reitner et al. 2000). Generally speaking, very little is known about the community fine structure or the specific role that the microorganisms play in iron oxidation and/or deposition.

The aim of the present paper is to summarize the research carried out with the iron-encrusted biofilm epibiotic on *Montacuta ferruginosa,* a marine bivalve that lives in the burrow of an echinoid (Gage 1966). This bivalve can be easlily sampled in the shore area and up to 18 bivalves may be found in the same burrow. To date, this biofilm has been investigated with three microbiological approaches, i.e. microscopy, molecular biology, and cultivation. The iron mineral as well as two iron precipitation pathways have also been studied in the biofilm.

2. MICROSCOPIC APPROACH

Under the scanning electron microscope, the biofilm of *M. ferruginosa* appears as a structured microbial mat with three separate layers (Gillan & De Ridder 1995, 1997). The inner layer is relatively thick (about 50 m) and made of densely packed mineral granules that adhere to the shell. Remains of filamentous bacteria (ferric iron-encrusted empty sheaths and extracellular material) are occasionally found in this layer. The intermediary layer is relatively thin (up to 10 m) and consists of various types of ferric iron-encrusted microorganisms. Filamentous bacteria related to Beggiatoaceae are predominant; they correspond to 2 morpho-physiological types that resemble the sulfide-oxidizing genera *Beggiatoa* and *Thiothrix*. Remains of lysed filaments frequently occur and are always heavily encrusted with ferric iron. Single rod-shaped bacteria are dispersed over the surface of the filamentous bacteria while microcolonies of coccoid bacteria and stalked bacteria locally occur within the layer. Dead stalked protozoans extensively covered with iron minerals are also typically seen : they correspond to 2 species of Ciliophora belonging to the Phyllopharyngea (Suctoria) and to 1 species of choanoflagellate. The outer layer consists of living microorganisms that are similar to those of the intermediary layer. The filamentous bacteria are particularly abundant; they extend up to 3 mm from the bivalve. The stalked protozoans and the filamentous bacteria are usually attached to the intermediary layer by one of their extremities; often they are partly encrusted with ferric minerals.

The genesis of this mineral-microbial mat could be partly explained by a three-step process : (1) ferric iron deposition within microbial exopolymeric substances (EPS) in the outer layer; (2) release and accumulation of heavily ferric iron-encrusted EPS after lysis of the bacteria in the intermediary layer; (3) degradation of EPS and accumulation of ferric iron minerals in the inner layer.

3. MOLECULAR BIOLOGY APPROACH

The genetic diversity of the biofilm of *M. ferruginosa* was assayed by denaturing gradient gel electrophoresis (DGGE) analysis of short (193-bp) 16S ribosomal DNA PCR products obtained with primers specific for the domain *Bacteria* (Gillan et al. 1998). This analysis revealed a diverse consortium : biofilms contained 9 to 21 DGGE bands and a total of 40 different band positions were found in the analysis of 20 individual biofilms. These 20 individual biofilms did not produce identical DGGE profiles (each profile is unique). However, different biofilms share common bands, suggesting that similar bacteria can be found in different biofilms. DGGE analysis also suggests that biofilms from bivalves living in the same burrow are more similar to each other than to biofilms from other burrows. This means that microorganisms can be transferred from one biofilm to another inside a burrow.

The phylogenetic affiliations of the sequence types were determined by cloning and sequencing the 16S rRNA genes. Close relatives of the genera *Pseudoalteromonas*, *Colwellia*, and *Oceanospirillum* (members of the *-Proteobacteria* lineage), as well as *Flexibacter maritimus* (a member of the *Cytophaga-Flavobacter-Bacteroides* lineage) were found in the biofilms. All these genera are aerobic heterotrophic microbes. One clone from *M. ferruginosa,* clone S3C16, has a sequence very similar to the PVB_OTU_4 sequence (S = 93.2%) retrieved from the bacterial mat community of Pele's Vents, an active deep-sea hydrothermal vent system in Hawaii (Moyer et al. 1995). Interestingly, this vent field bacterial mat, which is apparently dominated by filamentous bacteria encrusted by a mineral very similar to the mineral of *M. ferruginosa,* also contains bacteria related to the genera *Colwellia* and *Pseudoalteromonas*. It is not known if there is any relationship between the simultaneous presence of these three sequences (related to PVB_OTU_4, *Colwellia,* and *Pseudoalteromonas*) and the presence of the iron mineral.

4. CULTIVATION APPROACH

A classic cultivation approach was applied to the biofilm of *M. ferruginosa*. Only heterotrophic bacteria were cultivated. Viable counts and strain isolations were performed. Viable counts data have indicated that the number of colony forming units (CFU) per bivalve were significantly higher (t-test, P=0.045) when the culture media contains iron complexed to citrate (1.5 g/liter). This indicates the presence of bacteria that need citrate complexed to iron, noncomplexed citrate or only iron to grow. Such bacterial species could participate to the formation of the mineral typical of *M. ferruginosa* (microbial degradation of soluble organic Fe(III) complexes may cause iron precipitation - see below). The number of CFU per mg (dry weight) were 520 to 840 times higher for iron-encrusted biofilms than for the neighbouring sediments.

27 strains were isolated and characterized phenotypically. 17 strains were tentatively classified in the family Cytophagaceae (in the genera *Cytophaga, Flexibacter* and *Marinilabilia*); 3 in the family Vibrionaceae (in the genera *Photobacterium*); 3 in the genera *Leucothrix*; 2 in the genera *Micrococcus* or *Moraxella*; and 1 in the nocardioform actinomycetes.

5. CONJUNCTION OF THE THREE MICROBIOLOGICAL APPROACHES

The three microbiological approaches used to investigate the biofilm of *M. ferruginosa* have resulted in three different data sets which have now to be cross-checked and re-evaluated in order to get an insight into the composition of the real microbial community within the biofilm. Discrepancies between different approaches are frequently reported in microbiology. They simply reflect the many methodological biases of each approach, and particularly the facts that the most abundant heterotrophic marine bacterial species are not readily culturable by commonly used methods and that the cultured marine bacteria are not well represented in sequence databases (Suzuki et al. 1997).

The discrepancies between the three techniques used above may be parltly solved by the use of fluorescent in situ hybridization (FISH). FISH was tested with fluorescein-labelled probes and epifluorescence. However, no convincing results were obtained due to (i) the very low intensity of the fluorescent signal obtained (even with universal probes), (ii) the very low percentage of fluorescent cells, and (iii) the problem of autofluorescence. It is obvious that, despite the numerous advantages of fluorescent oligonucleotide probes, application of this technique to natural bacterial assemblages, particularly to marine bacteria, still faces many limitations (see also: Ouverney & Fuhrman 1997). It is possible that classic FISH does not work with the biofilm because probes are adsorbed to the iron mineral. Removal of iron and/or the use of confocal microscopy are probably needed here.

It is obvious, as shown by microscopy, that two morphotypes of multicellular filamentous bacteria (*Beggiatoa*- like and *Thiothrix*- like filaments) are always present and abundant in the biofilm. However, the identity of these filaments remains uncertain because *Beggiatoa* and *Thiothrix* were not retrieved by molecular and cultivation techniques, and because FISH was not successful. However, as *Leucothrix* strains were isolated here in pure culture, it is possible that the sulfide-oxidizing *Thiothrix*-like bacteria observed in the biofilm are *Leucothrix* bacteria (*Leucothrix* and *Thiothrix* share the same morphology but differ by the sulfur-oxidizing ability, *Thiothrix* being able to oxidize sulfide). This hypothesis is also supported by the recent discovery of sulfide-oxidizing *Leucothrix* species (*L. thiophila*) by Dul'tseva and collaborators (1996). The only way to confirm the identifications is to persevere with FISH, and sequence the rDNA of the bacterial strains isolated from the biofilm.

Microscopy has also shown that unicellular filamentous bacteria resembling Cytophagaceae were abundantly present in the biofilm (morphotypes 3 and 4). In this case only, the three approaches concur : Cytophagaceae are most probably present in the biofilm.

6. MINERALOGY OF THE IRON MINERAL OF *M. FERRUGINOSA*

X-ray diffraction and energy-dispersive X-ray (EDAX) analyses have indicated that the mineral is amorphous and rich in iron, phosphorus and calcium with traces of silicon and magnesium (Gillan & De Ridder 1995, 1997). The mineral forms granules whose diameter ranges from 0.05 to 1 m (in the colloidal range). Infrared absorption spectra of the mineral were obtained by applying the KBr disc method. Phosphate absorption bands due to P-O stretching vibrations were observed at 1,100 and 1,020 cm^{-1} (Gillan and De Ridder 2001). This result, as well as the similarity of spectra with other well known (bio)minerals, suggests that the mineral is an amorphous iron oxyhydroxide gel with phosphate sorbed on its surface rather than a pure ferric phosphate. Iron oxyhydroxide gels are minerals that may form abiotically in seawater (Byrne and Kester 1976). However, transmission electron microscopy observations have suggested that the mineral is not the result of a simple microbial bioaccumulation but the result of an in situ microbial biomineralization process (Gillan and De Ridder 2001).

7. EXPERIMENTS WITH IRON

The aim of these experiments was to study two possible microbial iron precipitation pathways in the biofilm, namely microbial iron oxidation and microbial degradation of organic Fe(III) complexes (Gillan et al. 2000). The iron-oxidizing activity was assayed spectrophotometrically by following the formation of the dye Wurster blue in biofilm extracts. An iron-oxidizing activity was effectively detected in extracts obtained by oxalic acid treatment of biofilm fragments. On the contrary, extracts obtained without oxalic acid treatment, heated extracts or extracts supplemented with $HgCl_2$ did not show any activity. This suggests that an iron-oxidizing factor (IOF), possibly an enzyme, was co-precipitated with the iron mineral (oxalic acid dissolves iron precipitates). Additional information gathered using SDS-PAGE, gel filtration chromatography and UV spectrophotometry indicate that the IOF would be a small peptide or glycopeptide (MW 1,350 D).

Microbial degradation of organic Fe(III) complexes was assayed with living biofilm fragments incubated in a media containing ferric citrate (1 mM). Analysis of the supernatants at various intervals of time using atomic absorption spectrometry revealed that the iron complex was degraded by living microorganisms much faster than in the heat-killed negative controls. We conclude that ferric iron precipitation in the biofilm, in addition to other processes, may proceed via microbial Fe(II) oxidation as well as microbial degradation of organic Fe(III) complexes.

8. CONCLUSIONS

It may first be concluded that the type of species detected in a Recent microbial community is highly dependent on the investigation method. FISH analysis is absolutely necessary to confront results of different approaches but this type of analysis is not always successful.

Secondly, it was shown that various pathways of microbial iron deposition may exist in the same microbial community. The simple observation of Fe(III) encrusted microbes cannot be taken as diagnostic for any microbial iron deposition process.

Understanding Recent iron-encrusted microbial communities can influence ideas about the environmental interpretation of fossilized iron-encrusted microbial communities that abound in the sedimentary rock record. For example it was concluded that the biofilm of *M. ferruginosa* could be a modern analog of the fossilized iron-encrusted microbial communities found in the Bajocian stratotype (Mid-Jurassic, northern Normandy, France) (Préat et al. 2000).

Many new iron-encrusted biofilms undoubtedly await description in the marine environment. For instance the epibiotic biofilm of *Urothoe poseidonis,* a symbiotic amphipod living in the same burrow as *M. ferruginosa,* and the biofilm of *Hydrobia ulvae,* a gastropod living at the surface of the sediments (Gillan & Cadée 2000). As the diversity of microbial consortia and populations is usually very high, the description of these systems may represent one of the major challenges of the future. The understanding of the ways that microbes interact with each other and with their mineral-rich environment is only now beginning to emerge.

REFERENCES

Baross, J.A. and Deming, J.W. (1985) The role of bacteria in the ecology of black-smoker environments, *Biol Soc Wash Bull* 6, 355-371.

Burnett, B.R. and Nealson, K.H. (1981) Organic films and microorganisms associated with manganese nodules, *Deep-Sea Res* 28, 637-645.

Byrne, R.H. and Kester, D.R. (1976) Solubility of hydrous ferric oxide and iron speciation in seawater, *Mar Chem* 4, 255-274.

Cowen, J.P. and Silver, M.W. (1984) The association of iron and manganese with bacteria on marine macroparticulate material, *Science* 224, 1340-1342.

Dul'tseva, N.M., Dubinina, G.A., and Lysenko, A.M. (1996) Isolation of marine filamentous sulfur bacteria and description of the new species *Leucothrix thiophila* sp nov., *Microbiology* 65, 79-87.

Gage, J. (1966) Observations on the bivalves *Montacuta substriata* and *M. ferruginosa,* "commensals" with spatangoids, *J mar biol Ass UK* 46, 49-70.

Ghiorse, W.C. (1984) Biology of iron- and manganese-depositing bacteria, *Ann Rev Microbiol* 38, 515-550.

Gillan, D.C. and Cadée, G.C. (2000) Iron-encrusted diatoms and bacteria epibiotic on *Hydrobia ulvae* (Gastropoda: Prosobranchia), *J Sea Res* 43, 83-91.

Gillan, D.C. and De Ridder, C. (1995) The microbial community associated with *Montacuta ferruginosa*, a commensal bivalve of the echinoid *Echinocardium cordatum*, in R.H. Emson, A.B. Smith, and A.C. Campbell (eds.), *Echinoderm Research 1995,* Balkema, Rotterdam, pp. 71-76.

Gillan, D.C. and De Ridder, C. (1997) Morphology of a ferric iron-encrusted biofilm forming on the shell of a burrowing bivalve (Mollusca), *Aquat Microb Ecol* 12, 1-10.

Gillan, D.C., Speksnijder, A.G.C.L., Zwart, G., and De Ridder, C. (1998) Genetic diversity of the biofilm covering *Montacuta ferruginosa* (Mollusca, Bivalvia) as evaluated by denaturing gradient gel electrophoresis analysis and cloning of PCR-amplified gene fragments coding for 16S rRNA, *Appl Environ Microbiol* 64, 3464-3472.

Gillan, D.C., Warnau, M., De Vrind-de Jong, E.W., Boulvain, F., Préat, A., and De Ridder, C. (2000) Iron oxidation and deposition in the biofilm covering *Montacuta ferruginosa* (Mollusca, Bivalvia), *Geomicrobiol J* 17, 141-150.

Gillan, D.C. and De Ridder, C. (2001) Accumulation of a ferric mineral in the biofilm of *Montacuta ferruginosa* (Mollusca, Bivalvia). Biomineralization, bioaccumulation, and inference of paleoenvironments, *Chem Geol* 177, 371-379.

Jannasch, H;W. and Wirsen, C.O. (1981) Morphological survey of microbial mats near deep-sea thermal vents, *Appl Environ Microbiol* 41, 528-538.

Juniper, S.K. and Tebo, B.M. (1995) Microbe-metal interactions and mineral deposition at hydrothermal vents, in D.M. Karl (ed.), *The microbiology of deep-sea hydrothermal vents,* CRC Press, New York, pp. 219-253.

Karl, D.M., Brittain, A.M., and Tilbrook, B.D. (1989) Hydrothermal and microbial processes at Loihi Seamount, a mid-plate hot-spot volcano, *Deep-Sea Res* 36, 1655-1673.

Konhauser, K.O. (1998) Diversity of bacterial iron mineralization, *Earth-Science Rev* 43, 91-121.

Moyer, C.L., Dobbs, F.C., and Karl, D.M. (1995) Phylogenetic diversity of the bacterial community from a microbial mat at an active, hydrothermal vent system, Loihi Seamount, Hawaii, *Appl Environ Microbiol* 61, 1555-1562.

Ouverney, C.C. and Fuhrmann, J.A. (1997) Increase in fluorescence intensity of 16S rRNA in situ hybridization in natural samples treated with chloramphenicol, *Appl Environ Microbiol* 63, 2735-2740.

Préat, A., Mamet, B., De Ridder, C., Boulvain, F., and Gillan, D. (2000) Iron bacterial and fungal mats, Bajocian stratotype (Mid-Jurassic, northern Normandy, France), *Sediment Geol* 137, 107-126.

Reitner, J., Thiel, V., Zankl, H., Michaelis, W., Wörheide, G., and Gautret, P. (2000) Organic and biogeochemical patterns in cryptic microbialites, in R.E. Riding and S.M. Awramik (eds.), *Microbial sediments,* Springer-Verlag, Berlin, pp. 149-160.

Suzuki, M.T., Rappé, M.S., Haimberger, Z.W., Winfield, H., Adair, N., Ströbel, J., and Giovannoni, S.J. (1997) Bacterial diversity among small-subunit rRNA gene clones and cellular isolates from the same seawater sample, *Appl Environ Microbiol* 63, 983-989.

Chapter 16

MICROBIAL ENDOLITHS AS INTERNAL BIOFILMS

Stjepko GOLUBIC
Department of Biology, Boston University, 5 Cummington St. Boston, MA 02215, USA golubic@bio.bu.edu

Jürgen SCHNEIDER
Geowissenschaftliches Zentrum der Universitaet Goettingen, Goldschmidtstr. 3, D-37077 Göttingen, Germany

INTRODUCTION

Microbial biofilms could be defined as living organic interfaces between inorganic physical states of matter. They act as arbitrators in interactions that occur at the interfaces between the lithosphere on one side and hydrosphere and atmosphere on the other. Some dissolve the substrate, and penetrate its interior. Microbial occupants of the lithobiontic ecological niches thus include both the exterior (epiliths) and interior (endoliths) of solid substrate surfaces. Euendoliths actively penetrate calcareous substrates and thereby create the endolithic microenvironment, whereas crypto- and chasmoendoliths take advantage of the existing internal microenvironments which they modify (Golubic et al. 1981, Schneider & Le Campion-Alsumard 1999). Microbial endoliths cope successfully with ecological conditions ranging from extremes of polar and alpine deserts to the most optimized environments in tropical reefs (Jaag 1945, Chazotte et al. 1995, Büdel 1999, Pentecost & Whitton, 2000, Vincent 2000, Wynn-Williams 2000). Their colonization involves pioneers of newly exposed substrates (Le Campion-Alsumard 1975), and successions of changing populations toward establishment of stratified microbial ecosystems. The distribution of endolithic biofilms is global, but the ecological settings under which they occur are extremely diverse and so are the assemblages of microbial endoliths that occupy them. Their constituents differ significantly from one another in appearance, complexity, taxonomic composition, in metabolic and ecological properties, as well as in the timing of their activities. Within the euphotic zone, autochthonous primary producers, mostly cyanobacteria and algae, usually dominate endolithic communities. Endolithic communities, which extend into the aphotic environment are composed of heterotrophs which depend on organic matter in the substrate they penetrate, such as skeletons and bioclasts (Golubic et al. 1975). Older literature is reviewed by Radtke et al. (1997).

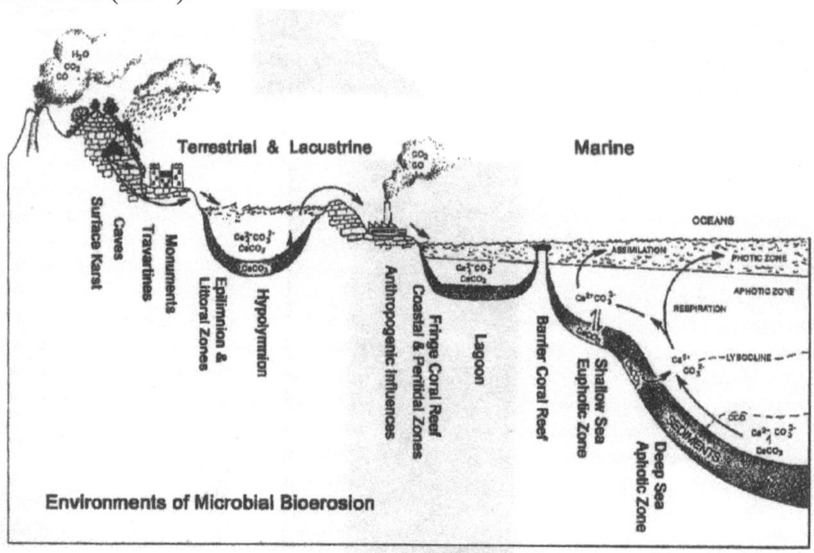

Figure 1. An idealized profile from high mountain tops to oceanic depths illustrating diversity of environmental settings occupied by endolithic biofilms as integrated into the global carbon cycling. Anthropogenic influences are added as possible modifying factors to natural processes (after Golubic et al. 1979).

Boring behavior of microbial endoliths is often specific, and the shape of the resulting borings conforms closely to the shape of the organism. Thus, many prokaryotic and eukaryotic microbial endoliths leave in the substrate recognizable and fossilizable traces. These remains contain information about paleoenvironments they once occupied. For this reason, microborings of modern and ancient endoliths are analyzed, compared, named and classified separately from the organisms that made them - as trace fossils (Vogel et al. 1999, Radtke 1991). If recognized in the fossil record, the assemblages of traces produced by phototrophic endoliths characterize sediments deposited within the photic zone. Phototrophic components are lacking in non-illuminated substrates deposited, for example, in the deep sea.

Endolithic communities composed of different microorganisms are important agents participating in the process of bioerosion in a variety of environments. As a group, they were able to overcome vast ecological differences in water supply, temperature, salinity, nutrient and energy supply. In this short review we present selected case histories illustrating the broad spectrum of endolithic communities as internal biofilms, from the high mountain tops to the depths of the ocean (Figure 1).

Exposed rock surfaces emerging under the retreating glaciers are among the most challenging frontiers for the expanding biosphere. Lichens are in the forefront of the occupation of this habitat. The process is exceedingly slow as it takes at least 10 to 15 years for an endolithic thallus to develop and about ten times longer for these rocks to become completely overgrown and colonized by euendolithic lichens, depending on the roughness and internal fabric of the rock (Pohl 2000). In contrast, freshwater and marine intertidal and subtidal biofilms develop rapidly within hours and the first endoliths penetrate the rock within days or weeks. The invasion of bare rocks by lichens starts usually with fungal penetration, whereas the cyanobacterial and algal settlers are initially free. The long-term survival of algae on drier habitats, however, depends on their association with resident fungi. The hyphae of endolithic lichens leave the rock surface largely intact. Only the perithecia are exposed to the rock surface. They penetrate several 100 μm deep into the substrate where they form a distinct layer. The phycobiontic layer of symbiotic algae develops later above the mycobiontic, hyphal layer (Figure 2).

Fully developed endolithic lichen is a stratified biofilm comprised of several distinct layers. This colonization strategy minimizes the erosion of the surface and leaves a protective layer for the lichen. A convenient, similarly protected endolithic environment is available in naturally porous rocks such as sandstone. Such cryptoendolithic habitats are populated by coccoid cyanobacteria dominated by the genus *Chroococcidiopsis* (Grilli-Caiola et al. 1996). These populations may become lichenized in dryer exposures and were able to populate sandstone in hot and cold deserts, from Negev and Sahara to Dry Valleys in the Antarctica (Friedmann & Ocampo 1976, Friedmann 1982). Similar assemblages known as chasmoendoliths inhabit fissures in the rocks, where the organisms are also protected from excessive radiation and water is retained by capillary action.

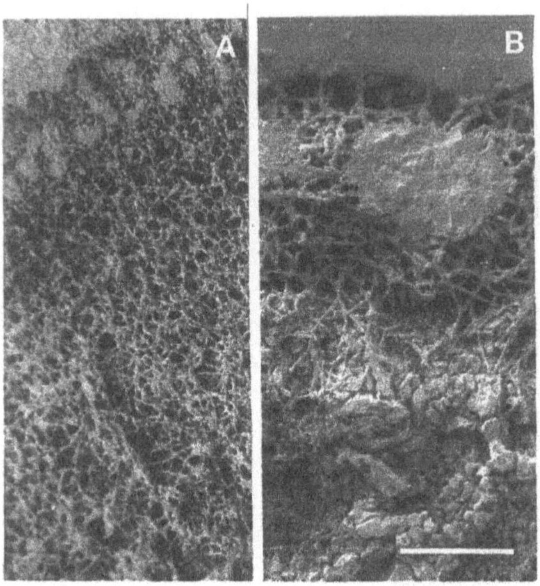

Figure 2. Section through endolithic lichen on exposed limestone cliff in Austrian Alps. A. Resin cast of hyphal borings penetrating a ca 1.5 mm into the rock. The smooth area upper left is the rock surface, followed by a 100 μm thick phycobiontic layer and a deeper mycobiontic layer. B. Detail of the phycobiontic layer. Scale bar is 50 μm for A and 200 μm for B.

Free cyanobacteria and algae settle as epiliths on rock surfaces, which are sporadically wetted by drainage of melting snow or rain. They form dark vertical lines on cliffs called "ink-streaks" (Tintenstriche; Jaag 1945). Epilithic and endolithic, free and lichenized cyanobacteria in this range are protected from excessive light and UV irradiation by extracellular mycosporine-like pigments scytonemin and gloeocapsin, which are embedded into their gelatinous envelopes and sheaths (Büdel et al. 1997, Garcia-Pichel 1998). Their existence is facilitated by the availability of water, and their diversity increases with increased wetting frequency and shade (e.g. cave entrances). A development of secondary endolithic habitats may be induced by carbonate precipitation, which results in porous rocks called travertine and tufa. Larger accumulations of carbonate deposits modify water flow in creeks and rivers into alternating pools and rapids, which become populated by different biofilm assemblages. Epilithic and endolithic lichens as well as free cyanobacteria and algae comprise external and internal biofilms over vast areas of the continent's rocky surfaces. Together with microbial desert crusts, these communities may have dominated the lands for billions of years during the Precambrian times, and contributed to bio-geo-chemical cycling of minerals and formation of mature soils long before the evolution of higher plants and animals (Campbell 1980, Garcia-Pichel 1998, Garcia-Pichel & Belnap 1996).

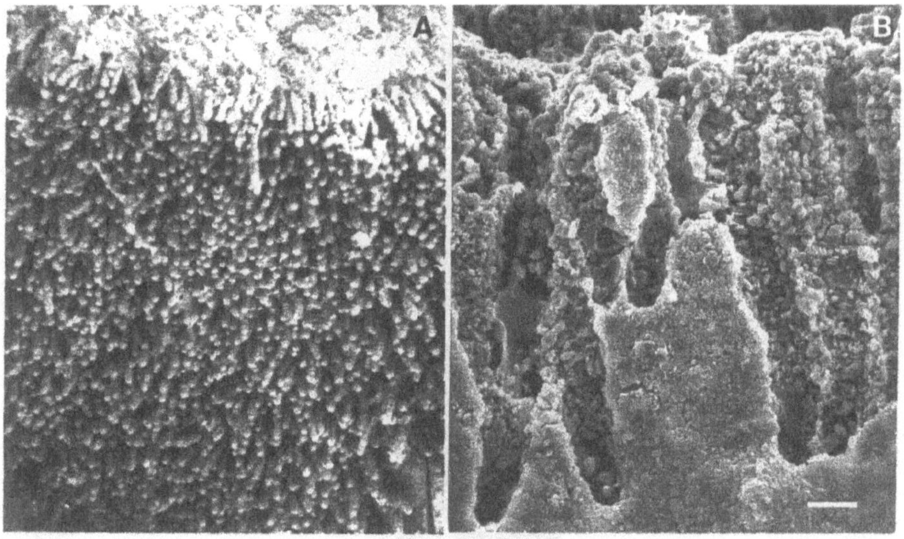

Figure 3. Microbial endolithic biofilms in the upper intertidal ranges. A. Resin casts of borings of dense population of the cyanobacterium *Kyrtuthrix dalmatica* inside limestone of the Mediterranean rocky coast. B. Fractured rock with empty boreholes of *Hyella caespitosa*. Micrite around the limestone grains are etched within the boreholes. Scale bar is 100 μm for A and 10 μm for B.

On the margins of every pool, in the littoral zone of every lake and along the sea coasts, microbial assemblages form external and internal biofilms which are arranged in characteristic color zones in accordance with different wetting frequencies by wave spray and changing water level (Le Campion-Alsumard 1966, Schneider 1976). In the littoral of alpine lakes, cyanobacteria-dominated biofilms are encrusted with calcium carbonate precipitates forming microreefs (Schneider et al. 1983). The micro-environment beneath the crust is characterized by a prevalence of corrosive processes, including activities of specialized microbial endoliths. Some of them (e.g. *Schizothrix lacustris*) are both endo- and epilithic as they span across the rock-water interface. Carbonate is dissolved at the lower part and precipitates at the upper end of the same organism (see figure 6, p. 421 in Schneider & Le Campion-Alsumard 1999). Similar mechanisms were described in marine coastal settings (Kobluk & Risk 1977) where microbial endoliths bore and remove carbonate, but the precipitates do not form as readily as in freshwater.

The intertidal ranges of rocky limestone sea shores offer the most dramatic examples of coastal bio-erosion. The process involves complex interactions between biological corrosion by microbial endoliths and biological abrasion carried out by their grazers (Schneider 1976). Dense populations of various endolithic cyanobacteria form internal biofilms (Figure 3). Population densities of 150 to 500 thousand individual borings/cm^2 are common, removing up to 50% of the rock surface (Schneider 1976).

Depth of penetration into the rock, of phototrophic microbial endoliths is light-limited. However, their growth and primary production output supports a variety of grazing animals, equipped with hard, rock-scraping mouthparts. Gastropods, placophors, echinoderms and scarid fish that feed on endolithic microorganisms, remove the uppermost layers of the rock, thereby displacing the illuminated layer and cyanobacterial boring deeper into the rock (Figure 4).

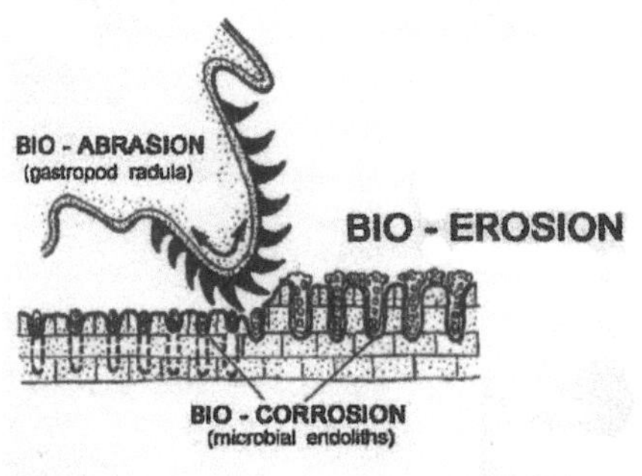

Figure 4. Mechanisms of coastal bioerosion

Combined activities of microbial bio-corrosion and bio-abrasion by grazers lead to formation of bio-erosional notches causing progressive destruction of limestone coasts (Torunski 1979, Schneider & Torunski 1982, Radtke et al. 1996). As a consequence of settling preferences, endurance and competitiveness of endolithic cyanobacteria, as well as of ecological requirements of grazing animals, coastal rocks are biodegraded selectively, resulting in peculiar geomorphologic features of bio-karst which include elaborate sharp-edged rock forms alternating with flat-bottom rock pools (Figure 5).

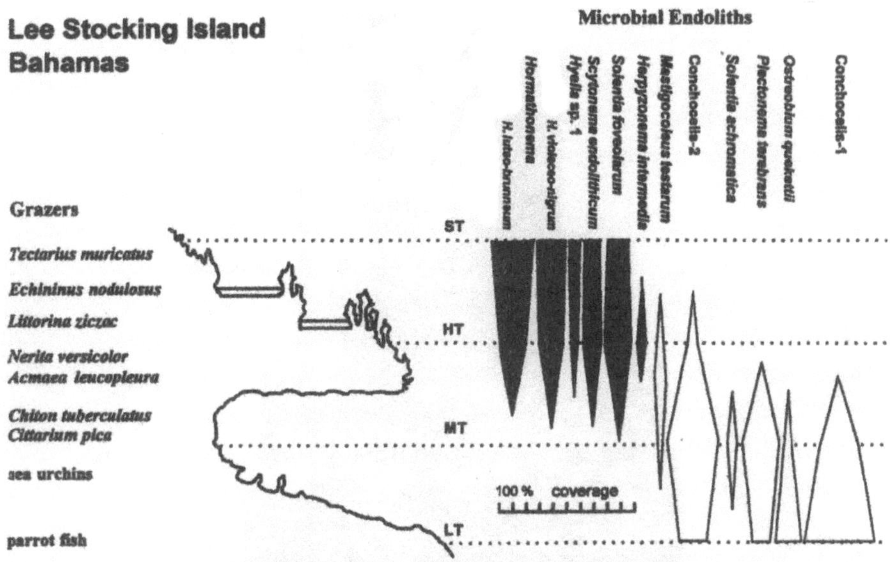

Figure 5. Distribution of dominant microbial endoliths (above) and their grazers (left) along a coastal profile with well developed bio-karst rock pools and a deep bio-erosional notch near Lee Stocking Island, Bahamas. ST, HT, MT and LT designate supra-, high, mean and low tidal levels (from Radtke et al. 1996).

These processes of mutual positive re-enforcement result in modifications of microbial microenvironments by providing sheltered niches with improved water retention, which foster microbial growth and, in turn, attract intensified bioerosion by grazers. Fragmentation of the rock due to removal of layers of carbonate substrate by grazers of microbial endoliths, together with the activities of macroscopic endoliths, such as clionid sponges contribute jointly to generation and accumulation of fine grained sediment in shallow marine environments (Schneider & Torunski 1982, Schneider & Le Campion-Alsumard 1999).

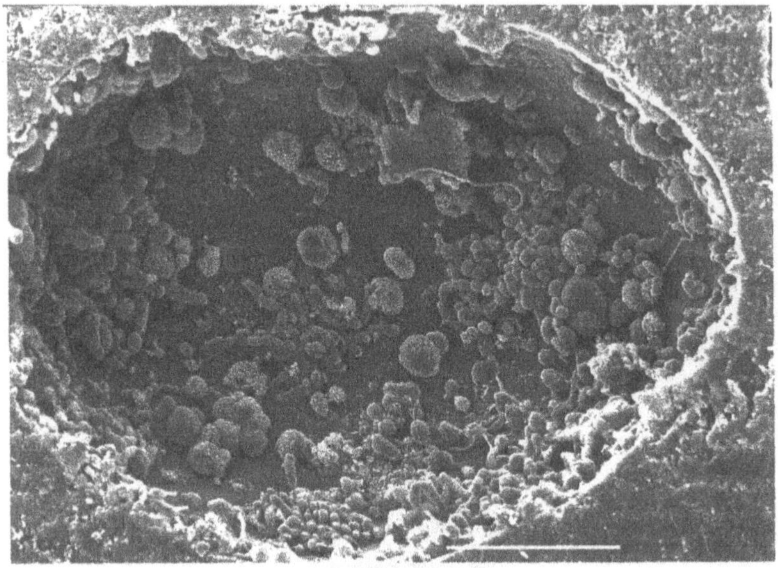

Figure 6. Assemblage of microborings in a single ooid sand grain from agitated shallow marine environment of the Bahama carbonate platform. The pattern is characteristic of the activity of phototrophic cyanobacterial endoliths. SEM of resin-casts boreholes, scale bar is 100 μm long.

Microbial endoliths are among few organisms able to settle on soft, shifting grounds. Entire assemblages of endoliths are often housed in a single sand grain (Figure 6). A specialized assemblage of phototrophic endoliths in ooid sands of the Arabian/Persian Gulf and the Bahama Carbonate Platform includes *Hyella gigas, H. salutans, H. reptans, H. racemus* and others (Al-Thukair & Golubic 1991, Al-Thukair et al. 1994, Gektidis & Golubic 1996, Gektidis 1997). A similar assemblage is found in bioclasts, but its composition changes with the stabilization of the sediment approaching the composition of carbonate hard grounds.

Ecological conditions along marine coasts are similar to those on alpine cliffs and lake littoral with respect to water supply limitations, but are additionally complicated by fluctuations in salinity. Isolated rock-pools experience evaporation and rapid increases in salinity up to a stage of crystallization of sea salts. As in the lacustrine settings, the epilithic and endolithic biofilms are dominated by cyanobacteria, but the species composition appears to be different. Cyanobacterial assemblages are arranged in zones in response to average wetting frequency, modified by microtopography of the limestone surface. In the top zone, which is only sporadically wetted by rain and wave spray, the rocks are coated with epilithic cyanobacterial biofilms. The rock beneath them is penetrated and occupied by an internal, endolithic biofilm composed of different cyanobacteria, including *Hormathonema luteo-brunneum*, *H. violaceo-nigrum*, *Scytonema endolithicum* and *Hyella balani* (Le Campion-Alsumard 1979, Le Campion-Alsumard & Golubic 1985). Species of *Hyella* are usually the pioneers that first occupy newly exposed rocks, however, on habitats with extreme water shortage *Hyella* is lichenized. The association with fungal hyphae provides additional protection to the phycobiont as evident by avoidance of marine lichens by grazers. Lower in the supratidal zone and deeper in the rock penetrate branched networks of *Lithonema adriaticum,* a species which is joined in tropical settings by the branched heterocystous cyanobacterium *Herpyzonema intermedium.* Contiguous carpets of short, parallel stubs of *Solentia paulocellulare* dominate small depressions on well-drained limestone rocks, while steel-blue colored endolithic covers of different *Solentia* species, mostly *S. foveolarum* and *S. achromatica*, line the floors of rock pools. The penetrating apical cells of these coccoid cyanobacteria are positioned at the end of a layered gelatinous stalk (Le Campion-Alsumard et al. 1996). The lower intertidal ranges are dominated by heterocystous cyanobacteria *Kyrtuthrix dalmatica* and *Mastigocoleus testarum*. The latter extends its distribution into subtidal ranges where it competes with other species of *Hyella* and with eukaryotic endoliths.

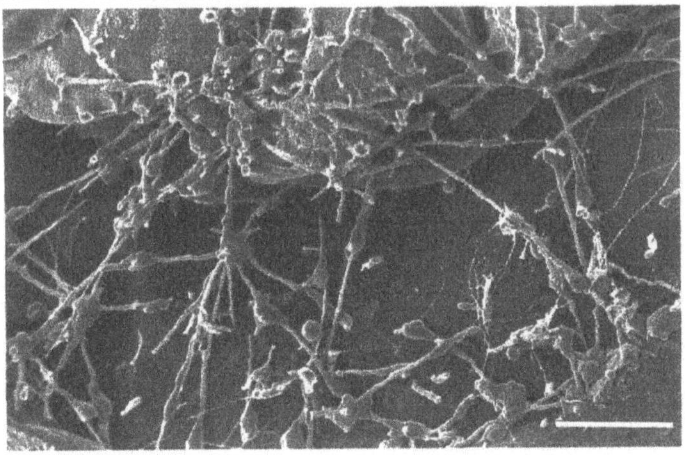

Figure 7. Assemblage of microborings from the deep sea aphotic zone, produced by heterotrophic microbial endoliths in a test of the foraminifer *Laticarinina.*. Scale bar is 100 μm long.

zonal distribution of subtidal taxa follows the gradient of light attenuation. Polyphotic chlorophytes *Phaeophila dendroides* and *P. tenuior* are replaced in greater depths and in shaded microenvironments by the oligophotic Conchocelis-stages of bangiacean rhodophytes and by the siphonal chlorophyte *Ostreobium quekettii.* Algal light harvesting dominates carbonate substrates down to a depth of 300 m in clear Caribbean seas (G. Radtke, personal communication, 1991). The same organism penetrates skeletons of virtually every coral (Lukas 1974, Le Campion-Alsumard et al. 1995b). The ubiquitous *Ostreobium* in coral skeletons is accompanied and attacked by endolithic ascomycotic fungi which also attack the polyps. The latter respond by producing pearl-like deposits to plug up fungal borings (Le Campion-Alsumard et al. 1995a). Thus, each coral is a dynamic system of at least 4 separate biological components, two of which are endolithic. These specialized endoliths, present in coral reefs of world's oceans may become a critical factor in coral health when their hosts are under stress from pollution and global warming (Bentis et al. 2000).

A different, heterotrophic community of microbial endoliths exploits organic compounds invested in shells and skeletal bioclasts. Heterotrophic microbial endoliths are biologically least explored. They are light-independent and, thus extend from the shallows into the abyssal depths. Their biological affinity has been attributed to fungi and lower fungi, specifically chytrids and thoustochytrids, but these assignments remain speculative. Their diversity was assessed by the shapes of their microborings, which are often characteristic (Figure 7). Carbonate skeletons of planktic as well as benthic organisms provide organic rich substrates targeted by heterotrophic endoliths which leave a benthic overprint in the fossil record (Golubic et al. 1984).

Endolithic biofilms are a geologically ancient ecological niche. The oldest fossil endolith was identified in silicified stromatolitic crusts of the 1,500-1,700 Ma old Dahong Yu Formation of China (Zhang & Golubic 1987). By its orientation and cell organization, it was comparable with modern endoliths of the genus *Hyella*. By the time of the Neoproterozoic, some 800 Ma ago, the ooid assemblages of several *Eohyella* species were found to be comparable with modern ones observed in the ooids of the Bahama Bank and Arabian Gulf (Knoll et al. 1986) Organically preserved body fossils of endolithic Conchocelis-stages of bangiacean rhodophytes were found in sediments 420 Ma old (Kazmierczak & Golubic 1976, Campbell 1980). Microborings are common throughout the Phanerozoic (Glaub et al. 1999). They are preserved as natural casts when filled with insoluble materials. However, a significant proportion of microborings remained preserved intact. These can be cast in polymerizing resin, studied by SEM and compared with modern endolithic traces. The fossil evidence supports an early occupation and diversification of various terrestrial, freshwater and marine endolithic niches, which may have well been established by Mesoproterozoic times, and achieved their modern diversity and composition before the dawn of Phanerozoic. Thus the value of most fossil microborings are expected to be in paleoecological rather than stratigraphic applications. A successful distinction between microborings originating from phototrophic vs. heterotrophic endoliths provides valuable paleobathymetric indication. Studies on distribution of modern endoliths and experimental exposure of carbonate substrates to microbial boring (Vogel et al. 2000) relate morphological characteristics of borings to biological properties of their makers. An ancient euphotic zone can thus be distinguished from deeper sedimentary settings on the basis of characteristic assemblages of microborings (Vogel et al. 1995, 1999, Glaub & Bundschuh 1997).

CONCLUSIONS

1. Endolithic biofilms have global distribution. They are comprised of specialized microbial assemblages which thrive and often dominate in various terrestrial and aquatic environments under conditions ranging from extreme to optimal.
2. The success of the endolithic mode of life is often granted by symbiotic integration and mutual support of consortial members as illustrated by lichens as the first colonizers of newly exposed terrestrial surfaces. Under more optimal conditions in tropical seas the consortial members in corals (polyps, zooxanthellae, endolithic algae and fungi) live in a state of precarious and often disturbed balance of powers. Coastal bioerosion is a consequence of integration of microbial endoliths and their grazers at the ecosystem level. Together they form a destructive force of geological dimensions.
3. Whereas each of the constituents of global microbial biofilms brings its own ecological requirements and activities, the environmental consequences can only be understood when their integrated impact is considered.
4. Distribution of microbial biofilms follows gradients of critical environmental determinants, such as water supply and light. They form

distinctive zones, which are compressed across sharp environmental boundaries, e.g. between terrestrial and aquatic realms and extended within each of these realms.

5. Microbial bioerosive activity has a long geologic history, which left a rich fossil record of boreholes and microboring patterns, which speak of ancient integrations and reveal the nature of paleoenvironmental conditions under which they lived.

REFERENCES:

Al–Thukair, A.A. & Golubic, S. (1991): Five new *Hyella* species from the Arabian Gulf. - Algological studies, 64, 167-197.

Al-Thukair, A.A., Golubic, S. & Rosen, G. (1994): New euendolithic cyanobacteria from the Bahama Bank and the Arabian Gulf: *Hyella racemus,* sp. nov..- J. Phycol., 30, 764-769.

Bentis, C.J., Kaufman, L. & Golubic, S. (2000): Endolithic fungi in reef-building corals (Order: Scleractinia) are common, cosmopolitan, and potentially pathogenic. - Biological Bulletin, 198, 254-260.

Büdel, B. (1999): Ecology and diversity of rock-inhabiting cyanobacteria in tropical regions. - Eur. J. Phycol., 34, 361-370.

Büdel, B., Karsten U. & Garcia-Pichel, F. (1997): Ultraviolet-absorbing scytonemin and mycosporine-like amino acid derivatives in exposed, rock-inhabiting cyanobacterial lichens. - Oecologia, 112, 165-172.

Campbell, S.E. (1980): *Paleoconchocelis starmachii*, a carbonate boring microfossil from the Upper Silurian of Poland (425 million years old): implications for the evolution of the Bangiaceae (Rhodophyta).- Phycologia, 19, 25-36.

Chazottes, V., Le Campion-Alsumard, T. & Peyrot-Clausade, M. (1995): Bioerosion rates of coral reefs: interaction between macroborers, microborers and grazers (Moorea, French Polynesia). - Palaeogeogr. Palaeoclimatol. Palaeoecol., 113, 189-198.

Friedmann, E.I. (1982): Endolithic microorganisms and the Antarctic cold desert.- Science, 215, 1045-1053.

Friedmann, E.I. & Ocampo, R. (1976): Endolithic blue-green algae in the dry valleys: Primary producers in the Antarctic desert ecosystem. - Science, 193, 1247-1249.

Garcia-Pichel, F. (1998): Solar ultraviolet and the evolutionary history of cyanobacteria. - Origins of Life and Evolution of the Biosphere, 28, 321-347.

Garcia-Pichel, F. & Belnap, J. (1996): Microenvironments and microscale productivity of cyanobacterial desert crusts. - J. Phycol. 32, 774-782.

Gektidis, M. (1997): Microborings in Bahamian ooids. - Cour. Forsch.-Inst. Senckenberg, 201, 109-121.

Gektidis, M. & Golubic, S. (1996): A new endolithic cyanobacterium: *Hyella vacans* sp. nov. from Lee Stocking Island, Bahamas. - Nova Hedwigia, 112, 93-100.

Glaub, I., Balog, S.-J., Bundschuh, M., Gektidis, M., Hofmann, K., Radtke, G., Schmidt, H. & Vogel, K. (1999): Euendolithic cyanobacteria/cyanophyta and their traces in earth history. - in: Charpy, L. & Larkum, A.W.D. (eds.): Marine cyanobacteria. - Bull. Inst. Océanogr. Monaco, no. Spec. 19, 135-142.

Glaub, I. & Bundschuh, M. (1997): Comparative study on Silurian and Jurassic/Lower Cretacious microborings. - Cour. Forsch.-Inst. Senckenberg, 201, 123-135.

Golubic, S., Campbell, S.E., Drobne, K., Cameron, B., Balsam, W.L., Cimerman, F. & Dubois, L. (1984): Microbial endoliths: a benthic overprint in the sedimentary record, and a paleobathymetric cross-reference with foraminifera. - J. Paleontology, 58, 351-361.

Golubic, S., Friedmann, I. & Schneider, J. (1981): The lithobiontic ecological niche, with special reference to microorganisms. - J. Sediment. Petrol., 51, 475-478.

Golubic, S., Krumbein, W. & Schneider, J. (1979): The carbon cycle, in: Trudinger, P.A. & Swaine, D.J. (eds.): Biogeochemical cycling of mineral forming elements, Stud. Environ. Sci., 3, 29-45.

Golubic, S., Perkins, R.D. & Lukas, K.J. (1975): Boring microorganisms and microborings in carbonate substrates. - in: Frey, R.W. (ed.): The study of trace fossils, 229-259, Springer, Heidelberg-Berlin-New York.

Grilli-Caiola, M., Billi, D. & Friedmann, E.I. (1996): Effect of desiccation on envelopes of the cyanobacterium *Chroococcidiopsis* sp. (Chroococcales). Eur. J. Phycol., 31, 97-105.

Jaag, O. (1945): Untersuchungen über die Vegetation und Biologie der Algen des nackten Gesteins in den Alpen, im Jura und im schweizerischen Mittelland. - Beitr. Kryptogamenfl. Schweiz, 9, 560 p.

Kazmierczak, J. & Golubic, S. (1976). Oldest organic remains of boring algae from Polish Upper Silurian. - Nature, London, 261, 404-406.

Knoll, A.H., Golubic, S., Green, J. and Swett, K. (1986): Organically preserved microbial endoliths from the Late Proterozoic of East Greenland. - Nature, London, 321, 856-857.

Kobluk, D.R. & Risk, M.J. (1977): Calcification of exposed filaments of endolithic algae, micrite envelope formation and sediment production. - J. Sedim. Petrol., 47, 517-528.

Le Campion-Alsumard, T. (1966): Contribution à l'étude des cyanophycées lithophytes des étages supralittoral et mediolittoral (région de Marseille). - Téthys, 1, 119-172, Marseille.

Le Campion-Alsumard, T. (1975): Étude expérimentale de la colonization d'éclats de calcite par les cyanophycées endolithes marines. - Cah. Biol. Mar., 16, 177-185, Roscoff.

Le Campion-Alsumard, T. (1979): Les cyanophycées endolithes marines: Systématique, ultrastructure, écologie et biodestruction.- Oceanologica Acta, 2, 143-156, Paris.

Le Campion-Alsumard, T. & Golubic, S. (1985): *Hyella caespitosa* Bornet & Flahault and *Hyella balani* Lehmann (Pleurocapsales, Cyanophyta): a comparative study. - Algological Studies, 38/39, 119-148.

Le Campion-Alsumard, T., Golubic, S. & Pantazidou, A. 1996. On the euendolithic genus *Solentia* Ercegovic (Cyanophyta/Cyanobacteria). - Algological Studies, 83: 303-329.

Le Campion-Alsumard, T., Golubic, S. & Priess, K. (1995a): Fungi in corals: symbiosis or disease? Interaction between polyps and fungi causes pearl-like skeleton biomineralization. - Mar. Ecol. Prog. Ser., 117, 137-147.

Le Campion-Alsumard, T., Golubic, S. & Hutchings, P. 1995b. Microbial endoliths in skeletons of live and dead corals: *Porites lobata* (Moorea, French Polynesia). - Mar. Ecol. Prog. Ser., 117, 149-157.

Lukas, K. J. (1974): Two species of the chlorphyte *Ostreobium* from skeletons of Atlantic and Caribean reef corals.- J. Phycol., 10, 331-335.

Pentecost, A. & Whitton, B. A. (2000): Limestones – in: Whitton, B. A. & Potts, M. (eds.): The ecology of cyanobacteria – their diversity in time and space.- Kluver Acad. Publ., 257-279.

Pohl, W. (2000): Wechselwirkungen zwischen endolithischen Biofilmen und Karbonatgesteinen in alpinen Gebieten Mitteleuropas. - Ph.D. - dissertation, University of Göttingen. URL: http://webdoc.gwdg.de/diss/ 2000/pohl/.

Radtke, G. (1991): Die mikroendolithischen Spurenfossilien im Alttertiär Westeuropas und ihre paläoökologische Bedeutung. - Cour. Forsch.-Inst. Senckenberg, 138, 307-340.

Radtke, G., Hofmann, K. & Golubic, S. (1997): A bibliographic overview of micro- and macroscopic bioerosion. - Cour. Forsch.-Inst. Senckenberg, 201, 307-340.

Radtke, G., Le Campion-Alsumard, T. & Golubic, S. (1996): Microbial assemblages of the bioerosional "notch" along tropical limestone coasts. - Algological studies, 83, 469-482.

Schneider, J. (1976): Biological and inorganic factors in the destruction of limestone coasts. - Contrib. Sedimentol., 6, 1-112 .

Schneider, J. & Torunski, H. (1982): Biokarst on limestone coasts, morphogenesis and sediment production. - Mar. Ecol., 4, 45-63.

Schneider, J., Schröder, H.-G., & Le Campion-Alsumard, T. (1983): Algal micro-reefs: coated grains from freshwater environments. In: Peryt, T. M. (ed.), Coated Grains, 284-298, Springer, Berlin.

Schneider, J. & Le Campion-Alsumard, T. (1999): Construction and destruction of carbonates by marine and freshwater cyanobacteria. - Eur. J. Phycol., 34, 417-426.

Torunksi, H. (1979): Biological erosion and its significance for the morphogenesis of limestone coasts and for nearshore sedimentation (Northern Adriatic). - Senckenbergiana maritima, 11, 193-265.

Vincent, W. F. (2000): Cyanobacterial dominance in the polar region - in: Whitton, B. A. & Potts, M. (eds.): The ecology of cyanobacteria – their diversity in time and space. - Kluver Acad. Publ., 321-340.

Vogel, K., Bundschuh, M., Glaub, I., Hofmann, K., Radtke, G. & Schmidt, H. (1995): Hard substrate ichnocoenoses and their relations to light intensity and marine bathymetry. - N. Jb. Geol. Paläont., Abh., 195, (Seilacher Festschrift), 46-91.

Vogel, K., Balog, S.-J., Bundschuh, M., Gektidis, M., Glaub, I. Krutschinna, J. & Radtke, G. (1999): Bathymetrical studies in fossil reefs, with microendoliths as paleoecological indicators. - Profil (Stuttgart), 16, 181-191.

Vogel, K., Gektidis, M., Golubic, S., Kiene, W.E. & Radtke, G. (2000): Experimental studies on microbial bioerosion at Lee Stocking Island, Bahamas and One Tree Island, Great Barrier Reef, Australia: implications for paleoecological reconstructions. - Lethaia 33, 190-204.

Wynn-Williams, D. D. (2000): Cyanobacteria in deserts – life at the limit - in: Whitton, B. A. & Potts, M. (eds.): The ecology of cyanobacteria – their diversity in time and space. - Kluver Acad. Publ., 341-366.

Zhang Yun & Golubic, S. (1987): Endolithic microfossils (cyanophyta) from early Proterozoic stromatolites, Hebei, China. - Acta Micropal. Sinica, 4, 1-12.

Chapter 17

EPIBENTHIC CYANOBACTERIAL COMMUNITIES INTERACTING WITH SEDIMENTARY PROCESSES IN SILICICLASTIC DEPOSITIONAL SYSTEMS (PRESENT AND PAST)

Nora NOFFKE
Old Dominion University; Department of Ocean, Earth & Atmospheric Sciences; 4600 Elkhorn Avenue; Norfolk, VA 23529; USA; nnoffke@odu.edu

1. INTRODUCTION

Siliciclastic depositional systems like sandy tidal flats or shelf environments are predominantly governed by erosion, deposition and deformation (Pettijohn & Potter 1964). But also microbiological parameters influence the sedimentary dynamics, because the sediments are widely colonized by a variety of epibenthic microorganisms like cyanobacteria. Epibenthic microbes attach to the surfaces of depositional grains by adhesive mucous secretions that are known as 'extracellular polymeric substances (EPS)' (Decho 2000 and literature therein). Such organic coatings around grains are termed 'biofilms' (see Charaklis and Wilderer 1989, Decho 2000; Stolz 2000; also Paterson and Black 2000). At sites of favorable ecological conditions, biofilms continue to grow to form thick and significant organic layers that are known as 'microbial mats' (compare Krumbein 1983, also Gerdes & Krumbein 1987, Cohen et al. 1984, or Cohen & Rosenberg 1989, Neu 1994, Stolz 2000, Noffke in press b). Microbial mats can cover large areas of a sedimentary surface. Within shallow-marine depositional environments, the sedimentary dynamics controlls abundance and distribution of epibenthic microorganisms, and also the formation of biofilms and mats (Noffke et al. 2001a, b, Noffke et al. 2002).

From the geological point of view, cyanobacteria are an important goup (Cameron et al. 1985, Gerdes & Krumbein 1987, Knoll & Bauld 1989, Golubic & Knoll 1993, and literature therein, Gerdes et al. 2000b). Epibenthic species construct 'stromatolites', laminated sedimentary structures, that are known since the earliest Earth history until today (e.g. Walter 1976, Reid et al. 2000, see Krumbein 1983 for general terminology). Significant, reef-like stromatolites are characteristic for carbonate environments, where early mineral precipitation during degradation of organic matter by heterotrophs fascilitates early diagenetic cementation (Krumbein 1979b, Reid et al. 2000). In siliciclastic environments, which are focus of this paper, such updomed 'chemical' stromatolites do not occur. In such physically governed environments, the epibenthic cyanobacterial assemblages interact syndepositionally with the sedimentary dynamics like e.g. erosion. This biotic-physical interaction forms characteristic 'Microbially Induced Sedimentary Structures (MISS)' that can be observed both in Recent and fossil (for overview see Gerdes et al. 1993, 1994, 2000b, Noffke et al. 2001a,b, 2003, Noffke in press a). MISS differ significantly from common stromatolites. Very common are 'wrinkle structures', fossilized microbial mats, that are widely distributed in consolidated rocks (Noffke 2000, 2002, Noffke et al. 2002, definition of wrinkle structures by Hagadorn & Bottjer 1997).

The purpose of this paper is to give an overview on strategies of epibenthic cyanobacterial communities in interacting with physical sedimentary dynamics, and on the significance of the biotic influence for the sedimentology of modern and ancient siliciclastics.

2. STUDY AREAS

2.1 Tidal flats of Mellum Island (southern North Sea)

Mellum Island is located in the southern North Sea (Fig. 1A). The sandy tidal flats are overgrown by epibenthic cyanobacteria forming thin biofilms in the lower intertidal zone, and thick microbial mats in the upper intertidal and lower supratidal zones (Stal 1985; Gerdes and Krumbein 1987; Gerdes et al. 1987, Villbrandt 1992; Noffke and Krumbein 1999b).

2.2 Lower arenigian siliciclastics, Montagne Noire, France

Siliciclastic shallow-marine rocks of Lower Arenigian age crop out in the Montagne Noire, southern France (Fig. 1 B). The succession includes mud- and siltstones, as well as fine to coarse-grained sandstones and record a depositional area that was located within a cool paleoclimate zone. The paleoenvironment resembles sedimentologically and geomorphologically the modern southern North Sea coast (compare Noffke 1992, Noffke & Nitsch 1994, Noffke 2000). A stratigraphic section comprising the Schistes et Gres de la Cluse de l'Orb – Formation (GCO-Fm.) was measured.

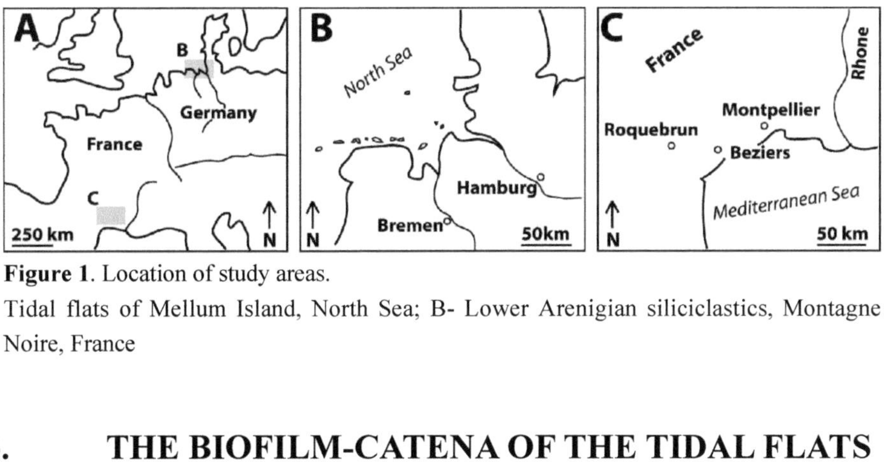

Figure 1. Location of study areas.
Tidal flats of Mellum Island, North Sea; B- Lower Arenigian siliciclastics, Montagne Noire, France

3. THE BIOFILM-CATENA OF THE TIDAL FLATS OF MELLUM ISLAND

The siliciclastic tidal flats of Mellum Island are colonized by different cyanobacterial populations of which each is dominated by one of the three species *Merismopedia punctata, Oscillatoria limosa,* or *Microcoleus chthonoplastes* (Stal 1985, Gerdes et al. 1987, Gerdes & Krumbein 1987, Noffke 1997, Noffke & Krumbein 1999). The dominance of the species within the cyanobacterial populations changes with the topographical height of the tidal flats. The reason for this is that from the mean low to the spring high water line the sedimentary conditions change and as a response to this, a lateral succession of distinct cyanobacterial populations establishes (Fig. 2). Within the lower intertidal zone, we can find *Merismopedia punctata*-dominated populations, whereas the upper intertidal area is overgrown by mats constructed mainly by *Oscillatoria limosa.* The bacterial community of the lower supratidal zone is dominated by *Microcoleus chthonoplastes.*

The lower intertidal zone is characterized by steady rework of the sediments by waves or tidal currents. How do photoautotroph cyanobacteria escape lethal burial in such an hydrodynamically active environment? *Merismopedia punctata*, a coccoid species that forms pillow-like clusters of spheroidal cells, attaches as biofilms on single mineral grains of the depositional surface. In experiment is shown, that if the bacterially overgrown sediment is disturbed by water agitation, the biofilm-coated grains float up more easily than 'sterile' quartz sand particles. As long as the water moves, the biofilm-coated grains stay in suspension, whereas the non-colonized sand grains mainly just roll back and forth on the sedimentary surface. Only after the water movement calmes down, the biofilm-coated grains sink to the floor and are thus deposited always ontop the sedimentary surface. This reaction of a biofilm-coated grain on water agitation is result of its small relative specific weight related to its larger diameter compared to the relatively high specific weight related to the smaller diameter of 'sterile' quartz grains. By this mechanism, *Merismopedia punctata*-overgrown grains stay atop the depositional surface. Its influence on the sedimentary dynamics is less important, e.g. no biostabilization effect (see paragraph 4) was measured (Noffke & Krumbein 1999).

The upper intertidal zone of Mellum Island is overgrown by *Oscillatoria limosa*. This organism forms single trichomes of high mobility, and constructs very thin, temporary mat layers (Villbrandt 1992). The upper intertidal area is characterized by diurnal, periodical rework by the flood current. *O. limosa* reacts on each of the regular burial events by rapid migration towards the new sedimentary surface, where it re-establishes a mat layer within a few hours (Villbrandt 1992, Noffke & Krumbein 1999). Because of the periodical inundation of the tidal surface, salinity values show moderate ranges compared to those of higher tidal areas, and desiccation events are rare. Therefore, *O. limosa* does not produce great amounts of EPS to protect itself against osmotic pressure or desiccation. In experiment in a flume chamber was shown that biostabilization effects on the sediments are about 3-5 times greater than that of non-colonized sands (compare paragraph 4). A specific surface morphology of the tidal flats document this effect (Noffke & Krumbein 1999).

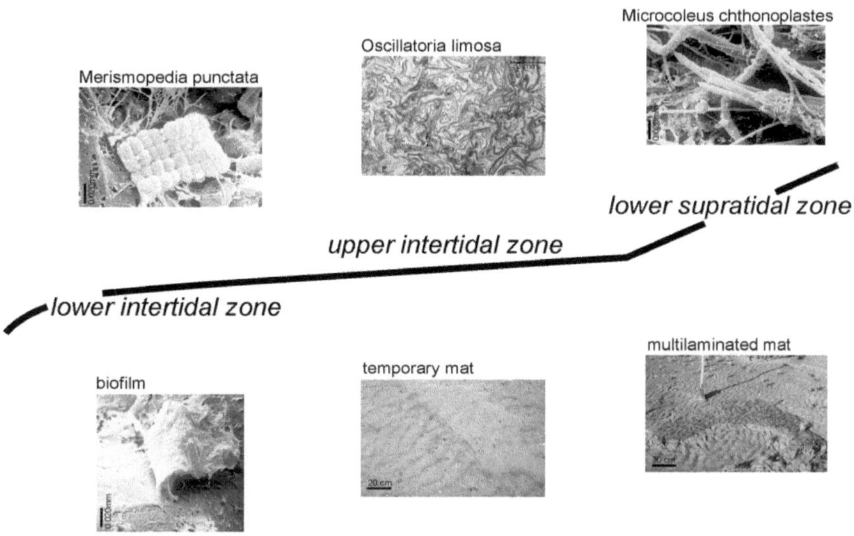

Figure 2. The biofilm-catena of the tidal flats of Mellum Island. Three species dominate cyanobacterial populations that colonize the main tidal zones. They form biofilms or mats in response to prevailing hydrodynamic conditions.

The lower supratidal zone of Mellum Island is covered by microbial mats of significant appearance (see figures in Gerdes & Krumbein 1987, Krumbein 1987, and older literature therein). The uppermost layer of these mats is predominantly constructed by *Microcoleus chthonoplastes* (e.g., Gerdes & Krumbein 1987, Krumbein 1987, Noffke & Krumbein 1999). This cyanobacterial species forms 'spaghetti-like' bundles of trichomes that are surrounded by an EPS-rich glycocalix. Together with heterotrophic bacteria, *M. chthonoplastes* generates a complex multilaminated and versicoloured mat system, that was termed 'Farbstreifen-Sandwatt - versicoloured tidal sand' by Schulz (1936). With its long lasting subaerial exposure in between the spring high tides the lower supratidal zone implies special ecological conditions the cyanobacteria have to respond to. During hot summer days, evaporation produces high values of salinity (Riege, 1994) and the mat-covered deposits may even dry out. After few hours of rain fall, the values decrease rapidly (Riege, 1994) and the mat surface is re-wetted. To protect itself against these extremely varying conditions, *M. chthonoplastes* secrets lots amounts of EPS and surrounds its trichomes by a robust sheath. The mucilages buffer salinity and provide a constant degree of moisture within the mat system, so a mat surface feels always wet. Because of reworking of the substrate take place only during spring high tides, it is sufficient for survival of the cyanobacterial population that only the hormogonia (pieces of trichomes falling apart) of *M. chthonoplastes* can move through sediments to escape lethal burial (Gerdes & Krumbein, 1987, see experiments by Gerdes et al. 1991). The thick mats contribute greatly to stabilization of the deposits, and the characteristic surface relief composed of erosional remenants and pockets evolves (Noffke & Krumbein 1999, Noffke 1999).

The example of the tidal flats of Mellum Island shows, how cyanobacterial species adapt to particular sedimentary conditions prevailing within each of the tidal zones. As a response to the different sedimentary dynamics, the species compositions of the epibenthic cyanobacterial populations changes from the low water to the high water line. Thus a lateral succession of different cyanobacterial populations establishes on the tidal flats. To describe such a succession, the term 'biofilm-catena' was proposed (Noffke 1997). 'Biofilm-catenae' can be observed also at other coastal environments (Noffke et al. 2001a).

4. CYANOBACTERIAL ACTIVITIES INTERACTING WITH SEDIMENTARY PROCESSES IN MODERN TIDAL FLATS AND THE FORMATION OF 'MICROBIALLY INDUCED SEDIMENTARY STRUCTURES'

Microbial assemblages interact with erosion or deposition, and this interference is recorded by particular 'Microbially Induced Sedimentary Structures' (Fig. 3). The main cyanobacterial activities processing the formation of the structures are leveling, biostabilization, imprinting, grain-separation, and baffling, trapping, and binding (Gerdes et al. 2000, Noffke et al. 2001b, Noffke in press a).

4.1 Leveling

Thick mats ovregrow a preceedingly rippled tidal surface and smoothens out the original surface relief, e.g. ripple marks. A mature microbial mat has a planar surface and the original tidal surface morphology of its substrate below is not longer visible (compare Noffke & Krumbein 1999a).

4.2 Biostabilization

The term biostabilization was first defined by Paterson (1994), and it describes the fixation of sediment by microorganisms such as diatoms, cyanobacteria, fungi, and others (compare also Neumann et al. 1970, De Boer 1981, Grant and Gust 1987, Dade et al. 1990, Gerdes et al. 2000, the volume on 'biostabilization of sediments' by Krumbein et al. 1994, or, Paterson 1997 for review, Noffke et al. 2001b, 2003). Biostabilization increases the resistance of microbial-interwoven sedimentary surfaces to erosion, but it reduces also the exchange of gases between sediment and water or sediment and atmosphere (Gerdes et al. 2000b, Noffke et al. 2001b). The reason for the immense stabilization effect of such filamantous cyanobacteria is threefold: (i) the copious and soft EPS fill in the original gruff tidal surface, so the frictional forces between the bottom water current and the depositional surface are reduced; (ii), additionally, the sedimentary grains are glued together and fixed by the adhesive bacterial mucilages, and (iii), the filaments themselves construct a dense network around the mineral particles (Noffke 1998, 1999, Noffke et al. 2001b). One can compare the construction of filaments (composed of several trichomes surrounded by a sheath) with the ropes of a suspension bridge. Bundled ropes (or trichomes) can react more flexible to tensile forces than single ropes (trichomes). Indeed, it was shown by Yallop et al. (1994) that a microbial mat of *Microcoleus chthonoplastes* reacts on drag forces by oscillation. It is assumable that this flexible reaction result from the specific construction of the filaments.

4.3 Imprinting

Biofilms grow over physically shaped surface structures, like ripple marks, without altering the original sedimentary relief. We can detect buried depositional surfaces in vertical sections through the sediment with the aid of their former biofilms. The biofilms line (=imprint) the surfaces (Noffke et al. 1996).

4.4 Grain separation

Grain separation is a process of upward transportation of sedimentary grains by a growing microbial mat. Over time, the biofilms around mineral grains grow and become thicker. The thickening biomass elevates the mineral grains and separates them from each other, and the grains 'float' without contact with each other within the organic mass (Noffke et al. 1997a,b).

4.5 Baffling, trapping and binding

Filamentous cyanobacteria like *Microcoleus chthonoplastes* accumulate actively sediment by baffling, trapping and binding (Black 1933, for tidal flats of Mellum Island: Gerdes & Krumbein 1987, Krumbein 1987, Noffke et al. 1997a, b, 2003, Noffke & Krumbein 1999). The cyanobacteria orient themselves perpendicularly to the mat surface and reach into the supernatant water, where they generate microzones of reduced flow velocity (‚baffling‘). The lowered current velocity triggers fall-out and settling of suspended mineral grains, a process which is termed 'trapping' (Black 1933). The baffled and trapped grains are incorporated into the mat fabrics during growth of the microbial layer ('binding' sensu Black 1933).

4.6 Primary sedimentary structures rising from cyanobactrial activities: ‘microbially induced sedimentary structures’

The term ‘primary sedimentary structures’ comprises all structures that are formed syndepositionally. This term includes on the one hand physical sedimentary structures that originate from erosion, deposition or deformation of sediments like e.g. ripple marks, parting lineation or dish structures. On the other hand, the term includes biogene structures like burrows and traces that originate from activities of macroorganisms like polychaets. Stromatolites are grouped therein (compare Pettijohn & Potter 1964). We proposed to introduce an additional group 'Microbially Induced Sedimentary Structures (MISS)' (Noffke et al. 2001b, 2003, Noffke in press a), which comprises phenomena that rise from interaction of bacterial communities with physical parameters (Fig. 3).

Biotic-physical structures are ‘leveled surfaces’, in the fossil record to be recognized as 'wrinkle structures', crinkled bedding planes (Hagadorn & Bottjer 1997, Noffke 2000, Noffke et al. 2003). 'Microbial mat chips' were torn from their parent mat during storms by water agitation (Logan 1961; Neumann et al. 1970; Gerdes and Krumbein 1987, Noffke et al. 1996, fossil examples reported by Pflüger and Gresse (1996). Characteristic tidal surface morphologies that document interactions of biostabilization and erosion are 'erosional remnants and pockets' (Reineck 1979; Gerdes et al. 1993; Noffke 1999), or 'multidirected ripple marks' (Noffke et al. 1995, 1996; Noffke 1998). Intrasedimentary gas entrapped underneath sealing microbial mat layers produce a characteristic 'sponge pore fabrics' (Noffke et al. 1996; Noffke et al. 1997a, b; Gerdes et al. 2000b). Preceeding ripple mark valleys can be traced as 'sinoidal laminae' in vertically cut sediment cores (Noffke et al. 1997b, Gerdes et al. 2000b). Grains incorporated into the mat fabrics often show an orientation with their long-axes parallel to the depositional surface. This texture in thin-sections are described as 'mat layer-bound oriented grains' (Noffke et al., 1997a,b). Baffling, trapping and binding is reflected by 'biolaminites' in the tidal sands of Mellum Island (Gerdes and

Krumbein 1987, or Gerdes et al. 1991, Gerdes et al. 2000b; compare also Grotzinger & Knoll 1999, and Reid et al. 2000).

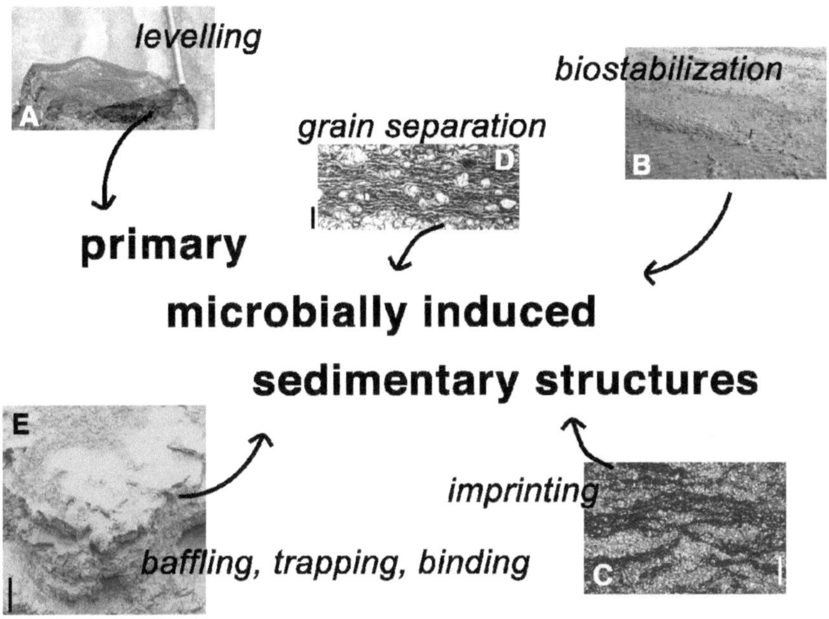

Figure 3. Cyanobacteria interact with physical sedimentary dynamics. This biotic-physical interference produces 'Primary Microbially Induced Sedimentary Structures', which are placed as own category into the Classification of Primary Sedimentary Structures.

5. THE SIGNIFICANCE OF BIOTIC-PHYSICAL INTERFERENCE IN THE FORMATION OF BEDDING PLANES

A bedding plane separates a single bed from adjoining beds. It represents either a period of non-deposition during sediment formation, or a sudden change of depositional conditions (e.g., Fuechtbauer 1988).

Within sediment that is composed of loose grains, e.g. tidal sands, one presupposition for the formation of a bedding plane is that the depositional surface was immobilized before a subsequent sediment layer was deposited. That is, the sedimentary surface was sheltered against primary destruction by erosion during placement of a fresh sand layer. Thus the preservation potential of a surface becomes higher. Loose surface sands of tidal flats can become fixed by different physical parameters. A thin salt crust can develop during ebb tides, but also wetting of a previously dry sandy surface by fog or rime can fasten the mineral particles (Reineck et al. 1995). Aside grain fixation, also a physically derived intra-sedimentary texture that imprints a former depositional surface may induce a bedding plane within an otherwise homogeneously composed sediment. Specific uni-directional grain orientation in a tidal surface is produced by particular current flow patterns during ascending flood (Allen 1964, compare Reineck et al. 1995). Drapes of grains of smaller (or coarser) grain sizes deposited during ebb tides (or during storms) onto the depositional surface may trigger the formation of bedding planes (Reineck et al. 1995).

The question is, if aside these physical effects, also microbial mats, by colonizing the interface between sediment and water, induce the formation of bedding planes in siliciclastics (Fig. 4).

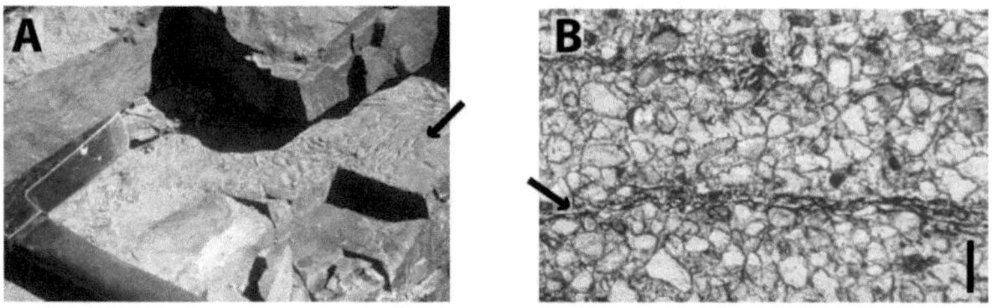

Figure 4. Do microbial mats contribute to the formation of bedding planes ? A- Bedding plane in consolidated rocks, Lower Arenigian, Montagne Noire, France; Note crinkled appearance of rock surface. B- Microbial mat covering ripple marks. Vertical section through sediment core from tidal deposits, Mellum Island.

5.1 Mat-related effects producing bedding planes in tidal sediments of Mellum Island

In relief casts of core samples and in thin-sections of artificially hardened tidal deposits, mat layer-bound oriented grains or enrichments of fine-grained particles can be observed. The layers separate the sandy deposits. As documented by field observations and measurements, such textures are first result of cyanobacterial leveling and stabilization of the sedimentary surface. Microbial leveling reduces the frictional forces between the sediment surface and moving bottom water, whereas the network of cyanobacterial filaments entangles the depositional grains and protects them from being ripped off by erosion (compare 4.1. and 4.2.). Grain separation and orientation originates during microbial mat growth and EPS-enrichment (compare 4.4.). This bacterial effect, as well as baffling, trapping and binding (active accumulation of fine sediment by microbial mats, compare 4.5.) leads to specific textures. The biogene textures line former depositional surfaces the same way like the physically induced ones. After consolidation of the sediment and rock formation, weathering may later hammer out ancient surfaces along these planes of differing, both biologically or physically derived textures.

5.2 Microbial mat-induced bedding planes in lower arenigian siliciclastics, Montagne Noire, France

We can find bedding planes that were formed not only by physical but also by biological factors in a stratigraphic section of the Lower Arenigian, Montagne Noire, in France (Noffke 2000). Within the stratigraphic section of the GCO-Fm., we can distinguish 4 groups of physically derived bedding planes, and 2 groups of bacterially induced ones (Fig. 5).

Nine bedding planes show wrinkle structures (= lithificated microbial mats, Hagadorn & Bottjer 1997) atop. In vertical sections through the wrinkle structures, we can see its internal biolaminated pattern, and a few single mineral grains floating within the ancient mat fabrics. The fossil mats contain also small-sized grains (silt) that lack in the sandy interlayers. The bedding planes show no original surface morphology, therefore it is concluded that the former mats were thick enough to level completely the relief of its substrate below. Because of the planar appearance of the bedding planes, they were termed 'non-transparent wrinkle structures' (Noffke 2000, Noffke et al., 2002).

Five wavy upper bedding planes show similar internal textures like the wrinkled surfaces, and 4 rippled rock bed surfaces are covered by crinkles that resemble a skin. Wavy upper bedding planes and skin-like crinkled bedding planes record former, very thin microbial mats that did not level the preceeding depositional surface and the surface structures. Because the original surface relief is still visible underneath the ancient mat, such bedding planes were termed 'tranparent wrinkle structures' (Noffke 2000, Noffke et al. 2002). In thin sections, they contain silt-size particles and a great amount of mica flakes.

Thirty-two of the bedding planes studied are covered by fine-grained sediments, but show none of the microbial mat-induced structures or textures. Therefore the fines were probably deposited during periods of lower hydrodynamic energies. Nineteen bedding planes are amalgamated, that is they record severe storms that eroded parts of the former depositional surface before fresh sediment was placed. Finally, we distinguish mud- and siltstone beds that contain no distinct bedding planes.

Biogenic bedding planes make 12.6 % of all bedding planes studied, and they thus represent a significant category within the total amount of bedding planes. Conversely, physically derived bedding planes make the greatest portion within the GCO-Fm., and record tidal areas of unfavourable hydrodynamic conditions that are avoided by cyanobacteria (compare Noffke et al., 2002).

The results show that bacterial activities contribute to the formation of bedding planes in siliciclastic tidal flats. Depositional surfaces are protected against erosion by biostabilization, and the bacterial communities produce distinct textures within the otherwise homogeneous sediments. Along these textures, weathering may later hammer out the depositional surfaces in the consolidated rock.

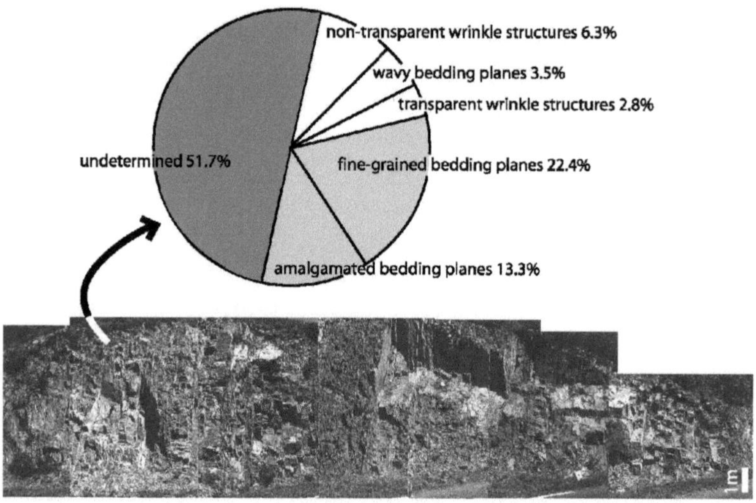

Figure 5. Types of bedding planes in the stratigraphic section of the Gres et Schistes de la Cluse de l'Orb, Lower Arenigian, Montagne Noire, France. We distinguish physically derived and bacterially induced bedding planes.

6. CONCLUSIONS

Shallow-marine siliciclastic deposits are colonized by various populations of benthic microorganisms, of those cyanobacteria are important parameters in sedimentary processes. The tidal flats of Mellum Island, southern North Sea, show how in correspondence with the tidal zones and the changing characteristic sedimentary dynamics, a lateral succession of different cyanobacterial populations establishes. We term such a succession 'biofilm-catena'. Biofilm-catenae interfere the physical parameters (erosion, deposition, deformation) that govern siliciclastic sedimentary systems. Such biotic-physical interaction produces 'Microbially Induced Sedimentary Structures (MISS)'. MISS are grouped as own category into the existing classification scheme of primary sedimentary structures. Another important aspect is the contribution of benthic cyanobacteria to the formation of bedding planes. Bedding planes in tidal sands or consolidated sandstones may not only result from physical factors affecting the former sedimentary surface, but could also be formed by cyanobacterial activities. Biogene and physical bedding planes are shown with the example of the Lower Arenigian, Montagne Noire, France. The knowledge of microbial activities in erosional and depositional processes within physical sedimentary systems should give rise to new perspectives in sedimentological investigations of siliciclastic environments of Earth, but also other planetary bodies.

7. ACKNOWLEDGEMENTS

I cordially thank my teachers Gisela Gerdes, Hansmartin Huessner, Andrew H. Knoll, Wolfgang E. Krumbein, and Dolf Seilacher. Financial support was granted by Deutsche Forschungsgemeinschaft (DFG) projects no. Ge 64/1-2, No 380/1-1 and No 380/1-2, and the Deutsche Akademie der Naturwissenschaften Leopoldina zu Halle/Saale (BMFT LPD 9901/8-2).

REFERENCES

Allen, J. R. L., 1964, Primary current lineation in the lower Old Red Sandtsone (Devonian), Anglo Welsh Basin: Sedimentology, v. 3, p. 98-108.

Black, M., 1933, The algal sediments of Andros Islands, Bahamas: Royal Society of London Philosophical Transactions, Ser. B, v. 222, p. 165-192.

Cameron, B., Cameron, D., and Jones, J.R., 1985, Modern algal mats in intertidal and supratidal quartz sands, northeastern Massachussetts, USA: Society of Economic Paleontologists and Mineralogists, p. 211-223.

Charaklis, W. G., and Wilderer, P. A., 1989, Structure and function of biofilms: Dahlem Workshop Reports, Life Sciences Research Report 46, Chichester, 386p.

Cohen, Y., Castenholz, R.W., and Halvorson, H. O., 1984, Microbial mats: Stromatolites: Alan R. Liss Press, New York, 498 p.

Cohen, Y. and Rosenberg, E., 1989, Microbial mats. Physiological ecology of benthic microbial communities: American Society of Microbiologists, Washington, D.C., 494p.

Dade, W.B., Davis, J. D., Nichols, P. D., Nowell, A. R. M., Thistle, D., Trexler, M. B. and White, D. C., 1990, Effects of bacterial exopolymer adhesion on the entrainment of sand. Geomicrobiological Journal, 8, 1-16.

De Boer, P.L., 1981, Mechanical effects of microorganisms on intertidal bedform migration: Sedimentology, v. 28, p. 129-132.

Decho, A.W., 2000, Exopolymer microdomains as a structuring agent for heterogeneity within microbial biofilms, *in* Riding, R. and Awramik, S., eds., Microbial Sediments: Berlin, Springer-Verlag, p. 9-15.

Gerdes, G. and Krumbein, W. E., 1987, Biolaminated deposits: Berlin, 183p.: Gerdes, G., Krumbein, W. E. & Reineck, H. E. (Eds.): Mellum – Portrait einer Insel. p. 62-76, 5 Figs., Kramer, Frankfurt/Main.

Gerdes, G., Krumbein, W.E., and Reineck, H.E., 1991, Biolaminations – ecological versus depositional dynamics, *in* Einsele, G., Ricken, W. and Seilacher, A., eds., Cycles and events in stratigraphy: Springer, Berlin, p. 592-607.

Gerdes, G., Claes, M., Dunajtschik-Piewak, K., Riege, H., Krumbein, W. E., and Reineck, H. E., 1993, Contribution of microbial mats to sedimentary surface structures. Facies, 29, 61-74.

Gerdes, G., Krumbein, W.E. and Reineck, H.E., 1994a, Microbial mats as architects of sedimentary surface structures, *in* Krumbein, W.E., Paterson, D.M., and Stal, L., eds., Biostabilization of sediments: Oldenburg, Bibliotheks und Informationssystem der Universität Oldenburg, 526 p .

Gerdes, G., Noffke, N, and Klenke, Th. 2000b, Microbial signatures in peritidal siliciclastic sediments: A catalogue: Sedimentology, v. 47, p. 279-308.

Golubic, S. and Knoll, A.H., 1993: Prokaryotes, *in* Lipps, J., ed., Fossil prokaryotes and protists: Blackwell, Boston, p. 51-76.

Grant, J. and Gust, G., 1987, Prediction of coastal sediment stability from photopigment content of mats of purple sulfur bacteria: Nature, 330, p. 244-246.

Grotzinger, J.P. and Knoll, A.H., 1999, Stromatolites in Precambrian Carbonates: Evolutionary Mileposts or Environmental Dipsticks?: Annual Review of Earth and Planetary Science, v. 27, p. 313-358.

Hagadorn, J.W. and Bottjer, D., 1997, Wrinkle structures: microbially mediated sedimentary structures common in subtidal siliciclastic settings at the Proterozoic-Phanerozoic transition: Geology, v. 25, p. 1047-1050.

Knoll, A.H. & Bauld, J., 1989, The evolution of ecological tolerance in prokaryotes: Transactions of the Royal Society of Edinburgh: Earth Sciences, v. 80, p. 209-223.

Krumbein, W.E., 1979, Photolithotrophic and chemoorganotrophic activity of bacteria and algae as related to beach rock formation and degradation (Gulf of Acaba (Sinai)). - Geomicrobiol. J., 1, 139-203

Krumbein, W.E., 1983, Stromatolites – The challenge of a term in space and time: Precambrian Research, v. 20, p. 493-531.

Krumbein, W.E., Paterson, D.M., and Stal, L.J., eds., 1994, Biostabilization of Sediments: BIS, University of Oldenburg, 526p.

Logan, B.W.,1961, *Cryptozoon* and associated stromatolites from the Recent, Shark Bay, Western Australia. *Journ. Geol.*, 69, no.5, 517-533.

Neu, T.R., 1994, Biofilms and microbial mats. In: *Biostabilization* (Ed. by W.E. Krumbein, D.M. Paterson and L.J. Stal), 9-16, Oldenburg.

Neumann, A.C., Gebelein, C.D., and Scoffin, T.P., 1970, The composition, structure and erodiability of subtidal mats, Abaco, Bahamas: Journal of Sedimentary Petrology, v. 40, p. 274-297.

Noffke, N. & Nitsch, E., 1994, Sedimentology of Lower Ordovician clastic shelf deposits, Montagne Noire, France: Geologie de France, v. 4, p. 3-19.

Noffke, N., Gerdes, G., Klenke, Th., and Krumbein, W.E 1995, Microbially induced sedimentary structures: Tagung Sediment 95 Freiberg, Abstract, p. 141.

Noffke, N., Gerdes, G., Klenke, Th., and Krumbein, W.E. 1996, Microbially induced sedimentary structures – examples from modern sediments of siliciclastic tidal flats: Zentralblatt für Geologie und Paläontologie Teil I, v. 1995, p. 307-316.

Noffke, N., 1997, Mikrobiell induzierte Sedimentstrukturen in siliziklastischen Wattablagerungen: PhD Thesis, University of Oldenburg, 127p.

Noffke, N., 1998, Multidirected ripple marks rising from biological and sedimentological processes in modern lower supratidal deposits (Mellum Island, southern North Sea): Geology, v. 26, 10, p. 879-992.

Noffke, N., 1999, Erosional remnants and pockets evolving from biotic-physical interactions in a Recent lower supratidal environment: Sedimentary Geology, v. 123, p. 175-181.

Noffke, N., 2000, Extensive microbial mats and their influences on the erosional and depositional dynamics of a siliciclastic cold water environment (Lower Arenigian, Montagne Noire, France): Sedimentary Geology, v. 136, p. 207-215.

Noffke, N., in press a, Microbially induced sedimentary structures: formation and application to sedimentology. - Middleton, C. (Ed.): Encyclopedia of sediments.

Noffke, N., in press b, Bacteria in sediments. - Middleton, C. (Ed.): Encyclopedia of sediments.

Noffke, N., 2002, ONLINE: The concept of geobiological studies: the example of microbially induced sedimentary structures in physical depositional systems: Palaios, v. 17, i-iii.

Noffke, N., Gerdes, G., Klenke, Th., and Krumbein, W.E., 1997a, A microscopic sedimentary succession of graded sand and microbial mats in modern siliciclastic tidal flats: Sedimentary Geology, v. 110, p. 1-6.

Noffke, N., Gerdes, G., Klenke, Th., and Krumbein, W.E., 1997b, The impact of sediment-stabilizing biofilms on the architecture of sedimentary structures (Mellum Island, southern North Sea): Courir Forschungs-Institut Senckenberg, v. 201, p. 297-305.

Noffke, N., Gerdes, G., Klenke, Th. & Krumbein, W.E., 2001a, Microbially induced sedimentary structures indicating small-scale facies zones within coastal environments of southern Tunisia (Recent and Pleistocene): Facies, v. 44, p. 23-30.

Noffke, N., Gerdes, G., Klenke, Th. & Krumbein, W.E., 2001b, Classification of primary microbially induced sedimentary structures: Journal of Sedimentary Research, v. 71, 5, 649-656.

Noffke, N., Gerdes, G., and Klenke, Th., 2003, Benthic cyanobacteria and their influence on the sedimentary dynamics of peritidal depositional systems (siliciclastic, evaporitic salty and evaporitic carbonatic).- *Earth Science Review,* 12, 1-14.

Noffke, N., Knoll, A.H. & Grotzinger, J., 2002, Sedimentary parameters controlling occurrence and preservation of microbial mats in siliciclastic depositional systems: Palaois, v. 17, 1-12.

Noffke, N. and Krumbein, W.E., 1999, A quantitative approach to sedimentary surface structures contoured by the interplay of microbial colonization and physical dynamics: Sedimentology, v. 46, p. 417-426.

Paterson, D. M., 1994, Microbiological mediation of sediment structure and behaviour, in Stal, L.J. and Caumette, P., eds., Microbial Mats, Berlin, Springer-Verlag, p. 97 – 109.

Paterson, D. M., 1997, Biological mediation of sediment erodiability: ecological and physical dynamics, *in* Burt, N., Parker, R. and Watts, J., eds., Cohesive sediments: Wiley, London, p. 215-229.

Paterson, D. M., and BLACK, K. S., 2000, Siliciclastic intertidal microbial sediments, *in* Riding, R., and Awramik, S., eds., Microbial sediments, Berlin, Springer, p. 217 – 225.

Pettijohn, F.J., & Potter, P.E., 1964, Atlas and glossary of primary sedimentary structures: Berlin, Springer-Verlag, 370 p..

Reid, R.P., Visscher, P.T., Decho, A.W., Stolz, J.F., Bebout, B.M., Dupraz, C., MacIntyre, I.G., Pearl, H.W., Pinckney, J.L., Prufert-Bebout, L., Steppe, T.F. and DesMarais, D.J., 2000, The role of microbes in accretion, lamination and early lithification of modern marine stromatolites: Nature, v. 406, p. 989-991.

Reineck, H.E., 1979, Rezente und fossile Algenmatten und Wurzelhorizonte: Natur und Museum, v. 109, p. 290-296.

Reineck, H. E., Gerdes, G. and Noffke, N., 1995, Physikalische Kraefte, die Rippelfelder erhalten, ehe sie versteinern: Natur und Museum, v. 125, p. 163-172.

Riege, H., 1994, *Untersuchungen zur Karbonatfällung in Mikrobenmatten.* PhD Thesis, University of Oldenburg, pp. 139.

Schulz, E., 1936, Das Farbstreifen-Sandwatt und seine Fauna, eine ökologisch-biozönotische Untersuchung an der Nordsee. *Kieler Meeresforsch.*, 1, 359-378.

Stolz, J. F., 2000, Structure of microbial mats and biofilms, *in* Riding, R., and Awramik, S., eds., Microbial sediments, Berlin, Springer-Verlag, p. 1-8.

Stal, L.J., 1985, Ökologie der Bakterien des Farbstreifen-Sandwattes. In: Gerdes, G., Krumbein, W. E., Reineck, H. E. (Eds.*), Mellum - Porträit einer Insel.* (Ed. by G. Gerdes, W.E. Krumbein, and H.E. Reineck), 188-202, Frankfurt am Main.

Yallop, M.L., De Winder, B., Paterson, D.M. and Stal, L.J., 1994, Comparative study on primary production and biogenic stabilization of cohesive and non-cohesive marine sediments inhabited by microphytobenthos: Estuarine, Coastal and Shelf Sciences, v. 39, p. 565-582.

Villbrandt, M., 1992, Interactions of nitrogen fixation and photosynthesis in marine cyanobacterial mats (Mellum, southern North Sea): PhD Thesis University of Oldenburg, 153p.

Walter, M.R., ed., 1976, Stromatolites: Developments in Sedimentology, v. 20, Elsevier, Amsterdam, Oxford, New York, 790p.

Chapter 18

THE ROLE OF BIOFILMS IN THE FOSSILIZATION OF NON-BIOMINERALIZED TISSUES

Derek E.G. BRIGGS
Department of Geology and Geophysics,
Yale University, P.O. Box 208109, New Haven, CT 06520-8109, USA
derek.briggs@yale.edu

1. INTRODUCTION

Deposits that preserve soft-bodied fossils are of fundamental importance in palaeontology because they provide much more complete data on ancient communities than does the normal shelly fossil record. Biofilms have long been recognized as playing an important role in the preservation of Konservat-Lagerstätten (i.e. conservation deposits, characterized by the quality of fossil preservation) (see Gall 2001). Seilacher *et al.* (1985) identified them as one of three major agents involved in promoting preservation, alongside stagnation and obrution (rapid burial). Biofilms are layers, a few mm thick, of spatially organized microbial cells within a matrix of exopolymeric substance (EPS), a glycocalyx secreted by the cells. Biofilms form the basis of microbial mats, i.e. interacting microbial communities consisting of phototrophic cyanobacteria, often with associated algae, fungi, protozoans, and heterotrophic bacteria, which produce extensive layered deposits in the geological record (stromatolites). Mats are normally stratified vertically downwards from aerobic to anaerobic in terms of the electron acceptors used in metabolism, and from phototrophic to chemotrophic.

The persistence of microbial mats is contingent on an absence of grazers (mats are consumed, *inter alia*, by gastropods, annelids, crustaceans and fishes) and minimal disturbance by bioturbators. For this reason they were widespread in the Precambrian, prior to the rise of invertebrates. In Phanerozoic sequences, on the other hand, microbial mats occur mainly in settings where metazoans were inhibited (e.g. by low oxygen, high salinity, low or high pH, high temperature). This contrast was highlighted by Seilacher and Pflüger (1994, see also Seilacher 1999) in their description of seabed conditions as 'matgrounds' in the Precambrian and dominantly 'mixgrounds' during the Phanerozoic. The prevalence of microbial mats in the Precambrian was a major factor in the widespread fossilization of early metazoans (the Ediacaran organisms) in spite of their soft-bodied nature. The rise of grazing and bioturbation during the Precambrian-Cambrian transition restricted this particular taphonomic 'window' (Gehling 1999) to more extreme environments where microbial mats continued to flourish.

2. THE ROLE OF MICROBIAL MATS

Microbial mats promote the preservation and recovery of soft-bodied fossils in both marine and lacustrine settings in a number of ways (see e.g. Seilacher *et al.* 1985, Allison and Briggs 1991).

1. *Capture.* Where microbial mats form on the surface of standing bodies of water they may trap carcases, particularly those of insects, which are transported to the sediment-water interface when the mat sinks (Gall 1995, Harding and Chant 2000). Insects may even live on or in the algal mats while they are floating on the water surface, as suggested for the famous Tertiary limestones of Riversleigh in Queensland (Archer *et al.* 1994). Where mats grow over a decaying carcase at the sediment-water interface, they may prevent it from floating
2. *Protection.* Microbial mats may form a shroud over a decaying carcase (Figure 1), protecting it from scavengers and burrowing animals (Gall 1990, 1995). Mats protect sediments from erosion, and prevent the removal and disarticulation of carcases that adhere to them, or are overgrown by them. The current velocity required to erode sediment bound by microbial mats is 3-20 times higher than for sediment alone (Krumbein *et al.* 1994). This sealing effect also enhances the preservation of tracks and trails (see Gall 2001).
3. *Decay inhibition.* Microbial mats may reduce decay rates by acting as a barrier to decay microbes, and by promoting conditions unfavourable to certain bacteria (Gall 1990, 2001).

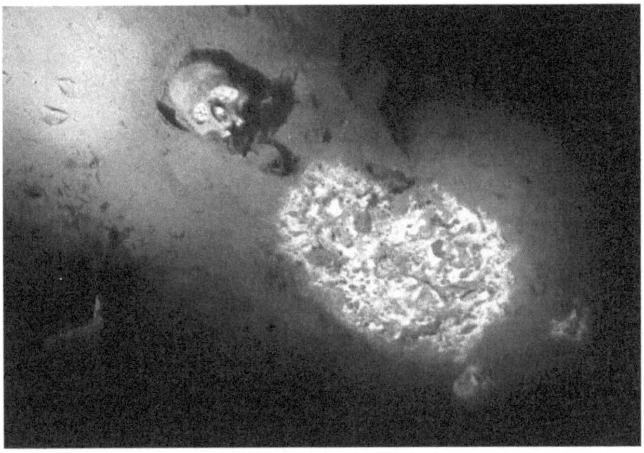

Figure 1. Human body covered with *Beggiatoia* from the Koenigsee, near Berchtesgaden in the Bavarian Alps. (Photograph taken by Dr. Fricke with the submersible Jago.)

4. *Formation of a microenvironment.* Microbial mats enclose a microenvironment in which the particular conditions (e.g. anoxia, low pH, high S^{2-}) required for soft-tissue mineralization may become established (see Sagemann *et al.* 1999).
5. *Mineralization.* Microbial mats are a barrier to the transfer of ions and may trap and concentrate them within a carcase. They are the site of abrupt changes in geochemical gradients. Ions may reach sufficient concentrations to promote replication of the soft tissues, or the microbial mat itself, by minerals (Briggs & Kear 1993). Mineralization is also instrumental in creating the 'sole veneer' that leads to the preservation of the Ediacara fossils as moulds and casts - so called 'death masks' (Gehling 1999: see below)
6. *Pseudomorphing.* In some cases the microbial mat forms on the surface of the carcase, corresponding faithfully to the outline of the tissues. If the mat then becomes mineralized (e.g. soft tissue of Las Hoyas dinosaur: Briggs *et al.* 1997; fossils from the Messel Shale and from Enspel: Wuttke 1983, Toporski *et al.* 2001) information on the gross morphology of the soft-tissues is preserved.
7. *Parting.* Microbial mats result in partings within siliciclastic and carbonate sequences which expose the fossils when the rock is split (e.g. Gehling 1999).

Identification of the former presence of microbial mats in the rock record may be problematic. The survival of residual carbon in laminae is not in itself diagnostic and biomarkers characteristic of microbes may not be present. Positive identification requires the discovery of microbial cells or filaments in life position (Schieber 1999). Textures on bedding plane surfaces may, however, provide a strong indication. Such features are found, for example, in the mid-Proterozoic Belt Supergroup of Montana, Idaho and British Columbia: domal buildups, cohesive behaviour, characteristic laminae, irregular curved-wrinkled bed surfaces, ripple patches, and laminae enriched in mica or early diagenetic minerals (Schieber 1999). They are rare, however, in younger rocks where they are confined to environments that inhibit metazoans (e.g. Pflüger 1999). Some of the most striking examples of large-scale surface textures, which can be attributed confidently to microbial mats, occur in the Jurassic (Kimmeridgian) lithographic limestones at Cerin in the Jura of France where evaporation led to elevated salinity in quiet lagoons (Bernier *et al.* 1991, Gall *et al.* 1985). Bedding plane surfaces exhibit filamentous structure, folding, tearing, and fragmentation into ribbons. Somewhat less dramatic evidence is exhibited by the Jurassic (Tithonian) Solnhofen Limestone (Seilacher *et al.* 1985)

Microbial mats may provide a sticky layer on the sea-bed to which carcases adhere prior to burial (Gehling 1999), or they may grow on the surface of a freshly deposited carcase. Microbial mats can also form within the sediment. They can grow rapidly enough to precede much decay, and they impact on soft-tissue preservation in three main ways:

1. by promoting conditions that lead to the replication of soft-tissues in minerals (e.g. Martill 1988, Wilby *et al.* 1996), or
2. by the mats themselves becoming mineralized and outlining the soft-tissues (pseudomorphing) (e.g. Wuttke 1983, Briggs *et al.* 1997), or
3. by promoting early diagenesis of the immediately overlying sediment to form a 'death mask' that leads to preservation as an external mould and corresponding cast, as in the Ediacara fossils (e.g. Gehling 1999).

3. PROMOTING CONDITIONS THAT LEAD TO MINERALIZATION

There is a clear association between microbial mats, high concentrations of phosphate, and phosphatized soft tissues in the limestones of Cerin (Wilby *et al.* 1996, Wilby and Briggs 1997). The sequence was systematically excavated over a period of about 20 years (Bernier *et al.* 1992) and all the observed macrofossils were collected, providing a basis for surveying the occurrence of exceptional preservation. The carcases were deposited on the upper surface of the beds after the sediment had settled from suspension; the surfaces were then rapidly overgrown by microbial mats (Figure 2).

Figures 2, 3. Microbial mats and phosphatized muscle tissue from the Jurassic of Cerin, France. 2. Section through fossilized microbial mats (M) which trap phosphorus resulting in concentrations toward the top of the layers of carbonate sediment. Scale bar 1 cm. 3. Three-dimensionally preserved muscle fibres (running NW-SE) from a fish from level 202 (concentration of 0.045 weight% P_2O_5). Scale bar 20μm. (After Wilby and Briggs 1997.).

Phosphatized soft tissues (Figure 3) are evident in 17% (n=283) of fishes and 30% (n=53) of crustaceans and only occur in the lower marine part of the sequence, where evidence of microbial mats is most prevalent. Phosphorus levels are significantly higher in parts of the sequence where soft tissues are preserved, than where they are absent (the average P_2O_5 concentration in the plattenkalk is 0.017 weight %; where phosphatized soft tissues occur recorded values range from 0.038 wt% to 0.067 weight %: Wilby *et al.* 1996). Concentrations vary on a mm scale, increasing upwards through an individual bed to the mat on its surface. The mat clearly acted as a barrier to the diffusion of dissolved phosphorus, perhaps by controlling the redox boundary, and/or promoting transitory fixation in the organic material (see Wilby *et al.* 1996). The role of microbial mats in preserving soft-tissues is also likely to have been critical in other deposits such as the Jurassic Solnhofen Limestone of Germany (Seilacher *et al.* 1985) and the Cretaceous Santana Formation of Brazil (Martill 1988), even where the evidence of their presence is less striking. The presence of microbial mats, and consequent elevation of phosphorus concentrations in sediment, results in phosphatization of soft-tissues in a wider range of taxa than in deposits where the main source for phosphorus is the carcases themselves (Wilby *et al.* 1995).

4. MINERALIZED MATS PRESERVING A CARCASE OUTLINE

This process, which occurs in a variety of Konservat-Lagerstätten, and involves a range of minerals, has been referred to as bacterial autolithification (see Davis and Briggs 1995). The unique specimen of the ornithomimosaur dinosaur *Pelecanimimus* from the Lower Cretaceous lacustrine sequence of Las Hoyas, Spain, preserves some of the soft-tissues in outline, including the throat pouch and a possible occipital crest (Pérez-Moreno *et al.* 1994). The soft tissues are preserved as a microbial mat (Figure 4), now mineralized in apatite, which consists of thin layers of coccoids embedded in films (Figure 5) that are thought to represent a glycocalyx (Briggs *et al.* 1997). The margin of this mat forms a defined line which presumably represents the body outline. Phosphorus released from the carcase was trapped by the mat in a manner similar to that described in the previous section (some details of muscle tissues are also preserved, but in iron carbonate) (Briggs *et al.* 1997). Similar mats formed of coccoid cyanobacteria within a glycocalyx become mineralized in aragonite in Lake Van (Turkey) today (Kazmierczak *et al.* 1996).

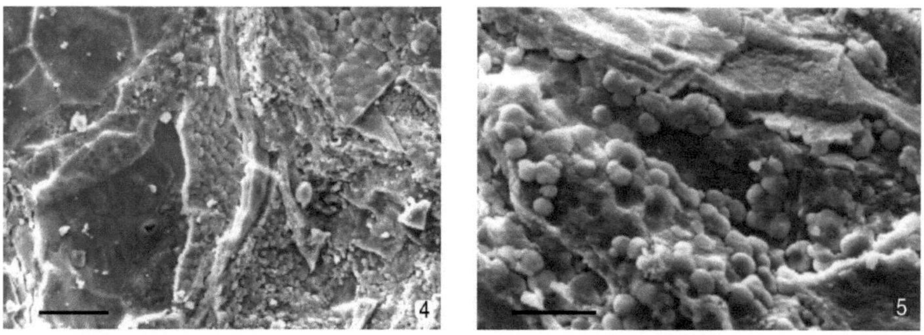

Figures 4, 5. Microbial mats overlying the dinosaur *Pelecanimimus* from the Lower Cretaceous of Las Hoyas, Cuenca, Spain (LH 7777, Museo de Cuenca). 4. Layers of mineralized microbial mats overlying the specimen. Scale bar 20µm. 5. Detail of the phosphatized mats showing mineralized coccoids and surrounding glycocalyx. Scale bar 10µm. (After Briggs *et al.* 1997.)

Bacterial biofilms have also been reported as the mode of preservation of large tadpoles from the Oligocene of Enspel, Germany, where they consist of spherical, coccoidal or rod-shaped forms 0.5 to 0.8 µm in diameter, which are composed of carbonaceous matter and calcium phosphate (Toporski *et al.* 2001). The outlines of vertebrate soft tissues in the Eocene Messel Shale have likewise long been known to be the result of a covering of coccoid and rod shaped bacteria up to 2µm in diameter which is preserved in siderite

(Wuttke 1983). Mats have also been shown to preserve the outline of soft-tissues in ichthyosaurs, most famously those of the Jurassic of Holzmaden, Germany, where they are made up of filamentous cyanobacteria reported to be composed of carbonaceous material in the fossils (Martill 1987). Analyses are necessary, however, to eliminate the possibility that authigenic mineralization has also taken place.

5. MATS FORMING 'DEATH MASKS'

The famous Precambrian Ediacaran biotas include a diversity of organisms, but the majority are preserved as impressions in relatively coarse sandstones, a phenomenon largely unknown in post-Proterozoic rocks (Gehling 1999). The nature and affinities of some of the fossils remain controversial and a number of experiments have been carried out (reviewed in Briggs 1995) to try to understand their preservation, but none has so far reproduced the process. Seilacher (1999) explained the lifestyles of many of the taxa in the context of a 'matground' - a layer of microbial mat on and/or within the upper part of the sediment. Gehling (1999) extended this concept to explain the unusual preservation of the fossils.

Specimens of *Dickinsonia*, for example, are preserved in sediment as a positive epirelief which has replaced the decayed organism and been cast by an impression (external mould) of the upper surface of the body in the overlying sand (Figure 6). Thus the sediment overlying the carcase must have been cemented prior to decay of the organism and the sediment below must have remained soft enough to form a cast. Other Ediacaran organisms (e.g. *Phyllozoon*) decayed more rapidly and collapsed leaving an impression of the lower surface. Gehling (1999) envisaged the organisms buried *in situ* on a mat surface by an influx of sediment during a storm event. The base of the introduced sediment then became a locus for early diagenetic cementation, forming a 'sole veneer' or death mask. All the organic material subsequently decayed and the fossils are preserved as external moulds and casts. In some cases there may be a partial fill of the body space (e.g. *Ernietta*, *Ventrogyrus*: see discussion of preservation of the latter in Burzin *et al.* 1998). The relief preserved by the sole veneer depends on the composition (and decay resistance) of the buried organism (see Gehling 1999 for further details).

Figure 6. The Precambrian organism *Dickinsonia costata* from the late Proterozoic Ediacaran interval, Flinders Ranges, South Australia, x0.64. (Photographed by DEGB at the University of Adelaide.)

6. CONCLUSION

The role of microbial mats in the fossilization of non-biomineralized tissues has been largely inferred to date on the basis of observations and analyses of fossils and their sedimentological context. More reliable deductions can be made about the controls on preservation where mineralization is induced in the laboratory and the taphonomic processes are monitored as they proceed (Briggs 1995). A range of experiments has been carried out on the controls on the replication of soft tissues in authigenic minerals (e.g. Sagemann *et al.* 1999) and there is no reason why this cannot be extended to microbial mats. It should even be possible to investigate the role of mats in preserving soft tissues in relatively coarse sandstone. As Gehling (1999, p. 54) pointed out 'Assuming that Ediacaran organisms were composed of familiar biological materials, and that both the water chemistry and microbial life of the terminal Proterozoic were similar to modern systems, the formation of sole-veneer, mineralized 'death masks' as a means of preserving buried soft-bodied organisms should be amenable to laboratory modelling.'

REFERENCES

Allison, P.A. and Briggs, D.E.G. 1991 Taphonomy of non-mineralized tissues. in Allison, P.A. and Briggs, D.E.G. (eds) *Taphonomy: releasing the data locked in the fossil record.* Plenum Press, New York, 25-70.

Archer, M., Hand, S.J. and Godhelp, H. 1994 *Riversleigh, the story of animals in ancient rainforests of inland Australia.* Second edition. Reed Books, Chatswood, NSW, 264 pp.

Bernier, P., Gaillard, C., Gall, J.-C., Barale, G., Bourseau, J.-P., Buffetaut, E., and Wenz, S. 1991 Morphogenetic impact of microbial mats on surface structures of Kimmeridgian micritic limestones (Cerin, France). *Sedimentology* 38, 127-136.

Bernier, P., Barale, G., Bourseau, J.-P., Buffetaut, E., Gaillard, C., Gall, J.-C., and Wenz, S. 1992 The paleoecological excavations at Cerin (southern French Jura Mountains), results and interpretation. *Geobios, Mémoire Spécial* 16, suppl. 2, 35 pp.

Briggs, D. E. G., and Kear, A. J. 1993 Fossilization of soft-tissue in the laboratory. *Science* 259, 1439-1442.

Briggs, D.E.G. 1995 Experimental taphonomy. *Palaios* 10, 539-550.

Briggs, D.E.G., Wilby, P.R., Pérez-Moreno, B.P., Sanz, J.L. and Fregenal-Martínez, M. 1997 The mineralization of dinosaur soft-tissue in the Early Cretaceous of Las Hoyas, Spain. *Journal of the Geological Society, London* 154, 587-588.

Burzin, M.B., Grazhdankin, D.V. and Bronnikov, A.A. 1998 The mystery of the Ediacaran organisms. *Journal of journals* 2, 47-53.

Davis, P.G. and Briggs, D.E.G. 1995 The fossilization of feathers. *Geology* 23, 783-786.

Gall, J.-C., Bernier, P., Gaillard, C., Barale, G., Bourseau, J.-P., Buffetaut, E., and Wenz, S. 1985 Influence du développement d'un voile algaire sur la sédimentation et la taphonomie des calcaires lithographiques. Exemple du gisement de Cerin (Kimmeridgien supérieur, Jura méridional français). *Académie des Sciences, Paris, Comptes Rendus* 301, ser. II, p. 547-552.

Gall, J.-C. 1990 Les voiles microbiens. Leur contribution à la fossilisation des organismes au corps mou. *Lethaia* 23, 21-28.

Gall, J.-C. 1995 Biofilms et mattes microbiennes; leur contribution à la sédimentogenèse. *Comptes rendus Academie Sciences, Paris*, sér. IIa 321, 823-835.

Gall, J.-C. 2001 Role of microbial mats. in Briggs, D.E.G. and Crowther, P.R. eds., *Palaeobiology 2*, Blackwell, p. 280-284.

Gehling, J.G. 1999 Microbial mats in terminal Proterozoic siliciclastics: Edicaran death masks. *Palaios* 14, 40-57.

Harding, I.C., and Chant, I.C. 2000 Self-sedimented diatom mats as agents of exceptional fossil preservation in the Oligocene Florissant lake beds, Colorado, United States. *Geology* 28, 195-198.

Kazmierczak, J., Coleman, M.L., Gruszczynski, M. and Kempe, S. 1996. Cyanobacterial key to the genesis of micritic and peloidal limestones in ancient seas. *Acta Palaeontologica Polonica* 41, 319-338.

Krumbein, W.E., Paterson, D.M. and Stal, L.J. eds. 1994 *Biostabilization of sediments*. Bibliotheks und Informationssystem der Carl von Ossietzky Universität Oldenburg, p. 97-105.

Martill, D. M. 1987 Prokaryote mats replacing soft tissues in Mesozoic marine reptiles. *Modern Geology* 11, 265-269.

Martill, D. M. 1988 Preservation of fish in the Cretaceous of Brazil. *Palaeontology* 31, 1-18.

Pérez-Moreno, B. P., Sanz, J. L., Buscalioni, A. D., Moratalla, J. J., Ortega, F. & Rasskin-Gutman, D. 1994 A unique multitoothed ornithomimosaur dinosaur from the Lower Cretaceous of Spain. *Nature* 370, 363-367.

Pflüger, F. 1999 Matground structures and redox facies. *Palaios* 14, 25-39.

Sagemann, J., Bale, S.J., Briggs, D.E.G. and Parkes, R.J. 1999 Controls on the formation of authigenic minerals in association with decaying organic matter: an experimental approach. *Geochimica Cosmochimica Acta* 63, 1083-1095.

Schieber, J. 1999 Microbial mats in terrigenous clastics: the challenge of identification in the rock record. *Palaios* 14, 3-12.

Seilacher, A. 1999 Biomat-related lifestyles in the Precambrian. *Palaios* 14, 86-93.

Seilacher, A. and Pflüger, F. 1994 From biomats to benthic agriculture: A biohistoric revolution: in Krumbein, W.E., Paterson, D.M. and Stal, L.J. eds., *Biostabilization of sediments*. Bibliotheks und Informationssystem der Carl von Ossietzky Universität Oldenburg, p. 97-105.

Seilacher, A., Reif, W.-E., and Westphal, F. 1985 Sedimentological, ecological and temporal patterns of fossil Lagerstätten. *Philosophical Transactions Royal Society of London* B 311, 5-23.

Toporski, J.K.W., Steele, A., Westall, F., Avci, R., Martill, D.M., and McKay, D.S. 2002 Morphological and spectral investigation of exceptionally well-preserved bacterial biofilms from the Oligocene Enspel formation, Germany. *Geochimica Cosmochimica Acta* 66, 1773-91.

Wilby, P.R., Briggs, D.E.G. and Pinna, G. 1995 Soft tissue preservation in the Sinemurian of Osteno, Italy - comparisons with Solnhofen and with experimental results. *2nd International Symposium on Lithographic Limestones, Cuenca, Spain, July 1995, Extended Abstracts.* 163-164. Universidad Autonoma de Madrid.

Wilby, P.R., Briggs, D.E.G., Bernier, P. and Gaillard, C. 1996 Role of microbial mats in the fossilization of soft tissues. *Geology* 24, 787-790.

Wilby, P.R. and Briggs, D.E.G. 1997 Taxonomic trends in the resolution of detail preserved in fossil phosphatized soft tissues. *Geobios, Mémoire spécial* 20, 493-502.

Wuttke, M. 1983 'Weichteil-Erhaltung' durch lithifizierte Mikroorganismen bei mittel-eozänen Vertebraten aus den Ölschiefern der 'Grube Messel' bei Darmstadt. *Senk. Lethaea* 64, 509-527.

Chapter 19

FOSSIL AND SUBRECENT FUNGAL COMMUNITIES IN THREE CALCRETE SERIES FROM THE DEVONIAN OF THE CANADIAN ROCKY MOUNTAINS, CARBONIFEROUS OF NORTHERN FRANCE AND CRETACEOUS OF CENTRAL ITALY

Possible selective fungal action on rocks

Alain. PREAT[1], K. KOLO[1], B. MAMET[1], Anna A.. GORBUSHINA[2] and David C. GILLAN[3]

1. Département des Sciences de la Terre et de l'Environnement CP160/02, Université Libre de Bruxelles, 50 av.FD Roosevelt, B-1050, Brussels, Belgium
apreat@ulb.ac.be
2. Geomikrobiologie, ICBM, Carl von Ossietzky Universität, Postfach 2503, D-2611, Oldenburg, Germany
3. Laboratoire de Biologie Marine CP160/15, Université Libre de Bruxelles, 50 av. F.D. Roosevelt, B-1050, Brussels, Belgium

1. INTRODUCTION

Paleosols and other types of subaerial exposures are present in ancient shallow water limestone sequences and continental facies (Wright, et al., 1988). They provide important information on sea-level changes, sedimentation rates and early diagenetic processes. Calcrete petrography is now well established, mainly from the soils on Recent and Pleistocene limestones (Freytet & Plaziat, 1982, Esteban & Klappa, 1983, Wright & Tucker, 1991). Although microbial organisms are frequently associated with calcretes, their role is still debated and poorly understood. Especially interesting is the association of surficial layers with microbial communities referred to as biofilms. In Recent subaerial environments such biofilms are ubiquitous and are dominated by microscopic fungi (micromycetes). Fungi are also dominant in modern soils and are therefore to be expected to occur in all buried environments. However, it is not clear, how subaerial biofilms and associated paleosols together with inhabiting organisms, are fossilized and preserved. These communities are particularly dominated by microscopical fungi, which were present in the Earth history since Precambrian (e.g. Tyler & Barghorn, 1954). Fungi are numerous in fossil record, however, their identification is difficult and prepares a lot of problems for mycologists and paleontologists (Sherwood-Pike & Gray, 1985). The aim of this paper is to describe fungal communities found in three distinct geological calcretes and to discuss the extent to which they participated in the calcrete carbonate factory.

2. GEOLOGICAL SETTINGS

2.1 Upper devonian of the Rocky Mountains (Canada)

The Mount Hawk, Ronde, Sassenach and Palliser formations form a predominantly carbonate succession of Late Devonian age (Geldsetzer et al., 1987). This sequence extends from the Roche Miette to the Cinquefoil areas and covers more than 20 km in the Jasper Basin (Alberta, north of Calgary). The Upper Devonian unit is composed of several hundred metres of reefal and peritidal limestones and dolomites belonging to a regional carbonate platform. The Roche Miette and Cinquefoil sections consist of stromatoporoid and algal wackestones-bindstones interbedded with loferites, fine-grained laminar dolomites and stromatolitic mudstones. The overall evolution is regressive from deeper subtidal biostromes in the lower part, to intertidal and supratidal lagoons in the upper part. At that level, calcite pseudomorphs after sulphates are present and interbedded with aeolian cross-bedded siltstones indicating a general sebkha environment. Various thin calcretes occur in the upper part of the section: they are pelleted ('black pebbles' sensu Strasser & Davaud, 1983) or laminar and contain mud-cracks, micritic coatings on bioclasts, *Palaeomicrocodium* algae (Mamet & Préat, 1985) and small-sized roots. Bituminous stylolites are also present. Filamentous fungal borings or endoliths are present in a few paleosols associated with silty stromatolitic mudstones and laminar dolocretes (Plate 1, Figures 1,2). They are sometimes observed inside the root systems and can be reddish. The porosity of the rocks is minimal.

Plate 1

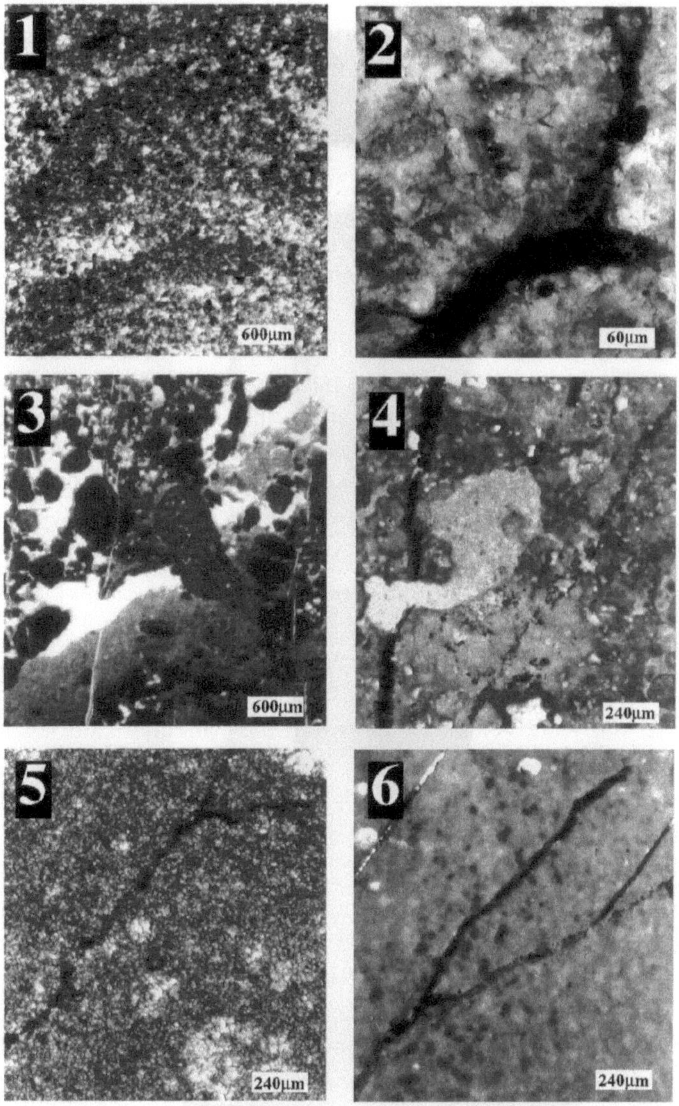

Figs 1, 2 : Stromatolitic calcrete (Fig. 1) with blackish dichotomous hyphae (Fig. 2). Thin-sections, sample CQF204. Roche Miette section, Canada, Devonian. **Figs 3-6** : Crystal vadose silt cavities (Figs 3, 4) and fine-grained calcite microspar (Figs 4, 5) in paleosol. The cavities intersect former dichotomic blackish hyphae (Figs 4, 6). Thin-sections, sample IB255 (Fig. 3), IB296 (Fig. 4), IB253 (Fig. 5), IB261 (Fig. 6). Avesnes-sur-Helpe section, France, Carboniferous.

2.2 Lower carboniferous (Viséan) of northern France

The Terwagne Formation is a semi-evaporitic restricted carbonate sequence of Early Carboniferous age (Viséan) outcropping in the Avesnes basin (Han et al., 2000) and mined in a quarry at Avesnes-sur-Helpe (northern France). The series is around 80 m thick and composed of foraminiferal packstones, ostracod-calciphaerid wackestones-packstones and algal-sponge bindstones interstratified with fine-grained idiotopic dolomites containing sulphate pseudomorphs. The formation consists of a succession of meter-scale peritidal cycles composed of very shallow subtidal fenestral wackestones and supratidal cryptalgal/stromatolitic laminites representing an inner shelf with evaporitic sebkhas. It is marked by numerous subaerial crusts and well-developed paleosols including crystal vadose silt cavities with pendant fibrous cements, rounded black pebbles, micritized ooids and pisoids (caliches), alveolar-septal structures, circum-granular cracks and ?*Palaeomicrocodium* aggregates. A large suite of geopetal textures on microcavities show multiple internal micro-graded infillings (Plate 1, Figures 3,4) suggesting pedogenetic 'pseudo-microkarst' vugs (sensu Freytet & Plaziat, 1982). Inter- and intragranular fungal networks and probable calcified faecal pellets are also present (Plate 1, Figures 5,6). As for the Devonian of Canada the rocks are tight as a consequence of important burial (Helsen, 1995; Han et al., 2000). As indicated above, the Avesnes quarry is currently active. This means that the collected samples are fresh and without surface contamination?

2.3 Cretaceous of central Italy

The margin of the Apulian platform in the Montagna della Maiella (Central Italy) is controlled by long term sea-level changes (Vecsei et al., 1998). The Fara San Martino succession, located 30 km SE of Pescara (Adriatic Sea), exposes a continuous 400 m thick profile of Lower Cretaceous well-bedded lagoonal to supratidal carbonates deposited in a low energy environment. Microfacies are composed of porous mudstones, wackestones with poorly preserved benthic foraminifers and various algae, and loferites with stromatolites. Laminar cryptalgal mats are commonly associated with mud-cracks. The series is composed of < 10 m-thick vertically stacked parasequences. These are capped by paleosols that contain a large suite of calcrete fabrics including rhizoconcretions, alveolar textures, 'clotted' fabrics associated with abundant calcite microspar ('sparmicritization' sensu Kahle, 1991), irregular calcareous crusts, geopetal crystal silts, biogenic laminar crusts, micritic coatings and angular pebble to cobble-sized algal chips (Plate 2, Figure 1). Dolocretes with sulphate pseudomorphs are also present. Extensive surfaces of exposure cover fenestral mudstones and contain abundant fungal networks and spores. These organisms associated with the top of the cycles (over more than 30 cm) are abundant, varied and well preserved. Most of the filaments are brownish to whitish, very long (up to 600 μm) with varying diameters (5 to 30 μm), straight and branched with many dichotomies in a same specimen (Plate 2, Figures 3,4). Other Fungi are present as very thin blackish filaments.

The porosity is important (20-25%), vuggy, fenestral, intergranular or, less frequently, moldic when associated with the roots and the compaction is low (no sutured contacts or grain stylolites). Pores are partially infilled with clear calcite as needle-fibers with thin brownish bands orthogonal to the elongation of the crystals (Plate 2, Figure 2). These crystals could be related to the activity of the Fungi (Verrechia, 2000) since long dichotomic filaments are associated with them. The brownish felting is probably related to the activity of microbial (bacterial?) mats as they form the same structures in Recent travertines and tufa (Janssen et al., 1999).

2.4 Sedimentology: synthesis

The systematic occurrence of *Fungi imperfecti* in the calcrete levels of three geological settings suggest that biomineralization was important in the formation of soil fabrics. The Fungi are the prominent species in these levels. Conditions were stressful for other organisms due to prolonged contacts with the atmosphere (subaerial exposure). In absence of competitors Fungi were able to proliferate in almost all calcrete levels. Their role in biomineralization was dominant as indicated by certain types of coated grains, bridging grains, rhizoconcretions, sparmicritisation, micritisation, perforations and secondary calcite formation from excreted fungal Ca-oxalates.

Microscopical analysis revealed two different fungal communities, an 'old' one, blackish (example of Plate 1, Figure 2), associated with grain boundaries and surfaces in all the three settings, and a Recent one, brownish, yellowish or colorless cross-cutting and perforating the grains and the matrix of the Devonian and Cretaceous rocks (example of Plate 3, Figures 3,4). These fungal communities are exclusively associated with the paleosols, whether these levels are tight (Devonian) or porous (Cretaceous). No Recent fungal communities were found in the Carboniferous for obvious reason. The outcrop was still too fresh for a fungal attack to develop any Recent subaerial biofilm.

Plate 2

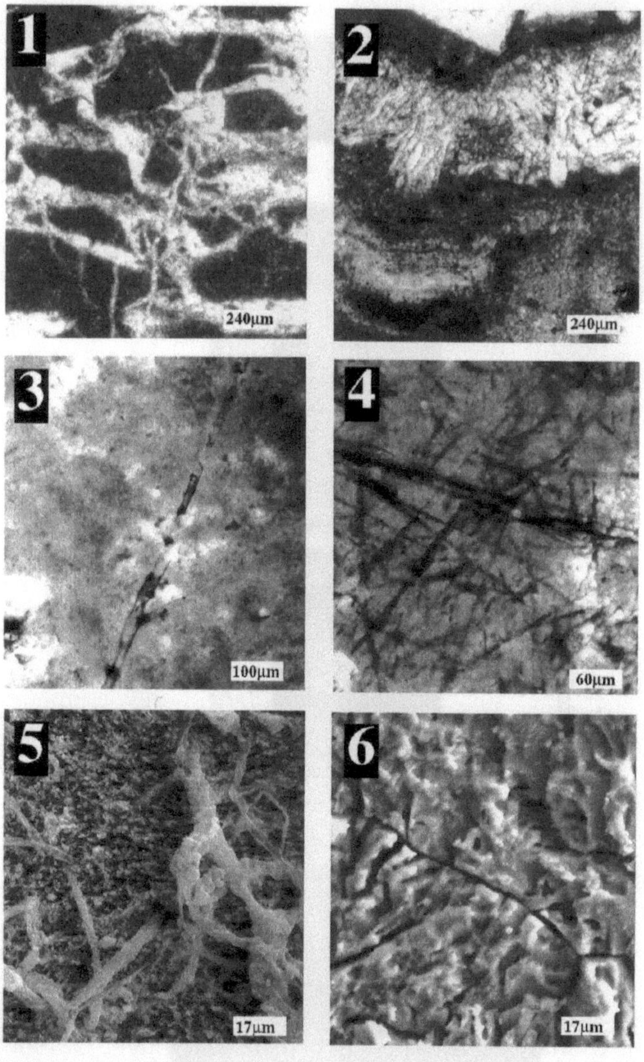

Plate 2: All figures from Fara San Martino section, Italy, Cretaceous. **Fig. 1**: Angular pebble to cobble-sized algal chips associated with a calcrete level. Thin-section, sample FSM79. **Fig. 2**: Pore, partially infilled with clear calcite as needle-fibers with orthogonal thin brownish bands. Upper part shows idiotopic dolomitic crystals. Thin-section, sample FSM3. **Figs 3, 4**: Long straight dichotomic and colorless (Fig. 3) or brownish (Fig. 4) fungal filaments through rock matrix. Thin-sections, samples FSM60 (Fig.3) and FSM6 (Fig.4). **Figs 5, 6**: Dichotomic hyphae coated with flaky calcite through the micritic matrix (Fig 5) and isolated calcite crystals showing 'old' fungal dichotomic perforations. The micritic matrix is in SEM pictures on slightly etched samples FSM60 (Fig. 5) and FSM 20 (Fig .6).

3. RELATIONSHIP BETWEEN FUNGI, SEDIMENTS AND ROCKS

3.1 Methodology

3.1.1 Extraction of OM

The Devonian and Carboniferous samples are represented by dark and very dark rocks relatively rich in black organic matter (OM). The Cretaceous samples are lightly to whitish coloured, generally with poorer content of lightly coloured OM. Traditional OM extraction techniques (Robinson, 1969; Durand and Nicaise, 1980) involve surface cleaning, crushing, extraction of soluble OM with soxhlet apparatus or ultrasound, alternate usage of HCl and HF acids for rock digestion (mostly carbonates and silicates), liberation of contained OM, and finally washing, filtration, centrifuge concentration and heavy liquid separation of solid non-soluble OM for petrography. Here, a slightly modified method was applied. The whole process was carried out manually and only whole rock samples were used to ensure the best undamaged OM recovery. Rock samples (150-200 g) were peeled (1-2 cm) using a high speed rock cutter to remove any extraneous material, washed several times, and for extra precaution dipped in concentrated HCl for 5-10 mn. The remaining samples were washed several times with distilled water and acid-digested using diluted HCl only. The recovered OM containing fungal material was filtered, washed several times, concentrated and separated manually.

3.1.2 Other preparations

Thin-sections were impregnated with slow penetration and setting resin from which the surface was later ground. Alternatively whole thin-sections were dipped in HCl under controlled dissolving process of exposed rock matrix. The resulting embedded-dissolved thin-sections revealed the *in situ* fungal network distribution. It was also possible to differentiate (fluorescence and light microscopy) between the grain or matrix-dependant fungal material, and the perforating and matrix independant fungal material. Additional sample preparations included etched thin-sections for SEM imaging. Rock-Eval pyrolysis (Tissot & Welte, 1984; Bordenave et al., 1993) was used to determine both the level of maturity of organic matter through the Tmax parameter and the quantity (weight %) of OM in the samples (Table 1). Leco furnace supplied additional analyses on TOC %. As the fungal material was the sole component of the OM, it was possible to draw some conclusions on the relative age of the fungal matter using the level of maturation.

3.2 Devonian fungi

Seventeen samples from calcrete levels yielded rich fungal material on extraction associated with very fine amorphous OM attached to fine filaments and sometimes covering them. The other 24 samples representing rock facies (peri-reefal packstones, lagoonal wackestones ...) unrelated to calcretes contained no fungal material. The fungal material composed of highly dichotomous blackish filaments of various sizes, generally 1-15 μm, represents a mycelium-like structure. Separate spores and hyphal conidiaphores were detected (Plate 3, Figure 1). Hyphal walls appeared partly or sometimes wholly mineralized under polarised light. They are interlaced with very dark-reddish rootlets-like OM (Plate 3, Figures 2,3) reflecting a mycorrhizal association. Both materials were blackish with reddish tint as observed with fluorescence microscopy using UV and blue filters 350-500 g. Such characteristics would reflect thermally stressed OM (Robert, 1985; Van Gijzel, 1981) suggesting a depositional origin. Furthermore, the fluorescence pattern revealed two different types of fungal networks. The first penetrated the rock as straight filaments either solitary or in bundles irrespective of the grain geometry, crossing them and giving strong green-yellowish fluorescence. The point of origin appeared to be the sample's surface suggesting a Recent origin. The second is always a part of grain geometry, as a coating and in-between grain boundaries. This second type was non-fluorescent. In embedded-and-dissolved thin-sections, the first type criss-crossed the second type which appeared as honeycomb structure reflecting the round shape of the dissolved grains (Plate 3, Figure 3-6). Rock-Eval pyrolysis gave maturity parameter of Tmax around 442 °C. (Table 1). This level is quite high and reflects thermally stressed OM that had a long burial history, thus 'old' OM which is consistent with fluorescence results. In embedded-and-dissolved thin-sections it was possible to confirm the presence of Ca-oxalate crystals (Plate 3, Figure 7).

Plate 3

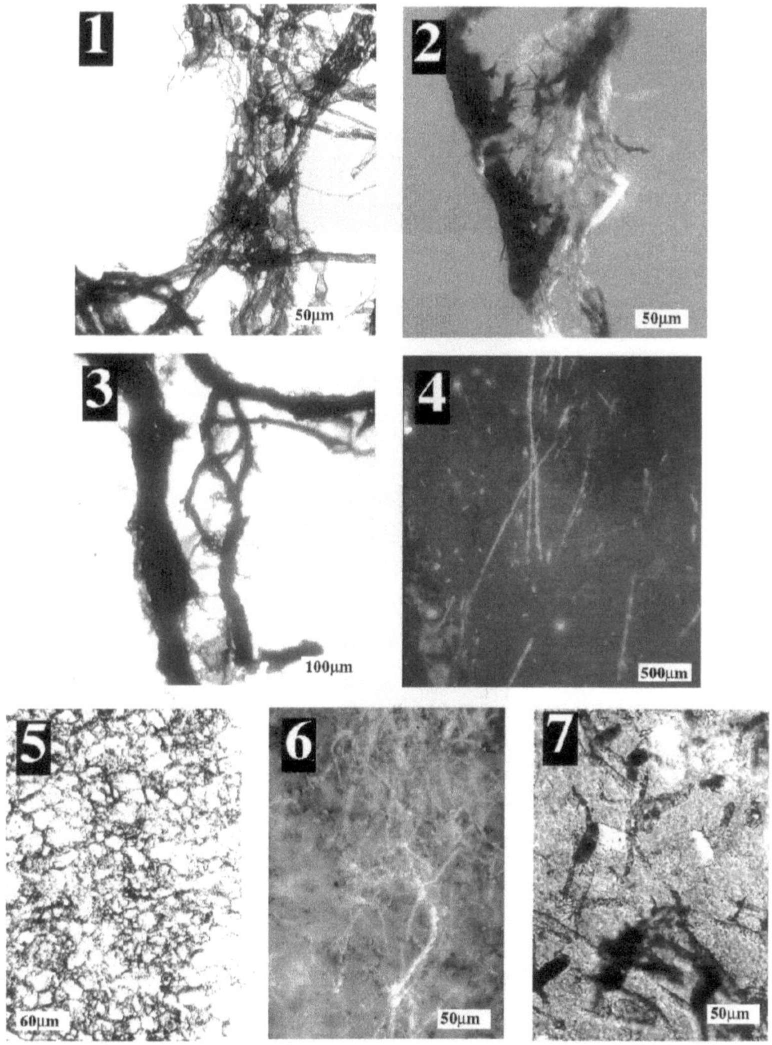

Plate 3: All the figures from Roche Miette section, Canada, Devonian. **Fig. 1**: Highly dichotomous mycelium, conidiophore and spores. Extracted material, sample CQF431, p.pl. **Fig. 2** :'Old' black non-fluorescent hyphae interlaced with Recent highly fluorescent hyphae and with black reddish rootlets-like OM. Extracted material, sample CQF204, fl.l. **Fig. 3**: Black non-fluorescent rootlets interlaced with hyphae. Extracted material, sample CQF418, fl.l. **Fig. 4**: Recent highly fluorescent hyphae penetrating the rock matrix. Some hyphae appear to be conidiophores. Etched thin-section, sample CQF406, fl.l. **Fig. 5**: Embedded dissolved thin-section showing 'old' black funglal material surrounding dissolved grains. Sample CQF406, p.pl. **Fig. 6**: Embedded dissolved thin-section showing Recent highly fluorescent hyphae penetrating rock matrix in all directions with 'old' black non-fluorescent fungal material in between the grains. Sample CQF418, fl.l. **Fig. 7**: Embedded dissolved thin-section showing Ca-oxalate crystals associated with Recent dichotomous hyphae. Sample CQF428, cr.p.pl.

Sample	Age	Tmax	TOC-1	TOC-2	OM%
FSM6	K	-	<0.1	-	0-1
FSM20	K	450	<0.1	<0.1	0.35
FSM35	K	-	<0.1	0.5	0.6
FSM36	K	-	-	<0.1	0.2
FSM60	K	-	-	<0.1	0.3
FSM66	K	-	-	<0.1	0.3
FSM69	K	-	-	<0.1	0.3
FSM79	K	-	-	-	0.2
FSM99	K	-	-	-	0.2
FSM124	K	453	<0.1	<0.1	0.2
FSM125	K	-	-	<0.1	0.2
FSM139	K	-	-	<0.1	0.4
FSM141	K	452	<0.1	-	0.4
FSM200	K	442	<0.1	-	-
FSM227	K	449	<0.1	-	-
CQF423	D	-	-	-	0.4
CQF424	D	-	-	-	0.4
CQF425	D	-	-	0.5	0.6
CQF427	D	442	<0.1	-	0.4
CQF428	D	442	<0.1	-	0.4
CQF429	D	-	-	-	0.2
CQF451	D	442	<0.1	-	0.3

Table 1. FSM Fara San Martino profile, K Cretaceous, CQF Cinquefoil profile, D Devonian, Tmax (°C) from Rock-Eval pyrolysis, TOC-1 % from Rock-Eval pyrolysis, TOC-2 % from Leco, OM% from manual separation

3.3 Carboniferous fungi

The 35 analysed rock samples of different rock facies and palaeoenvironments (subtidal lagoons, oolitic bars, algal mats, paleosols) returned fungal and fungal-like material only from calcretes levels. The fungal remains showed high fragility, typical of very old organic matter and special care was applied to ensure good retrieval. The material is composed of differently sized mostly rounded structures, which are sometimes resembling hyphae, moniliform hyphae, branching thalli, but predominantly spores or yeast-like round cells with budding patterns. Sometimes filamentous and round cells were obviously presenting interconnected stages of the life cycles (germinating spores, (Plate 4, Figures 1-3). Sizes of the structures were surprisingly different ranging from 5 to 500 μm (Plate 5,

Figures 1-4). Bodies similar to zygospores are present. Oogonium-like structure with antheridium attachment could also be observed (Plate 4, Figure 4). In one etched thin-section (Plate 4, Figure 5), a conidiaphore similar to present *Penicillium* was observed. All the retrieved fungal material was non-fluorescent. On the surface of these fungal remains it was possible to see under high magnification very fine hyphae inside or covering and cutting through these bodies (Plate 4, Figure 3). The distribution of these bodies is grain-related and no cutting or crossing were observed. The rock samples being gathered from a fresh section of an active quarry insured that Recent endolithic fungal attack was not possible and the retrieved material is non-contamined. Thermal evolution based on illite crystallinities (Han et al., 2000) or conodont alteration index (Helsen, 1995), gave a Tmax around 195°C.

3.4 Cretaceous fungi

The 65 analysed samples gave varied and abundant fungal material again from the calcrete levels only. Samples representing other rock types (reefs, lagoons, algal marshes) were sterile. All typical fungal features were present, including well developed mycelia of variably sized hyphae belonging apparently to different communities, spores and conidia, oogonia with hyphal and antheridial attachments (Plate 4, Figure 6). The general color is light, yellowish and the filaments are transparent and very long with diameters between 1-10 μm. Larger filaments are also present (15-20 μm), but are less abundant. Within this lightly colored Fungi is mixed and interwoven much darker (reddish brown) and mineralized filaments and/or plant rootlets. In one instance, a cellular plant structure was observed. Investigation with fluorescence microscope of isolated material, of embedded-and-dissolved thin-sections and of thin-sections showed fluorescence distribution patterns similar to the previous sections suggesting the same association of 'old' and sub-Recent Fungi. SEM imaging revealed filaments covered with flakey calcite, penetrating the matrix or residing in pores (Plate 2, Figure 5). Some isolated large grains surrounded by fine micrite showed 'old ' fungal perforations while the micrite was free from any perforations, suggesting a possible pre- or synlithification fungal attack associated with subaerial exposure and calcrete formation (Plate 2, Figure 6).

Plate 4

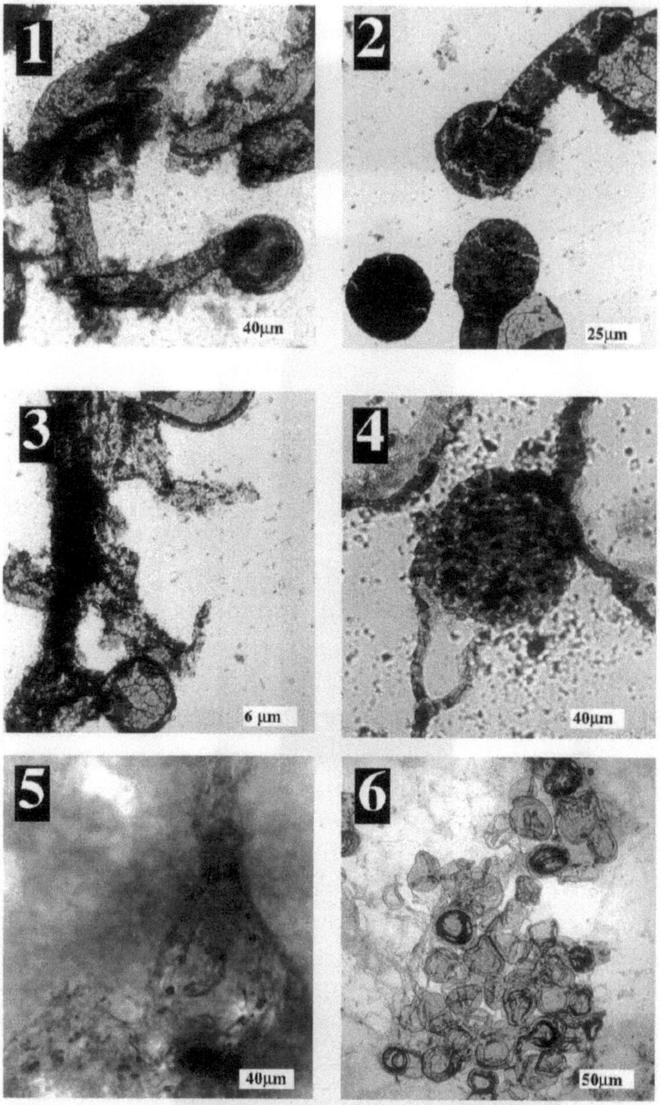

Figures 1-5 from Avesnes-sur-Helpe section, France, Carboniferous. **Fig. 1**: A sporangiophore with sporangium and degraded hyphae. Extracted material, sample IB223, p.pl. **Fig. 2**: Sporangium with the germinating spores. Extracted material, sample IB247,p.pl. **Figs. 3-4**: Fungal thallus with the fruiting bodies showing endobiotic? parasitic? hyphae (Fig. 3) and oogonium-like structure with antheridium attachment (Fig. 4). Extracted material, sample IB223, p.pl. **Fig. 5**: A conidiophore similar to modern *Penicillium*. Etched thin-section, sample IB218,p.pl. **Fig. 6**: A mass of spores, conidia, oogonia, with hyphal and antheridial attachments. Extracted material, sample FSM3, Fara San Martino section, Cretaceous, Italy, Fl.1.

Plate 5

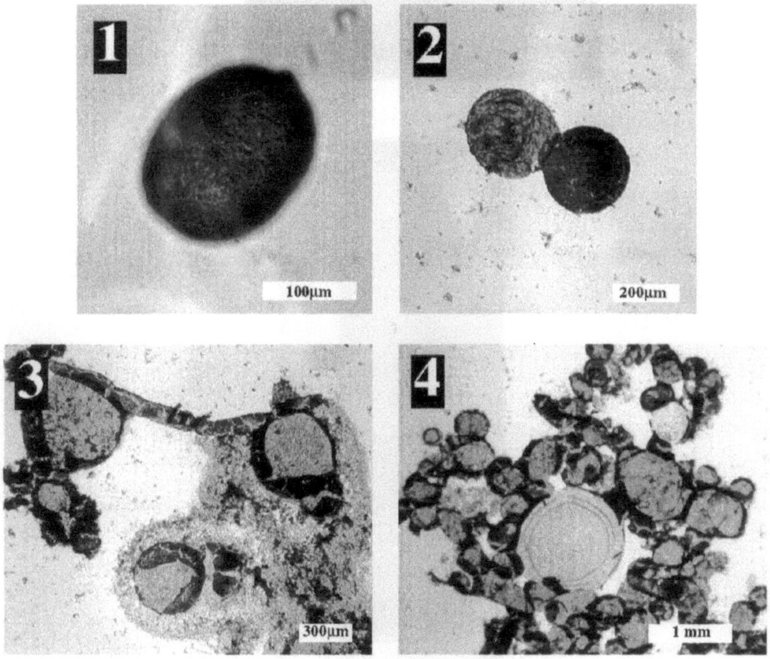

Figures 1-4 from Avesnes-sur-Helpe section, France, Carboniferous. **Fig. 1**: A spore with a typical attachment structure. Extracted material, sample IB219. **Fig. 2**: Two closely attached cells. Extracted material, sample IB223. **Fig. 3**: Various organisms (or various shapes of one organism) with EPS-like material. Extracted material, sample IB223. **Fig. 4**: Diversity of sizes, sample IB223.

4. DISCUSSION AND CONCLUSION

In general terms the idea is that *Fungi imperfecti* invaded at least during two periods the same calcrete layers, the first time during eogenesis (when calcretes were unlithified sediments) and the last time during telogenesis (when calcretes were lithified rocks). These layers constituted the top of thin shallowing-upward carbonate tidal sequences which were regularly exposed as a result of relative sea-level decrease. The fossilized fungal hyphae and spores are probably opportunistic since no other organisms are associated. Due to the stressful environmental conditions related to the subaerial exposition (contact with the atmosphere) and semi-evaporitic conditions (sulphate pseudomorphs), these Fungi did not suffer competition with other organisms. Detailed analysis of the shallowing-upward sequences has shown that the Fungi disappeared a few cm before or after exposure time, when other organisms (bacteria, algae, ostracods, foraminifers, mollusks...) were returned. These 'first' old fungal communities are quite well preserved and could not be confused with Recent ones which invaded only later the same levels. The 'old' filaments are indeed blackish, partly degraded and localized between the original sedimentary grains (pisoids, aggregates...). They are cut or displaced by early vadose cavities or bioturbation figures (Plate 1). The Recent (?or sub-Recent) fungal community is in contrast brownish, yellowish to colorless, well preserved and mainly perforating the original grains and structures (cavities, burrowing, laminations...).

Contamination is considered here as any organic material that is not related to the rock system. The procedure applied here insured that only organic matter related directly to the rock was extracted (3.1). A double check including random sample selection, numbering and analysis together with reversed analysis: sample selection > extraction of OM > rock facies analysis insured the validity of fungal material relationship to calcretes. The authenticity of fungal material was further proved by the absence of fungal material from all the other samples that were not calcretes.

Tmax from Rock-Eval pyrolysis and coloration of spores and pollens (Thermal Alteration Index- TAI, Robert 1985) are good indicators of the degree of thermal stress (duration and intensity) to which the OM was exposed to. Thus Recent OM can not show high maturation degree (Tmax > 436 °C, TAI > 2.5, brownish-reddish-black) except in cases where this OM was exposed to abnormal thermal stress (e.g. igneous activity). Therefore the OM (fungal material) had a very long thermal exposure related to burial. In contrast, the other immature part of OM (lighter colors and strong fluorescence) is related to Recent contamination. The unstable Ca-oxalates (weddelite, from XRD) were observed only in Devonian and Cretaceous samples where Recent and 'old' Fungi are associated but not in the Carboniferous samples (deep fresh samples) where no Recent fungal material was recovered. Association of calcretes with old depositional fungal material poses no interpretation questions as these rock types are active sites for microorganisms bio-activities.

It is widely accepted that Fungi show no specific rock type preference for attack and infestation (Verrecchia 2000; Gorbushina & Krumbein 2000). On the other hand some experimental work (Silverman & Munoz, 1970) has implied a selective behaviour. At the time being, the question is unresolved.

Experiments in progress aim to explain why these former calcretes were selectively invaded by these modern Fungi. One major reason could be a higher organic carbon content precisely due to the activity of former microbes (bacteria and Fungi). Also when the 'modern' invasion started is a question of debate. Their absence from the Carboniferous in a fresh outcrop seems to indicate that they are very young.

ACKNOWLEDGEMENTS

The authors wish to thank Roberta Bruni (University of Ancona), Dr. Devleeschouwer (Belgian Geological Survey) for help on sampling and collecting data. Thanks also due to Prof. Verrechia (University of Neuchâtel) for stimulating discussions and Rock-Eval analyses, and Prof. Tribovillard (University of Lille) for TOC analyses. We tank also M. Pinte for the access of the Avesnes-sur-Helpe quarry. The investigations were supported by the Fonds National de la Recherche Scientifique de Belgique (project FRFC n°2-4515-99) and The Communauté Française de Belgique (project 01.3.1).

REFERENCES

Bordenave ML., Espitalié J., Leplat P., Oudin JL. and Vandenbrouke M. (1993) Screening techniques for source rock evaluation. In: Bordenave ML. (Ed.) Applied Petroleum Geochemistry Editions Technip, *Paris, 217-278.*

Durand, B., and Nicaise, G., 1980, Procedures for kerogen isolation. *In: Durand, B., (Ed.), Kerogen-Insoluble organic from sedimentary rocks. Edition Technip, pp. 37-53.*

Esteban M. and Klappa CF. (1983) Subaerial exposure environment. *In: Carbonate Depositional Environments, Scholle PA., Bebout DG. and Moore CH. (Eds)., Am. Ass. Petrol. Geol. Mem., 33, 1-54.*

Freytet F. and Plaziat JC. (1982) Continental carbonate sedimentation and pedogenesis -Late Cretaceous and Early Tertiary of Southern France. *In: Contribution to Sedimentology, Stuttgart, Purser BH. (Ed.), **12**, 213pp.*

Geldsetzer HHJ., Goodfellow WD., McLaren DJ. and Orchard MJ. (1987) Sulfur-isotope anomaly associated with the Frasnian-Famennian extinction, Medicine Lake, Alberta, Canada. *Geology, **15**, 393-396.*

Gorbushina A. and Krumbein W.E. (2000) Subaerial microbial mats and their effects on soil and rock. *In: Microbial Sediments, Riding RE. and Awramik SM. (Eds.), Springer Verlag, 161-170.*

Han GM., Préat A., Chamley H., Deconinck JF. and Mansy JL. (2000) Palaeozoic clay mineral sedimentation and diagenesis in the Dinant and Avesnes Basins (Belgium, France): relationships with Variscan tectonism. *Sedimentary Geology, **136**, 217-238.*

Helsen S. (1995) Burial history of Palaeozoic strata in Belgium as revealed by conodont colour alteration data and thickness distributions. *Geol. Rundsch.*, ***84***, *738-747.*

Janssen A., Swennen R., Podoor N. and Keppens E. (1999) Biological and diagenetic influence in Recent and fossil tufa deposits from Belgium. *Sedimentary Geology,* ***126***, *75-95.*

Kahle CF. (1991) Origin of subaerial Holocene calcareous crusts: role of algae, Fungi and sparmicritisation. *In: Calcretes, Wright VP. and Tucker ME. (Eds.), Blackwell Sc. Publ.,* ***2***, *207-230.*

Mamet, B. and Préat A. (1985) Sur la présence de *Palaeomicrocodium* (*Algue?, Incertae sedis*) dans le Givétien Inférieur de Belgique. *Géobios,* ***18***, *389-392.*

Robinson WE. (1969) Isolation procedures for kerogen and associated soluble organic materials. *In: Organic Geochemistry, Methods and Results, Eglinton G. and Murphy MTJ. (Eds), Springer Verlag, Berlin-Heidelberg, 181-195.*

Robert P.(1985) Histoire géothermique et diagenèse organique. *Bull. Centres Rech. Explor.-Prod., Elf-Aquitaine, Mém.* ***8***, *Pau.. 345p..*

Silverman MP. and Munoz E.F. (1970) Fungal attack on rock: solubilization and altered infrared spectra. *Science,* ***169***, *985-987.*

Sherwood-Pike, M.A. and Gray, J. (1985) Silurian fungal remains: probable records of the Class Ascomycetes. *Lethaia,* ***18***, *1-20.*

Strasser A. and Davaud E. (1983) Black pebbles of the Purbeckian (Swiss and French Jura): lithology, geochemistry and origin. *Eclogae geol. Helv.,* ***76***, *551-580.*

Tissot B. and Welte D. (1984) Petroleum Formation and Occurrence. *Second Edition. Springer Verlag, Berlin, .699p.*

Tyler, S.A. and Barghorn, E.S. (1954) Occurrence of structurally preserved in Pre-Cambrian rocks of the Canadian shield. *Science,* ***119***, *606-608.*

Vecsei A., Sanders DGK, Bernouilli D., Eberli, G. and Pignalti, JS (1998) Cretaceous to Miocene sequence stratigraphy and evolution of the Maiella carbonate platform margin, Italy. *In*: *SEPM Special Publication, Mesozoic and Cenozoic sequence stratigraphy of European basins, de Graciansky PC., Hardenbol, J., Jacquin T. and Vail P.R. (Eds), No****60***, *53-74.*

Verrecchia E. (2000) Fungi and sediments. *In: Microbial Sediments, Riding RE. and Awramik SM. (Eds.), Springer Verlag, 68-75..*

Wright VP. and Tucker ME. (1991) (Eds.) Calcretes. *Reprint Series Volume* ***2*** *of the International Association of Sedimentologists, Blackwell Scientific Publications, 352pp.*

Wright VP, Platt NH. and Wimbledon WA. (1988) Biogenic laminar calcretes: evidences of calcified root-mat horizons in paleosols. *Sedimentology,* ***35***, *303-620.*

Van Gijzel P. (1981) Applications of geomicrophotometry of kerogen, solid hydrocarbons and crude oil to petroleum exploration. *In: Brooks J. (Ed.) Organic Maturation Studies and Fossil Exploration. Academic Press, London and New York, 351-377.*

Chapter 20

ARCHEAN BIOFILMS PRESERVED IN THE SWAZILAND SUPERGROUP, SOUTH AFRICA

Maud M.WALSH
Department of Agronomy
104 Sturgis Hall
Louisiana State University
Baton Rouge, LA 70803, USA
evwals@lsu.edu

Frances WESTALL
Lunar and Planetary Institute
3600 Bay Area Boulevard
Houston, TX 77058
USA

1. INTRODUCTION

The discovery of microfossils and stromatolites in Archean rocks of South Africa and Australia respectively (Awramik et al., 1983; Walsh and Lowe, 1985; Byerly et al., 1986; Walsh, 1992; Schopf, 1993; Westall and Gerneke, 1998; Hofmann et al., 1999; Westall et al., 2001) indicate that life was present on Earth as 3.5 billion years ago, and geochemical evidence suggests that it may have arisen as early as 3.8 billion years ago (Schidlowski, 1988; Mojzsis et al., 1996; Rosing, 1999). Preserved microfossils, however, are rare in Archean rocks; our information on occurrence of biota is from our examination of other evidence of life, including the presence of preserved microbial mats or biofilms. Fossilized biofilms are contained within carbonaceous cherts, rocks composed almost entirely of microcrystalline quartz that are black or dark gray in color because of the presence of kerogen, degraded organic matter. The carbonaceous cherts formed by the early silicification of biogenic deposits of organic matter (Lowe, 1999). Total organic carbon in the cherts examined in this study ranges from 0.10 to 14.6 mg C/ g (Hayes et al., 1983; Walsh and Lowe, 1999). This paper describes the results of a petrographic investigation of the carbonaceous cherts of the Barberton Greenstone Belt, South Africa. Over 400 petrographic thin sections were examined using light microscopy in an effort to characterize the mode and environment of formation of the sediment precursors to the carbonaceous cherts. Scanning electron microscopy has also been employed in examining some of the samples and the results reported in Westall et al. (2002—this volume).

2. GEOLOGIC AND ENVIRONMENT AL SETTING

The Barberton Greenstone Belt is located in the eastern part of the Kaapvaal Craton, South Africa. The predominantly volcanic supracrustal sequence that makes of up the belt is the Swaziland Supergroup, which includes three major lithostratigraphic subdivisions: the Onverwacht, Fig Tree, and Moodies Groups. Single zircon evaporation ages at 3445 +/- 4 Ma have been determined for the felsic volcanics of the uppermost Hooggenoeg Formation (Kroner et al., 1991). The lowermost Kromberg chert has been dated at 3416 +/- 4 Ma (Kroner et al., 1991; Byerly et al., 1996) and the upper chert of the Kromberg Formation has also been recently dated at 3334 +/- 6 Ma (Byerly et al., 1996). The uppermost dated unit of the Onverwacht Group is a chert within the Mendon Formation dated at 3298 +/- 6 Ma. Ages for the overlying Fig Tree Group range from 3258 +/- 3 Ma to 3225 +/- 3 Ma (Kroner et al., 1991).

The Swaziland Supergroup is made up of a lower predominantly volcanic succession, the Onverwacht Group, and an upper, mainly sedimentary succession, the Fig Tree and Moodies Groups. The Onverwacht Group is subdivided into the Sandspruit, Theespruit, Komati, Hooggenoeg, Kromberg and Mendon Formations (Lowe and Byerly, 1999). The depositional settings for the cherts of the Onverwacht Group are of two major types: platformal and basinal (Lowe, 1999). The platform deposits of the Hooggenoeg, Kromberg and lower cycles of the Mendon Formation were deposited either in shallow-water and subaerial environments, where they tend to be asociated with komatiitic or dacitic volcanic units, or under quiet, slightly deeper-water conditions associated with basaltic volcanism (Lowe, 1999). The upper cycles of the Mendon Formation represent fine-grained basinal sediments deposited mainly in deep, quiet water. The Fig Tree Group preserves sediments representing basinal, fan delta, and shallow water deposits and contains both volcaniclastic, terrigneous and orthochemical units. The Moodies Group is a quartz-rich sedimentary sequence deposited in alluvial to marginal marine to shallow-marine settings on stable continental crust (Heubeck and Lowe, 1999; Heubeck, 2001).

Asteroid impacts were also an important factor in the early Archaen environment. Calculations indicate that large asteroid impacts may have periodically extinguished the Earth's surface biota prior to approximately 3.8 Gyr (Sleep et al., 1989; Sleep and Zahnle, 1998; Ryder et al., 2000). Several deposits of impact-generated spherules indicate that impacts of substantial size occurred during during the time period represented by the Onverwacht Group (Lowe and Byerly, 1986; Lowe et al., 1989).

3. PLANAR MICROBIAL MATS

Fine kerogenous laminations that alternate with pure silica laminations are a common form of fossilized biofilm in the black and white banded cherts that cap ultramafic volcanic rock sequences. The kerogenous laminations that comprise the mats are 1-20 microns thick and the layers of pure siica (now represented by microcrystalline quartz) are slightly thicker (Figure 1A). The laminations are slightly wavy, with relief on the order of a few millimeters. On the micron scale the laminations are commonly discontinuous, but on a larger scale many can be traced across sample surfaces several cm wide. They are wavy or crinkly on a small scale, with maximum relief of a few mm. The gross structure of the layers is similar to that of modern microbial mats, with the alternation of organic-rich layers with layers of detritus or precipitate. Evidence for intermittent wave activity and subaerial exposure, including the presence of fold-over structures and rip-up clasts (Figure 1B), suggests that the planar mats formed in shallow subaqueous environments.

The resence of preserved filamentous microbes within fine kerogenous laminations further supports the contention that such laminations are preserved microbial mats. Filamentous fossils were reported from one location in the Hooggenoeg Formation, a 4m-thick section of upper Hooggenoeg chert in a locality on the east limb of the Onverwacht Anticline (Walsh, 1992). The microstructures are present in an approximately 60-cm-thick black and white banded chert overlying silicified volcaniclastic sands. The chert containing filamentous fossils is made up mainly of fine kerogenous laminations interlayered with thin accumulations of detrital kerogenous grains. Pyrite grains 5-25 μm in size are scattered throughout the thin section. Solid threadlike filaments composed of kerogen and fine pyrite grains occur maily in clusters within both the fine laminae and the detrital layers. The filaments are have cross- sectional diameters ranging from less than 0.2 μm to 2.5 μm and lengths up to 200 μm.

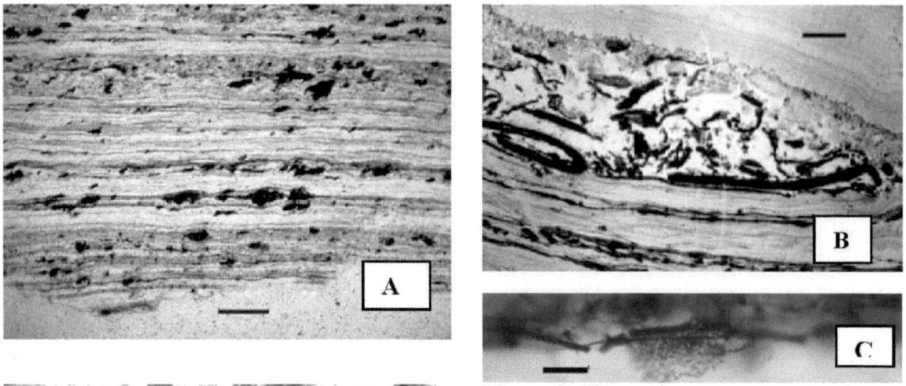

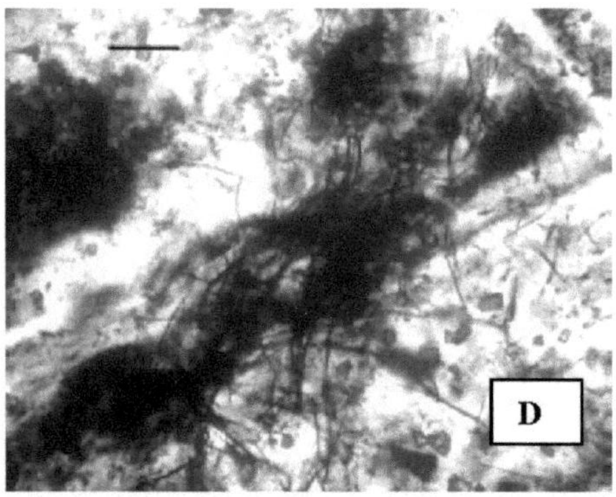

Figure 1. Planar kerogenous mats. A. Flat kerogenous laminations, Hooggenoeg chert (MW48-12; 26°00'30"S, 31°00'05"E). Scale bar =500µm. B. Folded laminations and rip-up clasts (MW15-1; 25°55'08", 30°58'44"). Scale bar =1mm. C. Fossil filament parallel to planar laminations in Kromberg chert (SAF 23-57; 26°01'42"S, 30°59'26"E). Scale bar =10µm. D. Tangle of fossil filaments in Kromberg chert (MW45-7; 26°01'42"S, 30°59'26"E). Scale bar =20µm

A more complex community of filamentous microbes is preserved in a sample from the Kromberg Formation (Walsh, 1992). The microfossils are preserved within layers of fine carbonaceous laminae that contain some lithic (intraclast?) fragments. Hollow cylindrical filaments range from 1.4 to 1.2 µm in diameter and 10-150 µm in length and solid threadlike filaments range from less than 0.2 than 0.2 µm to 2.5 µm and lengths up to 200 µm . Most are non-septate, with a few exhibiting slight constrictions at intervals of approximately 1 µm, or breakage at intervals of several µm. The walls of the filaments are composed of carbonaceous matter and fine pyrite grains. The filaments are commonly oriented subparallel to bedding (Figure 1C), but in some cases extend downward between layers, or radiate from a tangle of filaments (Figure 1D). The filamentous microbes may represent only a portion of the original community, but seem to have had a role in the construction of the mats. Morphologically the preserved microbes are similar to both filamentous cyanobacteria and to filamentous sulfur bacteria, but their size range is closer to that of filamentous bacteria.

Fine kerogenous laminations are common in the carbonaceous cherts of the Hoogenoeg and Kromberg formations and are associated with shallow water deposits that formed over volcanic flows (Walsh and Lowe, 1999).

They are rare in the upper Mendon and Fig Tree formation, which generally represent deeper water deposits with more clastic input.

Wrinkly laminations that appear to be kerogenous have been reported associated with shallow-water quartzose sandstone of the Moodies Formation (Heubeck, 2001). The laminations, less than 1 mm in thickness, drape ripple structures. Some contain domal or bulbous cm-size upward protrusions that appear to have had cohesive strength. They may represent the remains of microbial mats that formed on ripples and sand waves in a shallow subtidal environment during quiet intervals between sediment deposition.

4. COMPOSITE KEROGENOUS PARTICLES

Commonly associated with the flat-lying mats are lobate particles that are made up of smaller carbonaceous particles connected by a silica film. Lobate, usually more-or-less equant clumps of kerogen that are made up of smaller, commonly rounded carbonaceous particles are here termed composite grains (Fig. 2). They range from about 100 μm to over 1000 μm in diameter. Individual component carbonaceous particles, ranging from 25 to 100 μm in diameter, are commonly surrounded and cemented by botryoidal, coarsely crystalline microquartz that, in rare cases, shows relict fibrous texture. Isolated flakes of sericite are common within the particles.

Composite grains are most common in bands of carbonaceous laminations and in detrital layers composed of mixtures of simple and composite grains. In some beds, composite grains are mixed with sand-sized lithic grains. In others, composite grains are so loosely packed within pure translucent chert that there are few grain-grain contacts. Their occurrence in sedimentary layers with chert intraclasts suggests that the silica coating formed while the particles were in the water column. Unlike simple carbonaceous grains, composite grains are typically associated with lithic grains of approximately the same size, indicating that the density of the composite grains was similar to that of the lithics, possibly because of a mineral coating. A modern analogy may be the flocculent material found in the vicinity of the Juan de Fuca hydrothermal vents which contains cells uniformly coated with iron and silica and has been interpreted to represent microbial cells that inhabitated the sub-seafloor hydrothermal system and were coated with mineral precipitates as they made contact with cooling fluids (Juniper et al., 1995). Fossil microbes have not been located within composite grains. Like the fine kerogenous laminations considered to represent planar microbial mats, composite grains are common in the Hooggenoeg and lower Kromberg Formations. The presence of biofilms is suggested by the cohesion of the kerogenous particles, the eivdent trapping of kerogenous and lithic particles within the layers, and the close environmental association of the composite particles with the microbial mats.

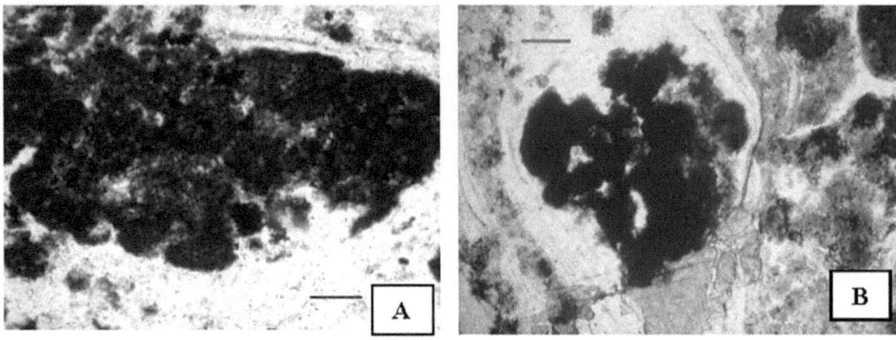

Figure 2. Composite kerogenous grains. A: Hooggenoeg Formation (MW55-3; 25o57'12"; 30o52'35") B. Kromberg Formation (MW45-26; 26o01'45"S, 30o59'21"E . Scale bar equals 100 microns in each.

5. STROMATOLITES?

The third type of probable biofilm preserved in the Swaziland Supergroup, rare in comparison with the other two, is flat-laminated and domal structures resembling stromatolites (Byerly et al., 1986; for discussion see Lowe, 1994, Buick et al., 1995). The stromatolites are variable in morphology, occuring as low relief laterally linked domes 0.5 to 6 cm in height or, rarely, as pseudocolumns (Figure 3A). The individual laminae range in thickness from 50 to 100 μm. The stromatolitic layers, which range from less than 1 to about 20 cm thick, directly overlie volcanic flows and are in many place overlain by a subsequent volcanic flow. Intraclasts of the stromatolitic layers are commonly found in association with the intact layers, suggesting that disruption, possibly by desiccation, and local transport of the particles has occurred. The stromatolites are made up of mainly microcrystalline quartz, with tourmaline, fine rutile, and rare carbonaceous matter (Figure 3B). No microfossils have been located in the stromatolites with light microscopy (for a fuller discussion of the fine detail of the structures, see Westall et al., 2002-- this volume). Boron isotopic analyses of the tourmaline suggest that a two-stage boron enrichment occurred, in which marine evaporitic boron was hydrothermally remobilized (Byerly and Palmer, 1991). The combination of the evidence of erosion and local transport of stromatolite fragments and the hydrothermal isotopic signature suggest that the stromatolites most likely formed in the vicity of hydrothermal vents in shallow water environments.

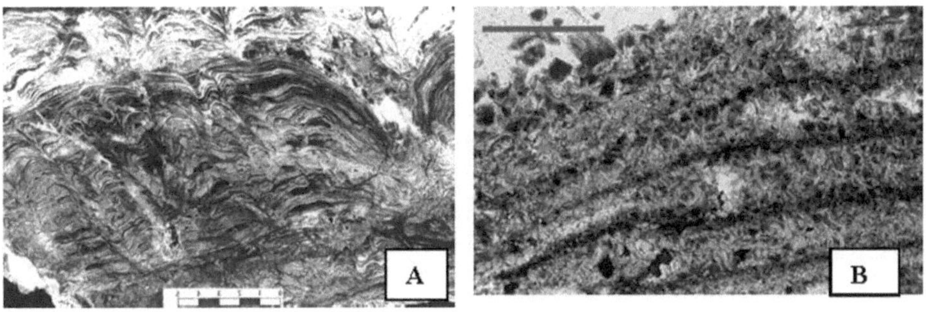

Figure 3. Stromatolite-like structures in the Mendon Formation. A. Cut surface of stromatolite-like structure. Scale is in cm. B. Detailed view of tourmaline-rich layers of stromatolite-like structure in petrographic thin section (25°54'10.5"S, 31°02'41"E). Scale bar equals 40 μm.

6. DISCUSSION AND CONCLUSIONS

A variety of morphologies of probable fossilized biofilms is present in the Swaziland Supergroup. The biofilms are found in shallow-water deposits that formed in the intervals between volcanic eruptions and asteroid impacts. The morphological similarities between the planar kerogenous laminations and modern microbial mats, as well as the presence of fossils microbes in some examples, strongly argues for their interpretation as preserved biofilms/microbial mats. The resemblance of composite kerogenous particles to flocculent particles containing colonies of microbes and associated with deep-sea vents suggests that the Archean structures may have formed in a similar manner, though precipitation of silica on cells in the water column. This implies that Arhcean life may have had a planktonic component as well as a substrate-bound component Stromatolite-like structures lack the internal feature and associated microbes of the planar mats. Their interpretation as biofilms requires further study, focusing on chemical signatures that might constrain conditions of formation. The variety of probable biofilm types suggests that microbes adapted to diverse ecological niches on the early Earth.

7. ACKNOWLEDGEMENTS

Maud Walsh gratefully acknowledges support from the Louisiana State University Council for Research for a summer stipend and Louisiana Space Consortium and the Board of Regents Support fund, and the NASA Space Grant Program under NGT5-40035.

Frances Westall acknowledges M.de Wit, J. Högbom, the NRC and the NAI (through D. McKay) for partial support of this research.

REFERENCES

Awramik, S.M., Schopf, J.W., and Walter, M.R. (1983) Filamentous fossil bacteria from the Archaean of Western Australia. *Precambrian Research* 20: 357-374.

Buick, R., Groves, D. I., and Dunlop, J. S. R. 1995. Abiological origins of described stromatolites older than 3.2 Ga: Comment. *Geology* 23: 191.

Byerly, G., Lowe, D.R., and Walsh, M.M. (1986) Stromatolites from the 3300-3500 Myr Swaziland Supergroup, Barberton Mountain Land, South Africa. *Nature* 319, 489-491.

Byerly, G. R., Kroner, A., Lowe, D. R., Todt, W., and Walsh, M. M. 1996. Prolonged magmatism and time constraints for sediment deposition in the early Archean Barberton greenstone belt; evidence from the Upper Onverwacht and Fig Tree groups. *Precambrian Research* 78: 125-138.

Byerly, G.R. and Palmer, M.R. (1991) Tourmaline mineralization in the Barberton greenstone belt, South Africa: Early Archaean metasomatism by evaporite-derived boron. *Contrib. Minerol. Petrol.* 107, 387-402.

Hayes, J. M., Kaplan, I. R., and Wedeking, K. W., 1983, Precambrian organic geochemistry, preservation of the record, In Schopf, J. W. (Editor) *Earth's Earliest Biosphere: its origin and evolution*: Princeton University Press, Princeton, p. 93-133.

Hofmann, H.J., Grey, K., Hickman, A.H., Thorpe, R.I. (1999) Origin of 3.45 Ga coniform stromatolites in the Warrawoona Group, Western Australia. *Geol. Soc. Amer. Bull.* 111: 1256-1262.

Heubeck, C. 2001. Microbial films of the Archean Moodies Group (~3.1Ga), Barberton Greenstone Belt, South Africa. In: Fossil and Recent Biofilms, Programme, Abstracts and Participants, ed. By W. E. Krumbein, T. Dornieden and M. Volkmann. University of Oldenburg, 2001.

Heubeck, C. and Lowe, D. R. 1999. Sedimentary petrography and provenance of the Archean Moodies Gorup, Barberton Greenstone Belt, in *Geologic evolution of the Barberton Greenstone Belt, South Africa*, edited by D.R. Lowe and G.R. Byerly, Geol. Soc. Am. Sp. Paper 329, pp. 259-286.

Juniper, S. K., Martinequ, P., Sarrazin, J. and Gelinas, Y., 1995, Microbial-mineral floc associated with nascent hydrothermal activity on CoAxial Segment, Juan de Fuca Ridge: *Geophysical Research Letters* 22: 179-182.

Kroner, A., Byerly, G. R., and Lowe, D. R., 1991, Chronology of early Archean granite-greenstone evolution in the Barberton Mountain Land, South Africa, based on precise dating by single zircon evaporation: *Earth Planet. Science Letters* 103: 41-54.

Lowe, D. R. 1994. Abiological origin of described stromatolies older than 3.2 Ga. *Geology*, 22, 387-390.

Lowe, D. R. 1999. Petrology and sedimentology of cherts and related silicified sedimentary rocks in the Swaziland Supergroup in *Geologic evolution of the Barberton Greenstone Belt, South Africa*, edited by D.R. Lowe and G.R. Byerly, Geol. Soc. Am. Sp. Paper 329, pp. 83-114.

Lowe, D. R. and Byerly, G. R. 1999. Stratigraphy of the west-central part of the Barberton Greenstone Belt, South Africa, in *Geologic evolution of the Barberton Greenstone Belt, South Africa*, edited by D.R. Lowe and G.R. Byerly, Geol. Soc. Am. Sp. Paper 329, pp. 1-36.

Lowe, D. R., Byerly, G. R., Asaro, F. and Kyte, F. J., 1989. Geological and Geochemical Record of 3400-Million-Year-Old Terrestrial Meteorite Impacts. *Science* 245: 959-962.

Mojzsis, S.J., Arrhenius, G., Keegan, K.D., Harrison, T.H., Nutman, A.P. and Friend, C.L.R., 1996. Evidence for life on earth 3,800 million years ago. *Nature* 384: 55-58.

Rosing, M.T (1999) 13C depleted carbon microparticles in >3700-Ma seafloor sedimentary rocks from Westa Greenland. *Science* 283: 674-676.

Ryder, G., Koeberl, C., and Mojzsis, S.J. (2000) Heavy bombradment of the Earth at ~3.85 Ga: The search for petrographic and geochemical evidence. In *Origins of the Earth and Moon*

Schidlowski, M. (1988) A 3,800-million-year isotopic record of life from carbon in sedimentary rocks. *Nature*, 333, 313-335.

Schopf, J.W. (1993) Microfossils of the Early Archean Apex Chert: new evidence of the antiquity of life. *Science*, 260, 640-646.

Sleep, N. H., Zahnle, K. J., Kasting, J. F. and Morowitz, H. J., 1989. Annihilaton of ecosystems by large asteroid impacts on the early Earth. *Nature*, 342: 139-142.

Sleep, N.H. and Zahnle, K.I (1998) Refugia from asteroid impacts on early Mars and early Earth. *J. Geophys. Res.*, 103, 28,529-28,544.

Walsh, M.M., 1992. Microfossils and possible microfossils from the Early Archean Onverwacht Group, Barberton Mountain Land, South Africa. *Precambrian Res.*, 54, 271-293.

Walsh, M.M. and Lowe, D.R. (1985) Filamentous microfossils from the 3,500 Myr-old Onverwacht Group, Barberton Mountain Land, South Africa. *Nature*, 314: 530-532.

Walsh, M.M. and Lowe, D.R. (1999) Modes of accumulation of carbonaceous matter in the early Archaean: A petrographic and geochemical study of the carbonaceous cherts of the Swaziland Supergroup, in *Geologic evolution of the Barberton Greenstone Belt,*

South Africa, edited by D.R. Lowe and G.R. Byerly, Geol. Soc. Am. Sp. Paper 329, pp. 115-132.

Westall, F. and Gerneke, D. (1998) Electron microscope methods in the search for the earliest forms of life on earth (in 3.5-3.3 Ga cherts from the Barberton greenstone belt, South Africa): applications for extraterrestrial life studies. *Proc. SPIE, Intl. Soc. Opt. Eng.*, 3114, 158-169.

Westall, F., de Wit, M.J., Dann, J., de Ronde, C.E.J., Gaast, S.J. and Gerneke, D. (2001) Early Archaean fossil bacteria and biofilms in in hydrothermally-influenced, shallow water sediments, Barberton greenstone belt, South Africa. *Precambrian Research* 106: 93-116.

Westall, F., Walsh, M.M., Toporski, J. and Steele, A. (2002) Fossil biofilms and the search for life on Mars. (this volume)

Chapter 21

IS THE FEEDBACK BETWEEN GENETIC STRUCTURE AND COMMUNITY STRUCTURE THE COMPUTATIONAL MECHANISM OF BIOFILM EVOLUTION?

Douglas E. CALDWELL
Dept. of Applied Microbiology and Food Science
51 Campus Drive
University of Saskatchewan
Saskatoon, Saskatchewan S7K 5A8 Canada
doug..caldwell@home.com

1. INTRODUCTION

In his 1859 volume titled "The Origin of Species By Means of Natural Selection or the Preservation of Favoured Races in the Struggle For Life" Darwin stated that natural selection is not the only means of evolution. In his words, "Furthermore, I am convinced that Natural Selection has been the most important, but not the exclusive means of modification" (Darwin 1859). Zavarzin (2000) has referred to this alternative aspect of evolution as the "non-Darwinian domain". This alternative approach may be necessary to understand the emergence of community-level organization in multispecies biofilms.

One example of non-selective evolution may be the Earth itself. Gaia theory postulates that Earth is a self-regulating system (Levit and Krumbein 2000) and the Daisy World computer simulation shows that this is theoretically possible (Watson and Lovelock, 1983). However, Gaia theory does not postulate the mechanism by which global selfregulatory mechanisms could have originated. There is only one Earth and it is not the result of diverging lines of planetary descent from which nature might have selected fit and unfit planets. Hence global regulatory systems probably emerged through nonselective mechanisms of modification. However, due to the difficulty of performing empirical experiments on a planetary scale, it is difficult to resolve this question experimentally.

Continuous culture is another example of modification with no apparent mechanism of selection. Unlike the Earth, it is a laboratory system that is inexpensive, easily controlled, and well understood in mathematical terms. Continuous culture thus provides an ideal system in which to study non-selective evolution and it has led to the development of nested proliferation theory as an alternative to selection theory (Caldwell et al 1997). In addition, the movement of biofilm communities from sphere to sphere in a dual-dilution continuous culture might be analogous to the movement of biospheres from planet to planet. Continuous culture might thus shed some light on the evolution of the Earth's biosphere if microorganisms are capable of moving between planets.

2. NON-SELECTIVE EVOLUTION IN CONTINUOUS CULTURE

During the continuous culture of microorganisms, fresh medium continuously flows into the culture. Spent cells and medium continuously flow out through an overflow device to ensure that the inflow equals outflow. This maintains a culture of actively growing cells without having to subculture on a daily basis. Another advantage of continuous culture, as opposed to batch culture, is that the kinetics of growth are easily defined and controlled.

To gain an intuitive understanding of continuous culture kinetics, imagine holding a glass of water under a slowly running faucet and letting it overflow into a sink. The glass of water contains a food coloring that is gradually diluted out by the water overflowing the glass. If instead of food coloring, the glass contained colored microorganisms that must use nutrients in the tap water to grow, the color of the growing organisms would be diluted out more slowly than the food coloring. This would be due to the ability of the colored organisms to proliferate. The growth of the organisms prevents them from being diluted out until the flow rate of the tap is increased to the point at which the dilution rate of the culture (flow of tap water divided by the volume of the glass) exceeds the maximum specific growth rate (μmax) of the microorganisms. Setting the dilution rate (D) of a continuous culture at a value below the maximum specific growth rate, automatically sets the growth rate of the organisms to the dilution rate of the culture (growth rate now being controlled by the restricted inflow of nutrient rather than the maximum speed of growth under ideal conditions).

Bacteria are removed from continuous culture by non-selective dilution, rather than by differential selection. However, they still undergo evolutionary modification as has been demonstrated by numerous investigators and this work has been summarized in a review by Caldwell et al (1997). During the modification of individual bacteria, the cells reproduce by binary fission and thus they do not age and die (as in the case of most speciated plants and animals). Consequently organisms are not differentially selected by survival.

Lethal mutations do occur but have a negligible effect on the growth of the survivors and thus they have no appreciable effect on the outcome of evolutionary modification. This is due to the low rate of lethal mutations, (k)(c), compared to the high rate at which viable cells are lost from the culture through dilution, (D)(c), as per the following equation.

$$\partial c/\partial t = (\mu)(c) - (D)(c) - (k)(c) \qquad (1)$$

Where:

$\partial c/\partial t$= rate of change in cell density
$(\mu)(c)$ = change in cell density due to growth
$(D)(c)$ = change in cell density due to dilution
$(k)(c)$ = change in cell density due to spontaneous lethal mutations
c = cell density or biomass
t = time
k = specific death rate
μ = specific growth rate
D = dilution rate

Thus selective death can be reduced or eliminated as a factor in the evolution of continuous cultures. Similarly, it is possible to eliminate or reduce the selective dilution of organisms from the culture. This requires that the outflow of the culture be designed so that organisms will not be differentially selected by the overflow mechanism. Cells leave the culture non-specifically (all cells and groups of cells have an equal probability of leaving) without regard for size, shape, rate of growth, metabolic efficiency, or any other individual or collective attribute. This sometimes requires the use of a Feuillade siphon overflow (Schiefer and Caldwell, 1982), or a similar device, to ensure that clumped cells and filaments of cells are diluted at the same rate as individual rod-shaped cells. In other situations, silicone coatings may be needed on the interior surfaces of the culture to ensure that adherent cells are not selectively retained during dilution. Regardless of the specific characteristic or trait in question, the central principle in continuous culture is that the removal of cells must be as non-selective as possible.

This makes it possible to reduce all potential sources of selection until they have no appreciable effect on the outcome of evolutionary modifications occurring in continuous culture. This allows the investigator to focus almost exclusively on non-selective evolutionary mechanisms. Evolution in these non-selective cultures requires the removal of organisms by attrition (non-selective dilution), but it does not require selective removal (it does not require natural selection, artificial selection, k selection, r selection, individual selection, group selection, or multilevel selection). Non-selective removal is required to sustain constant proliferation. Constant proliferation then enables the continual emergence of new forms through multiscalar recombination (through simultaneous recombination both on the scale of genetic structure and on the scale of community structure).

3. CULTIVATION OF BIOFILMS AND BIOAGGREGATES IN CONTINUOUS CULTURES

Continuous cultures are normally used to cultivate isolated lines of planktonic cells. However, they are also used to cultivate multispecies planktonic communities, as well as biofilms and bioaggregates. Changes in the environment of these cultures result in a corresponding modification of genetic structures (genetic sequence) and community structures (species composition, distribution, and abundance). This occurs in the absence of any appreciable selective pressure due either to selective mortality or to the selective removal of organisms (Caldwell et al 1997, Caldwell 2000).

Bioaggregate communities are grown in a Feuillade continuous culture as described above and consist of multispecies flocs that tend to foul and clog the overflow rather than being diluted uniformly. The aggregates increase in size until the shear forces within the culture (caused by aeration or stirring) are sufficient to fragment them into smaller pieces. This is in contrast to the proliferation of individual cells that divide by binary fission. The mean size of aggregates can be controlled by controlling the agitation of the culture, while the mean size of individual cells remains relatively unaffected.

Biofilm communities are grown in a dual-dilution continuous culture. Dual-dilution cultures consist of a round-bottomed flask containing either standard glass beads or neutral density glass beads (Korber et al. 1994). A multispecies biofilm community forms through simultaneous attachment and growth on the surface of the beads and the walls of the culture vessel. The aqueous dilution rate is equal to the flow rate divided by the volume of the aqueous phase. However, the surface area of the culture is also diluted and its dilution rate is controlled by replacing the culture vessel twice each day, while retaining the glass beads and aqueous medium. Increasing or decreasing the surface area of the glass beads in the culture increases or decreases the surface dilution rate as opposed to the aqueous dilution rate. It is interesting to note that the movement of biofilm communities from one glass sphere to another in continuous culture is analogous to the movement of a biosphere from one planet to another. In this analogy the size of a planet would be analogous to the size of a word in a cosmic language, as explained below.

4. THE INHERITANCE OF CONTEXTUAL INFORMATION

Microorganisms evolve in continuous culture despite the lack of selection via either cell mortality or cell removal, and this anomaly has led to the development of nested proliferation theory as an alternative means of evolutionary modification (Caldwell 2000). The central concept of proliferation theory is that information can not be contained in isolation within either a molecule or an organism. Instead, it always resides jointly within context as well as content. Thus when an organism reproduces it inherits information residing jointly within its internal environment and within its external environment. The external information affects both the magnitude and meaning of the information within its internal microenvironment. The external information may seem less important to an object than the information residing within it. However, both types of information are inherited and are integral components of a single geometric calculator. The need to postulate that evolving information resides both externally and internally (i.e. simultaneously among objects as well as within objects) can be better appreciated by considering the following examples: bits in a CD, three-way light switches on a wall, words in a sentence, *Anabaena* sp. and *Zoogloea* sp. in continuous culture, and bits in a byte.

4.1 Bits on a CD

When data is recorded on a compact disk (CD) a laser is normally used to record and read a spiral series of dots (pits or spots created by a laser beam). Each dot represents a bit and is either on or off depending upon whether it is present or absent. As is shown in Fig. 1 a bit on a compact disk can, in theory, be toggled on or off either by toggling the bit itself or by toggling the background. For example, a black dot on a white background can be turned off in one of two ways. The first is to make the black dot the same color as the background by making the dot white. The second is to make the dot the same color as the background by making the background black. It is easier to change the internal environment (content) of the dot than it is to change the external environment (context) of each dot. Consequently, this is normally what is done to toggle bits on and off. This does not mean that the background lacks information. The contextual information is simply taken for granted as a matter of expedience. However, ecology is the study of relationships between objects and their environment. Thus the ecology of informational objects is always a matter of describing changes in informational interrelationships between contextual information and information content (with each object being the interface between context and content).

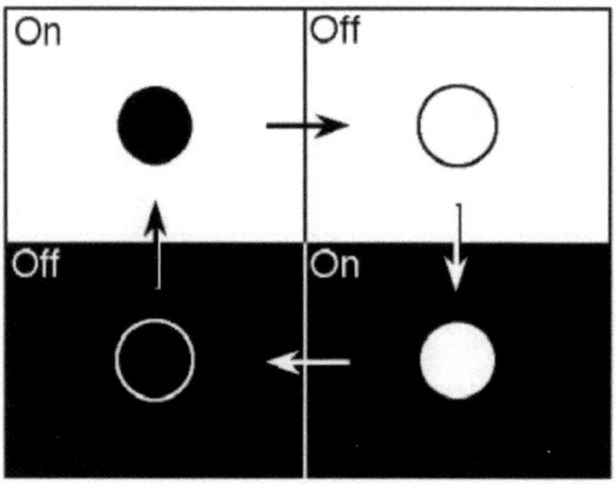

Figure 1 The potential life cycle of a bit. A bit can be turned on or off either by toggling the external environment of the bit (vertical arrows) or by toggling the internal environment of the bit (horizontal arrows).

4.2 Light Switches on a Wall

Informational relationships like the one above may seem to have no practical significance. However, one practical application is the use of three-way light switches at the bottom and top of a flight of stairs. The lights can be turned on using the bottom switch and then turned off using the top switch. These switches are never labeled on (up) and off (down) like normal light switches. Their meaning changes depending upon the setting of the other switch. If a series of three-way switches and four way switches are used together, then any number of switches can be used to turn a single light on and off. In this situation the meaning of up and down for any individual switch is unknown until the settings of all the other switches have been determined. In the case of these light switches, contextual information has been given functionality and thus becomes more meaningful because it is useful. It is useful because it can be controlled, although in most ecological situations information content is easier to control than the contextual information. However, meaning and significance is always derived mutually from both content as well as context.

4.3 Words in a Sentence

To clarify the importance of both content and context in determining meaning, consider the words in a sentence. The meaning of each word can not be determined from its spelling alone. Meaning is derived both from the positioning of the letters within a word and the positioning of the words around it.

For example, consider the following sentences:

Time flies like an arrow.
Fruit flies like a banana.

In the first sentence, "flies" is a verb. In the second sentence, "flies" becomes the plural form of a noun. When reading the first sentence the meaning of "flies" is derived from the context of the words around it. When reading the second sentence it is logical to assume that the same meaning should apply, but then the last word in the sentence, "banana", redefines the meaning of "flies" and the meaning of the entire sentence must be reinterpreted. This is a confusing effect associated with a paradigm shift, during which one must migrate from one frame of reference to another.

It is also important to note that when words are spoken, information (meaning and significance) also resides in the inflection of the voice. This means that the external environment responsible for influencing the meaning of words is not restricted to the sequence of words but is also dependent upon other aspects of environmental context. In many cases inflection is unintentional, and yet it is strategic and can be interpreted. In the case of a skilled actress, it is often intentional. However, in both cases it affects proliferation and is strategic information. The error may be in assuming that space-time positioning geometry which is not intended (subconscious) is not informative or of strategic significance. It should also be noted that inflection and oral language involve timing, and hence it is necessary to postulate that it is space-time geometry, rather than just spatial geometry, that is potentially informative.

Margalef (1963) may have been the first to make the connection between the similarities in the information content of both words and organisms (although Darwin noted similarities in the evolution of words and of species). In Margalef's words, "Let us imagine the structure of a community in terms of a message, written in a language with a number of symbols equal to the number of species, and where individual symbols stand for individuals". He also discussed the possibility that elements of an organism's environment, such as "dead wood" and "rabbit burrows" are "elements of organization" and products of a 'series of decisions" influencing the propagation of ecological systems.

4.4 Anabaena sp. and Zoogloea sp. In Continuous Culture

The relationships among groups of organisms changes depending upon environmental context, just as the meaning of words depends upon their internal and external environment. As shown in Table 1, the effect of a *Zoogloea* sp. on an *Anabaena* sp. varies from a 2,700% increase to a 13% decrease in *Anaebaena* sp. cell density, depending on changes in environmental context. The functional relationship between the organisms changes although there is no internal change in the organisms.

Table 1. Effect of environmental context on the percentage stimulation of an *Anabaena* sp. by the presence of a *Zoogloea* sp. in carbon dioxide limited continuous culture. Data from Schiefer and Caldwell (1982).

Environmental Context		Percentage Stimulaton		
gas phase	glucose	*Anabaena sp.*	chl a	phycocyanin
Carbon dioxide free air	1.5	710%	790%	760%
Air (0.03% carbon dioxide)	1.5	700%	410%	640%
5% carbon dioxide in air	1.5	-13%	-124%	4%
Carbon dioxide free air	7.5	2,700%	1,800%	2,900%

4.5 Bits in a Byte

Context determines the quantity of information as well as its meaning. This can be demonstrated by considering the effect of the number of bits in a byte. A bit is a single on/off signal and this is the smallest possible unit of digital information. A byte is a linear series of 8 bits. While a single bit can represent only two possible codes, two bits can represent four codes as shown in Table 2. The four codes that can be created by using two bits are: on-on, on-off, off-on, and off-off. Three bits in sequence can produce 8 different codes. Note that there is a synergistic effect as bits are combined to create information as shown below.

Table 2. Bits, nibbles, bytes and the synergistic role of context in determining the quantity of information as well as the meaning of information.

Bits (dots on a CD)	Information Potential (total possible number)	Information Content (per bit)
1 (a bit)	2	2.0
2	4	2.0
3	8	2.7
4 (a nibble)	16	4.0
5	32	6.4
6	64	10.6
7	128	18.3
8 (a byte)	256	32

This type of informational synergy is inherent in the geometry of information. It is thus a part of the causal mechanism of evolution. However, the cooperation and competition between organisms, as illustrated in the example of the *Anabaena* and *Zoogloea* sp. associations above, is defined by environmental context. Cooperation and competition are thus most effectively understood as the products of informational calculations, rather than being the primary mechanism itself. They are effects rather than causes. They are calculated, but they are not the mechanism of calculation.

If this is so, then it is possible that Darwin mistook effect for cause, noting that whenever evolution occurred, competition also occurred. He might have therefore concluded that competition was a cause when it was the consequence, just as cooperation is an inevitable consequence of evolution rather than a causal mechanism (and perhaps it too has been mistaken as cause rather than effect). Thus cooperation and competition always occur when evolution occurs. However, this is because they have a common cause and not because they are necessarily the causes of evolution in and of themselves. The primal cause of evolution is more likely to be the reciprocal control of growth by two or more nested geometric (informational) structures. If so, then the universe is a calculating mechanism that encodes information in the space-time geometry of each object. Each object being dependent on proliferation for its continued existence, and proliferation being dependent upon positioning geometry (on multiple scales of space and time).

5. NESTED PROLIFERATION THEORY AND MULTISCALAR RECOMBINATION

Postulating the existence of nested information systems makes it possible to consider whether feedback between the various levels of geometric nesting might provide a computational control mechanism sufficient to explain the emergence of complex adaptive systems in situations where there is no apparent physical mechanism of selection (continuous culture of biofilm and bioaggregate communities). Using the growth of *Anabaena* and *Zoogloea* sp. as an example (as discussed above) it is apparent that community structure and genetic structure are nested geometries with feedbacks that link the proliferation (specific growth rate) of genetic structures with that of community structures as shown in Fig. 2.

The rearrangement of genetic structures (through conjugation, transduction, and transformation) produces novel genetic structure (genetic sequences). Innovation in genetic structure has either a positive or a negative effect on the growth rate of the community structures within which the genetic structures are enclosed (nested). If the effect of genetic innovation (on the growth rate of the community) is positive then the innovative bioaggregate is retained in the culture. Those community aggregates lacking this innovation are lost through attrition (dilution) due to their lower growth rate (although they are not selected individually or collectively by either selective dilution or selective death). Note, that even though the specific growth rate of the genetic structure itself might be reduced by this genetic innovation, the improved specific growth rate of the community structure can compensate for this loss if the density of the genetic structures residing within the community structure increases. This enables the growth rate of the genetic sequences to keep pace with faster growing community networks. That is similar to the effect observed in rapidly proliferating bacteria, in which cells can replicate more quickly than does the circular bacterial genome. In this situation each bacterial cell must contain multiple copies of the genome if the net rate of DNA synthesis is to keep pace with the net rate at which cells are produced.

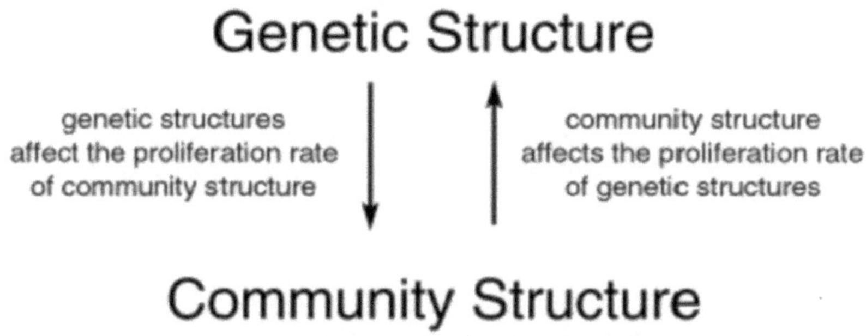

Figure 2 Genetic structure affects the growth rate of community structures and community structure affects the growth rate of genetic structures. This reciprocal feedback provides a computational mechanism for biofilm modification. More specifically, the restructuring of communities through the emigration and immigration of groups and individuals (recombination of community structures), affects the growth rate of the enclosed genetic structures. The reciprocal effect is that the restructuring of the enclosed genetic structures through genetic recombination (recombination of genetic structures) affects the growth rate of the enclosing community structures. This reciprocal feedback mechanism calculates compromises between the proliferation of community structures and genetic structures so that they are jointly optimized rather than being optimized individually. Continuous culture results in the attrition of any genetic innovation that might impair the proliferation of essential innovations in community structure. Thus, within the context of continuous culture, community structure constrains genetic structure through the external physical mechanisms of the dilution system. Similarly, genetic structure constrains community structure but this occurs through the internal biochemical mechanisms of transcription and translation.

Similarly, the recombination of community structures (through immigration and emigration) produces innovative communities (community-level innovation). These innovations in community structure then affect the specific growth rate of the proliferating genetic material which they enclose. Negative effects on the specific growth rate of genetic material would require that the density (copy number of genomes and/or cells) of the necessary coding be increased so that its net rate of propagation equals that of the community structure. This would then reduce the specific growth rate of the community due to costs associated with maintaining additional redundant copies of genetic material. If the liabilities associated with maintaining these additional copies are greater than the assets derived from the direct increases in the growth rate of the community structures then the innovation would be lost through attrition (non-selective dilution).

Changes in genetic structure that compromise critical innovations in community structure are lost through attrition (lost by dilution during continuous culture) and hence community structure constrains genetic structure. Changes in community structure that compromise critical innovations in genetic structure are also lost through attrition, and hence genetic structure constrains community structure. Consequently, it is likely that the reciprocal (and reiterative) feedback between genetic structure and community structure is the primary computational mechanism responsible for the evolutionary modification of biofilm communities. This system of checks and balances might be integral to the proliferation and modification of any set of nested geometries.

Both community structure and genetic structure are heritable if the lifetime of each exceeds the lifetime of the organisms involved. However, it should be noted that neither genetic structure nor community structure is inherited without modification. Both genetic structures and community structures are normally reshuffled (recombined) whenever they propagate and this is a source of innovation. Thus neither genetic innovation nor community innovation is heritable without modification, and all forms of innovation are continuously evaluated through attrition rather than differential selection.

6. THE POWER EQUATION

Combining the mathematics of exponential growth with the mathematics of information gives a relationship describing the growth of information capacity for proliferating chemical, physical, and biological systems (Caldwell 2000). This relationship was derived by assuming that the relative positioning of each physical object affects the proliferation of that object and thus embodies strategic information.

The exponential growth equation is:

$$c = (co)\,(e^{(\mu)(t)}) \qquad (2)$$

Where:

c = the final number of objects resulting from growth, replication, proliferation, or

propagation

co = the initial number of objects

e = the base of the natural logarithms

μ = the instantaneous fractional rate of increase (specific growth rate)

t = the time period over which the increase occurs

The information equation is:

$$C = (((\,(c_1)^{c_2})^{c_3})^{\ldots c_n}) \qquad (3)$$

Where:
n = maximum number of geometric levels at which objects within a system can recombine (e.g. in an alphabetic system there would be seven levels (n=7), if letters can be recombined within words (first level), if words can be recombined within words (second level), if words can be recombined within sentences (third level), sentences within paragraphs (fourth level), paragraphs within chapters (fifth level), chapters within books (sixth level), and books within libraries (seventh level)).
C = the total number of unique configurations or codes that are possible by repositioning the objects existing within all levels of an information hierarchy
c = the number of unique configurations or codes that are possible by repositioning the objects existing within each level of an information hierarchy Each set of parentheses represents one geometric level in an information system containing n levels, numbered 1, 2, 3, 4, etc. up to the nth level (cn).

Assuming that the position of each proliferating object is of strategic importance in ist proliferation and that it therefore embodies information, allows substitution of the expression in Eq. 1 (for c, the number of objects formed at any point in time) into Eq. 2 (that gives C, the number of codes that could be produced from a linear series of physical objects that are either present or absent in a geometric series). The result is the power equation (Eq. 4). This derivation tacitly assumes that mind (geometric computation) and matter (chemical, physical, and biological objects) can not be separated. Consequently each physical movement of an object is presumed to be a computational event (thought) as well as a physical event (movement).

$$C = [[[[\,(c1)\,(e^{(\mu)(t)})\,]\,[(c2)^{\,(e(\mu)(t))}\,]\,]\,[\,(c3)\,(e^{(\mu)(t)})\,]\,]\ldots[\,(cn)\,(e^{(\mu)(t)})\,]\,] \qquad (4)$$

7. CONCLUSIONS

Proliferation theory leads to the conclusion that evolution results from the growth of a nested series of proliferating objects, arranged much like a set of Russian nested dolls. An example is the proliferation of nucleic acids within cells, of cells within organisms, of organisms within communities, and of communities within ecosystems. Each level in the series is both habitat and inhabitant and this provides the computational mechanism for calculating compromise necessary to optimize proliferation across all these levels of spatial and temporal geometry. Thus each layer can grow somewhat independently, but the form and function of each is a calculated response to the form, function, and growth rate of the others. Each life form not only competes and/or cooperates but it is also continuously tuned so that it proliferates in harmony with the proliferation of the evolving forms that are nestled within it and that it is nestled within.

This principle of nesting may apply from the subatomic to the cosmic scale of existence. If so, then the proliferation of all physical, chemical and biological objects is controlled through calculations carried out by reciprocal feedbacks between nested space-time geometries, and the idea of nested computational geometry is necessary and sufficient to explain physical evolution as well as biological evolution. This makes physics a secondary consequence rather than a primary cause, and inverts the central precept of natural science.

It thus seems likely that the properties and behaviors of all chemical, physical, and biological objects may be due to geometric calculations encoded by the relative spacetime positioning for each of those objects. This would make evolutionary ecology and/or information ecology the core of both natural science and social science; with sociology, economics, chemistry, physics, and biology being the consequences of a universal computational mechanism. The "natural laws" of chemistry, physics, and biology thus now seem more likely to be secondary effects rather than primary causes.

As Albert Einstein wrote in "Evolution of Physics" published in 1938, "Physical concepts are free creations of the human mind, and are not, however it may seem, uniquely determined by the external world."

As Galileo Galilei wrote in Il Saggiatore, published in 1623. "Philosophy is written in this grand book - I mean the Universe - which stands continually open to our gaze, but it cannot be understood unless one first learns to comprehend the language and interpret the characters in which it is written. It is written in the language of mathematics, and ist characters are triangles, circles, and other geometrical figures, without which it is humanly impossible to understand a single word of it; without these, one is wandering about in a dark labyrinth."

REFERENCES

Caldwell, D. E. (2000) The origin of communities. Gaia Circular. 3(1), 27-32.

Caldwell, D. E., Wolfaardt, G. M., Korber, D. R., Karthikeyan, S., Lawrence, J. R., and Brannan, D. K. (2001) Cultivation of microbial consortia and communities. In: Manual of Environmental Microbiology. ASM press. Washington. In Press.

Caldwell, D. E., Wolfaardt, G. M., Korber, D. R. and Lawrence, J. R. (1997) Do bacterial communities transcend Darwinism? Advances in Microbial Ecology. Ed. By Jones. Plenum Press, New York 1997. pp. 105-191.

Cairns-Smith, A. G. (1982). Genetic Takeover and the Mineral Origins of Life. Cambridge University Press. 477 pp.

Darwin, C. (1859) The Origin of Species By Means of Natural Selection or the Preservation of Favoured Races in the Struggle for Life. New American Library. New York. 479 pp.

Gardner, M. (1970) The Fantastic Combinations of John Conway's new solitaire game "Life". Sci. Amer. 223, 120-123.

Korber, D. R., Lawrence, J. R., and Caldwell, D. E.. (1994) Effect of motility on surface colonization and reproductive success of *Pseudomonas fluorescens* in dual-dilution continuous culture and batch culture systems. Appl. Environ. Microbiol. 60, 1421-1429.

Levit, G. S., and Krumbein, W. E. (2000) The biosphere theory of V. I. Vernadsky and the Gaia theory of James Lovelock: A comparative analysis of the two theories and traditions. Journal of General Biology. 61, 133-144.

Margalef, R. (1963) On certain unifying principles in ecology. The American Naturalist 117, 357-373.

Schiefer, G. E., and Caldwell, D. E.. (1982) Synergistic interaction between *Anabaena* and *Zoogloea* spp. in carbon dioxide limited continuous cultures. Appl. Environ. Microbiol. 44, 84-87.

Watson, A. J. and Lovelock, J. E. (1983) Biological homeostasis of the global environment. Tellus. 35B, 284-289.

Zavarzin, G. A. (2000) The Non-Darwinian Domain of Evolution. Herald of the Russian Academy of Sciences. 70, 252-259.

Chapter 22

IS THERE AN ADEQUATE TERMINOLOGY OF BIOFILMS AND MICROBIAL MATS?

George S. LEVIT, Wolfgang E. KRUMBEIN
Friedrich-Schiller-University of Jena
Institut for the History of Medicine, Natural Science and Technology
(Ernst-Haeckel-Haus), D-07745 Jena, Germany
george.levit@uni-jena.de

1. INTRODUCTION

Clear and rigid terminological structure is an important prerequisite of any scientific communication and, after all, of the development of any scientific discipline. Yet, being carried away by their empirical investigations, scientists often neglect the terminological difficulties accumulating little by little in the theoretical apparatus. Sometimes an exterior look at a growing scientific discipline can be useful for detecting such difficulties.

Our purpose here cannot be to provide a complete analysis of the terminological structures of biofilm/microbial mat studies. We wish instead to give a glance at some crucial concepts of the field from the viewpoint of philosophy of science.

Because this book addresses a purely geobiological readership, some introductory words about philosophy of science are needed. Philosophy of science established itself through the 20th century as a well recognized discipline with its own methodological apparatus, journals and „classical„ agenda. Philosophy of science deals with the historical development and structure of scientific theories, methodological and terminological issues, and the philosophical consequences of current scientific achievements. The theoretical statements, which are too abstract and general to be immediately empirically proved, are also subject to logical analysis by the philosophers of science. In the recent decades some clearly specialized branches of the philosophy of science appeared exemplified by the philosophy of physics, philosophy of chemistry and philosophy of biology and medicine. These sub-disciplines have been in their turn ramified further in many new directions like the philosophy of molecular biology or the philosophy of quantum mechanics. Our attempt to draw attention of the theoreticians and philosophers to Biofilms and Microbial Mats Studies [henceforth BMMS] follows this tradition of the specialized philosophies.

The primary purpose of this paper is to analyze some aspects of the terminological structure of BMMS. From the viewpoint of methodology we will proceed rather as philosophers of science than scientists. This means that we will *not* examine the correspondence of scientific definitions to the empirical data. We will also *not* provide any new terms or definitions. We will instead briefly examine the structure of existing definitions and their potential of mutual terminological compatibility.

2. FORMULATING THE PROBLEM

It is commonly accepted now that microbial biofilms (MB) and microbial mats (MM) have played an important role in the Earth history by influencing the composition of the atmosphere and the hydrosphere, and by participating in stromatolite and biolaminite formation. Krumbein (1996) insists that both biofilms and microbial mats can be interpreted as sub-systems of the global organization of living matter or the biosphere. Considering the momentous importance of MB and MM for the whole structure of Earth and life sciences one would expect massive efforts directed to the clarification of the basic concepts. Actually inestimable work is made for describing the structure and matching the terminology of the microbial biofilm and mat studies. At the same time, reading the special literature on microbial biofilms and mats a sideliner blunders against an interesting problem: there is no clear definition of the difference between these two terms. Consequently there is no clarity about the relationships of theses two concepts, although they are often applied in the same texts and contexts. The term “Microbial Mat“ in historical perspective can be itself confusing, because it has emerged only recently from the formerly used term algal mat. Further the terms Bacterial Mat, Algal Mat and Microbial Mat were used synonymously for one and the same system in a publication with high impact factor (Doemel & Brock, 1974).

The problem is not up-to-the-minute news and was many times mentioned in the literature. 16 years ago Stal (1985) pointed out in his Doctoral thesis that „the concept of Microbial Mat is confusing". Despite of much research work that has been done upon MM in the recent years, there is still no clarity in this relation until now. Moreover, we have an impression that this issue is not regarded as a serious problem, because we could not find in the literature an extensive discussion of it. However, there are a lot of attempts to define these notions separately. Below we give some examples of MB and MM definitions and analyze these definitions briefly. All existing definitions can be divided into structural and functional definitions. *Structural* definitions describe the „mechanical" structure of MB and MM. They are insensitive to the peculiar kinds of micro-organims but provide a detailed description of the physical state and mechanical connections of the constituting parts (liquid, gaseous, solids). On the contrary, *functional* definitions are insensitive to the mechanical structure of MB and MM and describe their subjects in term of the functions different species play in the various biologically stabilized systems.

W. Characklis & P. Wilderer (1989) provide a classic example of a structural definition of biofilm, accepted by many other biofilm researchers: „*The biofilm* is, according to our definition, the sum of the biofilm volume elements bounded by the surface of the substratum and the interface between biofilm and liquid film. [...] The biofilm system consists of the biofilm, the substratum, and the medium in which it is immersed. [...] A biofilm accumulates when microorganisms colonize a surface and form mono- or multilayers of cells."

At first glance the above definition seems to be unproblematic. However, compare this definition of MB with Stal's definition of MM (Stal, 2000): „*Microbial mats* are generally formed by filamentous, entangled organisms that produce a microscopically 'mat-like' structure. In some cases such mats can indeed be peeled off from the sediment as a large, coherent piece. [...]. More generally the term microbial mat is used for multilayered microbial communities growing on sediments in a variety of different environments."

As one can see even these very comprehensive and instructive definitions do not separate MM and MB clearly. Apart from the metaphorical description of MM as 'mat-like' structures, we are confronted in both cases with *multilayered* microbial communities. Following these definitions one could come to the conclusion that the property of to be *filamentous* is crucial for distinguishing MM from MB. However, even a casual literature investigation shows that this is not the case.

Characterizing sub-aqueous biofilms Burns & Walker (2000) write: „Stable water levels and turbidity associated with flow regulation in the River Murray have promoted the growth of *filamentous* green algae and Cyanobacteria in *biofilms* on submerged wood". Another example: „The microbial *biofilm* community developed from attached single cells and distinct micro-colonies via a more confluent structure characterized by various *filamentous* bacteria to a mature biofilm rich in polymeric material with fewer cells on a per-area basis after 56 days" (Manz et.al., 1999). At the same time Stal (2000) points out that „benthic microbial communities of unicellular organisms, that usually do not form coherent structures, are also termed microbial mats." So, the property of to be filamentous cannot serve as a commonly acceptable characteristic for distinguishing MB and MM.

Some authors claimed that the typical property of MM is their *laminated* structure. However, there are laminated benthic microbial communities which we do not usually call MM. Besides, taking in consideration the *genesis* of microbial mats we must accept that a property 'to be laminated' corresponds only to the mature stages of MM development. In other words there is an urgent need of a genesis based definition of MM.

A characteristic example of a *functional* definition we found in an electronic textbook for MB students (Jean Douthwright, Gloria Delisle, Betty Eidemiller: http://www.personal.psu.edu/faculty/j/e/jel5/biofilms/.): „Microbial Mats are specialized microbial communities composed mainly of photosynthetic prokaryotes. Thus the principle distinction between *microbial mats* and *other biofilms* is their dependence on photosynthetic primary productivity as their source of energy." This definition contains a premise that MM is a kind of biofilm and points out an important feature of MM as opposed to other biofilms. However this definition cannot be seen as satisfactory. First, there are microbial mats (i.e. bioinert systems which are identified as MM by the majority of MM researchers) where prokaryotes do not play a key role, e.g., microbial mats composed almost exclusively by diatoms, *Dunaliella* and green alga (Asche et. al, 1985). There are also MM which are mainly composed of heterotrophic, lithotrophic and chemotrophic bacteria. Second, there are subaerial biofilms (again, we mean bioinert systems, which would be identified as biofilms by the majority of MB researchers) formed by rock inhabiting photosynthesizing cyanobacteria (Gorbushina et al., 1999).

Thus there are neither structural nor functional widely accepted and uncontroversial definitions of MM and MB. There is no unity about neither structural nor functional characteristics of MB and MM among the experts. What can be the reason of that situation? There are some answers possible, which, however, do not necessarily exclude each other.

1. We do not have enough empirical data for distinguishing between MM and MB.
2. There is no impassable border between these phenomena.
3. We just failed to agree about the difference of these two concepts.
4. There is no theory in the frame of which we could distinguish MM and MB.

There is also a historical explanation possible. One could assume, that pure biofilm investigations were performed by biologists and engineers in technologically oriented centers, whereas the study of microbial mats and stromatolites originates from geological departments. But we will not consider this possible explanation, because in the contemporary scientific world, in the Internet era, there are no historical excuses.

3. SOME PHILOSOPHICAL CONSIDERATIONS

In the philosophy of science one distinguishes two kinds of definitions. (1) „Descriptive definitions" attempt to describe common meanings of uses of terms. (2) „Prescriptive definitions" make recommendations for how to use the terms (Resnik, 2000). Prescriptive definitions are based not on the description of a common usage of a certain term but on the deeper theories that explicate terms. One can also approach the distinction between descriptive and prescriptive definitions from another side. The concept of a „descriptive definition" in the „descriptive natural sciences", like geology and some parts of biology, reflects not only the common usage of a term but also the descriptive character of this kind of definition itself. That means that in its pure form (which probably does not exist) a descriptive definition depicts a phenomenon without placing it in the general theoretical context. If in defining the biofilm one claims that „the biofilm is..." implying „the biofilm consists of..." we are faced with a descriptive definition. Thus almost all structural definitions of MM and MB we found in the literature are descriptive definitions.

On the contrary, prescriptive definitions usually define a term placing it in the broad theoretical context where this term makes sense. Note, that our distinction between descriptive and theoretical definitions has little to do with the distinction between empirical and theoretical terms by logical positivists (school of the philosophy of science), because our theoretical definitions describe objects that can be sometimes *directly observed* but can be only understood in the framework of a certain theory. The term 'theory' is used in its broadest sense. Theory is said to be a complex system of concepts, which organizes a certain class of phenomena. Theory is created for describing the essential characteristics and laws of the related field of knowledge. It is generally agreed that prediction (retrodiction) and explanation are central functions of a scientific theory. We also suppose that a sound scientific theory usually implies the existing of an alternative theoretical explanation. In the abstract theoretical modeling all terms are structured according the interior logic of a theory.

It is this possibility to directly observe some MBMM makes an illusory feeling that we do not need precise definitions, because it is *evident* what we mean by applying the terms MB or MM.

Classic examples of theoretical prescriptive definitions in biology delivers, f. e., the evolutionary theory, which is perhaps one of the most theoretically satiated parts of biology. In 1982 Gould & Vrba coined a neologism „*exaptation*“ to distinguish structures whose current functional roles are no part of the explanation of their origin from those whose roles are. The flippers of turtles are not specially modified for burying eggs in sand, but they are successfully used for this purpose. Although you can observe the flippers of turtle and it is evident that they are not designed for sand-digging, we cannot understand the meaning of the term „*exaptation*“ without accepting the whole theoretical structure of the Darwinian evolutionary theory and its interpretation by Gould & Vrba.

Almost all structural definitions of MB and MM, which are found in the literature, are descriptive definitions. Functional definitions look more like theoretical prescriptive definitions but are not really prescriptive, because there is no widely accepted theory (if any) placing MB and MM in a unite theoretical context.

The predominance of descriptive definitions usually points out the premature character of a scientific discipline. In the well developed parts of physics (which is a classic example of a mature scientific discipline) biology you will not find uncertain definitions, albeit there is enough uncertainty in the phenomena which are studied by physicists. Another reason for neglecting precise definitions is probably an unconscious orientation of many representatives of the descriptive scientific disciplines on multiplication of empirical data without paying attention to the theoretical fundamentals and hypothesis, which sometimes seem to be too speculative and redundant.

4. CONCLUSIONS

There is no clarity in the basic definitions of MB and MM, namely, there are no definitions clearly describing relationships between two basic concepts: microbial biofilms and microbial mats. As we stated already, different explanations of this situation are possible:

1. We do not have enough empirical data for distinguishing between MB and MM.
2. There is no impassable border between these phenomena.
3. We just failed to agree about the difference of these two concepts.
4. There is no theory in the frame of which we could distinguish MB and MM.

Figure 1. View of a hot Spring biofilm (Oregon)of cyanobacteria positioned on top of a *Chloroflexus* biofilm. These individual layers of biofilms interact with each other.

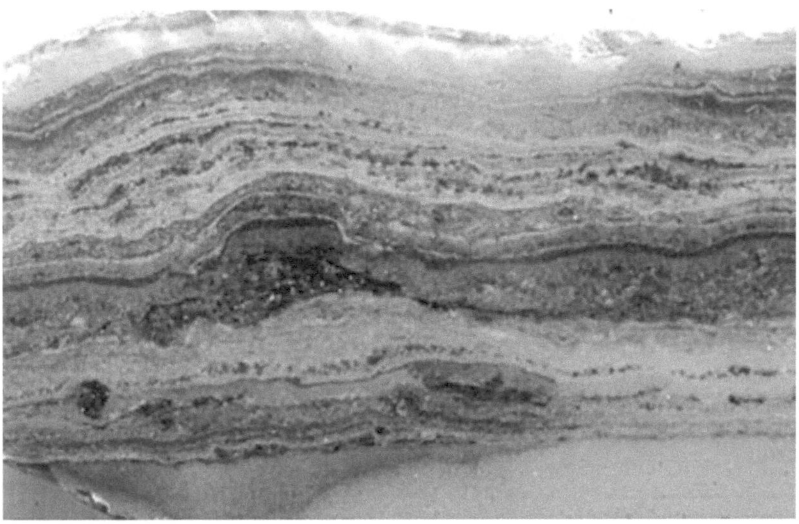

Figure 2. Serial fossilized biofilms making up a fossil Microbial Mat system (stromatolite). This iron depositing biofilm series consisted of fungal networks in a karst cave depositing the iron minerals goethite and hematite.

We do not believe that an increasing amount of empirical data would radically change the situation. As the diversity of already existing definitions and descriptions of MB and MM shows, there is enough empirical knowledge for making basic definitions. Although there is truly no impassible border between MB and MM (see Figs.1 & 2) this is not an excuse. As we know, nature often escapes sharp borders and the theoretical separation, description and classification of natural phenomena is usually connected with speculative operations. No mathematically based science would be possible, if we would not be able to work with idealized objects. Closely related phenomena can be separated only on the basis of a formalized scientific theory.

Obviously there is no agreement about the basic terminology of MB and MM. We claim that the depicted terminological confusion indicates the lack of a general theory (some alternative theories) which would comprise the concepts of MB and MM and place them in an unified theoretical context. With great scientific regret we state that the terminological confusion in BMMS must be connected with the predominating descriptive and technological character of research.

REFERENCES

Asche W., Ohadi H., Sander S., Schimmel B. & Krumbein W. „Mikrobielle Matten und Organismen." In: *Mikrobenmatten in den Salinen bei Guerande (Bretagne, Frankreich)*. Exursionsbericht. Universität Oldenburg, 1985.

Burns A, Walker K.F. Biofilms as food for decapods (Atyidae, Palaemonidae) in the River Murray, South Australia. Hydrobiologia 2000; 437 (1-3): 83-90

Characklis, W.G.,Wilderer, P. A. *Structure and function of biofilms*. Wiley: eds Chicester, 1989.

Doemel, W.N., Brock, T.D. Bacterial Stromatolites: origin of laminations. Science, 1974; 1984: 1083-1085

Douthwright, J., Delisle, G., Eidemiller, B. *A Manual of Biofilm Related Exercises*. American Society for Microbiology. http://www.personal.psu.edu/faculty/j/e/jel5/biofilms/

Gorbushina, A.A., Palinska, K.A., Krumbein, W.E. „Poikilotroph Growth Patterns in Rock Inhabiting Cyanobacteria." In: *The Phototrophic Prokaryots*, Peschek et al. (eds), Kluwer Academic Publishers/Plenum Publishers, New York, 1999.

Gould, S.J, Vrba, E. Exaptation – a missing term in the science of form. Paleobiology, 1982 8(1): 4-15

Krumbein, W. E. Stromatolites - Challenge of a Term through space and time. Precambrian Research 1983, 20, 493-531.

Krumbein, W.E. „Auf den Schultern des Riesen.“ In: *Mitteilungen Geol.-Paleont. Institut Universität Hamburg*, 1984, 56: 435-460 .

Krumbein W.E. „The year of the slime“. In: *Biostabilization of Sediments*, ed., Krumbein, W.E., Paterson D.M., Stal L.J., BIS-Verlag, Oldenburg, 1994.

Manz, W., Wendt-Potthoff, K., Neu, T.R., Szewzyk, U., Lawrence, J.R. Phylogenetic composition, spatial structure, and dynamics of lotic bacterial biofilms investigated by fluorescent in situ hybridization and confocal laser scanning microscopy. Microbial Ecology. 1999, 37 (4): 225-237

Resnik, D. A pragmatic approach to the demarcation problem, Stud. Hist. Phil. Sci. 2000, 31(2): 249-267.

Stal, L. Nitrogen-fixing bacteria in a marine microbial mat, Doctoral Thesis, Groningen University, 1985.

Stal, L. „Cyanobacterial Mats and Stromatolites.“ In: *The ecology of cyanobacteria : their diversity in time and space*, ed. by Brian A. Whitton. Kluwer Academic Publishers: Dordrecht, 2000.

Chapter 23

HIDDEN IN PLAIN SIGHT

A Macroscopic approach to biofilms and other visible manifestations of bacteria in the teaching of field microbiology

Betsey Dexter DYER
Department of Biology
Wheaton College, Norton, Massachusetts O2766 USA
BDyer@Wheatonma.edu

1. INTRODUCTION

Biofilms are manifestations of bacteria with access (even if temporary) to the specific parameters, such as chemicals, temperature, pH, and light that allow them to reproduce abundantly and to thrive. Large numbers of active bacteria often produce colors, bubbles, slimes, scums, odors, and flavors on an easily detectable scale. These may be considered the macroscopic field marks by which many bacteria (in fact most major taxonomic groups of bacteria) may be identified in the field. Individual microbiologists with expertise in the field are usually quite familiar with the macroscopic characteristics of their particular bacterial group. Nevertheless, the traditional approach to teaching microbiology seldom takes students to the obvious, user-friendly level of simply using unaided senses to make field identifications. A more typical approach is to treat bacteria as though they were tiny dots and dashes under a microscope or little colonies of monocultures. An inadvertent consequence is that many biologists (even ecologists) who take pride in knowing a great diversity of organisms, know almost nothing of bacteria and cannot make identifications on any level in the field. Yet bacteria are the most abundant organisms on Earth and the most diverse and as such, really should be incorporated into almost any biology course in which organismal diversity is a focus. Fortunately the invisibility of bacteria is an illusion and one that is easily remedied with a focus on their many macroscopic field marks. For example, biofilms of all types are outstanding sources of distinctive characteristics and their identification could be readily incorporated into any course in which students are learning to identify a diversity of organisms.

2. BACTERIA ARE THE MOST NUMEROUS OF ALL ORGANISMS

There are more individual bacteria on Earth than of any other organisms, 5×10^{30} according to calculations by Whitman et al (1998). Included in this number are the microbes hypothesized to fill the labyrinthian microcrevices deep in the Earth's crust. Also included are those that cover the surface area of nearly every other organism such that a simple count of their numbers might be multiplied by some much larger factor to get the number of symbiotic bacteria.

3. THE NUMBER OF BACTERIAL SPECIES IS EITHER OVER OR UNDERESTIMATED BUT EITHER WAY THEY ARE THE MOST DIVERSE OF ALL ORGANISMS

The number of bacterial species is either greatly overestimated or greatly underestimated The problem goes far beyond the usual "lumping vs. splitting" that is the foundation of most bickering in taxonomy. In fact the concept of species, especially as refined by Mayr (e.g. 1963) may not apply to bacteria at all. Bacteria may all be participants in a free-flowing, constant, and promiscuous exchange of DNA such that no particular genotype could ever clearly define a species. On the other hand, bacteria have a 4 billion year history and an extraordinary repertoire of metabolisms ranging from all possible autotrophies to the most discriminating of heterotrophies. Perhaps there are many more species (thousands-millions, based on these many trophies) than what is acknowledged in a typical microbiology text-a mere 2000-3000, based mostly on the willingness of a few opportunistic microbes to cooperate in the lab.

4. BACTERIA TYPICALLY HAVE SIMPLE MORPHOLOGIES; EVEN UNDER THE MICROSCOPE THEY MAY BE OVERLOOKED

Most bacteria look like a collection of punctuation marks: ".-,..--,,,,....~``~...", nearly featureless as well as nearly invisible. Even under a good microscope they may be easily missed and nearly impossible to interpret, to the dismay of beginning microbiology students. Yet it is our own size that is parochial (Gould, 1996) The "basic" animal is not a primate, nor a mammal, nor even a fellow vertebrate. It is the beetle for which (according to Haldane) God has apparently showed an inordinate fondness having made so many. J.B.S. Haldane (1892-1964), a mathematical biologist, noted that there are 8000 species of mammals but 400,000 species of beetles and commented, "the Creator if He exists, has a special preference for beetles."

Furthermore throughout most of the space of Earth and through most of the time of Earth, tiny organisms (bacteria) have always predominated, being the sole inhabitants for about 2.5 billion years and the majority inhabitants thereafter. Indeed, we (eukaryotes) are essentially cobbled together symbioses of bacteria, composed as we are of archaeal cells, proteobacterial mitochondria and (for some of us) cyanobacterial chloroplasts. We are walking, swimming, flying, and photosynthesizing bacterial communities. (Dyer, 1989) The question is not how can bacteria be so tiny but rather how did we get to be so large and multicellular (a topic not to be addressed here).

Quite possibly all of this matters to the ways in which we choose to study bacteria. They are not tiny versions of ourselves but rather we are anomalously large versions of them and our hypotheses concerning "their" world might be constructed accordingly.

5. TININESS IS NORMAL; WE ARE ANOMALOUS

The tininess of bacteria should inform our experimental approaches to and appreciations of their activities in the environment. For example, biofilms are often macroscopic, often slimy, scummy, at the very least filmy assemblages of bacteria taking advantage of a (perhaps temporary) concatenation of nutrients, light, pH and other factors of keen interest to microbes. Biofilms can be surprisingly well organized, with heterogeneous communities arrayed and juxtaposed for optimal interactions and exchanges. Therefore one question might be how can something so tiny as a bacterium become part of something so large and complex as a biofilm? The answer might be: precisely because they are so tiny. Diminutive size confers the following advantages:

1. Bacteria have an intimacy with the environment that we humans (insulated as we are behind a thick skin of dehydrated kerogen) can only imagine. Every nuance of ions or photons is detected, processed and resolved. The immediate environment is a chemical world, "visualized" molecule by molecule, reaction by reaction. Therefore it should be no surprise that bacteria arrange themselves with such precision at interfaces and between each other to form a more complex whole.

2. Bacteria have "hair-trigger" mechanisms for regulating their genes and by extension their metabolisms. Bacteria can change their strategies ad hoc according to the latest input from their environment. The classic well-known example (of many) is the "lac operon" (Jacob and Monod, 1961) by which at a moment's notice E.coli can take advantage of an unexpected influx of lactose to its environment and just as quickly shut down its lactose-specific enzymes when the supply is depleted or if something better (glucose) becomes available. Therefore the relationships of bacteria in a biofilm and their responses to each other and to environmental factors might be expected to be both quick and subtle. This is in marked contrast to the way in which animals like us respond to environmental changes. Our size allows us to maintain a sort of homeostasis such that most of our cells are likely to be oblivious of any change at all. Furthermore our multicellularity confers an opportunity for permanent commitment of each cell to a particular phenotype. Although we have all of the genes we need to produce an entire human being in any given cell of our elbows, in fact the dedication of our elbow cells to their task is so firm that extra eyes and noses will never sprout there no matter what environmental pressures there might be. Fascinating exceptions may be concocted in molecular biology labs such as the case of fruit flies sprouting eyes all over their bodies in response to gene regulation gone awry. (Halder et al. 1995) Furthermore many birth defects of humans are the results of errors in gene regulation.

3. Bacteria have a greater tendency than most organisms to be each other's immediate environments. Bacteria packed and stacked in a biofilm respond to each other as much as to any physical parameter. Perhaps our multicellularity is as close as we can get to that state but bacteria do it better by being in heterogeneous groups, interacting well across species boundaries. These interactions can and do include the promiscuous exchange of DNA with chimeric genotypes resulting. In contrast, the practices of most sexual organisms (such as animals) seem to prevent cross species activity but rather seem selected to preserve the cohesiveness and dignity of the genome. No claims are made for identification of bacteria on the species level which still requires lab work of the sort that fits the apprenticeship mode of acquiring knowledge. However considering that most biologists cannot

recognize (or name) one single bacterium on the "phylum" level, identification of major groups is a worthy accomplishment.

Biofilms have some symbiosis-like (or at the very least consortium-like) properties. Close symbioses often select strongly for reductions in genome size. Gene loss might be tolerated if missing functions are supplemented by a close neighbor. Biofilms, especially stable ones, may provide opportunities for coordinate use of some gene products and consequent loss of redundant genes, among members of the community. I thank Patrick Gunkel, an independent scientist in Woodshole, MA., for this suggestion. As a founder of the science of ideas, "ideonomy", Patrick felt compelled to offer an opposite scenario as well: The dynamic spatial and temporal relationships in a biofilm may actually preclude the transfer and loss of genes.

4. Bacteria reproduce rapidly by simple cell division. If nutrients are abundant, replication could occur more than once per hour. The evolutionary time scale at that rate is compressed. A biofilm of a month old has perhaps undergone as many as 5000 generations. Compare that with the number of generations in the entire history of Homo sapiens. Assuming 100,000 years of evolution and about 20 years per generation we have only just managed 5000 generations. The usual text book images of an evolving system, slowly shifting in response to isolation and selection may not be directly compared to the evolution of prolific, promiscuous microbes who have been at it for 4 billion years.

If a human were included into the membership of a biofilm, it would be like adding a stone statue, static and insensitive and effectively unreproductive, to a busy village square. Calling to mind such an image is a not entirely frivolous exercise. We humans are indeed slow-moving giants dwelling within the milieu of our symbiotic biofilms. Bacteria cover our skin surfaces favoring our moister spots and they are absolutely teaming throughout the nutrient rich environment of our gastro-intestinal tracts. Furthermore, bacteria (as mitochondria) live in nearly every cell. From a bacterial point of view we are a haphazardly defended mountain of nutrients.

6. BACTERIA CAN BE IDENTIFIED IN THE FIELD WITHOUT MICROSCOPY

It is well known to field microbiologists that many bacteria are identifiable in the field without microscopy. Indeed enrichment cultures often begin with a sample carefully chosen from an environment in which some bacterial manifestation is detectable. What is less available is access to that field knowledge. Information on bacterial field marks can be difficult to find unless one is getting some training under the tutelage of a field microbiologist or delving into technical literature on specific bacterial groups. The guild-like specialties of microbiologists and the apprentice-like training of the aspirants can lend itself to a cautious refusal to generalize about recognizing any bacteria in the field. Unlike the field components of nearly every other area of biology (e.g. botany, ornithology) field microbiology is not considered an amateur activity. Therefore we may be missing out on the opportunity to bring microbiology more inclusively into general biology curricula. Bacteria are an integral part of every biome, every biogeochemical cycle, every food web. Interpretation of (and appreciation for) bubbles, scums, slimes, flocs, films, odors, flavors, and other microbial manifestations should be within the realm of any field biologists (with or without the micro prefix).

7. THE RULES OF THUMB MAY BE READILY TAUGHT TO MICROBIOLOGY AND BIOLOGY STUDENTS

For nearly every major bacterial group there are "rules of thumb" by which reasonable identifications may be made in the field. This paper is not intended to be a complete list but just a sampling of the sort of information that can be gleaned from primary literature, textbooks, and reference tomes such as *The Prokaryotes* (Balows et al, 1991). No claims are made for identification of bacteria on the species level, which still requires lab work of the sort that fits the apprenticeship mode of acquiring knowledge. However considering that most biologists cannot recognize (or name) one single bacterium on the "phylum" level, identification of major groups is a worthy accomplishment. A more exhaustive and detailed compilation of macroscopic field marks may be found in *A Field Guide to Bacteria* (Dyer, 2003).

1. Look at extremes: In general bacteria display their field marks most readily in environments considered "extreme" by multicellular eukaryotes. Therefore boiling springs, salt flats, and anoxic environments of all kinds are wonderful introductions to the visible microbial world.

2. Sniff for sulfury compounds: Eukaryotes are minor participants in the sulfur cycle. Most interesting emissions are interpretable as bacterial.

3. Look beyond the eukaryotes: For example bright green (eukaryotic) scum may overgrow a wet photic zone but in the shadowy crevices or periodically dry areas, cyanobacteria show their blackish pigments.

4. Learn to decipher colors. Many pigments are eukaryotic and many others are geological. However bacteria often can form distinctive, highly visible pigments as in the case of bacterial communities of sulfur-rich environments in photic zones. Here is one example, easily shown and interpreted on a field trip in search of bacteria. On a stable, marine mud flat it may be possible to observe thick blue-black felts of cyanobacteria, underlain by scums of pink, the "purple sulfur bacteria", these pinks coming to the surface in especially sulfury areas. Beneath the pink may lie a thin layer of green, the "green sulfurs", followed by black depths of iron sulfide produced by *Desulfovibrio*-type activities. Here and there a surface milky-white film or cobweb of *Beggiatoa*-types may be seen.

The ornithologists with their long history of amateur participation call it "jizz": the sum total of all of the characteristics of a bird (sometimes glimpsed fleetingly or at a distance) that render it identifiable. These include location, activities, size, shape and markings. Professional ornithologists can recognize scores of birds by their "jizz" but so can competent amateurs. Such inclusiveness is within the realm of possibility in microbiology too.

8. THE EXAMPLES ARE OFTEN VIVID AND MEMORABLE AND EVEN VISCERAL

Gram-positive *Brevibacteria* reveal themselves as the producers of foot and dirty sock odor. Feasting upon loosened skin cells and often assisted by athletes foot fungus, they release a panoply of distinctive volatiles including S-methyl thiopropionate. Interestingly these same *Brevibacteria* are essential in the ripening of surface-washed cheeses. As with many cheeses and other bacterially enhanced foods, one must often acquire a taste for them and there is a fine line between "delicious" and "putrid". Cheese after all is milk gone bad in a more or less controlled manner resulting in something surprisingly edible and stable (even preserved). It is also typical that the community of bacteria found in and on us are also major participants in the production of fermented foods of all kinds. Poet Leon-Paul Fargue clearly made the connection between the odor of toes and of fine cheese when he described it as "les pieds de Dieu".

A case in point (for memorability) was a certain surface ripened Muenster cheese, purchased for me while I was working on my book, *A Field Guide to the Bacteria* and described here from an excerpt:

To assist me in researching this book, friends provided interesting cheeses with bacterial pedigrees. Dr. Jon Herskovitz had a very ripe wheel of Muenster cheese that he sent to me by mail. During transit, it became even riper so that the post office, wrapped the box in plastic in order to make delivery more tolerable. Nevertheless, the package reeked. At first, before I commenced unwrapping, it smelled like a gas stove leak and then increasingly like a full diaper pail as I removed the five layers of plastic bags and two layers of plastic wrap in which Jon had attempted to conceal the cheese. During the unwrapping I took the cheese out of doors to complete its unveiling. I then telephoned my friend Jon and he dared me to eat some. So I had two crackers full. The flavor was surprisingly mellow and creamy albeit quite complex and quite aggressive in its after-flavors, which lingered in my mouth and digestive system for hours. Delicious as it was, I rewrapped it tightly and discarded it in a public dumpster. Hours later the house still smelled of it. Packaged with the cheese and enclosed in a plastic bag had been several neckties. These had been rendered completely unwearable and had to be discarded. Days later the kitchen wastebasket continued to emit an odor. We discovered a completely dry outer wrapping of the cheese which had retained the smell and had transferred it to everything else in the trash.

9. HIDDEN IN PLAIN SIGHT

Field Microbiology can confer a completely different (and quite enlightening) point of view on the world. It is well worth indoctrinating our students whether or not they are planning to be microbiologists. For those willing to walk with head bent down, sniffing for tell tale odors and looking for spots of moisture, traces of slime and scum, and distinctive colors, the microbial world is entirely visible. Indeed eukaryotes can seem faded into the background serving as supports and structures for bacterial films and scums or acting as containers for fermenting communities. Bacteria are always present, just "Hidden in Plain Sight". The signs are all there for those willing to interpret them.

ACKNOWLEDGMENTS

I was fortunate to have been on several (very influential) microbiological field trips with Stjepko Golubic, Lynn Margulis, Ken Nealson, Wolfgang Krumbein and others of the NASA "Planetary Biology, Microbial Ecology" 1990 course.

REFERENCES

Balows, A, Truper, H, Dworkin, M.,Harder, W, and Schleifer, K-H (eds) (1991) *The Prokaryotes*, Springer-Verlag.

Dyer, B. (1989) Symbiosis and organismal boundaries *American Zoologist* **29**,1057-1060.

Dyer, B. (2003) *A Field Guide to the Bacteria*, Cornell University Press, New York.

Halder, G., Callaerts, P. and Gehring, W. (1995) Induction of ectopic eyes by targeted expression of the eyeless gene in *Drosophila. Science* 267: 1788-1792.

Gould, S.J. (1996) Microcosmos *Natural History* March, 21-23,66-68.

Jacob, F. and Monod, J. (1961) Genetic regulatory mechanisms in the synthesis of proteins, *J. Mol. Biol.* **3**,318-356.

Mayr, E. (1963) *Animal Species and Evolution* Harvard U. Press.

Whitman, W., Coleman, D., and Wiebe, W. (1998) Prokaryotes: the unseen majority *PNAS* **95**,6578-6583.

Chapter 24

IS IT POSSIBLE TO QUANTITATIVELY ASSESS THE ROLE OF ALGOBACTERIAL FILMS IN A WATER BODY?

Nickolai V. SHADRIN
Institute of Biology of Southern Seas,
Nakhimov Ave., 2, Sevastopol, 99011, Ukraine.
lib@ibss.iuf.net

Algobacterial communities or biofilms are among factors determining the evolution of biosphere. The role they had played can be quantitatively assessed from the thickness of sedimentary rocks composed with the involvement of them. At present biofilms also perform an active role in functioning of water bodies. The question of interest is whether it is possible to quantitatively estimate this role for a specific water body. The frank answer would be "no" rather than "yes". By now the notion about the inner structure of different biofilms and processes going in them has been drawn in plain outline as other sections of this book demonstrate. This allows to make more or less exact quantitative assessment of potential functional possibilities of the various types of biofilms. Much more incomplete is today's understanding of what characteristics and in what way may account for realization of these possibilities in a specific water body. In its turn, this insufficiency of the knowledge obstructs quantitative assessment of ecological significance of biofilms. Let us examine in brief some possible approaches to resolving of this problem.

One way endeavour to analyze the impact of each of the factors taken separately and then to integrate the results. However, this might be a deadlock because of the involved multidimension problem – a damnation of multi-measurements. An alternative would be the employment of some dimensionless integral characteristics, criteria of similarity. This proved to be a successful tool in disciplines affiliated to physics and technology. Probably, it is preferable to apply three groups of criteria, which would describe:

- general integrated abiotic characteristics of water bodies;
- specific geographical features and the intensity of solar energy flux reaching the earth;
- the structure and inner relations of the biotic component of ecosystem of the water body.

The three sets of factors work in co-operation. Let us take a simple mechanical analogy as an example and consider abiotic and biotic factors interrelation. It is like inflating a balloon (biota) inside bottles (water bodies) of different size. In some cases the balloon's capacity is larger than that of the bottle, then the balloon would not reach its maximum capacity. In other cases the balloon snaps long time before it would reach the bottle limits. Apparently, sustainable functioning of a community is possible only with the adjustment to a variety of external limiting factors.

In all probability, neither biota nor its integral parts, e.g. biofilms, will ever realize their potential in full. The diversity of limitations can be divided into two groups:

- factors which determine the possible total area of biofilms in a water body, and
- factors which determine the intensity of functioning of biofilms.

What may restrict development of biofilms in a water body? To answer this question it is essential to look into the term "biofilm". Leaving aside analysis and comparison of the definitions available by now, in out case we understand biofilms as organized aggregations of unicellular organisms found in the boundary zone between the different physical media. This definition implies that the main factor limiting development of biofilms in a water body is total area of the boundaries between physical media.

By now several criteria of similarity between water bodies have been proposed. Dimensionless similarity criterion - morphometric index of water body or a part of a water body - is the most convenient for serving the purpose under the present discussion (Shadrin, 1985):

$$\mu = \frac{L}{H_{av}} \tag{1}$$

where is morphometric index, L – the length of the coastline of a water body and H_{av} – average depth, this term a number of authors give as S/V, where S is the surface area of a water body and V is the volume of water in it (Khailov, 1982).

Physical meaning of the proposed index may be interpreted as the quantity of external boundaries in a water body or as the specific quantity of interfaces between media in it.

Physico-chemical and biological features of a water body depend on its morphometric index. For instance, using data cited in G.M. Bagrova (1983) the reliable relationship between the albedo of a lake and the morphometric index (r=0,91) was defined, and from data given in I.A. Arnoldi and K.A. Mosevich (1934) – the relationship between oxygen and temperature vertical gradients and - (r=0,98). Results of the analysis have been presented earlier (Shadrin, 1985).

In a series of biotic characteristics which reliably depend on only two are of special interest to us:

- the increase of total species number of unicellular algae with the increase of estimates in a lake. G.F. Zagorenko (1987) had revised data obtained for 142 lakes of the Soviet Union and reported about a reliable (= 0,001) correlation between species number in and morphometric index of water bodies (r = 0,64);
- the increase of bacterial content in water column of lakes with the growing . This conclusion was based on the analysis of two lake groups – one in Poland (Chechuga, 1961) and th other in near – Baikal region (Topolov, 1974). It was found that relation between bacterial concentration and is approximated by the equation (2) (Shadrin, 1985; Shadrin, Teryoshina, 1986):

$$N = a + b\mu \qquad (2)$$

where *N* is bacterial concentration in lake water and *a* and *b* are coefficients.

It should be noted that angular factor (b) was almost the same for the two lake groups, and coefficients of correlation between (N) and (μ) were 0,78 and 0,76, correspondingly. In one of the near – Baikal lakes the correlation between interannual variations in N and μ was even closer (r = 0,99). There is a simple explanation of this phenomenon. In natural water bodies and in artificial microcosms bacterial concentration measured under the surface (neuston film) and in near-bottom water layer is usually 1000 – 1500 times as large as that in the water column (Shadrin, 1985). Such high bacterial content is owing to cell removal from the neuston and bottom biofilms. In their turn, high values of morphometric index result from the increased specific number of interfaces in the water body and, correspondingly, the greater total surface area of the biofilms. This leads to a logical conclusion about positive relationship between the morphometric index and the role biofilms perform in ecosystem of the water body. In summer time the depth of some Crimean salt lakes decreases to 2 cm and less; then biofilms are factually the only active biotic element (community) left in those lakes – the share they contribute is close to 100%.

Analysis of the equation (2) permits to quantitatively estimate the part biofilms play in production – destruction processes by employing the following equation:

$$F = \left(\frac{a}{b\mu} + 1\right)^{-1} 100\% \tag{3}$$

where F is the specific portion of biofilms in production – destruction processes going in the water body. In as much as reliable estimates of coefficients used in the equation (2) are not available at present, comparison between the lakes can be made only at semi-quantitative level.

Several approaches can be applied for computing numerical estimates of the coefficients; one of them is finding cut total absolute values of surface area of interfaces in water bodies. Is it possible to carry out the task today? What difficulties we may face going this way?

The question is how large is the actual area of interfaces in a water body. Assessment of total area for water/air interface is an easy task while for water/hard substratum it is a hard nut to crack. The difficulty roots in high fractality of this boundary. The importance of fractality as a characteristic of biofilm inner structure has been stressed elsewhere (Krumbein, 1996), and we are focusing on the analysis of fractality of the surfaces on which biofilms grow.

In regarding this boundary integrally it is essential to distinguish two basic constituent parts:

- - interface water/solid matter as an abiogenous substrate (e. g. bottom, rocks, cliffs etc.), and
- - interface water/solid matter as a biogenous substrate (external surface of aquatic organisms).

Now let us examine the coastline as the triple boundary where all the three media meet. In evaluating its fractality maps of different scale and data of the field observations are to be used. The length of coastline between two specifies points was measured on the maps having 8-9 scales by means of curvimeter. Five sites of the Black Sea shoreline were under the study. Even without citing all the data obtained, it is evident that the range of distinctions is impressive. For instance, the coastline between Skadovsk and Bakalskaya spit was measured to be 12 km long using one map (1:120 000 000 scale) and 260 km long using the other (1:400 000 scale). In accordance with theory (Mandelbrot, 1982), relationship between the spot's measured length and the map's scale is well approximated with the equation

$$L = cK^{d} \tag{4}$$

were L is the length of the coastline, K – the scale of the map (km in cm), and c and d are factors. Table 1 gives equation parameters and determination factors (R^2) computed by the method of least squares. Field measurements of the coastline length as composed by 0.1, 1 and 10-m long stretches yielded results within the same range of power factor (d) variability. Values of the factor (d) are different for different sections of the coastline; in further computations we use $d_{\text{average}} = 1.96$. Then what would be the error in evaluating the surface area of a narrow strip of supralittoral by consulting a map with the scale 1:100 000 (1 cm for 1 km)? Given the scale of collecting biofilm samples is 1 cm (1:1), the error will make about 10 orders of magnitude – that is how much we underestimate the length of the examined area on the scale applicable to collecting samples. Measuring linear characteristics of the hard substratum directly supralittoral and littoral revealed the same relationship which has been approximated by the equations calculated based on the obtained data. The parameter d were lesser, varying from – 0.040 to – 0,46 with the average – 0.15. Values of coefficients of correlation between the length of the stretches (on the scale of the measurement) and the distance between points were from 0.95 to 0.99. Proceeding from these exponents, it is easy to figure out that using maps with the easy 1:100 000 will result in as huge underestimation of total area of the littoral (supralittoral) locations on the scale 1 x 1 cm as from 2.5 to 43 000 times. Bottom surface area at these locations would be 34 times as less on the average as it should actually be.

Therefore, in computation of total surface area of substrate fit for biofilm colonization it is essential to take into account fractality of the surfaces. This is also applicable to computation of total area of the biofilms. Employing equation (4) and average exponents it is possible to make estimates obtained from the maps conform to the scale we are interested in. Having more precise estimates would require field measurement of the sea bottom and coastline fractality.

Even more complicated task is assessment of total surface area of biotic substrates fit for development of biofilms. Two details make the problem difficult to cope with: first, evaluation of total substrate surface area, and second the peculiarity of the substrate surface area because of special interaction between host organism and unicellular epibionts settled on the surfaces and forming biofilms. Leaving aside the second aspect, we shall briefly touch upon the first one disregarding plankton and nekton as a part of the problem.

In formation of the interfaces on the sea bottom benthic macroorganisms play special role. (Their significance is usually increasing with the increase of morphometric index values in the water body, but this is beyond the scope of this article). Today it is hardly possible to evaluate total surface area of all macroorganisms populating some benthic community or other. What we can do yet is to compute total surface area in some benthic phytocenoses using data at hand. In Black Sea populations total surface area of the brown alga *Cystoseira* sp. makes up from 25 to 77 m^2 per m^2 of the sea bottom (Kovardakov et al., 1985, A. Prazukin, oral presentation). Employing data given by Dr. N.Milchakova (1989), total surface area of the leaves of Zostera spp. growing in the Black Sea was tentatively evaluated by author as about 20 m^2 per m^2 of the sea bottom.

At lost depth underwater potion of cane (*Phragmites* sp.) growth may contribute as much as 5-9 m^2 per m^2 of the bottom (Shadrin, unpublished). Large macroorganisms are usually covered by smaller ones adds to the total surface area used by settling unicellular creatures. An illustrative example are hydroids which abundantly grow on both animals and underwater vegetation (Grishicheva, Shadrin, 1999), pertinent computations have been made for the hydroid *Laomedia exigna* (Grishicheva, Shadrin, unpublished). One square centimeter of brown algae may provide shelter to about 20-30 hydroid module branches with the average surface area 4,5 mm^2 each ($4,5 \times 10^{-6}$ m^2). Assuming that this estimate is applicable to 20% of the macrophytes surface area, the result would be that hydroids on the macrophyte growth account for 9 m^2 extra surface per m^2 of the sea bottom. Moreover, 2500 and more diatom algal cells may inhabit 1 mm^2 (1×10^{-6} m^2) of the surface area of hydroids (other unicells also occur there but they were not accounted) on one square metre of the sea bottom only macrophytes and hydroids alone would form substrate surface area exceeding 85 m^2. Hydroids are not only alone animal epibiont of macrophytes. There are other animals epibionts, too. Multicellular epiphytic algae found growing on *Cystoseira* may add 25-400% to the total surface area of phytocenoses (Kovardakov et al., 1987). These data clearly show how huge may the surface area of biogenic substrate be compared with the area of the sea bottom.

This collection of scientific evidence leads to unencouraging conclusion that with our today's knowledge we cannot either reliably assess total surface area of the substrates and the biofilms or integrally quantity the role of biofilms in a water body. What appears to be indisputable, it is the present underestimation of biofilms significance which manifests itself in proposed models and conceptions. Correct understanding of the role biofilms play in water bodies would lead to principal revision of currently admitted notions about the structure of production – destruction processes. For instance, as the literature claims, the share microphytobenthos, in other words biofilms, contribute to the primary production of ecosystem of the Black Sea is about 3% while that of phytoplankton 94% (Prodanov et al., 1997). It follows then that the computations underestimate total area occupied by microphytobenthos about 34 times. Having made the corresponding adjustment, the re-computation will result in nearly the same estimates of the production yielded by microphytobenthos and by phytoplankton. This is also proved by that benthic diatoms washed out from the hard substrata in tidal zone often predominate in coastal phytoplankton (Shadrin, Gubanov, 2002).

It has been stated above that integral characteristics of a water body are the factor determining both total surface area of biofilms and the its rate of production – destraction processes in the water body. In modern ecology it has been acknowledged as banality that in populations production rate is limited by one of two integral factors which are the rate of resource influx into the population and the rate of produced biomass withdrawal. This has been repeatedly discussed in general outline and with the involvement of models. However, objective analysis has not been endeavoured as yet because of the insufficient scientific evidence. With our present knowledge the only evident fact is that total structure of biotic interrelations in water body and the extent to which biofilms fit these interrelations are of primary importance. Therefore it would be reasonable to place all biofilms on a scale two extremeties of which are:

- biofilms with nearly closed substance cycles, and internally regulated rate of processes;
- biofilms in which incoming and outcoming substance influxes by far exceed the inner turnover.

Biofilms of the first variety would most probably be found in oligotrophic water bodies (e.g. mount rivers) and those of the second variety in eutrophic waters.

Quantitative assessment of the role biofilms play in functioning of water ecosystems poses the problem which would hardly be coped with in the nearest future. In their advancement researchers should be guided by complementary principle and apply two approaches – one experimental and the other field study. Though understanding of the underlying molecular – physiological mechanisms and relevant potentialities is to be gained through experiments only, the researcher should be aware that a natural water body is not a laboratory flask.

This research was supported through grant N 97-30776 from INTAS.

The author expresses his gratitude to O. Yeryomin, E. Mikusheva and N. Grishicheva for rendering technical assistance, and to O. Klimentova for the help in translating the text into English.

Table. Parameters of the equation (4) $L = cK^d$ determined for different locations of the Black Sea coastline (*L* is the coastline's length and *K* – the scale of the map).

Location	c	d	R^2
Skadovsk – Bakalskaya spit	10,703	-1,491	0,952
Sevastopol – Yevpatoria	17,031	-2,357	0,889
Sochi – Batum	16,786	-2,328	0,915
Samsun – Erdemli	11,047	-1,388	0,983
Konstanza - Iliyechevsk	17,224	-2,205	0,958

REFERENCES

Arnold, N.N., N.A. Mosevich 1934. Bonitation of water bodies.-A reference book on fish management in small water bodies. Moscow-Leningrad OGIZ:80-122 (in Russian).

Bagirova G.M. 1983. Radiation balance assessment depending on the latitude and water body characteristics.-Water resources, N1:94-102 (in Russian).

Chechuga B. 1961. Seasonal changes of total bacterial number and generation time in lakes of different types.-In:Primary production of the seas and inland water bodies. Minsk:262-263 (in Russian).

Grishicheva N.P., N.V.Shadrin 1999. Hydroids as epibionts of mussel and *Cystoseira*.- In: The coasts of and the sea near Sevastopol: ecosystem processes and services to the society. Sevastopol, Aquavita: 229-238 (in Russian).

Khailov K.M. 1982. Near – boundary events in water bodies and their use in biotechnology.- Ecologiya, N6: 3-8 (in Russian).

Kovardakov S.A., A.V. Prazukin, Yu.K. Firsov, A.E. Popov. 1985. Coraplex adaptation of *Cystoseira* to gradient conditions. Kiev. Naukova Dumka:-216 pp. (in Russian).

Kovardakov S.A., Yu.K. Firsov, S.E. Zavalko. 1987. Cystoseira and its epiphytes in a recreation seawater area. The dynamics of biomass and absorption surface.- Current problems of algology. Abstr. 1-st AU-Union Conf., Cherkassy: 112-113 (in Russian).

Krumbein W.E. Geophysiology and parahistology of the interactions of organisms with the environment. P.S.Z.N.I: Marine Ecology, 1996, v.17, N1-3, p.1-21.

Mandelbrot B.B. The fractal geometry of Nature. Freeman, San Francisco, 1982.

Milchakova N.A. 1989. Comparative description of *Zostera* spp. Population structure in pure and mixed phytocenoses of sea grass (Kazachiya bay, Crimea).-Ekologiya Morya, 31: 26-30 (in Russian).

Milchakova N.A. 1989. Statistical analysis of the effect of bottom sediment granulometric composition of the abundance and size structure of *Zostera marina* populations in the Black Sea.- Ekologiya Morya. 32: 59-63 (in Russian).

Prodanov K., K. Mikhaiov, G. Daskalov, C. Maxim, A. Chashchin, A. Arkhipov, V. Shlyakhov, E. Ozdamar Environmental management of fish resources in the Black Sea and their rational exploitation. Studies and Reviews General fisheries Council for the Mediterranean. N. 68. Rome, FAO, 1997. 178 p.

Shadrin N.V. 1985. The dependence of production characteristics on morphomentric parameters of the water body.- In: The hydrobiology and hydroparasitology of Near – Baikal region. Novosibirsk, Nauka: 201-205 (in Russian).

Shadrin N.V., Gubanov V.V. 2002. Churning circulation as a special biotope.-Secon Int. Conf. "Oceanography of the Eastern Mediterranean and Black Sea: Similarities and differences of two interconnected basins", Abstracts, Ankara: 138.

Shadrin N.V., E.V.Therjoshina. 1986. About some factors influencing aquatic microorganisms community structure.- In: The structure and functionning of aquatic microorganisms communities. Novosibirsk, Nauka, 1986, 70-75 (in Russian).

Topolov A.A. 1974. Seasonal and interannual changes of bacterial abundance in Ivano-Arakhleik Lakes.- Gydrobiol. Zhurnal, N5: 85-87 (in Russian).

Zagorenko G.F. 1987. The dependence of planktonic flora diversity on morphometric characteristics of the water body.- Current problems of algology. Abstr. 1-st AU-Union Conf., Cherkassy: 106-107 (in Russian).

Chapter 25

THE ROLE OF FUNGAL BIOFILM AND NETWORKS IN THE TERRESTRIAL CALCIUM CARBONATE CYCLE

Eric P. VERRECCHIA
Institut de Géologie, Université de Neuchâtel
Rue Émile Argand 11, 2007 Neuchâtel, Switzerland
Eric.Verrecchia@unine.ch

Corinne LOISY
Université Michel Montaigne, EGID Bordeaux III
1 allée Daguin, 33607 Pessac cedex, France

Olivier BRAISSANT
Institut de Géologie, Université de Neuchâtel
Rue Émile Argand 11, 2007 Neuchâtel, Switzerland

Anna A. GORBUSHINA
ICBM, Laboratory of Geomicrobiology, Carl Von Ossietzky Universität
P.O. Box 2503, D-26111 Oldenburg, Germany

1. INTRODUCTION

In terrestrial environments, fungi constitute one of the most important microbial agents, but also due to their ability to decay organic matter but also because of their role in various biogeochemical cycles (Gadd, 1999; Verrecchia, 2000). Fungal involvement in carbonate rock destruction, dissolution and precipitation has been suspected for over 150 years (Braconnot, 1825) but convincing hypotheses to explain the processes of carbonate dissolution and precipitation were not proposed before the W's.

Nevertheless, their role in the calcium carbonate cycle remains largely underestimated. The aim of this chapter is to emphasize the great potential of fungi in $CaC0_3$ redistribution and sequestration at the surface of continents.

2. DISSOLUTION PROCESSES

Carbonate rocks are a frequent habitat for symbiotic fungi (lichens) and free-living fungi (Sterflinger and Krumbein, 1997; Gorbushina and Krumbein, 2000). Biological mechanisms of carbonate weathering are various and numerous: carbonate destruction and dissolution are caused by both bacteria and fungi (lichens included). Environments in which this process is most active are soils and sub-aerial rock surfaces (including caves).

The main characteristics of the destructive microbial mat are:

← the microflora of rocks emerging from soil or freshly exposed to the atmospheric environment drastically changes from the normal soil flora to very specialized poikilophilic micromycetes as well as other microorganisms;

← the destructive effects of the microflora finally established on and in such rocks (marbles, limestones, dolomites) results from an equilibrium between mechanical destruction, segregation and chemical solution by metabolic products. Hydrolytic enzymes and chelating compounds, including pigments, are taking part in these destructive processes;

← the organisms themselves form clusters, networks and diffuse biofilms, rather thaii a real microbial film or mat as in sub-aquatic environments. Their efforts and life strategies are concentrated on survival and complex interactions with the environment. The surface of interactions of rock substrate and environment (including living microorganisms) is considerably and continuously increased by these interactions;

← rock destruction has been demonstrated for marbles of Crimea, Delos and Israeli sites to be faster than any kind of weathering by atmospheric influences or acid rain. Degradation rates of up to 10 mm per 100 years have been documented by field and laboratory experiments in which physical measurements of heat transfer have showed the consequence of irreversible swelling and expansion of the rock crevice or cavity populated by poikilophilic black fungi.

3. PRECIPITATION PROCESSES

The two principal minerals of geological interest involved in precipitation inside fungal network are oxalates and carbonates. Oxalates are related to the fungal metabolism: excreted oxalic acid reacts on metal-bearing minerals to produce metal-oxalate inside or outside the thallus, a mean of immobilizing soluble metal ions (Sayer and Gadd, 1997; Ramsay et al., 1999). In carbonate rocks, oxalate crystals precipitate mainly as Ca-oxalates such as weddellite and whewellite (polyhydrate and monohydrate Ca-oxalate respectively). These crystals form inside the fungal hypha (Arnott and Webb, 1983; Clémençon, 1997), leading to spiky sheaths (Fig. 1. A, B). Precipitation of Ca-oxalates allows the removal of Ca (which can be toxic in high concentrations) from Ca-rich micro-environments. At the geological scale, oxalates are metastable: they transform into carbonates.

Two main processes lead to calcium carbonate precipitation by fungi: (1) an indirect process, involving a complex cycle starting with oxalic acid excretion, oxidation of oxalate minerals and secondary calcite precipitation in relationship with bacteria and (2) a direct process leading to hyphae biomineralization. This kind of precipitation of calcite crystals occurs directly inside the living fungal matter, mainly in the micro_brile of the chitinal sheath.

Oxalic acid can be considered as a toxic by-product of citric acid production, and its synthesis appears to depend on whether glucose or citric acid is used as a carbon source (Gadd, 1999). Oxalate can be produced directly during fungi's life as a result of its consumption of glucose. This leads to oxaloacetate production that is hydrolized into oxalate and acetate (Verrecchia, 1990), a reaction catalyzed by cytosolic oxaloacetase (Wolschek and Kubicek, 1999). In the case of a citric acid source, oxalic acid is produced through the glyoxylate cycle (Dutton and Evans, 1996;Wolschek and Kubicek, 1999). In carbonate environments, the former process seems more likely. If Ca-oxalates are easily observed on fungal sheath as raphidic or styloidic crystals (Fig. 1. A, B), the future of acetate molecules remain unclear (Gadd, 1999). A possible explanation has been explored: acetate can be rapidly used as a direct carbon source by soil bacteria. The use of acetates leads to calcium carbonate precipitation by bacteria. Therefore, the distinction between sedimentary micrite or microsparite and calcitic bacterial by-products resulting from acetates is dif_cult to make, leading to an underestimation of the role of fungi and associated bacteria in the$CaCO_3$ cycle.

In the laboratory, many soil bacterial strains easily transform acetate sources into calcite crystals (Fig. 1. C, D). In addition, soil bacteria also use Ca-oxalates precipitated by fungi as another carbon source leading to the precipitation of secondary carbonates. The transformation of oxalates into carbonates by bacteria associated with fungi is probably related to two different and/or complementary processes: (1) the role of an oxalate and formate metabolism enzymes (Lengeler et al., 2000) and (2) to an increase of medium pH resulting from oxalate consumption, and consequently an enhancement of the precipitation of calcite (Tamer and Aragno, 1980).

Figure 1. (A, B) Ca-oxalate minerals precipitated on fungal sheaths as raphidic (A) or styloidic (B) crystals. (C, D) Calcite crystals obtained in soil bacteria cultures on acetate (C) and citrate-rich (D) media

One of the contributions of fungi to the carbonate cycle is their ability to precipitate $CaC0_3$ in their sheath as needle fibre calcite (Callot et al., 1985a; Callot et al., 1985b; Verrecchia and Verrecchia, 1994). The fungal origin of needle fibre calcite is well demonstrated by observations of mineralized fungi showing the close relationship between crystals, microtubules and hyphal walls in both scanning electron and light microscopy (Fig. 2). Isotopic data (^{18}O and ^{13}C) of various types of needle fibre calcite associated with fungi show the complexity of the precipitation process. There is no specific isotopic signature (Fig. 3). The carbon source is probably fungal activity as well as the proximal environment of the fungal strains.

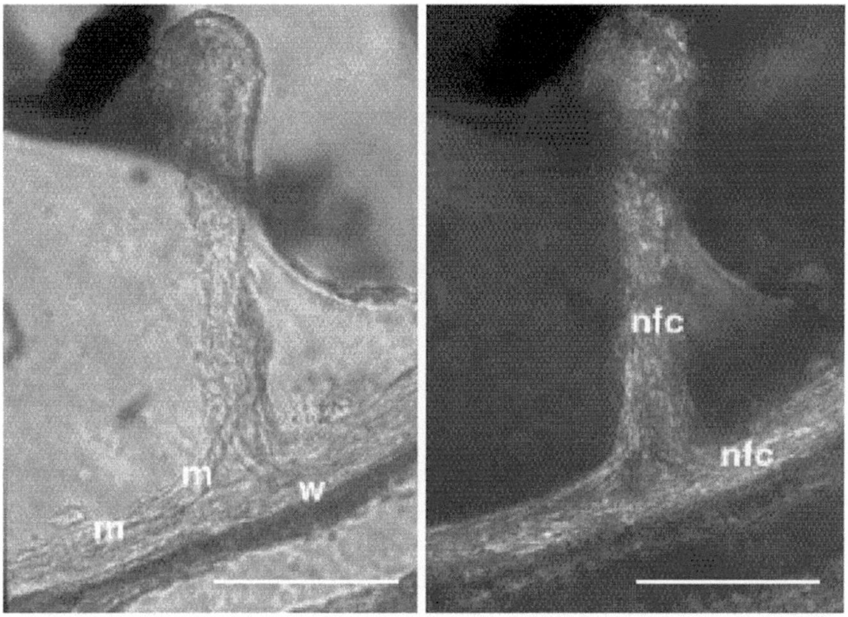

Figure 2. (A) Needle fibre calcite (nfc) precipitated inside fungal walls (w; m: microfibriles). Plane polarized light. (B) Same view in cross polarized light emphasizing the calcitic nature of fibers. Bar 30pm.

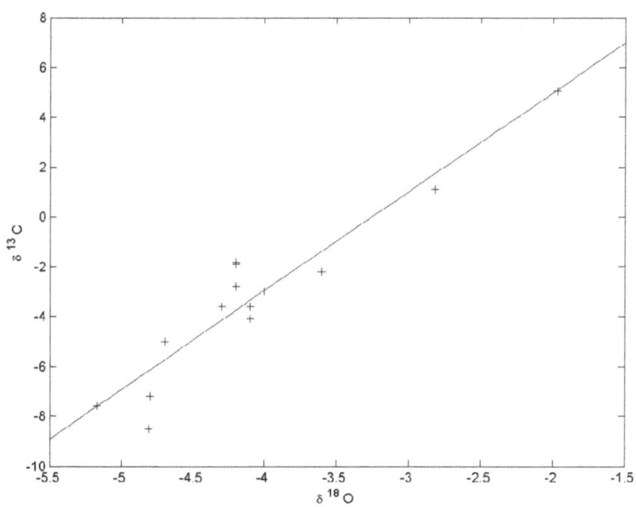

Figure 3. Isotopic signature of needle fibre calcite. There is a clear trend shown by the mixing line but no distinct isotopic signature for needle fibre calcite. This can result from the complex nature of NFC involving various carbon sources such as organic matter, former carbonates, $C0_2$, etc.

4. CONCLUSIONS

The role of fungi is crucial in the calcium carbonate cycle at the surface of continents. Fungi have a huge potential participation in carbonate weathering as well as precipitation. They are undoubtedly at the origin of much of the calcium carbonate accumulation in paleosols and in CaC03 enrichment of terrestrial surficial sediments throughout geological time. Their association with bacteria enables transformation of their by-products such as citrate, acetate and oxalate, reinforcing their role in calcium carbonate dynamics at the surface of continents.

REFERENCES

Amott, H.J. and Webb, M.A. (1983). The structure and formation of calcium oxalate crystal deposits on the hyphae of wood rot fungus, Scan. Eleetron. Mierosc. IV, 1747{1758.

Braconnot, H. (1825). De la présence de l'oxalate de chaux dans le monde minéral: existence du même sel en quantité énorme dans les plantes de la famille des lichens, et moyen avantageux d'en extraire de l'acide oxalique, Ann. Chim. Phys. 28, 318{322.

Callot, G., Mousain, D. and Plassard, C. (1985a). Concentration de carbonate de calcium sur la paroi des hyphes mycéliens, Agronomie 5, 143{150.

Callot, G., Guyon, A. and Mousain, D. (1985b). Inter-relations entre aiguilles de calcite et hyphes mycéliens, Agronomie 5, 209{216.

Clémençon, H. (1997). Anatomie der Hymenomyceten, F. Flück-Wirth Verlag, Teufen, Switzerland.

Dutton, M.V and Evans, C.S. (1996). Oxalate production by fungi: its role in pathogenicity and ecology in the soil environment, Can. J. Microbiol. 42, 881{895.

Gadd, G.M. (1999). Fungal production of citric and oxalic acid: Importance in metal speciation, physiology and biogeochemical processes, Adv. in Microbial Physiol. 41, 47{92.

Gorbushina, A.A. and Krumbein, W.E. (2000). Subaerial microbial mats and their effects on soil and rock, in R.E. Riding and S.M. Awramik (eds.), Microbial Sediments, Springer Verlag, Berlin, Heidelberg, pp. 161{170.

Lengeler, J.W., Drews, G. and Schlegel, H.G. (1999). Biology of the Prokaryotes, Blackwell Science, Thieme, Stuttgart, New York.

Ramsay, L.M., Sayer, J.A. and Gadd, G.M. (1999). Stress response of fungal colonies towards toxic metals. In: The Fungal Colony (N.A.R. Gow, G.

Robson and G.M. Gadd eds), pp. 178{700. Cambridge University Press, Cambridge.

Sayer, J.A. and Gadd, G.M. (1997). Solubilization and transformation of insoluble inorganic compounds to insoluble metal oxalates by Aspergillus niger, Mycological Res. 101, 653{661.

Sterfinger, K. and Krumbein,W.E. (1997) Dematiaceous Fungi as a major agent for biopitting on Mediterranean marbles and limestones, Geomicrobiol. J. 14, 219{230.

Tamer, A.Ü. and Aragno, M. (1980). Isolement, caractérisation et essai d'iden-tifcation de bactéries capables d'utiliser l'oxalate comme seule source de carbone et d'énergie, Bull. Soc. Neuchâtel. Sci. Nat. 103, 91{104.

Verrecchia, E.P. (1990). Lithodiagenetic implications of the calcium oxalate-carbonate biogeochemical cycle in semi-arid calcrete, Nazareth, Israel, Geomicrobiol. J. 8, 89{101.

Verrecchia, E. (1985a). Fungi and sediments, in R.E. Riding and S.M. Awramik (eds.), Microbial Sediments, Springer Verlag, Berlin, Heidelberg, pp. 68{75.

Verrecchia, E.P. and Verrecchia K.E. (1994). Needle-fibre calcite: a critical review and a proposed classification , J. Sedim. Res. A64, 650{664.

Wolschek, M.F. and Kubicek, C.P. (1999). Biochemistry of citric acid acumulation by Aspergilus niger. In: Citric acid biotechnology (B. Kristiansen, M. Mattey and J. Linden eds), pp. 11{31. Taylor and Francis, London.

Chapter 26

PAST AND PRESENT BIOFILM FORMATION IN DEEP FENNOSCANDIAN SHIELD GROUNDWATER

Karsten PEDERSEN
Göteborg University, Department of Cell and Molecular Biology, Microbiology Section, Box 462, SE-405 30 Göteborg, Sweden
pedersen@gmm.gu.se

1. INTRODUCTION

Intra-terrestrial life has been found at kilometre depths in deep subsea-floor sediments and in the basement crust beneath the sediments, up to 2800 m deep in continental sedimentary rocks, in igneous rock aquifers down to 3500 m and in fluid inclusions in ancient salt deposits from salt mines (Pedersen 2000). The biomass of these organisms may be equal to the total weight of all marine and terrestrial plants (Whitman et al. 1998). The intra-terrestrial microbes generally seem to be active at very low but significant rates, and several investigations indicate chemolithoautotrophs to form a chemosynthetic base (Stevens and McKinley 1995; Kotelnikova and Pedersen 1998). Hydrogen, methane and carbon dioxide gases are continuously generated in the interior of our planet (Apps and van de Kamp 1993) and seem to constitute sustainable sources of carbon and energy for deep intra-terrestrial biosphere ecosystems.

Intraterrestrial environments generally offer a tremendous surface area to microorganisms present in the liquid phase. Every particle in sedimentary formations that is surrounded by water offers a surface for attaching and growing microorganisms. The attachment and colonization of microorganisms on particles in porous media have recently been reviewed (Fletcher and Murphy 2001). Aquifers in hard, igneous rock offer an environment to the microbes very different from sedimentary formations. Typically, fractures of various diameters intersect the rock mass. Sometimes, regional fault zones that can be up to several meters appear. Many of those fractures are hydraulically conductive and transport water. Moving water, also very slowly moving, surfaces and microbes always generate biofilms in surface ecosystems. This presentation clearly demonstrates that the intraterrestrial environment does not make an exception.

2. FENNOSCANDIAN SHIELD AQUIFER ENVIRONMENT

The Fennoscandian Shield (also called the "Baltic Shield") is the slightly vaulted Precambrian rock area comprising parts of Norway, most of Sweden, all of Finland and most of the Karelian and Kola peninsulas. The Fennoscandian Shield mostly consists of granite and gneiss rocks (Gàal and Gorbatschev 1987). The age of the Fennoscandian Shield rock spans from 3100 million years in north to 1700 million years in the south and west. Cycles of heating and cooling and glaciations of the rock mass have created a huge three-dimensional intraterrestrial system of aquifers of which many are hydraulically conductive. A total of 75 specific borehole positions in 55 different boreholes down to 1700 m under ground, distributed over 11 sites in Finland and Sweden have been investigated for geology, chemistry, numbers of microorganisms, and microbial diversity and activity (Pedersen 2001). Those extensive investigations have revealed active microbial life. As the igneous rocks were too hot to host any life when formed, this life must have migrated via groundwater conducting fractures into cooled rock mass and adapted to the environmental conditions prevailing there. The life found represent a large physiological and genetic diversity, which reflects the variability in groundwater chemistry and rock composition of the Fennoscandian Shield. Those deep microbial communities may be independent of solar energy. A range of independent results have demonstrated a hydrogen dependency of the deep dwelling populations, which implies that a hydrogen driven biosphere may exist in depth of Fennoscandian Shield rocks.

The groundwater environment in igneous rock aquifers is generally anaerobic, normally with a pH of between 7 and 8. The pH is seldom below 7, but may approach 10 in some water. The presence of ferrous iron and sulfide attests to reduced and anoxic conditions. The content of organic carbon is generally low, being in the range of 0.1–10 mM. The content of dissolved solids is usually dominated by chloride, sodium and calcium, but these vary in concentration by four orders of magnitude. Such variability in groundwater composition makes it difficult to interpret chemical information. Multivariate mixing and mass balance calculations, so-called "M3 modeling", can be used to trace the effect from present and past groundwater flow with fair accuracy (Laaksoharju et al. 1999). This technique has identified five end-members, so-called "reference waters", that mix to form most types of groundwaters found in the Fennoscandian Shield aquifers. These waters have meteoric (i.e., through precipitation), glacial, marine (from the Baltic Sea), altered marine or deep brine characteristics.

3. ATTACHING AND GROWING MICROORGANISMS

Laminar flow reactors (Fig. 1) have frequently been used by the author for the study of biofilms (Pedersen 1982a, 1982b, 1990; Pedersen et al. 1986)

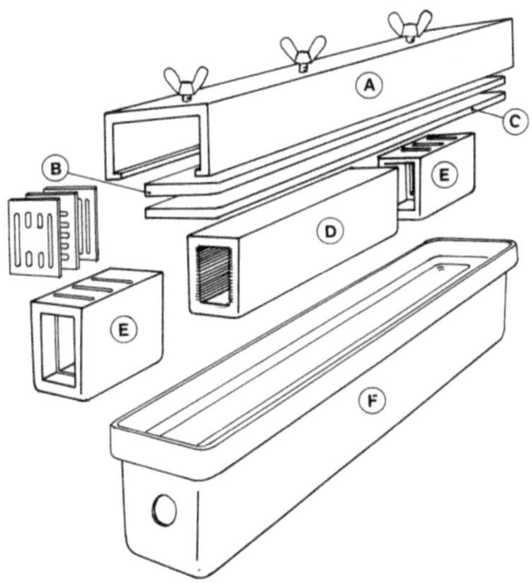

Figure 1. The biofilm reactor, made of polycarbonate. Aluminium sliding lid A; aluminium lid B; lid C; test pile D; diffusors E; reactor box F. Overall dimensions are 340 x 85 x 73 mm. The test pile takes 20 pieces of 140 x 24 x 0.17 mm glass surfaces. Alternatively, a test pile for 12 pieces of 140 x 24 x 0.17 mm surfaces of different materials can be installed

Biofilm reactors were at different times connected to slowly flowing groundwater from boreholes in the Stripa mine (812-1240 m) (Ekendahl and Pedersen 1994; Ekendahl et al. 1994; Pedersen et al. 1996)and to boreholes from ground on Laxemar (830-1078 m) (Pedersen and Ekendahl 1992) close to the Äspö hard rock laboratory (HRL) tunnel. From the Äspö HRL tunnel, five boreholes (68-192 m) were connected (Pedersen et al. 1996). Total numbers of attaching and growing biofilms were analysed. Metabolic activity was assayed with radio-labelled compounds, liquid scintillation and microautoradiography techniques. Some biofilms were studied using scanning electron microscopy (SEM) and microbial species diversity was investigated with 16S rDNA sequencing methodology.

3.1 Numbers

The number of attaching and growing microbes on surfaces exposed to slowly flowing groundwater, in the range of 1 to 20 mm sec^{-1}, show and initial phase of attachment, followed by slow growth (Fig 2). Colonies of microbes developed after about one month of exposure. It is obvious that biofilms develop also in intraterrestrial environments, albeit very slow. As a comparison, laboratory experiments with pure cultures, using the same type of biofim reactors, reached 10^7 cells cm^{-2} after 17 h^2 (from Pedersen et al. 1986); it took almost half a year in deep groundwater. The rate of processes in deep rock environments is on a very slow scale.

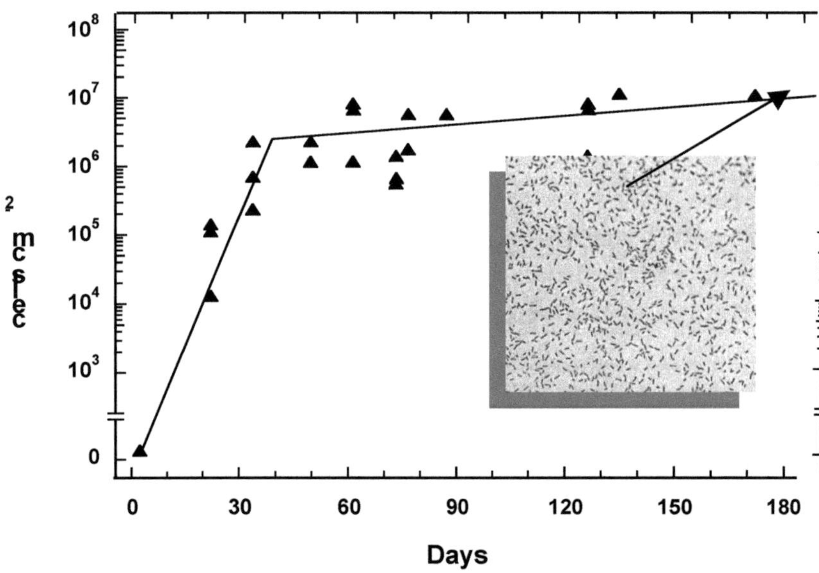

Figure 2. Total number of microorganisms on glass surfaces exposed to slowly flowing groundwater (1-20 mm sec^{-1}) from three different sites, the Stripa research mine, Laxemar and tha Äspö hard rock laboratory tunnel (see text for references). The photo illustrates a surface with just above 10^7 cells cm^{-2} (from Pedersen et al. 1986).

3.2 Diversity

The 16S rDNA diversity of attached microorganisms has been studied at the Stripa (Ekendahl et al. 1994) and Äspö HRL (Pedersen et al. 1996) sites. Representatives for Proteobacteria beta, gamma, delta, epsilon groups, and Gram positive bacteria was found. The diversity on the surfaces was generally different, and smaller, than what was found in the groundwater. Commonly, only a few species appeared. For example, one borehole had 90 % of the clones closely (98.6 %) related to *Bacillus megaterium* (Pedersen et al. 1996). This observation correlated nicely with the appearance of very large rods with a size typical for *B. megaterium* in this borehole (Fig. 3).

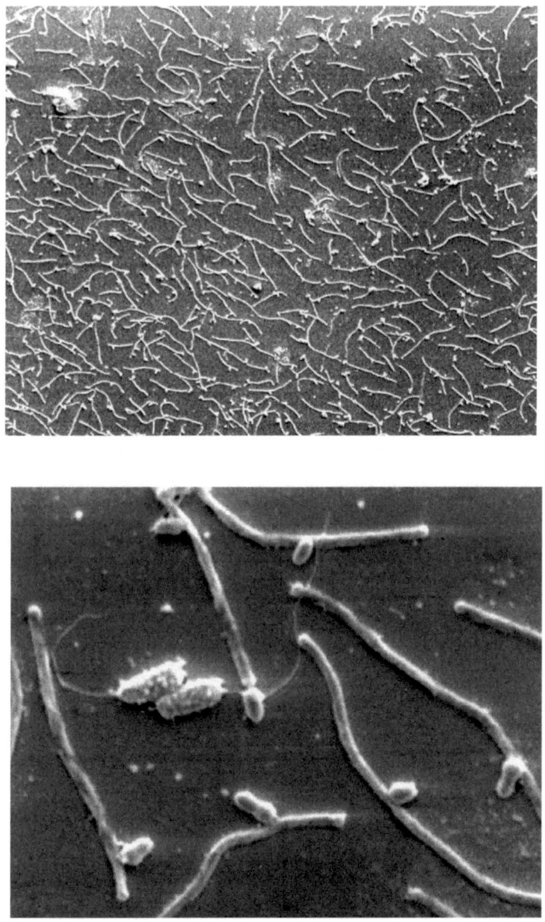

Figure 3. A glass surface that has been exposed to slowly flowing groundwater for 67 days. The water came from a borehole denoted KR0013, 68 m below ground. The top figure gives an overview and the lower figure shows a close up. The filamentous microbes on the images are about 5 – 7 m long.

As specific archeal amplification primers were not used in those studies, the probability to detect organisms belonging to the domain *Archaea* was strongly reduced. Therefore, *in situ* probe methodology using archaean primers was applied later on rock surfaces exposed to groundwater from KR0013 in the Äspö HRL tunnel. Cultivable methanogens had been found in this borehole earlier (Kotelnikova and Pedersen 1998). The hybridisation indeed demonstrated attached archaean cells that were growing in chains on the exposed rocks (see fig. 4 in Pedersen 1997). It therefore seems clarified that *Archaea*, possibly methanogens, participate in deep groundwater biofilm formation.

3.3 Activity

The metabolic activity of the investigated biofilms was studied using radio-labelled substrates. Formate, acetate, lactate, glucose, carbon dioxide and leucine were used (Pedersen and Ekendahl 1992; Ekendahl et al. 1994; Ekendahl and Pedersen 1994; Pedersen et al. 1996). Total numbers of attaching and growing biofilms were analysed. Leucine assimilation is virtually specific to bacteria if low (nM) concentrations are applied. This amino acid is used for protein synthesis by many bacteria during growth (Kirchman et al. 1985). It can also be used as a carbon and energy source and can be fermented by proteolytic clostridia via the Strickland reaction. High percentages, up to 98%, of most populations studied assimilated leucine, as measured using microautoradiography. The attached populations generally showed higher percentages of active cells that did the unattached microbes. This assimilation indicates that major portions of the studied communities were viable. The preferred carbon source was lactate, and the attached microbes generally showed a much larger metabolic activity per cell. Overall, it became clear that the attached populations were more metabolically active, as compared to the unattached populations. This may indicate that intraterrestrial microbial activity generally occurs on solid surfaces rather than in the goundwater, if there is a flow which can be very slow, of groundwater.

4. A FOSSIL BIOFILM

Many old fractures in hard rock are no longer open, because they have been filled with precipitating minerals such as calcite, dolomite, pyrite, epidote and chlorite, to mention some of the more common ones. This fact presents interesting possibilities for the search for microbial fossils, which have been captured in fluid inclusions in hard rock aquifers, as has been reported for halophilic bacteria (Vreeland et al. 1999). Two boreholes in the Äspö HRL tunnel (SA813B and SA923A) carried groundwater that was over-saturated with carbonate and calcium. They were explored for attaching and biofilm-forming microorganisms (Pedersen et al. 1996). Glass surfaces exposed to flowing groundwater at flow rates of 8–14 mm s^{-1} for 67 days were subsequently observed through light and scanning electron microscopes. Some 1.2 x 10^6 cells per cm^{-2} were found more or less buried in dense precipitates of calcite. Although partly artificial, this experiment supports the idea that microbes attached to fracture minerals in deep hard rock may become buried or trapped in fluid inclusions during growth of a fracture mineral.

An unusually long fracture core sample from 207 m borehole length (200 m depth) in KAS02. was extensively investigated with respect to stable isotopes (Tullborg et al. 1999). Obvious signs of biological activity in the form of light ^{13}C signatures in the calcite carbon were found. It was therefore assumed that fossil microorganisms may be detectable in the fracture mineral. Investigating the fracture with electron microscopy did indeed reveal bacterium-like structures in parts of it and X-ray analysis demonstrated these structures to be enriched with carbon (Pedersen et al. 1997). The found fossil microorganisms in the deep igneous rock fracture minerals were a good indication that microbial life was present deep under the island of Äspö long before drilling and excavation of the Äspö HRL.

5. IRON- AND SULPHUR-OXIDISING BACTERIA IN BIOFILMS

Whenever reduced, anaerobic ferrous iron bearing groundwater reaches an oxygenated environment, iron-oxidizing bacteria appear and grow. They oxidise ferrous iron and precipitate as ferric iron oxi-hydroxides together with large amounts of cell material. They eventually may form biofilm mats that can be several cm thick. In a similar way, sulphide-oxidising bacteria will grow and form biofilm mats.

5.1 Gallionella ferruginea biofilms

The genus *Gallionella* is characterized by its chemolithotrophic growth (Hallbeck and Pedersen 1991) with ferrous iron and its production of a twisted stalk consisting of a bundle of numerous fibers that makes *Gallionella* very easy to identify (Hallbeck and Pedersen 1990; Hallbeck et al. 1993). The formation of biofilms of *Gallionella* on surfaces has previously been reported (Hallbeck and Pedersen 1995). Water from 40–60 m deep wells was pumped over test surfaces and in less than 48 hours, dense biofilms of stalks and cells developed (Fig. 4). They were up to 30 m thick, showing the biofilms growth potential of this bacterium. The Äspö Hard Rock Laboratory (HRL) tunnel, is heavily populated with mats of *Gallionella* that can reach 10 cm or more in thickness. Stones and rock surfaces in pools and ditches in the tunnel are commonly covered with moss-like biofilms of *Gallionella*, demonstrating underground tunnels, mines, etc to be excellent sites for studies of BIOFILM processes related to this organism, such as metal accumulation (Ferris et al. 1999, 2000).

The stalks of *Gallionella* attract large amounts of iron hydroxides and also, of other metals. The processes governing this coating are not well established, although several intriguing theories exist. It can be anticipated that enzymatic ferrous iron oxidation by *Gallionella* in its extraction of electrons for energy production takes part in the build-up of metal oxides on the stalks. By this process, oxidized, ferric iron would be expelled from metabolic processes and adsorbed on the stalk. In addition, as suggested elewhere (Hallbeck and Pedersen 1995), the stalk may act as a protective agent against the creation of toxic levels of oxygen radicals. The stalk would then have a ferrous iron oxidation capability and a sorption capacity for ferric iron hydroxides and also, for other metals.

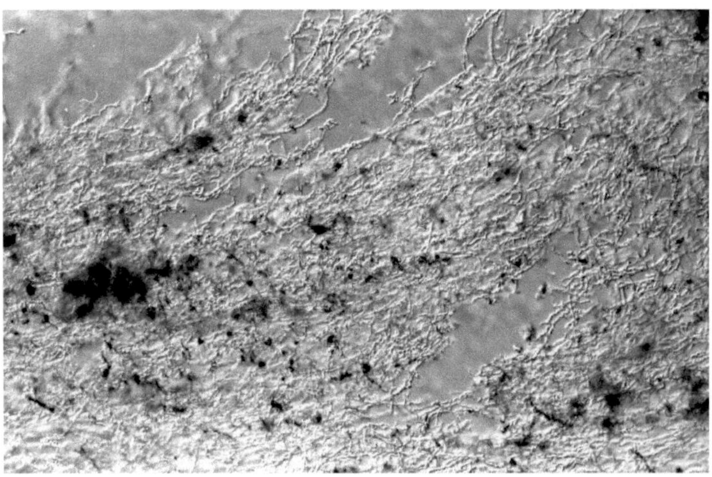

Figure 4. A biofilms of *Gallionella ferruginea* formed on glass surfaces in the biofilm reactor after 48 h. The long, filamentous stalks are visible as is dark spots of precipitated iron. The cells cannot easily be detected without staining with fluorescent stains. (Scale 1 cm = 40 m).

REFERENCES

Apps J A, van de Kamp P C, 1993. Energy gases of abiogenic origin in the Earth's crust. In: Howell G (ed) In The future of Energy gases, U.S. geological Survey Professional Papers. United States Government Printing Office, Washington. pp 81-132.

Ekendahl S, Arlinger J, Ståhl F, Pedersen K, 1994. Characterization of attached bacterial populations in deep granitic groundwater from the Stripa research mine with 16S-rRNA gene sequencing technique and scanning electron microscopy. Microbiology 140, 1575-1583.

Ekendahl S, Pedersen K, 1994. Carbon transformations by attached bacterial populations in granitic ground water from deep crystalline bed-rock of the Stripa research mine. Microbiology 140, 1565-1573.

Ferris F G, Hallberg R O, Lyvén B, Pedersen K, 2000. Retention of strontium, cesium, lead and uranium by bacterial iron oxides from a subterranean environment. Applied Geochemistry 15, 1035-1042.

Ferris F G, Konhauser K O, Lyuvén B, Pedersen K, 1999. Accumulation of metals by bacteriogenic iron oxides in a subterranean environment. Geomicrobiology Journal 16, 181-192.

Fletcher M, Murphy E, 2001. Transport of microorganisms in the subsurface: The role of attachment and colonization of particle surfaces. In: Fredrickson JK, Fletcher M (eds) In Subsurface microbiology and biogeochemistry. Wiley-Liss Inc., New York. pp 39-68.

Gàal G, Gorbatschev R, 1987. An outline of the Precambrian evolution of the Baltic shield. Precambrian Research 35, 15-52.

Hallbeck L, Pedersen K, 1990. Culture parameters regulating stalk formation and growth rate of *Gallionella ferruginea*. J Gen Microbiol 136, 1675-1680.

Hallbeck L, Pedersen K, 1991. Autotrophic and mixotrophic growth of *Gallionella ferruginea*. J Gen Microbiol 137, 2657-2661.

Hallbeck L, Pedersen K, 1995. Benefits associated with the stalk of *Gallionella ferruginea*, evaluated by comparison of a stalk-forming and a non-stalk-forming strain and biofilm studies in *situ*. Microbial Ecology 30, 257-268.

Hallbeck L, Ståhl F, Pedersen K, 1993. Phylogeny and phenotypic characterization of the stalk-forming and iron-oxidizing bacterium *Gallionella ferruginea*. J Gen Microbiol 139, 1531-1535.

Kirchman D, K'nees E, Hodson R, 1985. Leucine incorporation and its potential as a measure of protein synthesis by bacteria in natural aquatic systems. Appl Environ Microbiol 49, 599-607.

Kotelnikova S, Pedersen K, 1998. Distribution and activity of methanogens and homoacetogens in deep granitic aquifers at Äspö Hard Rock Laboratory, Sweden. FEMS Microbiology Ecology 26, 121-134.

Laaksoharju M, Skårman C, Skårman E, 1999. Multivariate mixing and mass balance (M3) calculations, a new tool for decoding hydrogeochemical information. Applied Geochemistry 14, 861-871.

Pedersen K, 1982a. Method for studying microbial biofilms in flowing-water systems. Appl Environ Microbiol 43, 6-13.

Pedersen K, 1982b. Factors regulating microbial biofilm development in a system with slowly flowing seawater. Appl Environ Microbiol 44, 1196-1204.

Pedersen K, 1990. Biofilm development on stainless steel and PVC surfaces in drinking water. Water Research 24, 239-243.

Pedersen K, 1997. Microbial life in granitic rock. FEMS Microbiol Rev 20, 399-414.

Pedersen K, 2000. Exploration of deep intraterrestrial life – current perspectives. FEMS Microbiol Lett 185, 9-16.

Pedersen K, 2001. Diversity and activity of microorganisms in deep igneous rock aquifers of the Fennoscandian Shield. In: Fredrickson JK, Fletcher M (eds) In Subsurface microbiology and biogeochemistry. Wiley-Liss Inc., New York. pp 97-139.

Pedersen K, Arlinger J, Ekendahl S, Hallbeck L, 1996. 16S rRNA gene diversity of attached and unattached groundwater bacteria along the access tunnel to the Äspö Hard Rock Laboratory, Sweden. FEMS Microbiology Ecology 19, 249-262.

Pedersen K, Ekendahl S, 1992. Assimilation of CO_2 and introduced organic compounds by bacterial communities in ground water from Southeastern Sweden deep crystalline bedrock. Microbial Ecology 23, 1-14.

Pedersen K, Ekendahl S, Tullborg E-L, Furnes H, Thorseth I-G, Tumyr O, 1997. Evidence of ancient life at 207 m depth in a granitic aquifer. Geology 25, 827-830.

Pedersen K, Holmström C, Olsson A-K, Pedersen A, 1986. Statistic evaluation of the influence of species variation, culture conditions, surface wettability and fluid shear on attachment and biofilm development of marine bacteria. Arch Microbiol 145, 1-8.

Stevens T O, McKinley J P, 1995. Lithoautotrophic microbial ecosystem in deep basalt aquifers. Science 270, 450-453.

Tullborg E-L, Landström O, Wallin B, 1999. Low-temperature trace element mobility influenced by biogenic activity- Indications from ^{18}O, ^{13}C, ^{34}S and trace element analyses of fracture calcite and pyrite in crystalline basement. Chemical Geology 157, 199-218.

Vreeland R H, Piselli Jr A F, McDonnough S, Meyers S S, 1999. Distribution and diversity of halophilitic bacteria in a subsurface salt formation. Extremophiles 2, 321-331.

Whitman W B, Coleman D C, Wiebe W J, 1998. Prokaryotes: The unseen majority. Proc Natl Acad Sci U S A 95, 6578-6583.

Chapter 27

MELANIN AND CHROMATIC CHANGES IN BLACK CRUSTS OF SANDSTONE MONUMENTS

Jesus M. VALERO, J. Iñaki ALAVA
INASMET. Department of Chemical Tech. and Environ. Mikeletegi Pasealekua 2. 20009. San Sebastian. Spain.
jvalero@inasmet.es

1. INTRODUCTION

It is widely accepted that rates of decay and soiling of stone monuments have accelerated in the last century. Soiling is an optical effect, a darkening of the surface that can be measured as a change in light reflectance (Newby et al 1991). Black crusts are among the most studied patterns of deterioration in historical buildings.

Black crusts are usually formed in areas sheltered from intensive rain water runoff. These crusts are composed by gypsum crystals mixed with the original stone materials and dust, aerosols and particulated materials of various origins. The chromatic alteration in black crusts has been attributed to soot deposits (airborne carbonaceous particles, uncombusted hydrocarbons, etc) on the surface of monuments. However, several authors have attributed a biological origin to some chromatic alterations present in monuments. Krumbein (1992) and Urzi et al (1992) stated that black stains on stone monuments have a biological origin. According to these authors, black fungi an their metabolic pigments called melanins have caused most of the blackening and brown patinas on monuments.

Although the possibility to use organic compounds of anthropogenic origin as carbon source has been pointed out by various authors, there is a lack of knowledge about the ability of black fungi to grow on these compounds.

At the same time, studies performed using Py-GC-MS comparing the organic compounds found in black crusts and those of melanin, have tried to demonstrate that melanin can not be responsible of the chromatic alteration, because no biological marker for melanin can be found in black crusts. There is also a lack of knowledge about the melanin structure of black fungi isolated from black crusts. In the present article our attention has focused on the chemical structure of melanins from black fungi.

2. METHODOLOGY

Black crusts were taken from sandstone monuments in the city of San Sebastian (Spain). Black fungi were isolated and identified. Their pigments were extracted and characterised using different techniques (FTIR, Azure A absorption and tryciclazole inhibition).

3. RESULTS

3.1 Fungi identification and acid production

Several fungi were isolated from samples of black crusts. The black fungi belonged to Cladosporium, Penicillium and Phoma genera (Table 1), all of them producing a dark green mycelium. The testing for acid production was negative for all the strains (Table 1), therefore deterioration bu acids is not likely to occur in these strains of black fungi.

Table 1. Identification and acid production of black fungi from black crusts

Strain	Identification	Acid production
T1	*Cladosporium spp.*	Negative
T3	?	Negative
T4	*Cladosporium spp.*	Negative
T5	*Cladosporium spp.*	Negative
T6	?	Negative
T7	*Penicillium corylophilum*	?
T9	*Phoma glomerata*	?

3.2 Pigment characterization

Pigment characterization was performed by FTIR analysis, Azure A absorption change and tryciclazole inhibition. Pigment extracted of strain T4 showed the FTIR spectrum that can be seen in figure 1. The IR spectrum confirm that the pigmentation of this fungus in indeed melanin.

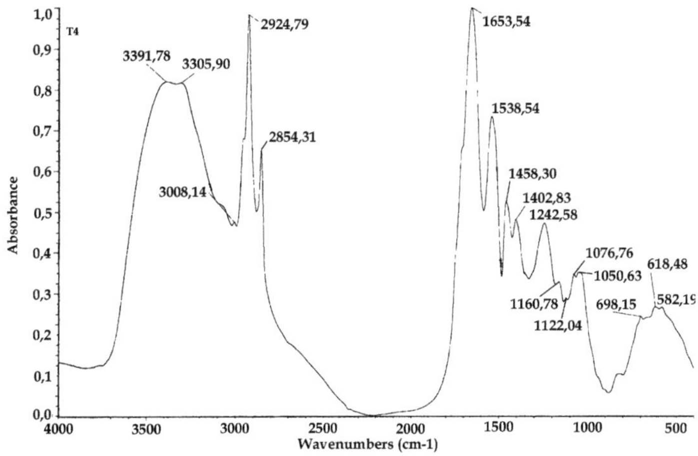

Figure 1. FTIR analysis of pigment extracted from Cladosporium spp.

Azure A analysis of extracted pigment also confirm that the pigment is melanin, showing (figure 2) the characteristic decrease on absorbance associated to melanin-Azure A binding.

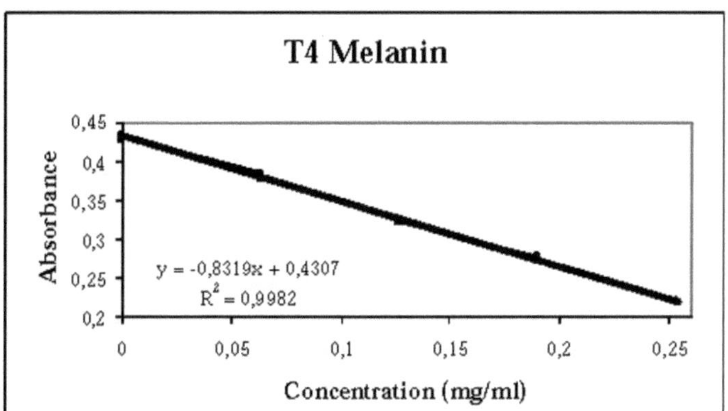

Figure 2. Azure A identification of melanin from T4

The tryciclazole assay demonstrate that melanin of five strains of black fungi (all the tested strains) was indeed DHN melanin. Tryciclazole inhibit the enzyme reductase of the DHN synthesis pathway thus producing a change in the colour of hiphae, that lose their pigment becoming white (table 2). Growth of fungi was not altered by tryciclazole concentration.

Table 2. Growth and inhibition of DHN melanin by tricyclazole

Strain	[tryc] (mg/l)	Growth	Hiphae colour
T1	0	+++	Dark green
T1	10-20-30	+++	White
T3	0	+++	Dark green
T3	10-20-30	+++	White
T4	0	+++	Dark green
T4	10-20-30	+++	White
T5	0	+++	Dark green
T5	10-20-30	+++	White
T6	0	+++	Dark green
T6	10-20-30	+++	White

4. CONCLUSIONS

Black fungi isolated from black crusts in sandstone are classified in the *Cladosporium*, *Phoma* and *Penicillium* genera. All the strains produced dark green colonies in culture but none of them produced acid. These results are in accordance with previous results in black fungi isolated from calcareous rocks (Wollenzein et al 1995). The pigment produced by these fungi is DHN melanin. The fact that DHN melanin is the only melanin present in these black fungi could explain why it has not been found an appropriate fungal marker for Py-GC-MS.

Saiz-Jimenez et al (1995) analysed various fungi using Py-GC-MS, but only one of them was a DHN melanin producer. In the pyrolisate of this fungi they found a large amount of naphthalene derivatives. Most of these derivatives were also found by Hermosin (1995) in black crusts around Europe, but this author indicated that there were due to atmospheric pollution. We think that some naphthalene derivatives could be used as fungal melanin markers. In this sense PY-GC-MS analysis of black crust should be re-evaluated accordingly.

The inherent characteristics of melanins make them difficult to analyse. Melanins are not soluble in most solvents, they have a heterogeneous structure, and are formed by a large range of associated molecules. Their breakdown on simple monomers, as other complex biomolecules, is impossible, thus making difficult the study of their structure. These facts can explain why the biological hypothesis of chromatic alteration has been difficult to prove. Work is in progress regarding the analysis of black crust to determine the presence of fungal DHN melanin in black crusts.

REFERENCES

Saiz-Jimenez C., Ortega-Calvo J.J., and De Leeuw J.W. 1995. The chemical structure of fungal melanins and their possible contribution to black stains in stone monuments. The Science of the Total Environment 167: 305-314.

Hermosín B. 1995. phD Thesis. University of Sevilla. Newby P.T., Mansfield T.A., and Hamilton R.S. 1991. The sources and economical implications of building soiling in urban areas. Sci. Total Environ., 100: 347-366.

Krumbein W.E. 1992. Colour changes of building stones and their direct and indirect biological causes. Proc. 7th Int. Cong. Deterioration and Conservation of Stone, LNEC, Lisbon, ed. J. Delgado Rodrigues, F. Henriques, and F. Telmo Jeremias, pp. 443-452.

Urzí C.E., Krumbein W.E., and Warscheid T. 1992. On the question of biogenic colour changes of Mediterranean monuments (coating – crust – microstromatolite – patina – scialbatura – skin – rock varnish), in D. Decrouez, J. Chamay and F. Zezza (Eds.), La Conservation des Monuments dans le Basin Méditerranéen, Musée d´art et d´histoire. Genève, p.397.

Wollenzein U., de Hoog G.S., Krumbein W.E., and Urzi C. On the isolation of microcolonial fungi occurring on and in marble and other calcareous rocks. The Science of the total Environment 167 (1995) 287-294.

Chapter 28

QUANTIFICATION AND SPATIAL RELATIONSHIP OF MICROORGANISMS IN SUB-AQUATIC AND SUB-AERIAL BIOFILMS

Karsten Rodenacker
GSF-National Research Center for Environment and Health,
Institute of Biomathematics and Biometry, 85764 Neuherberg, Germany
Karsten@Rodenacker.de

Martina Hausner
Institute for Water Quality Control and Waste Management,
Technische Universität München, Germany
M.Hausner@bv.tu-muenchen.de

Anna A. Gorbushina
Institute for Chemistry and Biology of the Marine Environment (ICBM),
Carl von Ossietzky Universität Oldenburg, Germany
a.gorbushina@uni-oldenburg.de

1. INTRODUCTION

Similarity is the one and only measurement of qualities. Explanations and ideas cannot be transferred without images. The transition from qualitative observation, comparison and description to quantitative and objective findings by measurements is rarely done for properties of objects represented by images or volumes. The extraction of quantitative information from images is either obscure, out of scope or, even worse, misused and misinterpreted. In contrast, the state of art of imaging, generation of digital image data, is highly developed.

Technical progress allows almost all laboratories to apply elaborate digital imaging methods to microscopic specimens. These are namely all methods to obtain digital data by classical and fluorescent light microscopy, as well as by confocal laser scanning microscopy, X-ray microtomography for (slice-wise) 3D digitization and techniques from electron microscopes and magnetic resonance and PET based devices. Digital data of the highest quality can be derived and visualized using most of these devices. However, visualization of 3D data is still under development for adequate, problem related representation.

Most of the collected data are only observed and used for qualitative consideration. The application of objective quantitative methods to digital data is done only to a very limited degree (Krumbein et al., 1994). The measuring methods applied to digital data are still not in focus and often considered as mysterious. Typically black (software) boxes containing a multitude of methods are attached to imaging devices and are often misused or misunderstood. From application of wrong scales through misuse of statistical methods to lacking consideration of dimensional data, no possible way of data misinterpretation is left out.

In section 2 we try to present image analysis (IA) methods from scratch. We outline a measuring hierarchy called *measurement continuum* for isolated, hence defined, interacting and replicated objects in two and three dimensions. Section 3 about quantification in experiments with biofilm and fungus growth and some concluding remarks (section 4) will finish this chapter. Accordingly, due to the space limitations this presentation can neither be complete nor fully comprehensive. However, the authors hope to help in clarifying this promising field of digital image analysis for further and broader usage.

2. THE MEASURING CONTINUUM

2.1. Definition of objects or segmentation

Based on the idea of the scientist, material is examined under a device (e.g. a microscope) and digital data are gathered and stored. In digital space images and volumes are represented by picture (*pixel*) or volume elements (*voxel*), respectively. An object is a set or heap of those pixels. In Fig. 1 the *raster* and the *pixels* of an object are shown in sketch form. The most important (and most cumbersome) step is now the definition of the above mentioned objects. Certain areas have to be identified as *region of interest* (ROI). This is called *delineation* or *segmentation* in the field of pattern recognition and image anaylsis (IA). After solving

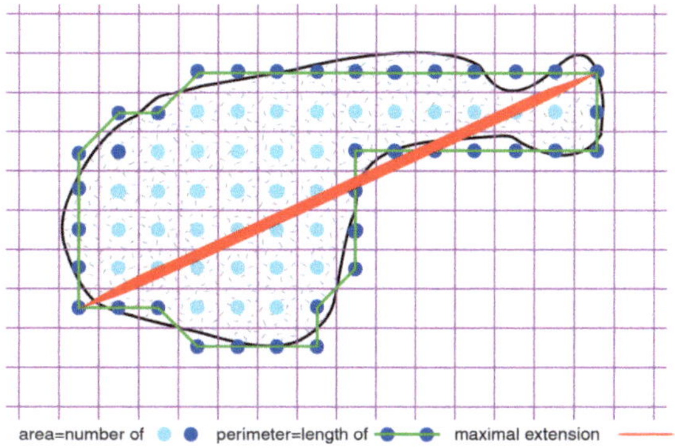

Figure 1. Digitisation scheme of a simple object

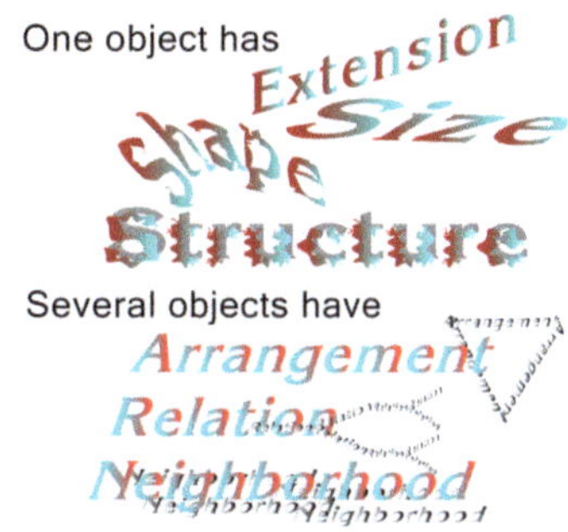

Figure 2. Measurable properties of objects

this step by hand or by semi-automatic methods the resulting regions are ready for measurement (Rodenacker, 1988). Fig. 3 shows a digitized image by successive increasing enlargement to determine the *pixels* and the *pixel values* and their use for segmentation by a threshold.

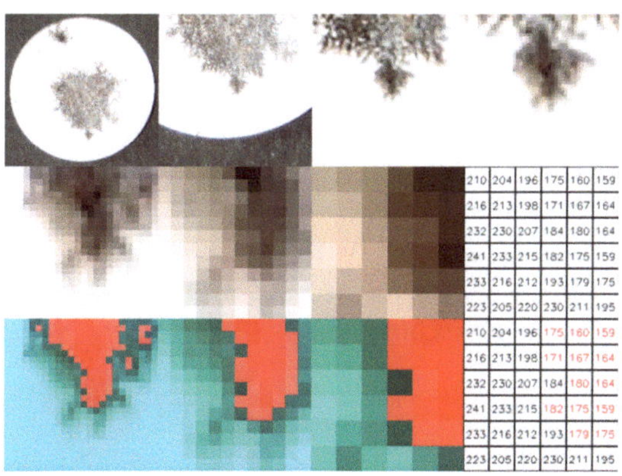

Figure 3. Sequence of different enlargements of the image of a fungus culture showing an example of segmentation by a threshold

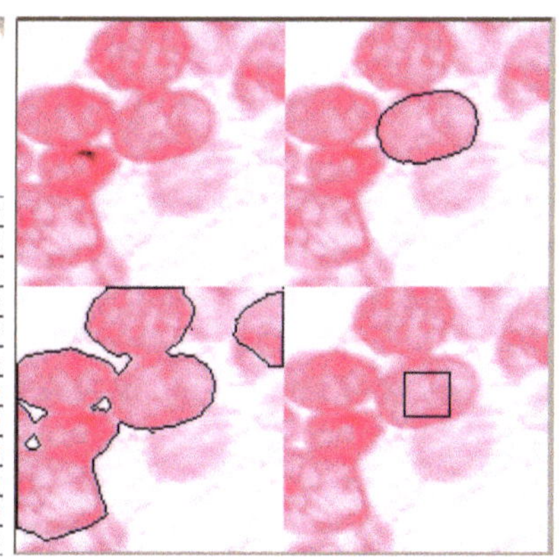

Figure 4. Different segmentations of a cell nucleus

In case of images, hence 2D data, we typically deal with projections of material. In 3D these data are the volume extensions of the material represented in digital form. Although the 3D representation is a more complete one with respect to images, it is more difficult to handle. We are not used to drawing and interacting in space only shown on screen. However, material has properties and in any dimension these properties can be used to differentiate regions of interest and to relate them to (external) experimental questions.

Pixels or voxels have properties like the location (*coordinates*) and *pixel values* resulting of the imaging process and, after segmentation, *labels*. A quite famous label is the one of *background*. For example a digital image from a transmission light microscope consists of pixels carrying some information about the amount of light transmitted through the material and possibly about its spectrum, the color. It might be possible that objects are detected where the transmitted light is below a certain *threshold* (Fig. 3). This leads us already to the measuring part since the amount of transmitted light is already a measurement of pixels in digital form. In fact for identification purposes the pixels are the objects of interest.

It should be clear that the regions defined or selected are highly dependent on the measuring task, hence the interest. In Fig. 4 a digital image of stained cell nuclei is displayed. Depending on the measuring task, different regions are outlined: one nucleus, all nuclei or a part of the central nucleus. The latter two possibilities might serve for quantitative analysis of the variability of transmitted light inside the nuclei. Fig. 2 illustrates how the segmentation of one nucleus alone can be considered for all properties of one object. All nuclei may deliver information about (internal) structure and (external) neighborhood.

The most important (qualitative) properties for given objects are illustrated in Fig. 2. Measurements will have to be related to one of these descriptive fields or terms. The measured *values* will have to be related to the hypotheses of the respective experiment or the external information. In Tab. 1 we try to establish an hierarchical structure reflecting the needs for differentiation of an actual experiment.

Table 1. Measurement continuum: Features of objects

Level	*Type*	*Feature*
0	pixel,voxel	value (color), location
1	set of pixels	e.g. number (sizes), mean of values, moments, diameter (extensions), form (shape), orientation (direction), local similarity (texture, structure)
2	group of sets of pixels	the same features as above with additional measurements of relations: describing neighborhood, connectivity, orientation
...	group of ...	etc.

2.2. Measurement of objects

In the previous section we defined objects, described how they were defined by pixels or voxels and what information they may contain (Tab. 1). In this section we will shortly illustrate how quantitative features can be extracted or calculated from this information.

In general a measurement is a count of *sub*-objects. Mathematically speaking, this is the sum (or integral) over a function f inside a region R. f is representing the image with its pixels or any other derived function. If an extension has to be measured, the region R is usually the set of border points of the object. For size and shape, the required region is the whole object itself whereas for structures, sets of sub-objects are often defined.

One of the most frequently used concepts for quantitative features is that of *moments* for different dimensions defined as

$$m_i^f = \sum_{x \in R \subset \mathbf{Z}} f(x) x^i, \qquad m_{ij}^f = \sum_{(x,y) \in R \subset \mathbf{Z}^2} f(x,y) x^i y^j \tag{28.1}$$

and m_{ijk}^f accordingly. Let $f(x,y)$ a binary image of an object with values 0 and 1, than m_{00}^f is equal to the area A of the region and $(m_{00}^f/m_{10}^f, m_{00}^f/m_{01}^f)$ are the centroid coordinates (kx, ky). With $f(x)$ the frequency distribution of pixel intensities, a histogram, m_0^f is the total integrated intensity, m_1^f the mean and m_2^f the variance of the pixel intensities. This is only to illustrate how simple the definition of measurements is or can be. The problem is more on the side of the identification of the region of interest and the interpretation of measurements for the respective experiment.

A general measurement procedure involves *data*, the *region of interest* (ROI), a *transformation* and the *measurement*. The up to now not mentioned transformation represents an enhancement of the data, such as frequency distribution, which dismisses all local properties and represents only statistical properties of the pixel values, or gradient transformation which enhances only local changes of intensity. Typically the transformation has a non-scalar result. The subsequent applied measurement results always in one scalar.

This distinction allows to define procedures for the detection of certain properties difficult to measure. For example shape in all possible expressions is not directly measurable. However, transformations can be designed extracting the property of interest. In addition to the works in mathematical morphology (Serra, 1982) Sir D'Arcy Thompson (Thompson, 1966) carried out invaluable work in this field.

2.3. Measurement of relations between objects

Objects carry now features, can be (re-)located (Sec. 2.2) and related to others by certain *rules*, e. g. neighborhood, distance or properties like *being similar in size, ... surrounded by ..., ... a recognized object type* ... or other inherent spatial and non-spatial object properties. With other words *intra- and inter-relationships* of the objects under focus are considered. Applications from the field of cyto- and histometry and introductions can be found in Rodenacker and Bengtsen, 2003. Fig. 5 shows an example for the definition of subregions (objects) and their use for estimation of growth spread in radial direction (Fig. 6). In Fig. 7 a

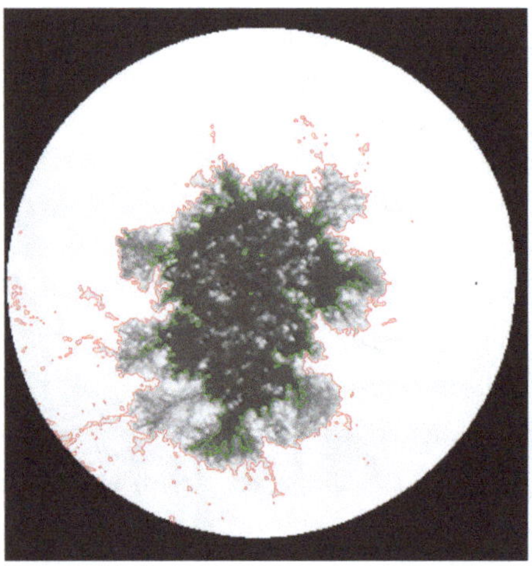

Figure 5. Fungus with two differently pigmented hyphal areas segmented

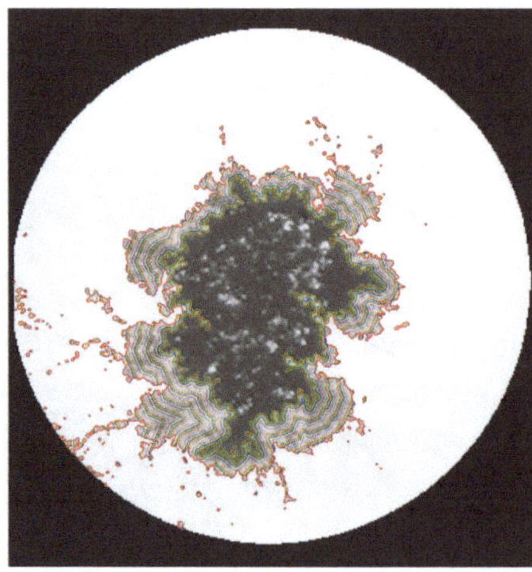

Figure 6. Same fungus with growth front simulated from strong pigmented hyphal area to the outside

region is devised into a set of patches (objects) that are related to each other by certain geometric criteria. Some arbitrarily chosen relations are plotted by joining lines. The formerly defined objects generate now new (higher level) objects with new interpretation. To explain this a bit more thoroughly, consider Fig. 7 *(left)*. Every object x, marked by a yellow star, has a set of outgoing joins. These joins with the set of related objects S define a neighborhood around the object x. $x^1 = S \cup x$ is a new object for further analysis on the base of all features of $\forall y \in S \cup x$. They are considered for further feature extraction. Typical spatial relationships are the above mentioned neighborhoods and others. The Delaunay triangulation (Fig. 7 *(left)* in blue) is very useful for questions

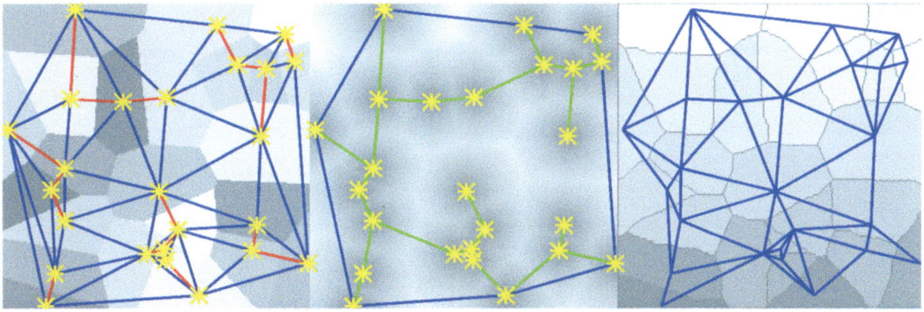

Figure 7. Partitioned region with different spatial neighborhood relationships: (*left*) Delaunay triangulation blue, nearest neighbor red, (*centre*) minimum spanning tree green, convex hull blue, (*right*) skeleton neighborhood

of surface coverage. The nearest neighborhood (Fig. 7 *(left)* in red) and the minimum spanning tree (Fig. 7 *(centre)* in green) deal with questions of reachability. The skeleton neighborhood (Fig. 7 *(right)* in blue) is directly applicable for the analysis of interface reaction and correlation. In this illustrative and artificial example the skeleton neighborhood is very similar to the Delaunay triangulation and differs only at the borders of the field. This is generally not the case for arbitrary shapes of the generating objects.

Distance based methods using (co-)localization and distribution of different labelled objects (possibly differently stained or fluorescently marked, or in general, with different feature values) serve also for quantification and characterization. In Fig. 8 *right* each pixel of one object from one class has a minimum distance to (any object of) the other class. The resulting distribution of distances can be statistically evaluated (Rodenacker et al., 2000) and used for characterization.

3. EXAMPLES

All calculations and illustrations are performed on normal PC's under IDL[1], a programmable and interactive data manipulation language and program environment. For calculations on spatial or temporal-spatial data (3D or even 4D) PCs have to be equipped with sufficient RAM memory.

[1]Research Systems Inc., Boulder, USA

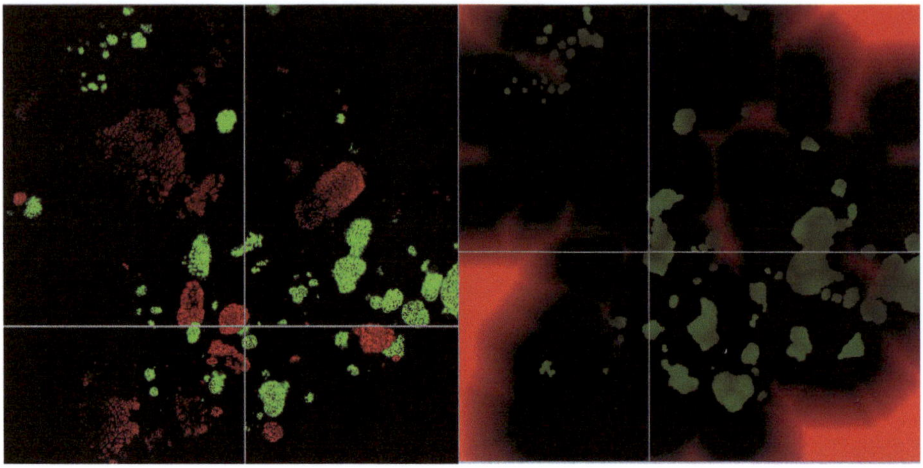

Figure 8. *left*: Optical section through a double FISH marked sludge flock (M. Wagner, Microbiology, TU München). *right*: Same section with green component and overlaying distances away from the red component in red. Bright red shows large distance.

3.1. Differentiation of wild and mutant bacteria *Pseudomonas aeruginosa*

Material and methods: *Growth of biofilms in flow cells:* Biofilms were grown on cover slips (0.2 mm thick) in continuous flow-through stainless steel cells ("flow cells"; 40 mm long, 4 mm wide, 8 mm in height) (Johnsen et al., 2000). Flow cells were inoculated with *Pseudomonas aeruginosa* strains PA01 and MW1 (Davies et al., 1998) as described by Wuertz et al., 2001. The strains were labeled with the red fluorescent protein (dsRed; Matz et al., 1999) using a pUT delivery plasmid (de Lorenzo et al., 1990) containing a dsRed cassette from Tolker-Nielsen et al., 2000 carrying a gentamycine resistance marker (Hentzer, 1999). The biofilm was irrigated with a mineral medium (Bastiaens et al., 2000) with 0.2% glucose as carbon source. Biofilm development took place at room temperature for 5 days with a continuous supply of the medium at a flow rate of 0.016 ml/min. *Microscopy:* The biofilm was observed on 5 consecutive days (Monday to Friday) directly in the flow cell with a LSM410 confocal laser scanning microscope[2], using a 63x 1.2 NA water-immersion objective. Upon excitation with the 543 nm laser line, cells emitted dsRed-conferred fluorescence in the orange-red spectral range (> 570 nm). Vertical stacks of images (xy-sections) were collected as described by Kühn et al., 1998. The total scanned

[2]Zeiss, Jena, Germany

depth was 20 μm (20 optical sections, $\Delta z = 1$ μm). The horizontal pixels size was calculated to 0.3053 μm. 10 vertical stacks in the x-direction were collected with the CLSM as single channel grey images and stored on a hard disk prior to image processing. *Measurement:* To characterize the bacteria several measurements were performed. To estimate their ability to build clusters a mathematical algorithm was applied to generate clusters (see Fig. 11 and Fig. 12) (Serra, 1982). A cluster was found if sufficient bacteria were located in a small neighborhood. The following sub-volumes were measured: whole cube, slice (per depth), cluster, cluster inner part (per cluster). Measured features were volumes, centroid coordinates of the clusters, substrate occupation area and density, mean growth height and spatial occupation (Rodenacker et al., 2000). The temporal changes were compared on feature level.

Results: Wild type bacteria have larger growth rate than the mutant (Fig. 9). The colonization of the substrate is more dense and the growth height or better biofilm thickness differs by about 2-3 μm between wild type and mutant bacteria. The relatively low biofilm thickness allows to consider the growth distribution and density in 2D by projection. The distribution of mutant bacteria is much more regular as compared to the more structured one of the wild type. In Fig. 10 the ratio between the morphologically closed occupation area and the original area is displayed. After day 3 (Wednesday) the bacteria started to grow in a different way which corresponded to the visual recognizable distribution in Fig. 11 and 12. The lower curve of the wild type shows a reduction in the colonization area. This was also observed by Davies et al., 1998, and is qualitatively described. This global shape feature does not characterize the specific shape of the wild type growth (Fig. 11, 9). It reflects only a deviation from the more regular distribution of the mutant.

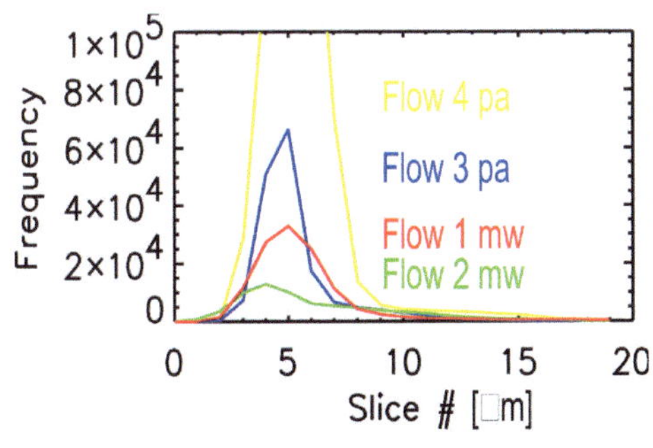

Figure 9. Volume distribution (count in units of 0.093 μm^3) per depth for two different flow channels per bacteria group

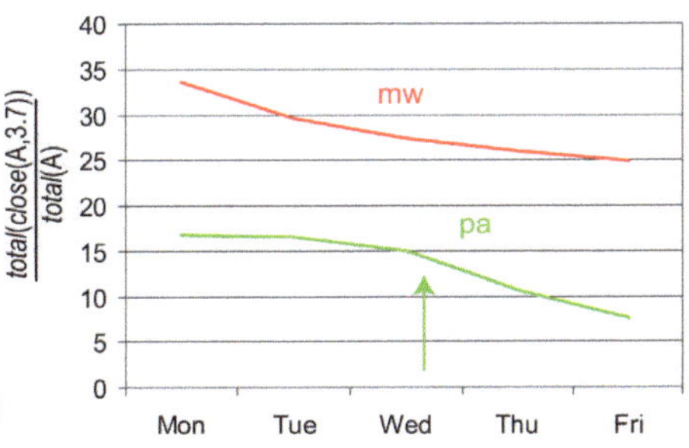

Figure 10. Growth density. The morphological closure was performed with a radius of 12 pixels corresponding to 3.66 μm

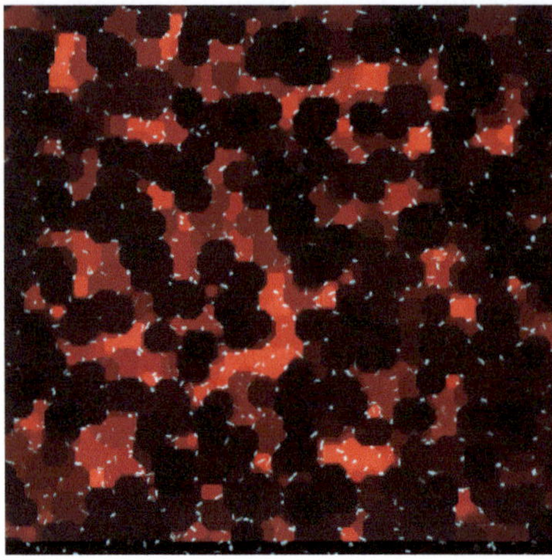

Figure 11. Wild type bacteria after four days of growth and morphological closure

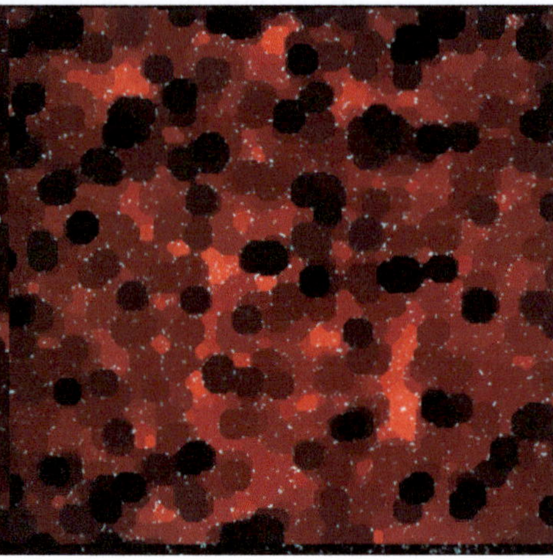

Figure 12. Mutant bacteria after four days of growth and morphological closure

3.2. Quantification of colony growth in fungi from subaerial biofilms

Material and methods: *Growth of fungi:* Fungal strains A28 *Coniosporium sp.*, A49 *Phaeococcomyces sp.*, A68 *Sarcinomyces sp.* and A95 *Phaeococcomyces sp.* were isolated from marble surfaces exposed in the atmospheric environment. These strains demonstrate a whole spectrum of adaptations to subaerial growth and are supposed to need atmospheric environment for successful development. Microcolonies of rock inhabiting black fungi were cultivated on nitrocellulose filter disks, placed on the surface of 2% malt agar plates. After 3 months the filter discs with well-developed colonies were covered by podzol soil (B) or sterile quartz sand (S) with a control cultured left open in the air (L). After three months of incubation with regular humidification of soil and sand the resulting colonies were taken out and used for measurement experiments. *Image collection:* Images were taken by two commercial digital cameras. Pixel sizes were about 0.07 mm per pixel and 0.09 mm per pixel which correspond to about 363 dpi and 282 dpi. *Measurement:* As all cells in these strains are dark pigmented, we were able to use a intensity of pigmentation for discrimination of the area occupied by the colony. It was distinguished between 2 growth zones with different occupation densities (loose and very compact or coalescing) (Fig. 13).

Results: Strains A28, A49 and A68 demonstrate similar growth behavior under cover independently of its quality (S or B) (Fig. 14). The hyphal area is reduced both under soil and sand compared to undisturbed growth in their natural environment - air (L). In A95 compact central colony part was seemingly unaffected by the substrate coverage, while the outgrowing hyphae (green zone, Fig. 13) were more influenced by the presence or absence of soil and sand. Our results demonstrate that

Figure 13. Illustration of the measurements: right the original fungus image, left the measured areas colored in green and red

the growth behavior of fairly complex fungal colonies can be characterized objectively by relatively simple measurement procedure. However, for further experiments it is necessary to include additional experimental stages thus presenting the process and not only the result.This addition would allow for the better understanding of the processes involved and for easier interpretation of the results. Image collection has to be further improved and standardized. Commercially available cameras also represent a constant source of the erroneous adjustments caused by inherent unsuppressible automatic programs.

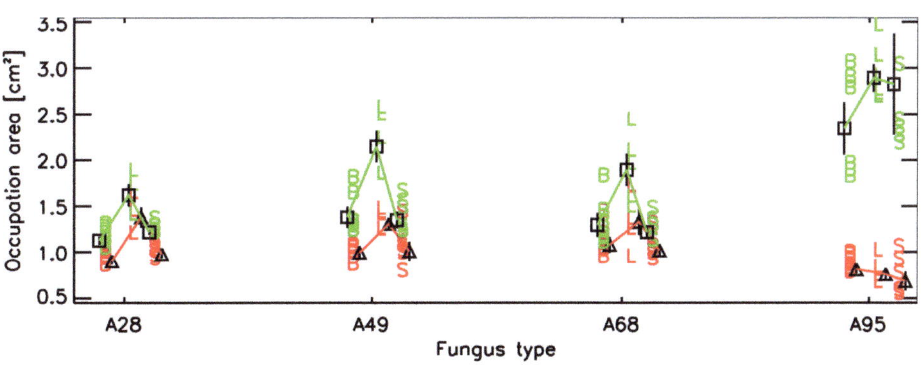

Figure 14. Hyphal area of four different fungi under different stressors (B=soil, L=air, S=sand) with mean values and standard deviation, green total hyphal area, red strong pigmented hyphal area

4. SUMMARY

An overview of methods in quantitative image analysis with emphasis to the measurement of objects properties as well as the measurement of object inter- and intra-relationships is given. Examples of a closed biofilm from *P. aeruginosa* and of complex zonated fungal colonies are used for growth quantification analysis. It was attempted to maintain comprehensibility for geo-micro-bio-myco-logists as well as acceptability in presentation for mathematician-physicist-engineer-computer-scientists.

REFERENCES

Bastiaens, L., Springael, D., Wattiau, P., Harms, H., deWachter, R., Verachtert, H., and Diels, L. (2000). Isolation of adherent polycyclic aromatic hydrocarbon (pah)-degrading bacteria using pah-sorbing carriers. *Appl Environ Microbiol.*, 66:1834–43.

Burger, G., Oberholzer, M., and Gössner, W., editors (1988). *Morphometrie in der Zyto- und Histopathologie.* Springer-Verlag, Berlin.

Davies, D., Parsek, M., Iglewski, B., Costerton, J., and Greenberg, E. (1998). The involvement of cell-to-cell signals in the development of biofilms. *Science*, 280:295–298.

de Lorenzo, V., Herrero, M., Jakubzik, U., and Timmis, K. (1990). Mini-Tn5 transposon derivatives for insertion mutagenesis, promoter probing, and chromosomal insertion of cloned dna in gram-negative eubacteria. *J Bacteriol.*, 172(11):6568–72.

Hentzer, H. (1999). unpubl. work.

Johnsen, A., Hausner, M., Schnell, A., and Wuertz, S. (2000). Evaluation of fluorescently labelled lectins for the noninvasive localisation of extracellular polymeric substance (eps) in sphingomonas biofilms. *Appl. Environ. Microbiol.*, 66:3487–3491.

Krumbein, W. E., Paterson, D. M., and Stal, L. J., editors (1994). *Biostabilisation of Sediments.* BIS-Verlag, Oldenburg.

Kühn, M., Hausner, M., Bungartz, H.-J., Wagner, M., Wilderer, P., and Wuertz, S. (1998). Automated confocal laser scanning microscopy and semiautomated image processing for analysis of biofilms. *Appl. Environ. Microbiol.*, 64:4115–27.

Matz, M., Fradkov, A., Labas, Y., Savitsky, A., Zaraisky, A., Markelov, M., and Lukyanov, S. (1999). Fluorescent proteins from nonbioluminescent *Anthozoa species. Nature Biotech*, 17:969–973.

Rodenacker, K. (1988). Vom Punktezählen zur Bildanalyse: Vorgehensweise und technische Möglichkeiten. In Burger et al., 1988, pages 25–38.

Rodenacker, K. and Bengtsen, E. (2003). A feature set for cytometry on digitized microscopic images. *Anal Cell Pathol*, 25(1):1–36.

Rodenacker, K., Brühl, A., Hausner, M., Kühn, M., Liebscher, V., Wagner, M., Winkler, G., and Wuertz, S. (2000). Quantification of biofilms in multi-spectral digital volumes from confocal laser-scanning microscopes. *Image Anal Stereol*, 19(1):39–43.

Serra, J. (1982). *Image analysis and mathematical morphology.* Academic Press Inc., London u.a.

Thompson, D. W. (1966). *On growth and Form.* Cambridge University Press.

Tolker-Nielsen, T., Brinch, U. C., Ragas, P. C., Andersen, J. B., Jacobsen, C., and Molin, S. (2000). Development and dynamics of *Pseudomonas sp.* biofilms. *J Bacteriol*, 182(22):6482–9.

Wuertz, S., Hendrickx, L., Kühn, M., Rodenacker, K., and Hausner, M. (2001). In situ quantification of gene transfer in biofilms. *Methods Enzymol*, 336:129–143.

Chapter 29

LIFE ON THE EDGE AND BEYOND

Joseph SECKBACH
P.O.Box 1132, Efrat 90435, Israel
seckbach@huji.ac.il

1. INTRODUCTION

Recent observations have shown that microbes of all three domains of life (Archaea, Bacteria, and Eukarya) are ubiquitously distributed all over the globe. Microorganisms are spread throughout many ecosystems: in land, water, and air. Recently our knowledge of the organisms living in very rigorous conditions (i.e., the "extremophiles") has greatly increased. Those severe habitats are considered anthropocentrically as "extreme", although for their inhabitants these places are equal to an oasis. Non-extremophilic organisms consider these harsh niches as an abyss or as infernal regions. On the other hand, some obligatory extremophiles are not able to tolerate the "normal" ambient conditions of growth. Among the extremophiles are members of the three domains of life, yet most of them are prokaryotes (Archaea and Bacteria) that were already found on Earth for 3.5 to 3.8 Ga (billion years before present). The Archaea, an ancient branch of microbial life, is the main group to thrive in extreme habitats (they consist of acidophiles, thermophiles, and the methanogens). Some of the prokaryotic cells can live without sunlight or organic carbon, and survive on sulfur, hydrogen and additional material that other microorganisms can not metabolize.

The extremophiles may represent pioneering cells, resembling those that occurred on young Earth. Several of them may be considered as potential candidates of microbes living in extraterrestrial bodies. In this chapter we discuss the main groups of the extremophiles and review aspects of Astrobiology with the possibilities for extraterrestrial life.

2. MICROBES ON THE EDGE: THE EXTREMOPHILES

The extremophiles developed several means to cope with the extreme habitats, such as DNA repair mechanisms, antifreeze proteins against ice formation, and other protective mechanisms. When these protection resources fail, some (bacterial) cells may produce resting spores which are long-term survival units. Spores are surrounded by heavy cell walls that enclose internal dehydrated cytoplasm and act with a minimum level of metabolism. Spores are thought to have survived in several niches, such as in the guts of frozen mammoths for long period. Cano and Borucki (1995) reported on revived bacteria isolated from insects which have been trapped in amber for 25 to 40 Myr (million years). Recently Vreeland and co-authors (2000) claimed to have resuscitated microbes isolated from underground surface salt crystals which have been dormant for a period of 250 Myr.

In many instances, combination of extreme conditions has been found, such as microbes growing under high hydrostatic pressure and very low temperature, or microorganisms thriving in hot springs and very acidic waters. Various microbes on Earth developed strategies to cope with extreme environments, such as the unicellular cyanobacterium *Chroococcidiopsis* sp. that survives a large variety of extreme conditions (Billi and Grilli Caiola, 1996). This prokaryote is found ubiquitously in various conditions of growth. It grows in hot springs, hot and cold deserts worldwide, in acidic, marine or fresh water, and in hypersaline habitats (Grilli Caiola et al., 1966). It survives extremely dry environments because of its ability to withstand long periods of desiccation (Billi et al., 2000). This cyanobacterium is present in several lichens as the cyanobiont. In arid deserts this prokaryote colonizes cryptoendolithic and hypolithic habitats (Grilli Caiola et al., 1993).

Recently it has been reported that fungi were growing on the Soviet space station Mir, which had been "out there" for a few years. Further, bacteria were left for about three years on the moon surface under harsh conditions, and have been revived upon their retrieval back to Earth.

Extreme environments harbor microbes that are challenged by utmost ranges of temperature (from psychrophiles to hyperthermopiles), various limits of pH levels (acidophiles and alkaliphiles), hypersaline solutions (halophiles), hydrostatic pressures (barophiles), desiccation, UV radiation, and growth under various compositions of atmospheres (such as under pure CO_2, ammonia, lack of oxygen, or anaerobic surroundings, etc.). For further discussion and references on extremophiles see: Horikoshi and Grant (1998), Madigan (in Seckbach, 2000b), Oren and Seckbach (2001), Seckbach (1997, 1999, 2000a, 2000b), Seckbach and Chela-Flores (2001), Seckbach and Oren

(2000, 2001), Seckbach and Walsh (1999). Table 1 exhibits some environmental conditions for the diversity of microorganisms.

2.1 Temperature effects

2.1.1 Thermophiles and Hyperthermophiles: Microbes Under High Temperatures

Thermophiles (45 – 75^{0}C) and hyperthermophiles (75 – 113^{0}C) grow in hot regions, in the depths of Earth (Onstott et al., in Seckbach, 1999), and deep in the submarine abyss of hydrothermal vents (Deming and Huston, in Seckbach, 2000b). For example, the archaeon *Pyrolobus fumarii* has a maximum known growth temperature of 113^{0}C (Stetter, 1998). This species is able to survive even at 121^{0}C in an autoclave for over 1 hour (Madigan and Oren, 1999). In the hot springs of Yellowstone National Park several species of Bacteria (e.g., *Thermus aquaticus)* and various cyanobacteria grow at high temperature around 70^{0}C, and archaeal cells grow close to boiling waters.

Thermophilic sulfate-reducing bacteria have been isolated from 75^{0}C oil–field water in the North Sea that originated from formations 2 to 4 km below the sea floor. Hyperthermophilic bacteria and Archaea grow fastest between 80 and 100^{0}C. They are generally unable to grow at temperatures below 60^{0}C. Thermophilic eukaryotes, in general, do not grow at temperatures over 60^{0}C; for example, the upper temperature range for *Cyanidium caldarium* is 57^{0}C (Seckbach, 1994) and for thermophilic fungi, around 60^{0}C. Some polychaetes live in burrows on hydrothermal vent chimneys with a strong temperature gradient that averages 68^{0}C (Table 1).

Since Darwin's "warm little pond," it has been widely accepted that the early life originated in warm or even hot environments. This proposal is consistent with what was commonly conjectured about the former warm temperature on Earth. Therefore it has been proposed that thermophiles were most probably among the first cellular pioneers on the early Earth (Seckbach, 1994, 1995, 1997). Various Archaea, such as many thermophiles, acidophiles and methanogens, cluster around the root of the phylogenetic tree of life. For further data see: Devis (1972), Seckbach (1995, 1999, 2000a), Stetter (1998).

2.1.2 The Psychrophiles: The Cold Lovers

Psychrophiles are microbes growing at _2^{0}C to < _20^{0}C. These cells are the cold lovers and are abundant in soils, waters or sea ice in frigid zones such as Antarctica, the Arctic regions, or the permafrost of Siberia. One can observe them as snow algae (e.g. the *Chlamydomonas nivalis, Chloromonas*, and certain dinoflagellates) that "paint" snow various bright colors (Hoham and Ling, in Seckbach 2000b). Organisms in cold environments may produce antifreeze proteins that inhibit ice formation and prevent internal damage to their cells (see Table 1).

Ice samples removed from deep (up to 3,600 meters) drills at Vostok station, Antarctica, have revealed various microorganisms (Karl et al., 1999; Priscu et al. 1999). Several articles compared the microorganisms growing in those cryohabitats to the potential of microbial life on and in the surface of extraterrestrial areas (Seckbach 1999, 2000a). For example, Abyzov et al. (1999) like several others, studied the Antarctic ice sheets as a model in the search of life on other planets and for the solution of some problems of exobiology. Galtier et al. (1999) suggested, on the base of their RNA studies, that the cradle of life was in cool environments rather than in high temperature habitats.

2.2 Halophiles; The salt lovers

Halophiles are microbes thriving in salt-rich environments (saline and hypersaline habitats), such as exist in the marine, salt lakes and in the Dead Sea (Israel). The reddish environmental coloration of these "purple salt bacteria" is due to the pigmented microorganisms (bacteriorhodopsin and carotenoids). Square, triangular, flat, gas vacuole-containing halophilic Archaean cells have been observed in hypersaline environments (Oren, in Seckbach, 1999). Table 1 demonstrates the halophilic range.

Recently, Vreeland et al. (2000) claimed that they revived Bacteria of the genus *Bacillus* from a tiny sample isolated in salt crystal from a depth of 570 meters. These spores were resuscitated after 250 Myr of dormancy in salt crystal (Vreeland et al. 2000). Genetically the isolate of Vreeland and co-authors is closest to bacterium *B. marismortui*, which was isolated from Dead Sea water sealed in bottles for 57 years.

When evaporating seawater reaches saturation, moderately halophilic bacteria are replaced by representatives from the archaeal domain of life, and these may reach concentrations of more than 10^8 cells per gram (Parkes, 2000). For further references on halophiles, see chapters by Grant et al., in Horikoshi and Grant (1998); Oren (2000); Oren, in Seckbach 1999, 2000b).

2.3 Living on the margins of pH limits

2.3.1 Acidophiles: Living in acidic environments

The acidophiles thrive in highly acidic habitats. Many prokaryotic acidophiles are known (for example, *Thermoplasma,* a heterotrophic Archaeon from hot coal tailings, or *Sulfolobus,* a chemoautotroph in hot springs). The acidothermophilic red alga *Cyanidium caldarium* (pH 0-4, 45-57^{0}C) tolerates even a solution of 1 N H_2SO_4 (Seckbach, 1994), while the green halophilic alga *Dunaliella acidophila* grows at pH values as low as 0.5 – 3.0 (Pick, in Seckbach, 1999). Most acidophiles maintain their intracellular pH near neutral (Seckbach, in Seckbach, 2000b pp. 107-116). Table 2 presents the internal vs. external pH values for some prokaryotes and eukaryotic acidophiles. From these data it is clear that the intracellular pH is slightly acidic, although much higher than the external acidic medium. The cells achieved this gradient through strong proton pumping activity with selective permeability properties preventing the uncontrolled entrance of protons. For further sources, see Seckbach (1994, 1999, 2000b).

2.3.2 Alkaliphiles; microbes in very high pH niches (of pH>10)

The soda lakes in Africa have higher range of pH and contain alkaliphilic microbes (such as the cyanobacterium *Spriulina platensis*). The intracellular pH in alkaliphilic *Bacillus* (Table 3) is rather higher than those in the acidophiles (Table 3) but still milder than their external medium. For further information on this habitat see: chapters on alkaliphiles in: Boussiba et al., in Seckbach (2000b); Horikoshi and Grant (1998); Oren and Seckbach (2001); Seckbach (1999, 2000b).

2.4 Desiccation: dormnacy of spores and cells

Davis (1972) recorded several algae that are able to survive very long periods. Spores are well known for their longevity – they were found in a 118-year old can of meat, and yeast has been cultured from a 166-year old bottle of porter ale (Parkes, 2000). Spores of bacteria have been revived from the gut of a bee stuck in amber for 25 to 40 million years old (Cano and Borucki, 1995). Life can exist inside a salt crystal, as claimed for *Bacillus* spores resuscitated after a period of dormancy for 250 million years (Vreeland et al. 2000). There have been other reports of bacteria in ancient salt deposits, varying in age from 195 Myr to 650 Myr (Parkes, 2000). Ancient bacteria from environmental samples, such as ancient rocks, coal and brick from 2,400 year-old Egyptian temples, have been previously reported (Postgate, 1999). Bacterial cells (*Streptococcus mitis*) were left inside a TV camera aboard Surveyor 3 on the surface of the moon for almost

Table 1. Diversity of microorganisms in the ranges of temperature, pH, and salinity

TEMPERATURE (^{0}C)

-20 **60** **115**

EUKARYOTES

Psychrophiles **Mesophiles** **Thermophiles** **Hyperthermophiles**

PERMAFROST

POLAR WATERS

MARINE WATERS

DESERTS

HOT SPRINGS

HYDROTHEMAL VENTS

ACIDITY (pH)

<1.0 **7.0** **12**

EUKARYOTES

ACIDOPHILES ALKALIPHILES

ACID MINES

& HOT SPRINGS

ACID RAIN

ACIDIC LAKES

LAKES & OCEANS

ALKALINE LAKES

SALINITY (g/l)

0 350

EUKARYOTES

FRESH WATER ORGANISMS EXTREME HALOPHILES

BRACKISH

MARINE

SALT LAKES

DEAD SEA

Adapted from K. H. Nealson (1999).

three years. There these cells were exposed to UV radiation, to vacuum, and to low temperature without nutrients. Upon retrieval of the camera and its transportation back to Earth by Apollo 12, these dormant cells were easily revived (Seckbach and Oren, in Seckbach, 2000b). All these cases indicate the broad spectrum of survivability of microorganisms.

3. THE NEW SCIENCE OF ASTROBIOLOGY

Astrobiology (synnonymous with Bioastronomy, Cosmobiology, Exobiology) covers fields of life on Earth, extremophiles and extraterrestrial Life (Chela-Flores, 2001). It is reasonable to believe that life, as we know it (which is based on carbon chemistry, moderated by liquid water and defined by a suitable atmosphere) may occur throughout the Universe. Some planets may possess Earth-like conditions that might favor microbial inhabitability. There is a promising model for extrapolation from terrestrial to extraterrestrial life–both types reside in the same universe, are constructed of the same elemental building blocks, and governed by the same natural laws.

Table 2. Values of acidophilic prokaryotes and eukaryotes showing internal vs. external pH data

	External pH	**Internal pH**
Archaea and Bacteria	**A**	**B**
Thermoplasma acidophilum	2.0	6.0
Sulfolobus acidocaldarius	2.5	6.0
Bacillus acidocaldarius	3.0	6.3
Algae		
Cyanidium caldarium	2.1	6.6
Chlorella saccharophila	4.0	7.1
Chlorella vulgaris Beij.	5.3	6.6.
Chlorella pyrenoidosa	3.1	7.0
Chara corallihna Klein ex Willd.	4.5	7.3
Scenedesmus quadricauda (Turp.) Breb.	3.1	6.9
Euglena mutabilis Schmitz	2.8	5.7
Dunaliella acidophila	(0.5) – 3.0	(7.2) – 6.2

For sources of the above data see Seckbach, in Seckbach (2000b) page 112.

Liquid water was present in the past and is present in some extraterrestrial bodies, such as Mars or the Jovian satellite Europa (and perhaps on Saturn's moon Titan). Mars and Europa are the most promising candidates for the discovery on their surface, or subsurface, living forms. Furthermore, McKay et al. (1996) studied the Martian meteorite found in Antarctica (ALH84001) and reported the presence of polycyclic aromatic hydrocarbons (PAHs) and mineral grains of crystals of supposed biological origin. In addition, these authors claim to have observed electron microscopically morphological structures of nanofossilized shapes resembling tiny bacteria. These observations prompted rebuttals on the authenticity of these findings. Oro (2000) argued against this "evidence", indicating several reasons against it. As of today the biogenic identity of those meteoritic structures has not been confirmed. Had it been, it might have been possible to say that life on Mars had flourished in the geological past.

Radar analysis of Vostok station (Antarctica) detected a liquid water lake present at a depth of ~ 4 km underneath the heavy icy layer. Microorganisms have been observed in drillings taken from thousand of meters beneath Vostok station (Karl et al., 1999; Priscu et al. 1999). Following analyses of the photos taken of Europa's surface, it is assumed that this Jovian moon contains a liquid ocean under its heavy icy layer. This Europan lake is presumably heated by geothermal volcanic sources, and it may harbor living organisms, similar to those found in various places on Earth (e.g., in Vostok Lake).

The appearance of microbial life on celestial bodies may be considered only if the past conditions on them were appropriate for generating biochemical compounds and for their polymerization into more complex structures. This notion brings some scientists to propose their thoughts of converting barren planets through terra formation for harboring life.

3.1 Origin of Life

The question of the beginning of life is still one of the most intriguing ones. It is assumed that the first (prokaryotic) cells appeared in hot habitats around 3.5 to 3.8 Ga ago as anaerobic heterotrophs or chemosynthetic microbes. Until that era, there were heavy bombardments of satellites from other celestial bodies on Earth. The primordial UV radiation did not permit any microbe to be exposed on the Earth's surface. Therefore, the first cells had to be protected and survived beneath the land surface or in the depth of the oceans (Davies, 1999). When the photosynthesis process took place, the first cyanobacteria released free oxygen into the atmosphere converting the anaerobic atmosphere into an aerobic one. The ozone accumulations eventually blocked the lethal UV radiation, enabling microbes to exit their aqueous habitats and to conquer the land. The stromatolites scattered all over the globe are the fossilized remnants of these photosynthetic pioneers. After the cyanobacteria, the eukaryotes (see further "Eukaryogenesis") and the multicellular organisms evolved.

Table 3. Extracellular vs. intracellular pH values in alkaliphilic *Bacillus*.

Microorgaisms	External pH	**Internal pH**
Bacillus alcaldophilus	8.0	8.0
	9.0	7.6
	10.0	8.6
	11.0	9.2
B. firmus RAB	7.0	7.7.
	9.0	8.0
	10.5	9.4
Bacillus strain Yn-2000	7.5	8.5
	8.5	7.9
	9.6	8.1
	10.2	8.4
Bacillus species C-125	7.0	7.3
	7.5	7.4
	8.0	7.6
	8.5	7.8
	9.0	7.9
	9.5	8.1
	10.0	8.2
	10.5	8.4

Based on Horikoshi, in Horikoshi and Grant (1998) page 159

3.1.1 Eukaryogenesis

The transition from the prokaryotic to eukaryotic phase (Eukaryogenesis) is still a biological mystery. There are two main schools of thought to explain this transformation from the primitive (anucleated) prokaryotic cell into the more advanced, complex nucleated cell. The "classical" view of this Pro-Eukaryotic phase (e.g., transformation from bacterium into a protozoa or algal cell) is explained by compartmentalization via autogenous pathways. The primeval bacterial cell enveloped different sections of its genome and each portion developed eventually into a specific organelle, see chapters by Jensen (1999), and in Seckbach, (1994); Nakamura (1999) and in Seckbach (1994); Seckbach et al. (1998). The more popular hypothesis is the endosymbiotic concept (supported by L. Margulis since the 1970s).

Natural selection, a mechanism for evolution of organisms first described by C. Darwin and now generally accepted by biologists, can be applied to the evolution of cells as well. This concept of eukaryogenesis via natural selection provides an alternative to the endosymbiotic origin of the eukaryotic phase (Seckbach et al. 1998).

3.2 Extraterrestrial Life

There is a view that precursors of life on Earth were "imported" from extraterrestrial sources via celestial objects such as meteorites. McKay et al (1996) claimed to observe Martian nanobacteria in a meteorite collected in Antarctica (ALH84001). Such meteorites arrived on Earth with the heavy bombardment of Earth from Mars prior to 3.8 Ga. Similar biological specimens have been observed in "Martian" meteorites found in Nakhla, Egypt, and Shergotty, India. McKay's report caused a wave of excitement that was followed by a sweep of skepticism, and neither side cares to back down. An abundance of objections followed the publication on nanobacterial "Martian worms" and the meteoritic extraterrestrial biogenetic material claimed to be present in the meteorites (Oro, 2000).

Recently the two Neapolitan scientists B. D'Argenio and G. Geraci (see: Abbott, 2001) claim to have found extraterrestrial bacteria in ancient rocks from a meteorite in one of the city's museums. In their data published by Italy's National Academy of Sciences, they report resuscitation of dried-up bacteria that had been locked in the rocks for millions of years. They believe that their results prove that life exists in outer space and evolved there. We have to take this report cautiously, since it was been published without peer review, and without any other laboratory repeating this experiment and observations.

3.2.1 Hypothetical attempt to make Planets inhabitable

A few decades ago Sagan (1961) proposed an extravagant engineering plan suggesting to seed the upper atmosphere of hostile planets (e.g., Venus) with cyanobacteria to make these extraterrestrial places habitable for life. Those microbes would grow on water and minerals available at the upper level of the planet, photosynthesize to produce oxygen that would eventually lower the hot temperature, and slowly make it comfortable for man. More recently Friedmann and Ocampo-Friedmann (1995) also proposed an idea for the propagation of Mars with terrestrial microbes by seeding it with the cyanobacterium *Chroococcidiopsis* for extraterrestrial "Terra forming". Other ideas proposed covering the cold planet with some greenhouses' gases for warming up their surfaces and ultimately making them inhabitable.

3.3 Life possibilities on Mars

Today the surface of Mars is inhospitable owing to the lack of liquid water, but fluids may exist inside the warmer interior of the planet. Martian photos demonstrated evidence of warmer and wet past conditions, with liquid running water. (occurring in lakes, rivers, canyons, and waterfalls). One can assume that if there was liquid water on Mars, then there was also microbial life. Today the Martian water exists as polar ice caps or under the surface. McKay et al. (1996) studied the Martian meteorite found in Antarctica (ALH84001) and reported the presence of polycyclic aromatic hydrocarbons and nanofossilized shapes resembling tiny bacteria (see above). Oro (2000) and others opposed this "evidence" and pointed out several reasons against these biogenetic sources. As of today the biological origin of those meteoritic structures has not been confirmed. It seems that the judgment will remain suspended out for some time until some samples will be brought from Mars on future missions.

If on the surface of red planet Life forms are not present, it does not mean that its subsurface may not harbor them. Recently, microbes from cryospheric habitats of Antarctica and the Arctic regions and the permafrost in Siberia have been observed. Those cryophilic organisms may serve as analogues for possible psychrophiles living on, or deep inside, the surface of Mars.

Another idea for the Martian terra-forming has been to bring bioengineered green plants to the red planet. The new plants' genome should enable them to grow under Martian ecological conditions, i.e., low O_2, low water, low temperature, UV radiation, and low air pressure. The oxygen released from those photosynthesizers should also provide a screen blanket against the UV radiation to the surface.

3.4 Life on Europa

Following surface analysis of the Jovian satellite, Europa, it has been proposed that this moon contains a salty liquid ocean under its heavy (~100 km thickness) ice layer. This Europan lake is presumably heated by geothermal volcanic sources, and it is assumed that this lake contains living organisms similar to those found in various places on Earth. Such theoretical thoughts of possible microbial life is like a "Ship in a bottle" which is cut off from the outside world. The above proposals are also based on the radar analysis of Vostok station (Antarctica) where a liquid water lake was detected to be present at a depth of ~ 4 km underneath the heavy icy layer.

Prokaryotes and Eukaryotes have been observed in drillings taken from thousands of meters beneath Vostok station

3.5 Other celestial habitable zones

In the past, planet Venus was a candidate for harboring life forms (Seckbach and Libby, 1970), but currently such a possibility seems remote because of its high temperature and pressures. Recently, the list of habitable celestial areas has included the Saturn moon Titan, where its atmospheric composition and the recent detection of water [Simatov, 2000] give some hope for exobiological life.

4. CONCLUSIONS

Life on Earth is ubiquitously present in many ecosystems, including very severe environments. The microbes of severe habitats (extremophiles) may serve as models for the possibilities for extraterrestrial life. The recent report on ancient Martian nanobacteria are promising but more proof is necessary. Only samples returned from Mars can prove whether this red planet had or has living organisms. Extraterrestrial life, should it be detected on planets, may hold the key to the origin of life on Earth. If life began on more than one world, it could be ubiquitous; if not, we may be alone. The answer is out there blowing somewhere the solar winds or in the "sands on that faraway but tantalizingly planet glowing in the southern sky" (Discover, 2001).

5. ACKNOWLEDGEMENT

I thank Professor **W. E. Krumbein** (University Oldenburg) for inviting me, and for the financial support, to attend the Symposium on **Fossil and Recent Biofilms**, organized by Hanse Institute for Advanced Study, Delmenhorst, Germany. Gratefulness is due to Professor **A. Oren** (Hebrew University) for critical proofreading of this manuscript.

REFERENCES

Abbott, A. (2001) Nature **411**: 229 (17 May, 2001).

Abyzov, S., Mitskevich, I.N., Poglazova, M.N., Brarkov, N.I., Lipenkov, V. Ya., Bobin, N.E., Kudryashov, B.B. and Pashkevich, V .M. (1999) Adv.Space Res. **23**: 371-376.

Billi, D. and Grilli Caiola, M. (1996). New Phytol. **133**: 563-571,

Billi, D., Friedmann, E.I., Hofer, K.G., Grilli Caiola, M. and Ocampo-Friedmann, R. (2000) Appl. and Environ. Microbiol. **66**: 1489-1492.

Cano, R. J. and Borucki, M. (1995) Science **268**: 1060-1064.

Chela-Flores, J. (2001) In: J. Seckbach (ed.) *The New Science of Astrobiology.* Volume III of "Cellular Origins, Life in Extreme Habitats and Astrobiology", J. Seckbach. Editor, Kluwer Academic Publishers, Dordrecht, The Netherlands.

Davies, P. (1999) The Fifth Miracle: The Search for the Origin and meaning of Life. Simon & Schuster. New York.

Davis, J.S. The Biologist **54**: 52-93.

Discover (2001) **22**(6): 33.

Friedmann, I.E. and Ocampo-Friedmann, R. (1995) Adv. Space. Res. **15**: 243-246.

Galtier, N., Tourasse, N. and Gouy, M. (1999) Science **283**: 220-221.

Grilli Caiola, M., Billi, D. and Friedmann, E.I. (1996) Eur. J. Phycol. **31**: 97-105.

Grilli Caiola, M., Campo-Friedmann, R. and Friedmann, E. I. (1993) Phycologia **32**, 315-322.

Horikoshi K. and Grant, W.D. (eds.) (1998) *Extremophiles: Microbial Life in Extreme Environments.* Wiley-Liss. New York.

Jensen, T.E.(1999) In: E. Wagner, J. Normann, H. Greppin, J.H.P. Hackstein, R.G. Herrmann, K.V. Kowallik, H.E.A. Schenk and J. Seckbach. (eds.) *From Symbiosis to Eukaryotism:* Endocytobiology VII. University of Geneva. pp. 63-74.

Karl, D.M., Bird, D.F., Björkman K., Houlihan, T., Shackelford, R. and Tupas, L. (1999) Science **286**: 2144-2147.

Madigan, M.T. and Oren, A. (1999) Curr. Opinion Microbiol. **2**: 265-269.

McKay, D.S., Gibson, E., Thomas-Keprta, K.L., Hogatolla, V., Romanek, C., Clemett, S., Chillier, S., Maechling, C.R. and Zare, R.N. (1996) Science **273**, 924-930.

Nakamura, H. (1999) In: E. Wagner, J. Normann, H. Greppin, J.H.P. Hackstein, R.G. Herrmann, K.V. Kowallik, H.E.A. Schenk and J. Seckbach (eds.) *From Symbiosis to Eukaryotism:* Endocytobiology VII. University of Geneva. pp. 53-61.

Nealson, K.H. (1999) Origin of Life and Evolution of the *Biosphere* **29**: 73-93.

Oren, A. (2000) In: B.A. Whitton and M. Potts (eds.) *The Ecology of Cyanobacteria.* Kluwer Academic Publishers, Dordrecht, The Netherlands. pp. 281-306.

Oren, A. and Seckbach, J. (2001) In: J. Elster, J. Seckbach, W.F. Vincent and O. Lhotsky (eds.) *Algae and Extreme Environments – Ecology and Physiology,* Schweizerbart'sche Verlagsbuchhandlung, Stuttgart, Germany, pp. 13-31

Oro, J. (2000) In: G. Lemarchan and K. Meech (eds.) *A new Era in Bioastronomy*, ASP Conference Series, vol. **213**: 285-299.

Parkes, R. J. (2000) Nature **407**: 844-845.

Postgate, J. (1999) New Scientist (21 July), 46-49.

Priscu, J.C., Adams, E.E., Lyons, W.B., Voyteck, M.A., Mogk, D.W., Brown, R.L., McKay, Ch. P., Takacs, C.D., Welch, K.A., Wolf, C.F., Kirshtein, J.D. and Avci, R. (1999) Science **286**: 2141-2144.

Sagan, C. (1961) Science **133**: 849-859.

Seckbach, J. (ed.) (1994) *Evolutionary Pathways and Enigmatic Algae: Cyanidium caldarium (Rhodophyta) and Related Cells.* Kluwer Academic Publishers, Dordrecht, The Netherlands.

Seckbach, J. (1995) J. Biol. Physics **20**: 335-345.

Seckbach, J. (1997) In: C.B. Cosmovici, S. Bowyer and D. Werthimer (eds.) *Astronomical and Biochemcial Origins and the Search for Life in the Universe.* Editrice Compositori, Bologna, Italy. pp. 511-523.

Seckbach, J. (ed.) (1999*) Enigmatic microorganisms and life in extreme environments.* Kluwer Academic Publishers, Dordrecht, The Netherlands.

Seckbach, J. (2000a). In: G. Lemarchan and K. Meech (eds.*) A new Era in Bioastronomy*, ASP Conference Series, vol. **213**: 379–386.

Seckbach, J. (ed.) (2000b). *Journey to Diverse Microbial Worlds: Adaptation to Exotic Environments.* Kluwer Academic Publishers, The Netherlands.

Seckbach, J. and Chela-Flores, J. (2001) In: P. Ehrenfreund, O. Angerer and B. Battrick (eds.) First Eurp. Workshop Ext-/Astro- Biology. Erup. Space Agency, ESA Pub. Div. Estec. Noordwijk, The Nederlands, pp. 255-260.

Seckbach, J. and Libby, W.F. (1970) Space Life Sciences **2**: 121-143.

Seckbach, J. and Oren, A. (2000) In: R.B. Hoover (ed.) *Instruments, Methods, and Missions for Astrobiology III.* Proceeding of SPIE – The International Soc. for Optical Engineering. **4137**: 89-95.

Seckbach, J. and Oren, A. (2001) In: J. Chela-Flores, T. Owen and F. Raulin (eds.) *First Steps in the Origin of Life in the Universe.* Sixth Trieste Conference on Chemical Evolution. Kluwer Academic Publishers, Dordrecht, The Netherlands. Pp. 241-245

Seckbach, J. and Walsh, M.M. (1999) In: E. Wagner, J. Normann, H. Greppin, J.H.P. Hackstein, R.G. Herrmann, K.V. Kowallik, H.E.A. Schenk and J. Seckbach. (eds.) *From Symbiosis to Eukaryotism:* Endocytobiology VII. University of Geneva. pp. 85-104.

Seckbach, J., Jensen, T.E., Matsuno, K., Nakamura, H., Walsh, M.M. and Chela-Flores, J. (1998) In: Julian Chela-Flores and F. Raulin (eds.) Exobiology: Matter, Energy, and Information in the Origin and Evolution of Life in the Universe, Kluwer Academic Publishers, Dordrecht, The Netherlands, pp. 235-240.

Simatov, M.B. (2000) In: G. Lemarchan and K. Meech (eds.*) A new Era in Bioastronomy*, ASP Conference Series, vol. **213**: pp. 333-338.

Stetter, K.O. (1998) In: K. Horikoshi and W.D. Grant (eds.) *Extremophiles: Microbial Life in Extreme Environments.* Wiley-Liss. New York. pp. 1-24.

Vreeland, R.H., Rosenzweig, W.D. and Powers, D.W. (2000) Nature **407**: 897-900.

Chapter 30

"ACTIVE PLANETARY COVER" CONCEPT AND LONG-TERM EVOLUTION OF PLANETARY CLIMATE

Yuri M. Svirezhev, Arthur Block and Werner von Bloh
Potsdam Institute for Climate Impact Research,
P.O.Box 60 12 03, D-14412, Potsdam, Germany
yuri@pik-potsdam.de

When we study the History of Science we discover two mutual contrary phenomena: either behind an apparent complexity a simplicity is hidden or, on the contrary, an evident simplicity conceals within itself an extraordinary complexity. (H. Poincaré, 1894)

1. INTRODUCTION

At the beginning of XIX. century J.-B. Lamarque had introduced the term "biosphere". In accordance with his definition the biosphere is a "scope of life" and it is an external cover for the Earth. In 1875 E. Süss, who also distinguished the biosphere as one of the Earth covers, introduced the same term in geology. But V. Vernadsky was the person who first created the modern concept of the biosphere as an active planetary cover, which does not only passively reflects its geological and geochemical environment but also transforms it. So that Lovelock (1979) has done only a next step when he has postulated that this transformation is an act of self-regulation creating optimal conditions for the biosphere existence.

If to formulate Vernadsky's concept as some axiomatic theory we have to understand: what kind of axioms lies at its basement? (Here they are presented in a more formal form than in Vernadsky's original work (1926))

1. In the course of all geological periods the living organisms have never been created directly from inorganic matter.

There is the analogue of this axiom in biology called Redi's Law ("life comes only from life").

2. The existing facts can not answer the question about the origin of the Earth life and its ancient history.

Certainly, we can use different speculations but having done that we go away beyond the framework of Vernadsky's Empirical Generalisation Method. There is only one way to resolve this contradiction, namely, to postulate that in spite of the fact that the possible pre-biosphere histories could be different, a result of the biosphere evolution in the course of all geological periods must be the contemporary biosphere. This is a typical ergodicity axiom. Mathematically this means that a topology of the system (biosphere) is sufficiently simple and it is "almost linear". May be this is true but it is now more and more clear that we are living in a "non-linear world". Therefore the "virtual biospheres" concept was suggested (Svirezhev, 1994): the contemporary Earth's biosphere is one of many possible (virtual) biospheres corresponding to multiple equilibria of some non-linear dynamical system, the "climate + biosphere + chemosphere + geosphere". In the course of planetary history and as a result of its own evolution and the evolution of its environment, the system passed through several bifurcation points, when external (in relation to the system) factors determined the direction the biosphere would take.

3. There were no lifeless geological epochs.

This means that the contemporary living matter is genetically connected with living matter of all previous epochs (*continuity axiom*). The next two axioms have, actually, a form of conservation laws. However, since they postulate some equilibrium properties of the biosphere, they can be called the axioms of *stationary state*.

4. The chemical composition of living matter was, on the average, the same as it is now.

5. The amount of living matter, on the average, was the same for all geological time.

These Vernadsky's axioms cause a lot of objections at the present time. And finally, the axioms, which determined the principles of functioning of the biosphere mechanisms, are:

6. The energy, which is stored and emitted by living organisms, is the solar one.

By means of them the energy is controlled the global chemical (in particular, the global biogeochemical cycles).

7. Vegetation plays the main role in the assimilation and allocation of the solar energy.

To agree with the axiom of the constancy of the total amount of living matter we have to assume that the biosphere evolution is followed by the way of only structural complication of the living matter.

Developed on the base of these axioms any model will be already the model of Vernadsky's biosphere, but any change of any axiom will give us some new model (of virtual biosphere). Let us try to consider an evolutionary tree of virtual biospheres instead of the almost linear graph of Vernadsky's biosphere. How to do this?

2. ACTIVE PLANETARY COVER AND POSSIBLE CANDIDATES TO PLAY THIS ROLE

If keeping in mind that life is a phenomenon, which can exist in some relatively narrow interval of temperatures, then immediately the following question arises: how is the planetary temperature kept within the interval? Is this a result of self-regulated interactions between the biosphere and climate (Lovelock, 1979; Schwartzman, 1999) or a purely random combination of different terrestrial and extra-terrestrial processes (Budyko et al., 1985)? Before to answer the question we introduce the following definition.

We define an active planetary cover ("planetary skin") as some system which directly affects the climate, forms it and, in turn, its dynamics depends on the climate. The system has to possess multiple equilibria.

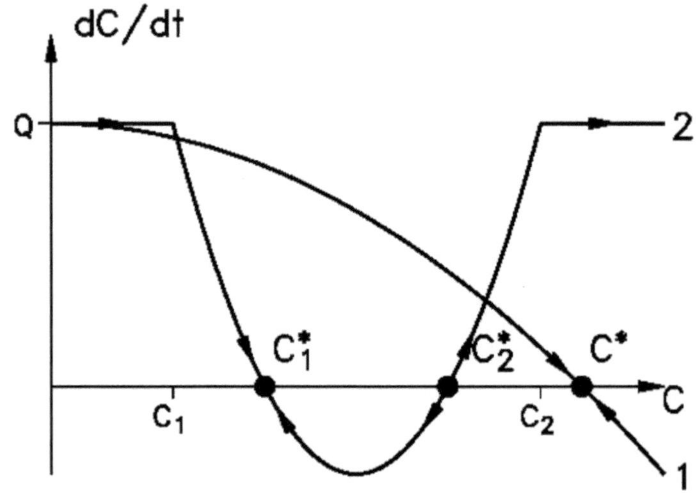

Figure 1 System phase portraits: (1) the planet surface is weathered, only weathering takes carbon out of the atmosphere; (2) the planet is covered by "green cover" with an unimodal temperature function.

Let $T(t)$ be the mean annual temperature of the Earth, $S(t)$ be the mean solar radiation, C be the total amount of atmospheric carbon, $(C)T^4$ the outgoing black-body radiation with $(C) = {}_0 (C)$ where ${}_0$ is the Stefan-Boltzmann constant and (C) is a monotonous decreasing function describing the greenhouse effect. The "skin" will be described by the variable p, which is a part of the Earth surface "covered" by skin with albedo ${}_p$; a remaining part of the surface has albedo ${}_q$. For instance, this can be the percentage of the Earth area taken over by the continents. We assume that $S(t)$ and $p(t)$ are the given functions of "slow" time. Then the simplest zero-dimensional model of such sort of planetary skin can be presented in the form:

$$T \approx \left(\frac{S\left[1 - \alpha_q - (\alpha_p - \alpha_q)p)\right](1+\rho)}{\sigma(C)} \right)^{1/4},$$

$$\frac{dC}{dt} = Q - \beta(T)pf_c(C) \qquad (1)$$

where is the coeffcient of spatial correlation between = 1 - (*x*) and *S*(*x*), *x* is spatial co-ordinate that reflect a spatial interposition of skin and "free" surface on the Globe. The value (*T*) describes dependence of the rate, with which the skin takes carbon out of the atmosphere, on the temperature. *Q*(*t*) is the inflow of carbon from the mantle (volcanism).

Let us consider different candidates for the role of active cover, which can actively regulate the Earth temperature.

a) **"planet with weathering":** For this process = $_w$ ~ exp(*T*); $f_c \sim \sqrt{C}$. The phase portrait of the system described by (1) is shown in Fig. 1. There is only one stable equilibrium C^*, which "floats" in slow time. It is the solution of equation [$T(C^*$, p)]$pf_c(C^*)$ = Q. It is obvious that corresponding to C^* the temperature T^* can get into the tolerable interval [T_1, T_2] only accidentally.

b) **"planet with vegetation" ("green cover"):** In this case = $_v$ is a finite unimodal function within the interval [T_1, T_2] and ≡ 0 outside of it. The function f_c is a monotonous increase function with saturation. The corresponding phase portrait is also shown in Fig. 1. There are two "slow floating" equilibria: stable with lower temperature, $C^*{}_1$, and unstable with higher one, $C^*{}_2$. Since then this means that the "green cover" can maintain the equilibrium temperature $T(C^*{}_1)$ belonging to the tolerable interval and lying in its low part, but the stability domain of the equilibrium is very small. Its upper boundary is $C^*{}_2 < C_2$ which corresponds to $T(C^*{}_2) < T_2$, i.e., it lies lower than the upper tolerable temperature.

c) **"planet with vegetation + weathering":** The best results in the sense of stability and maintenance of the tolerable interval for life can give the combination of "chemical" (like weathering crust) and "biological" (like vegetation) covers. Let us assume that chemical and biological pumps work simultaneously. Then the equation for atmosphere carbon is written as

$$\frac{dC}{pdt} = Q/p - \left[\beta_v(T)f_c^v(C) + \beta_w(T)f_c^w(C)\right] \qquad (2)$$

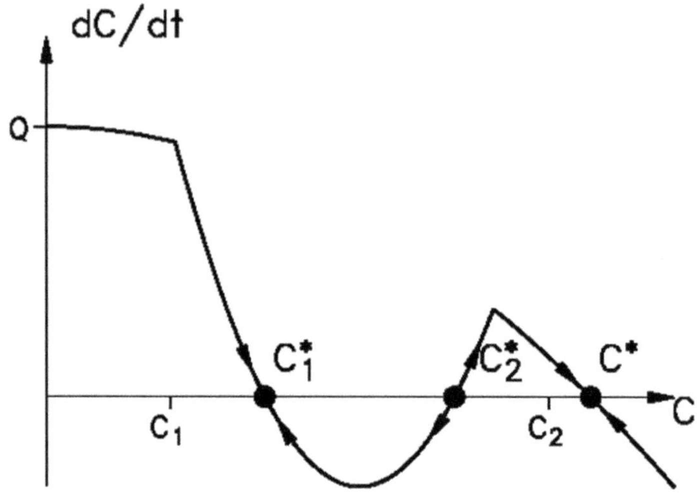

Figure 2 A system phase portrait: the planet cover is a combination of "chemical" (weathering) and "biological" (vegetation) covers. It is assumed that chemical and biological pumps work simultaneously.

From its phase portrait (Fig. 2) one can see that the system behaviour became more complicated, where this complexity depends on the value of bifurcation parameter Q/p. Let us consider its slow evolution from large to small values. When Q/p was large (high volcanism, small continental area that corresponds to the early Earth history), only one already stable equilibrium exists. A leading role plays here the weathering. The area of continents increases, the volcanism decreases, the value of bifurcation parameter Q/p also decreases. In the system arises an interesting transition regime with three equilibria: two are stable and one (intermediate) is unstable. If the first stable equilibrium corresponds to lower temperature lying within the tolerable interval then the second stable equilibrium corresponds to higher temperature lying outside of it. In order to shift it even if to the high border of the tolerable interval we additionally have to manipulate the cover albedo and its relative area. At last, let us consider the case with low value of Q/p when the volcanism is small and the continental area is large (present time). One equilibrium with tolerable temperature is only stable. The main role in the regulation of the Earth temperature plays the vegetation. Further decrease in Q/p takes the stable temperature outside of tolerable limits. The biosphere would be finished, the Earth would be transformed into an "ice desert".

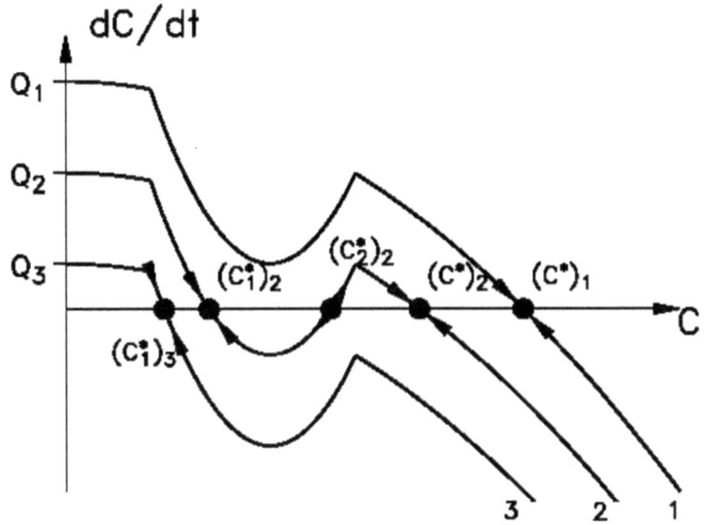

Figure 3 The planet cover is a combination of "chemical" (weathering) and "biological" (vegetation) covers. Phase portraits are considered for three values of carbon inflow from the mantle into the atmosphere: $Q_1 > Q_2 > Q_3$

d) "planet – chameleon": At the beginning the skin of planet had been "lunar" colour with = 0.07 that did not differ from the oceanic albedo (0.07 – 0.10). The radiation temperature of this planet was equal to 275 K. The mantle degassing can rise the planetary temperature up to 60°C – 70°C. If the new atmosphere contains H_2O, CO_2, NH_3, H_2S, S_2, HCl then an optical (not chemical) result of interaction between these gases and the matter of "lunar" surface was, as a rule, the increase in albedo. The analogous result will be assuming that the atmosphere contains a sufficient amount of oxygen to oxygenate surface rocks. For instance, the albedo of red sands is equal to 0.2. The finale stages of weathering are carbonates and sands with = 0.3 – 0.4. A rise of living skin of green colour (cyanobacterial mats, vascular plants, etc.) also increased the albedo up to = 0.12 – 0.22. All these processes led to one result: drop of initial temperature of the planet shifting it to tolerable values for life. On the other hand, if the planet was initially covered by snow then the vegetation arosed and its further propagation increased the temperature also shifting it to tolerable values.

Revenons á nos moutons and keeping in mind our question we can answer that the truth lies, as usually, in the middle: in order to get into the

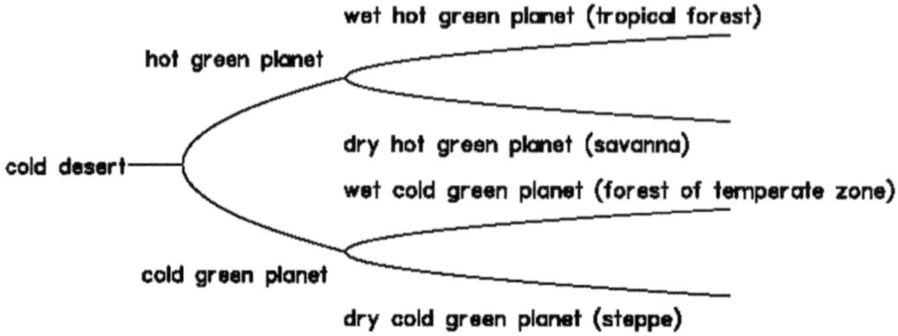

Figure 4 A bifurcation diagram of the system "climate + biosphere"

tolerable life interval we need a suitable combination of abiotic factors. But after the life has been arisen, it can maintain the global thermostat for its self-existence.

3. COMPLEX NON-LINEAR BEHAVIOUR OF PLANET COVERED BY "GREEN COVER"

Let us consider the following model (in details it is described in Svirezhev & von Bloh, 1998), when a typical example of active cover is the vegetation (surface biota), which is a main component of the biosphere. It changes the planetary albedo and influences the atmosphere composition trough the global carbon and hydrological cycles that leads to the change of climate. In turn, the dynamics of vegetation depends on the climate and the concentration of atmospheric CO_2. The system has the following bifurcation parameters:

1. The total amount of carbon. In the course of a "slow" planetary evolution (characteristic time 10^5 - 10^6 years) this value, involved in the "crust – biosphere" cycle, has been changed by the processes of silicate-rock weathering and volcanism. The latter is directly connected with the tectonic activity and continental dynamics.
2. The total amount of water.
3. The product P_m , where P_m is a "potential" productivity of global vegetation and is a residence time of carbon in biota. A "slow" dynamics of this parameter is closely connected with the evolution of carbon transformers and oxygen producers: from microorganisms to vascular trees.

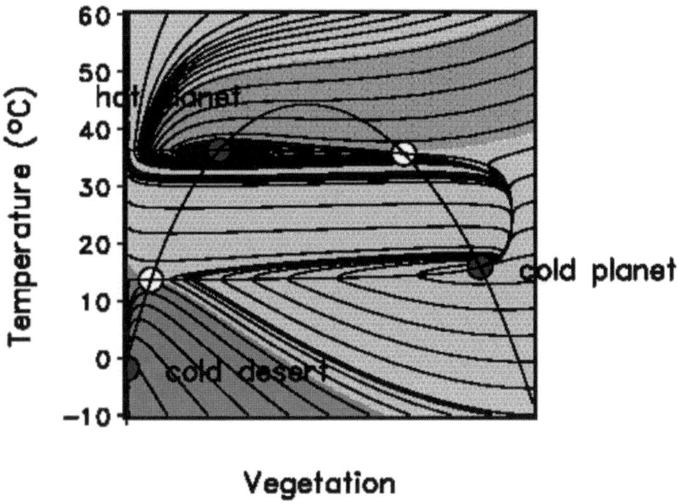

Figure 5 A phase portrait of the system "climate + biosphere" (projection onto the plane (temperature, vegetation)

What kind of bifurcation is possible in this system (Svirezhev, 1999)? We can observe the change of the planet's "status" from "a cold (ice) desert" either to "a cold green planet" or to "a hot green planet" (first bifurcation). Then either "a wet hot" planet, covered by tropical rain forest, or "a dry hot" planet (savanna) develops from hot green planet as a result of second bifurcation. Analogously, either "a wet cold" (temperate forest) or "a dry cold" (steppe) planet arise from green cold planet (Fig. 4). In this figure the bifurcation diagram is presented as a branching structure depending on one general value. Really, it depends on all three bifurcation parameters.

We shall consider the system with two causal loops: (1) vegetation albedo temperature vegetation, (2) vegetation atmosphere carbon temperature vegetation. In this case the system can have up to five different equilibria: three of them are stable and two others are unstable. The projection of its phase portrait onto the plane $(T;N)$ where T is the temperature and N is the density of biomass is shown in Fig. 5.

There are four attractive domains corresponding to the following equilibria:

1. "cold desert" when the planet has no vegetation and the planetary temperature is less than –3°C.

2. "cold green planet" with a rich vegetation and relatively low temperature, approximately equal to the current Earth's temperature, 15°C.

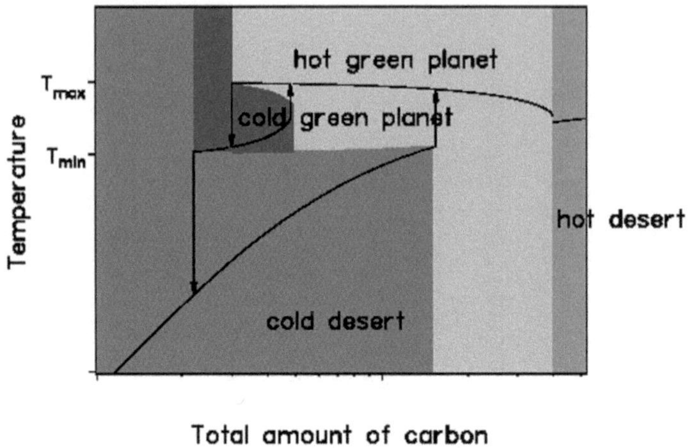

Figure 6 Bifurcation diagram of the "climate + biosphere" system. Black lines indicate trajectories of the evolution for either decreasing or increasing carbon. Gray-shaded areas denote the basis of attraction of the stable equilibria.

3. "hot green planet" with a poor vegetation and relatively high temperature, higher than 30°C; that is typical for Earth's hot deserts.

4. "hot desert" is a lifeless planet, its temperature is higher than the upper boundary of the tolerable interval (> 60°C).

Let us consider the evolution of the system topology when the total amount of carbon, A, slowly decreases. Fig. 6 plots the temperatures of the different equilibria as a function of A where the black lines indicate trajectories of the evolution for either decreasing or increasing carbon. Gray-shaded areas are the so-called basins of attraction, i.e. the set of initial conditions, starting in which trajectories come to the same stable equilibrium. The area denoted as the "cold desert" is the solution without vegetation on the planet. The equilibrium temperature lies below the lower tolerable temperature T_1, while the "hot desert" marks a solution with a temperature, which is too hot for the vegetation. The "hot green planet" and "cold green planet" are the non-trivial solutions with low and high vegetation densities, respectively.

Two hysteresis loops can be identified. Starting with the "hot desert" the planet evolves in the following way: hot desert hot green planet cold green planet cold desert. The transition "hot green planet cold green planet" is accompanied by almost an explosive increase in vegetation biomass 8-10 times. Starting from the cold desert, however, a different path is realized: cold desert hot green planet hot desert. This means that a current state of the coupled climate-biosphere system depends on its history. The change from one to another equilibrium is done in a non-continuous way. The arrows in the diagram indicate non-continuous transitions. The rise of the hot green planet from the hot desert only is a continuous process.

4. CYANOBACTERIAL MATS AND ZAVARZIN'S EARLY BIOSPHERE APPROACH

Let us consider our planet at early stages of its history (2 - 3 aeons ago) when one large and flat continent only was surrounded by mountains (Zharkov, 2000), so there was no contact between the continent and the slightly acid (pH = 4 - 5) ocean. This entire "saucer" was fulfilled by shallow water-body systems of intercontinental seas and lakes, which were ideal niches for Cyanobacter microbial mats (CBM). All this forms a special type of active cover, which consists of photosynthesising prokaryotic communities: CBM. This system can be considered as some first stage of the biosphere succession: the biosphere of CBM or Zavarzin's biosphere (Zavarzin, 1984). Namely this system has played a main role (as a predecessor) in the formation of the contemporary Phanerozoic biosphere and atmosphere. If assuming that contemporary CBM are analogues of those ancient ones then we can say that the CBM were optimally adapted to their ecological niche. Using results of the work of Gerasimenko & Zavarzin (1994) we distinguish the properties, which may play an adaptive and regulating role.

1. Their productivity was very high (as by contemporary tropical rain forest). For instance, under sub-optimal conditions (salinity 16-21%, T^0 40-45°C), the annual gross production and destruction of CBM is equal to 1.6 kgC/m^2 and 1.1 kgC/m^2 , correspondingly. Their difference is equal to 0.5 kgC/m^2. Although the destruction in CBM are mostly anaerobic but the production of oxygen by photosynthesis is proportional to this difference so that one square meter of CBM produces 1.3 kg O_2 per year.

2. CBM were covered by mucopolysaccharide films ("mucus"). A sandwich from 2 m of water and 2 mm of mucus guaranteed practically the full defence from UV radiation. Diffusion in mucus is the same as in pure water, mucus fibers within CBM are very good light conductors, due to this the light and nutrients are transported to the whole volume of CBM.

3. Due to the high species diversity of organisms, which form the mat, a full use of visible spectrum was attained.

4. CBM create conditions for $CaCO_3$ formation, since over a mat the pH = 8.4 – 9.6 and the concentration of HCO_3 is equal to 280 mg/l, while under a mat the pH decreases to 7 and the concentration of HCO_3 is equal to 915 mg/l.

CBM have a perfect structure organisation: light conductors, calcium and gypsum quasi-skeleton, their "shark skin" adjusted external water flows increasing the nutrients transport, etc. .

Considering the scenario in which the area of intercontinental seas covered by CBM was 10^7 km^2 (~ 6% of the contemporary continental area), we assume that all water, which was degassed from the mantle, is concentrated in the contemporary hydrosphere, 1.64 x 10^{24} g. All degassing nitrogen is concentrated in the atmosphere, 4.1 x 10^{21} g. If we test the composition of volcanic gases of the different origin (Voitkevich & Bessonov, 1986) then the best approximation will be the emission of Iceland geysers and fumaroles: 99.4% H_2O, 0.33% C in the form of CO_2, 0.05% N_2. In accordance with these assumptions we get the estimations for N_2: 0.825 x 10^{21} g, and for carbon (in the form of CO_2): 5.445 x 10^{21} g. In order to remove all this carbon dioxide 10^7 m^2 of CBM have to work one million years. 1.45 x 10^{22} g oxygen will be produced by photosynthetic mats. The most part of oxygen is spent for the oxidation of FeO, CO, SO_2 and H_2. The contemporary atmosphere contains 1.1 x 10^{21} g. Certainly, this is some extreme estimation under conditions very close to optimal ones. Substantively, the process of the atmosphere CO_2 "grazing" by CBM occurred much slower.

In accordance with Ronov's estimation (Budyko et al., 1985) the stratisphere contain 9.8 x 10^{22} g C. This value is 18 times higher than the estimation, which was mentioned above. In order to get a better consistency of these estimations we can either increase the percentage of N_2 (~ 5 times) and C (~ 18 times) in volcanic emanations or decrease the total amount of degassed water.

Let us consider the following thermodynamic model: a flow of mixed gases with the concentrations corresponding to volcanic emanations passes through an active membrane with properties of CBM. The process is isothermic and isobaric: T^0 = 50°C, pressure = 5 bar. Using the methods of non-linear thermodynamics we calculated the equilibrium atmosphere, which was established behind the membrane: 1% of CO_2 and 8% of O_2.

When as a result of intensive orographic processes the depth of intercontinental seas has begun to increase, this was a reason of massive ruin of CBM since their density was ~ 1.39 g/cm^3. The biosphere of the CBM has perished as the contemporary (Phanerozoic) biosphere (the biosphere of vascular plants), which has appeared about 0.6 billion years ago, would perish about 0.6 billion years after present (Franck et al., 2000).

REFERENCES

Budyko, M.I., A.B. Ronov, A.L. Yanshin, 1985. *The history of the atmosphere*. Gidrometeoizdat, Leningrad.

Franck, S., Block, A., von Bloh, W., Bounama, C., Schellnhuber, H.-J., and Svirezhev, Y., 2000. Reduction of biosphere life span as a consequence of geodynamics. *Tellus* 52B, 94-107.

Gerasimenko, L.M., Zavarzin, G.A., 1994. Relict cyanobacterial communities. In: "*Problems of Pre-anthropogenic Evolution of the Biosphere*" (Eds: Sokolov, B.S., Rozanov, A.Y.), Nauka, Moscow, 222-253.

Lovelock, J., 1979. *Gaia*. Oxford University Press, New York.

Schwartzman, D., 1999. *Life, Temperature, and the Earth: The Self-Organizing Biosphere (Methods and Cases in Conservation Science)*, Columbia University Press, New York, 304 p.

Svirezhev, Y., 1994. *Simple model of interaction between climate and vegetation: virtual biospheres*. IIASA Seminar. Laxenburg. Austria.

Svirezhev, Y., 1999. Virtual Biospheres: Complexity versus Simplicity. In: "*Tempos in science and nature: structures, relations, and complexity*" (Eds: C. Rossi, S. Bastianoni, A. Donati and N. Marchettini). Annals of the NYAS, New York, v. 879, 368-382.

Svirezhev, Y. and von Bloh, W., 1998. A zero-dimensional climate-vegetation model containing global carbon and hydrological cycle. *Ecological Modelling* 106, 119-132.

Vernadsky, V.I., 1926. *Biosphere*. Gostekhizdat, Leningrad.

Voitkevich, G.V., Bessonov, O.A., 1986. *Chemical evolution of the Earth*. Nedra, Moscow.

Zharkov, M.A., 2000. Personal communication.

Zavarzin, G.A., 1984. *Bacteria and Atmosphere Composition*. Nauka, Moscow.

Chapter 31

BACTERIAL BIOFILMS IN ASTROBIOLOGY: THE IMPORTANCE OF LIFE DETECTION

Jan TOPORSKI
University of Portsmouth, School of Earth and Environmental Sciences, Astrobiology Group, Burnaby Building, Portsmouth, PO1 3QL, UK, jan.toporski@port.ac.uk.

Andrew STEELE
NASA JSC, Mail Code SN, 77058 Houston, Texas, USA.

David S. McKay
NASA JSC, Mail Code SN, 77058 Houston, Texas, USA.

Frances WESTALL
Lunar and Planetary Institute, 3600 Bay Area Blvd., 77058 Houston, Texas, USA.

1. INTRODUCTION

Experience gathered by researchers during their hunt for evidence of early life on Earth has shown the difficulties associated with the interpretation of possible microbial fossils (Schopf and Walter, 1983; Schopf, 1999a; Brazier et al., 2001). Indeed the controversy surrounding the earliest life on Earth is akin to the debate (e.g. Thomas-Keprta, 2000; Golden et al., 2000) on the nature of the nano-structures in Martian meteorite ALH84001 described by McKay et al. (1996). Both these examples emphasise the difficulties involved in (a) conclusively identifying structures as fossil bacterial cells and (b) establishing their indigeneity/syngenicity to the host matrix. Better understanding of biological signatures in rocks is, therefore, needed in order to identify traces of microbial life (Knoll, 1999).

These traces include not only the characteristic morphologies of potentially fossilised microorganisms, but also include organic biomarker molecules, isotopic fractionations, biological mineralisation and possibly trace element concentrations. It is thus crucial to tackle the problems emerging from the search for evidence of early life on Earth and in exobiological and -palaeontological research with a multi-disciplinary approach (Knoll, 1999). Techniques traditionally applied in the hunt for evidence of early life on Earth included light and electron microscopy for the detection of morphological biomarkers (Schopf, 1999a; Westall, 2000) and gas chromatography – mass spectroscopy (GC-MS) for molecular biomarkers (Peters & Moldowan, 1993). Neither approach however is capable of combining morphological, isotopic and chemical information from individual structures, which is crucial to obtain unambiguous data.

There is a need for new technologies that will enhance our ability to obtain chemical information from individual microstructures and fossils (Knoll, 1999; Briggs et al., 2000). Although new techniques will need to be developed to fulfil the requirements of astrobiological research, the utilisation and adaptation of existing techniques previously applied to other areas of science, will surely provide new insights. Considering the importance of the combination of morphological and molecular biomarkers, a crucial requirement for the respective techniques is the need for high resolution in determining molecular, isotopic and morphological biosignatures. High sensitivity for the detection of minute quantities of molecular biomarkers on very small samples is a priority especially when facing the possibility of conducting astrobiological investigations on returned Mars samples. With the increased interest in biological processes in extraterrestrial environments/materials, there is a requirement to include biological techniques in the field of space exploration, especially when searching for evidence of extinct or extant life.

The following sections will outline the approaches and findings of our current research, which is aimed at addressing the challenges associated with the search for evidence of microbial life in terrestrial and extraterrestrial materials, as well as the future of our research direction in biotechnology with an emphasis on robotic exploration of Mars and other solar system bodies.

2. THE DEBATE ON POSSIBLE EVIDENCE OF LIFE IN METEORITES

In the past and for many years to come, meteorites are and will be the only extraterrestrial materials available for the search for evidence that life may have existed elsewhere in our solar system. Researchers devoted their efforts to the "question of living bacteria in stony meteorites" as early as in the 1930's. Roy and Hudson (1935) refuted and disproved Lipman's claims (1932), that "stony meteorites had brought down with them from somewhere in space a few surviving bacteria". Roy and Hudson (1935) conducted microbial culturing experiments and concluded from their results that the detected bacteria were terrestrial contaminants. Later researchers made a microbiological examination of the Murchison carbonaceous chondrite using microscopy techniques (Claus and Nagy, 1961). They interpreted findings of small (ca. 10 m), "organized elements" as microfossils indigenous to the meteorite, although "microbiological contaminants (common airborne bacteria, algae, etc.) in carbonaceous as well as in non-carbonaceous meteorites" had been observed by these authors. Microfossils have also been suggested to be present in a number of other carbonaceous meteorites (e.g. Palik, 1962; Timofejev, 1963). The reported microstructures however were identified as instrumental, mineralogical or organic artefacts (e.g. Briggs, 1962; Gregory, 1962). More recent work (e.g. Hoover et al., 1998) reported on microorganisms and their remnants on freshly fractured surfaces in the Murchison, Allende, and Efremovka meteorites. These researchers interpreted cyanobacteria-like structures as indigenous to the meteorite and thus to the chondrite parent body, based on the observation that these structures are "tightly conjugated" with the mineral matrix. However, no further data to support this hypothesis has been reported, *i.e.* testing for recent terrestrial contamination by microbial culturing experiments, despite the fresh appearance of some of the organisms (Zhmur and Gerasimenko, 1999). These authors suggested that chemolithotrophic microorganisms (autotrophs) could live in these meteorites, as well as in lunar soils and igneous rocks. However, the presence of indigenous carbonaceous matter and introduced terrestrial organic chemical contamination in the same meteorites and rocks would also support the growth of heterotrophic microorganisms, since they metabolise organic carbon (Steele 2000).

McKay et al. (1996) presented the most comprehensive suite of evidence for the possible detection of traces of extraterrestrial life within Martian meteorite ALH84001. Their range of findings went far beyond morphology alone and indeed 4 independent lines of possible evidence for extraterrestrial life were presented. Scientific opinion concerning the interpretation of the findings varied widely and vociferous arguments ensued with many still raging today 5 years after the announcement (e.g. Frankel and Buseck, 2000; Treiman, 2001; Thomas-Keprta et al., 2000 and 2001).

The first line of evidence of McKay et al. (1996) was the presence of Fe-Mg-Ca-carbonates and sulphide and oxide minerals occurring together in discrete zoned carbonate concretions (globules) of Martian origin (Mittlefehldt, 1994) within fracture surfaces of the meteorite. McKay et al. (1996) suggested that the association of these minerals could be related to biogenic processes, although the contrary has been argued (Schopf, 1999). The zoning of the carbonates has been interpreted as possibly being a result of microbial influence (McKay et al., 1996) and indeed can be produced under biologic control (Zhang et al., 1999; Barrat et al., 1999). Chemical zoning in Fe-Mg-Ca-carbonates could also be produced abiotically at elevated temperatures in laboratory conditions (Golden et al., 2000), although zoning and oxygen isotope non-equilibrium values argue against prolonged heating of carbonates (Treiman and Romanek, 1998). Further complications may have arisen from impact shock melting of the carbonates (Scott et al., 1997; Scott, 1999). Clearly, zoning in carbonate globules can be both biogenically and nonbiogenically mediated.

Second, McKay's group described the presence of minute iron sulphide (pyrrhotite and greigite) and magnetite crystals associated with the Fe- and Mg-rich globule rims. Pyrrhotite has not been established as a biogenic product in terrestrial organisms (Posfai et al., 1998) and has been suggested to probably be derived from pyrite. Greigite on the other hand is a product known from certain marine magnetotactic and sulphate reducing bacteria (Posfai et al., 1998). The presence of greigite in the ALH84001 carbonates, however, has been not confirmed (Frankel and Buseck, 2000). Some portion (~ 27%) of the elongated prismatic magnetite crystals associated with the carbonate rims show striking similarities to those found in some magnetotactic bacteria from Earth and are interpreted as biogenic (Thomas-Keprta et al., 2000 and 2001). These researchers report that this subset of magnetite is virtually identical in size, elongation, crystal shape, chemical purity and the absence of structural defects to a marine strain of magnetotactic bacteria (MV-1). Friedman et al. (2001) reports chains of the magnetite crystallites in the carbonate globules, suggesting the possibility of intact magnetosome preservation. However, magnetite grains with a number but not all of these characteristics are reported to form from rapid thermal decomposition of siderite (Golden et al., 2000) and Treiman (2001) argues further that each of the above characteristics is also consistent with an abiogenic origin, leaving the discussion on the magnetites as biomarkers devoid of consensus. However, currently one crucial fact remains: no abiogenic process has been shown to produce the range of properties exhibited by the sub population of ALH84001 magnetites examined by Thomas-Keprta et al. (2001).

The third line of evidence was the presence of organic chemical compounds (polycyclic aromatic hydrocarbons (PAHs)) associated with the carbonate minerals, which are argued to have formed by diagenesis of microorganisms (McKay et al., 1996). Other researchers, however, argue that PAHs can also form abiotically by Fischer-Tropsch type reactions (Anders, 1996; Zolotov and Shock, 2000) as long as molecular hydrogen is available. The decarbonatisation from siderite to form magnetite releases CO and CO_2 gas which can react with H_2 below 300 °C to form PAHs (Zolotov and Shock, 2000). A further point of discussion is the initial source of the PAHs. PAHs can readily form abiotically in space (Léger et al., 1987) and quite possibly on the surface of Mars without any intervention from life (Zolotov and Shock, 1999). Becker et al. (1997) suggests that the PAHs may have been introduced by melted Antarctic ice and were preferentially scavenged by the carbonates within the meteorite. This argument is supported by the fact that the ^{14}C values of the organics in the meteorite indicate 80 % of the organic material in ALH84001 has a probable terrestrial source (Jull et al., 1998). Clemett et al. (1998), however, makes a good case against the contamination of ALH84001 with terrestrial PAHs. He has shown that the concentration of PAHs in the carbonate globules is not an effect of selective deposition, as the PAHs present in the ice (at very low concentration) differ from those found in the meteorite. Other meteorites collected in the Allan Hills (ALH) region did not show evidence of PAH contamination and finally, their concentration is greater inside the meteorite than its exterior. That Stefan et al. (1999) found the PAHs broadly distributed in the meteorite and not concentrated to the carbonate globules only shows that the argument over PAHs in ALH84001 as biomarkers is far from settled.

Finally, McKay and colleagues (1996) presented images of "features resembling terrestrial microorganisms" as possibly being fossil remains of Martian microorganisms. Bradley et al. (1997) claim that the possible nanofossils (less than 0.5 m in length, ca. 50 nm in diameter) are actually irregularities in the surfaces of mineral grains, which were accentuated by the metal coating for electron microscope examination. McKay et al. (1997) replied that they found the same surface irregularities, and those were not attributed to possible Martian nanofossils. Steele et al. (1998) showed that the metal coating would not cause artefacts in the same size-range of the supposed Martian nanofossils. Furthermore, it cannot be ruled out that some of the bacterium-shaped objects described by McKay et al. (1996) may be derived from terrestrial biota, as Steele et al. (1998a, 2000) have reported the presence of living organisms in ALH84001. Debates subsequent to the announcement of the findings in ALH84001 in August 1996 culminated in a special workshop in Washington D.C. (Workshop on the lower size limit for life, October 1998), where these features were considered too small to represent fossil microorganisms (Schopf, 1999). Since then several researchers have presented evidence for the existence of nano-scale organisms (e.g. Çiftçioglu et al., 1997; Kajander et al., 1998). More importantly, Gillet et al. (2000) described bacterium-shaped objects form the Tatahouine meteorite that matched the size of the structures in ALH84001. Bacterial cultures extracted form the soil surrounding the meteorite pieces revealed morphologies similar to those observed within Tatahouine. Transmission electronmicrographs of these cultures showing the presence of cell walls and 16S rDNA analysis confirmed a terrestrial bacterial origin (Gillet et al., 2000). These findings render the size-argument used to refute McKay et al.'s (1996) hypothesis incongruous, since it is quite possible for "nanobacteria" to become fossilised in the same way as "normal" micron-sized bacteria. Putative fossil nanobacteria indeed have been reported from terrestrial palaeo-environments (Folk, 1993), but, as with the supposed Martian nanofossils, these claims are subject to considerable debate, and alternative abiogenic explanations have been suggested (e.g. Kirkland et al., 1999; Vali et al., 2001).

Although challenged for more than three-quarters of a century, the question whether life existed (or exists) elsewhere in our solar system remains still unresolved. The methods applied in the quest for evidence of life in meteorites failed to provide conclusive and unambiguous data allowing the assignment of observed features to a biogenic origin. This fact clearly emphasises the need for new strategies to discover traces of microbial life.

3. MICROBIAL CONTAMINATION OF METEORITES

It clearly emerges from the debates described above that the detection of traces of life on extraterrestrial materials is associated with great difficulties. Two criteria must be fulfilled for its unambiguous detection: (1) establishment of biogenic origin (either viable or non-viable), and (2) proof of indigeneity to the respective extraterrestrial material. In a scenario where (1) is positive for viable life, surely the second step must include considerations to test for viable Earth microbiota, *i.e.* the search for evidence of microbial contaminants. Indeed, the scientific debate over evidence of Martian life in ALH84001 has had major implications for the search for life elsewhere in our solar system. The foremost being that the techniques used to look for life on ALH84001 are still the techniques of choice to search for life in Mars return samples or on robotic missions. These techniques failed to detect terrestrial bacteria on ALH84001 (Steele et al., 2000), preferring to allude to organic contamination from ice (Bada et al., 1998; Jull et al., 1998). Furthermore recent work has shown that Viking could have missed up to as many as several million bacterial cells per gram of soil (Glavin et al., 2001). There is thus a serious problem in the space science community with regards to the detection of life in extraterrestrial samples. Our recent research, however, has shown that terrestrial microbial contaminants are no rarity in meteorites. In addition to Martian meteorite ALH84001, morphotypes indicating the presence of microorganisms could be detected in the Martian meteorite Nakhla (Toporski et al., 2000a), in the Murchison carbonaceous chondrite, as well as in a number of ordinary chondrites (Steele et al., 2000b), primarily using light, electron and atomic force microscopy. It is considered important, however, to understand contamination pathways and identify the microbial species involved in contamination and therefore trace the principle sources of contamination and understand more fully terrestrial microbial metabolism within the meteorites.

There are two ways to characterise contaminating organisms: (1) culturing experiments on small chips of the meteorites with subsequent biochemical and 16S and 18S rDNA characterisation of any isolates and/or (2) the direct extraction of DNA from the samples followed by PCR (polymerase chain reaction) using 16S and 18S rDNA sequence analysis (Whitby et al., 2000). In either case, characterisation (so far only completed on Allende but in progress on ALH84001 and Nakhla) was achieved using common primers for the detection of eubacteria and fungi. The direct DNA extraction approach was chosen in attempt to classify all species present in the meteorites and not only those culturable on the small range of growth media used in these studies.

The biological studies on Murchison and Allende have so far confirmed the presence of the bacterial genus *Bacillus spp.* and more specifically the species *B. licheniformis*, *B. pumilis*, *B. subtilis*, as well as the genera *Streptococcus spp.* and *Micrococcus spp.* (also *M. luteus)* and the species *Corynebacterium minutissimum*, *Staphylococcus capitis* and *S. auricularis.* As for fungi, *Neurospora spp* and *Aspergillus spp.* could thus far be identified. All the bacterial and fungal species thus far detected are chemoorganotrophic and therefore must be obtaining nutrients from the organic matter within the meteorite (Holt et al., 1994; Carlile & Watkinson, 1994). Some species are of probable human origin such as *Corynebacterium spp*, *Streptococcus spp.* and *Staphylococcus spp.* (*S. capitis* is found in human eyebrows and hair; Holt et al.,1994). Members of the genus *Bacillus* are from a variety of habitats including human skin, dust and soil. The fungi *Neurospora* is a member of the *Ascomycetes* and is mainly found in soils. Members of the *Aspergillus spp*. are ubiquitous in soils and can be associated with common laboratory contaminants and human contact (Carlisle & Watkinson, 1994). It is furthermore noteworthy that several of the characterised bacteria and all fungal species are spore forming. From these observations three likely routes of contamination can be envisaged: collision of the meteorite with the Earth (soil inhabitants), airborne contamination by spore-forming organisms (during atmospheric entry, time in environment, collection, curation and in the laboratory), and from human contact by either direct handling or from dust/skin flakes. None of the above organisms are exclusively found in soils and therefore no conclusions can be drawn concerning the primary colonisers of the meteorite upon cooling.

Our results have shown that it is possible to use microbiological and molecular biological techniques on extraterrestrial materials to detect the presence of microorganisms by means of their DNA on both external (fusion crust) and internal (cracks and fissures) surfaces (Whitby et al., 2000). This may have major implications for the study of organics from meteorites as the degradation and alteration by the terrestrial contaminants may affect interpretations and implications inferred from organic analysis. Compared to the previous efforts to detect life in meteorites, the techniques described in this chapter allow a systematic approach to the classification of microbial contaminants and therefore to the dynamics of microbial colonisation of meteorites. Also these results show that current molecular biology techniques do not require large samples sizes (only ca. 300 mg) to be used as effective tools for what is essentially a process of life detection in extraterrestrial materials. By studying microbial contamination of meteorites, the skills needed to search for life in other parts of the solar system can be improved and new techniques developed and rigorously tested.

Biological techniques allow the identification and characterisation of life in meteorites and thus aid the general multi-disciplinary strategies required for astrobiological research. Clearly the application of these techniques is strongly biased towards the detection of viable life, implying that different strategies have to be tested for the detection of extinct life.

4. IN SITU BIOMARKERS DETECTION ON FOSSIL BIOFILM

Crucial requirements for the combination of molecular and morphological fossil biomarkers are non-destructive analytical techniques. The technique of Time of Flight – Secondary Ion Mass Spectrometry (ToF-

SIMS) fulfils this requirement and has already been successfully applied to a variety of organic and inorganic materials (Steele et al., 2001 and references therein). ToF-SIMS allows *in-situ* analysis with high mass resolution as well as molecular imaging of micron-sized structures. In contrast to GC-MS, no extraction or derivatisation of the sample is required for ToF-SIMS analysis. The surface of the analysed sample remains virtually undamaged ($< 1\%$ of the uppermost surface layer is removed) (Briggs et al., 1989), thus, microscopical investigation of the same samples following analysis is possible, allowing the combination of molecular and morphological information.

We analysed fossil bacterial biofilms associated with macrofossils from volcanoclastic lacustrine sediments from the Upper Oligocene (ca. 25 m.y.) Enspel formation, Germany (Toporski et al., 2001). Soft tissues and coprolites have previously been reported to be preserved by replacement of microbial mats (Liebig, 1998). Preliminary scanning electron microscopy (SEM) studies showed that a coprolite consisted purely of fossilised bacterial cells and associated fossilised polymeric substances (Toporski et al., 2000b). Further samples of this specimen were carefully prepared for the combined investigation and overlaying of chemical and morphological biosignatures.

ToF-SIMS analysis of this material has revealed the presence of a number of organic substances such as alkanes, alkenes, aromatic and polycyclic aromatic hydrocarbons. More tentative are the interpretations of alkyl pyrroles and pyridyl (precursor molecules of purine and pyrimidine bases) as well as saturated and unsaturated fatty acids (Toporski et al., 2001). Most importantly however, ToF-SIMS analysis indicated the presence of hopanoids, a class of fossil biochemicals (triterpanes) considered diagnostic of prokaryotes (Peters & Moldowan, 1993). Our studies conducted on purified hopane standards using high mass resolution have shown that in the high-mass resolution mode of the ToF-SIMS instrument, differentiation of diagnostic hopane peaks is possible. Furthermore, all characteristic GC-MS fragments of these molecules could be detected and differentiated (Steele et al., 2001). The spectra on the fossil biofilms were collected in a slightly lower mass resolution mapping mode and therefore unmistakable differentiation and identification was not possible (follow up studies are underway). The centroids of a number of possible characteristic hopane peaks obtained on the fossil bacterial biofilms, however, are well within the mass-range associated with bacterial hopanes. There is a strong possibility therefore that hopanoids may be directly associated with the fossilised bacterial cells although further work including GC-MS analysis and high mass resolution ToF-SIMS analysis are needed to verify these interpretations. SEM studies on the ToF-SIMS analysed sample confirmed the sole presence of bacterial fossils. For the first time, it could be demonstrated that the combination of high magnification and high resolution imaging techniques, with state of the art *in-situ* molecular surface analysis techniques, is feasible on a micron scale on samples of geological and astrobiological interest. This allows the combination of spectral and morphological biomarkers to be undertaken on the same features on a sample.

The advantages of highly sensitive non-destructive *in-situ* analysis techniques for biomarker detection and characterisation are invaluable, particularly with respect to the envisaged Mars sample return missions. Considering the small amounts of sample material that will be available (whether dealing with meteoritic or returned samples), the usefulness of this method is even more evident. The advantages of a well-tested ion probe instrument capable of detecting micro-quantities of substances on exceedingly small samples are inestimable.

5. MICROARRAY ASSAYS FOR SOLAR SYSTEM EXPLORATION

Life detection in terrestrial laboratories has developed in the last 30 years at a rate equalled only by that of the growth of the computer industry. These techniques employ labelled probes to detect and characterise cells and molecules ranging from human tumours to the expression of single proteins in large bacterial consortia. Central to these technologies have been the development of DNA/RNA hybridisation, monoclonal antibody production, and DNA amplification by PCR. The approach we are currently exploring is that of using antibodies. Immunological methods utilising the extreme sensitivity and specificity of vertebrate immune systems have been developed and antibodies have been successfully raised to individual proteins within bacteria or whole bacterial species, individual amino acids, organic pollutants, DNA, RNA and important prokaryotic proteins (Klein and Horejsi, 1999).

Immunological techniques have also been used to assay for traces of extinct life and are becoming an important organic chemical tool. As early as 1980, Lowenstein was using a sensitive radioimmunoassay to detect DNA and proteinaceous material in fossilized bone (Lowenstein, 1980). Since then this technique has become increasingly important in the search for organic molecules in fossilized material (DeJong et al., 1974; Lowenstein, 1981, Muyzer et al., 1984; Muyzer and Westbroek, 1989; Lowenstein, 1985; Collins et al., 1991 Nerlich et al., 1993; Schweitzer et al., 1999; Schweitzer and Horner, 1999). More recently, work by Schweitzer et al. (1999) has shown that antibodies raised against hemoglobin, collagen and keratin can produce strong reactions in fossilized bone and bird claw tissues. This work shows that it is possible to detect biomarkers using immunological techniques in palaeontological samples including T-Rex (Schweitzer and Horner, 1999b) from the Mesozoic period back at least 100 million years. Clearly, antibodies can be specific to biomarkers for viable life (*i.e.* DNA, or cytochromes), prebiotic molecules (*i.e.* individual nucleotides or amino acids), pollutants (such as PAH's) and non-viable biomarkers within fossil materials. The combination of the antibodies specific to these four groups of antibodies can be blotted together onto a single glass slide to form a microarray plate.

Microarray technology has developed as a tool to aid biologists unravel the secrets of the human genome. Using technology refined from inkjet printers, up to 10000 separate DNA/antibody tests can be blotted onto a chip 2.5 cm square. Commercial instruments are readily available that manufacture and read these chips based on a fluorescence response from each test spot.

It has recently been successfully proposed the European (ESA; A. Steele PI) and North American (NASA; D.S. McKay PI) Space Agencies to develop a rapid, automated, compact, instrument for detecting evidence of both extinct and extant life on Mars. The instrument, initially entitled Mars Immunoassay Life Detection Instrument (MILDI), will be designed using microarray technology to accurately detect and quantify a wide range of potential biomarkers. This instrument will be hopefully used in conjunction with other life detection techniques to confirm the presence of extinct or extant life. However, it will be further extended to test for prebiotic molecules and introduced pollutants. Furthermore, as antibodies can be raised to whole bacterial species, MILDI will carry antibodies to known terrestrial microbial space-craft contaminants to act as a control for all other life detection techniques onboard a possible robotic mission.

The basic process is as follows: (1) Identify and choose relevant extinct and extant biomarkers. (2) Raise antibodies to these molecules (many are already described in the scientific literature). (3) Combine the antibodies with various fluorescent stains. (4) Place these onto a microarray. (5) Produce an extract of the sample of interest using water or a solvent. (6) Spread the extract onto the test plate and allow it to react. (7) Illuminate the test plate with UV light and image the plate with a CCD camera to record reaction. If the molecules for which the antibodies are designed are present, the antibodies will react with the molecules, the stain will be activated/quenched and the associated well will fluoresce. Rigorous controls for both cross-reactivity and false positives/negatives are being incorporated into the testing phase of these arrays. Proof of concept for this instrument will be blind testing of initially Martian relevant environments *e.g.* Antarctic rocks, hotsprings samples, as well as known contaminated Martian meteorites.

The search for life on Mars (and other planetary bodies) is complicated by several factors and must be based upon certain assumptions. There are three possibilities to consider: 1) life is currently extant in some form; 2) life did exist but is now extinct; 3) life never existed. In considering possibilities one and two, we must assume that life elsewhere follows some basic chemical parameters that life on Earth shares, otherwise we may not be able to recognise it. We must therefore assume that the component biomarkers common to life on this planet are also present as signals of life on other planets. One advantage of the use of antibodies is that the antibody/antigen reaction is based not on chemical recognition but on shape. If life evolved on Mars it need only have molecules with similar shapes to react with the antibodies. This is not the strange concept that it may seem considering that the shape of a component molecule is of paramount importance to the function of many biological systems including membranes, genetic material, catalytic enzymes and other proteins. Therefore the chemical composition/isomeric excess of a Martian organism may differ widely from that of an Earth organism.

The best case scenario is that life is currently present on Mars and indeed uses concepts similar to those on Earth. If so, we may be able to observe organisms and/or their immediate metabolic by-products directly. If life is extinct, chemical biomarkers may still be present but they may be altered over time and may be fragmentary. It may be difficult or impossible to prove that life never existed on Mars, but our confidence in that conclusion would build as we accumulate more negative evidence, particularly the total lack of chemical biomarkers which have been previously shown to last for billions of years on Earth (Summons and Walter, 1990).

6. SUMMARY

Modern Exo- and Astrobiology is on the verge to be able to discover evidence of life elsewhere in the solar system with a higher degree of confidence, but many of the techniques previously included in this effort have failed to provide unambiguous evidence, and least of all conclusive proof. New concepts and technologies are being tested to design integrated exo-/astrobiological strategies able to complete the challenging quest of the detection of extraterrestrial life. Including highly sensitive *in-situ* surface analysis techniques and biological/immunological methods in with other newer techniques, will help to increase the odds of finding a second statistical observation of life other than life on earth.

REFERENCES

Anders E. (1996) Evaluating the evidence for past life on Mars (letter). *Science* **274**, 2119-2121.

Bada J.L. Glavin D.P., McDonald G.D. and Becker L. (1998) A search for endogenous amino acids in Martian meteorite ALH84001. *Science*. **279**, 362 – 365.

Barrat J.A., Gillet Ph., Lesourd M., Blichert-Toft J., and Popeau G.R. (1999) The Tatahouine diogenite: Mineralogical and chemical effects of sixty-three years of terrestrial residence. *Meteorit. Planet. Sci.*, **34**, 91-97.

Becker L., Glavin D. P. and Bada J. L. (1997) Polycyclic aromatic hydrocarbons (PAHs) in Antarctic Martian meteorites, carbonaceous chondrites, and polar ice. *Geochim. Cosmochim. Acta* **61**, 475-481.

Bradley J.P., Harvey R.P. and McSween H.Y. (1997) No 'nanofossils' in Martian meteorite. *Nature*, **390**, 454.

Brasier, M.D., Green O.R., Lindsay J.F., Steele A. (2001) Earth's oldest fossils: questioning the evidence. *Nature*, in press.

Briggs D.E.G., Evershed R.P. and Lockheart M.J. (2000) The biomolecular paleontology of continental fossils. *In*: Erwin D.H. and Wing S.L. (eds.) Deep time: Paleobiology's perspective. *Paleobiology*, **26**, 169-193.

Briggs D., Brown A. and Vickerman J.C. (eds.) (1989) Handbook of static secondary ion mass spectrometry. John Wileys & Sons, Chichester, UK.

Briggs M.H. (1962) Properties of the organic microstructures of some carbonaceous chondrites. **Nature**, **195**, 1076-1077.

Carlile, M. J. & Watkinson S.C. (1994) The fungi. Academic Press London, Gen. Ref.

Çiftçioglu N., Pelttari A. and Kajander E.O. (1997) Extraordinary growth phases of nanobacteria isolated from mammalian blood. *Proceedings of SPIE*, **3111**, 429-435.

Claus G. and Nagy B. (1961) A microbiological examination of some carbonaceous chondrites. *Nature*, **192**, 594-596.

Clemett S.J., Dulay M.T., Gilette J.S., Chillier X.D.F., Mahajan T.B., and Zare R.N. (1998) Evidence for the extraterrestrial origin of polycyclic aromatic hydrocarbons (PAHs) in the Martian meteorite ALH 84001. *Faraday Discussions (Royal Soc. Chem.)*, **109**, 417-436.

Collins M.J., Muyzer G., Westbroek P., Curry G.B., Sandberg P.A., Xu S.J., Quinn R. and MacKinnon D. (1991) Preservation of fossils biopolymeric substances. *Geochim. Cosmochim. Acta*, **55**, 2253 – 2257.

DeJong E.W., Westbroek P., Westbroek J.F. and Bruning J.W. (1974) Preservation of antigenic properties of macromolecules over 70 Myr. *Nature*, **252**, 63 – 64.

Folk R.L. (1993) SEM imaging of bacteria and nannobacteria in carbonate sediments and rocks. *J. Sed. Petrol.*, **63**, 990-999.

Frankel R.B. and Buseck P.R. (2000) Magnetite biomineralisation and ancient life on Mars. *Current Opinion on Chemical Biology*, **4**, 171-176.

Friedmann E.I., Wierzchos J., Ascaso C. and Winklhofer M. (2001) Special Feature: Chains of magnetite crystals in the meteorite ALH84001: Evidence of biological origin, *Proc. Natl. Acad. Sci.*, **98**, 2176-2181.

Gillet Ph., Barrat J.A., Heulin Th., Achouak W., Lesourd M., Guyot F. and Benzerara K. (2000) Bacteria in the Tatahouine meteorite: nanometric-scale life in rocks. *Earth and Planetary Science Letters*, **175**,161-167.

Glavin D.P., Schubert M., Botta O., Kminek G. and Bada J.L. (2001) Detecting pyrolysis products from bacteria on Mars. *Earth Planet. Sci. Lett.*, **185**, 1-5.

Golden D.C., Ming D.W., Schwandt C.S., Lauer H.V.., Socki R., Morris R., Lofgren G.E. and McKay G.A. (2000) Inorganic formation of zoned Fe-Mg-Ca carbonate globules with magnetite and sulphide rims similar to those in Martian meteorite ALH84001. *LPSC XXXI*, LPI, 13-17 March 2000, Houston, Texas, #1799 (abstr.).

Gregory P.H. (1962) Identity of organised elements from meteorites. *Nature*, **194**, 1065.

Holt J.G., Krieg N.R., Sneath P.H.A., Staley J.T. and Williams S.T. Eds. (1994). *Bergey's Manual of Determinative Bacteriology*. Williams and Wilkins, Baltimore, USA.

Hoover R.B., Rozanov A.Yu., Zhmur S.I. and Gorlenko, V.M. (1998) Further Evidence of Microfossils in Carbonaceous Chondrites. *Proc. SPIE*, **3441**, 203-216.

Jull A.J.T., Courtney C., Jeffrey D.A. and Beck J.W. (1998). Isotopic evidence for a terrestrial source of organic compounds found in Martian meteorites ALH84001 and EETA 79001. *Science*. **279**, 365 – 369.

Kajander O.E., Björkland M. and Çiftçioglu N. (1998) Mineralization by nanobacteria. *Proc. SPIE*, **3441**, 86-94.

Kirkland B.L., Lynch F.L., Rahnis M.A., Folk R.L., Molineux I.J. and McLean R.J. (1999) Alternative origins for nanno-bacteria-like objects in calcite. *Geology*, **27**, 247-250.

Klein, J. & Horejsi, V. (1999) *Immunology*. Blackwell Scientific Publications, Oxford UK

Knoll, A. (1999) Recognition of a biological signature in rock. Discussion summary. In: *Size Limits of very small Microorganisms – Proceedings of a Workshop*. National Academic Press, Washington D.C., USA, pp. 85-87.

Léger A., d'Hendcourt L. and Boccara N. (1987) *Polycyclic Aromatic Hydrocarbons and Astrophysics*. Reidel, Boston, USA.

Liebig K. (1998) Fossil Microorganisms from the Eocene Messel Oil Shale of Southern Hesse, Germany. *Kaupia – Darmstätder Beiträge zur Naturgeschichte*, **7**, 1-95.

Lipman C.B. (1932) *Amer. Mus. Nov.*, **No. 588.**

Lowenstein J.M. (1985) Molecular approaches to the identification of species. *Am. Sci.*, **73,** 541 – 546.

Lowenstein J.M. (1981) Immunological reactions from fossil material. *Phil. Trans. R. Soc. Lond,* **292,** 143 – 149.

Lowenstein, J.M. (1980) Species specific proteins in fossils. *Naturwissenschaften,* **67**; 343 – 346.

McKay D. S. Gibson E. K. Jr., Thomas-Keprta K. L., and Vali H. (1997) No 'nanofossils' in Martian meteorite: reply. *Nature* **390**, 455-456.

McKay D.S., Gibson E.K., Thomas-Keprta K., Vali, H., Romanek C., Clemett S., Chillier X.D.F., Maechling C.R. & Zare R.N. (1996) Search for past life on Mars: possible relic biogenic activity in Martian meteorite ALH84001. *Science*. **273**, 924-930.

Mittlefehldt D. W. (1994) ALH84001, a cumulate orthopyroxenite member of the SNC meteorite group. *Meteoritics* **29**, 214-221

Muyzer G., Westbroek P., DeVrind J.P.M., Tanke J., Vriheid T., DeJong E.W., Bruning J.W. and Wehmiller J.F. (1984) Immunology and Organic Geochemistry, *Org Geochem.*, **6**, 847 – 855.

Muyzer G and Westbroek P. (1989) An immunohistochemical technique for the localisation of preserved biopolymeric remains in fossils. *Geochim. Cosmochim. Acta,* **53**, 1699 – 1702.

Nerlich A.G., Parsche F., Kirsch T., Wiest I. and von der Mark K. (1993) Immunohistochemical detection of interstitial collagens in bone and cartilage tissue remnants from an infant Peruvian mummy. *Am. J. Phys. Anthro*, **91**, 279 – 285.

Palik P. (1962) Further life-forms in the Orgueil meteorite. *Nature*, **194**, 1065.

Peters K. & Moldowan J.M. (1993) *The biomarker guide*. Prentice Hall, New Jersey, USA.

Posfai M., Buseck P.R., Bazylinski D.A. and Frankel R.B. (1998) Reaction sequence of iron sulphide minerals in bacteria and their use as biomarkers. *Science*, **280**, 880-883.

Roy S.K. and Hudson N.P. (1935) The question of living bacteria in stony meteorites. *Geological Series of Filed Museum of Natural History*, **6**, 179-198.

Schopf J.W. and Walter M.R. (1983) Archaean microfossils: New evidence of ancient microbes. Pp. 214-239 In: *Earth's Earliest Biospheres, Its Origin and Evolution.* J.W. Schopf (ed.). Princeton, New Jersey: Princeton University Press, USA.

Schopf, J.W. (1999a) Fossils and pseudofossils: lessons from the hunt for early life on Earth. In: *Size Limits of very small Microorganisms – Proceedings of a Workshop.* National Academic Press, Washington D.C., USA, pp. 88-93.

Schopf J.W. (1999b) Breakthrough Discoveries. In *Evolution! Facts and Fallacies* (ed. J.W. Schopf) Academic Press, N Y pp. 91-117.

Schweitzer M.H. & Horner, J.R. (1999) Intravascular microstructures in trabecular bone tissues of Tyrannosaurus Rex. *Annales de Paleontologie*, **85**, 179-192.

Schweitzer M.H., Watt J.A., Avci R., Forster C.A., Krause D.W., Knapp L., Rogers R.R., Beech I.B. and Marshall M. (1999) Keratin specific immunoreactivity in a late cretaceous bird from Madagascar. *Accepted Journal of Vertebrate Paleontology.*

Scott E. R. D., Yamaguchi A., and Krot A. N. (1997) Petrological evidence for shock melting of carbonates in the Martian meteorite ALH 84001. *Nature,* **387**, 377-379.

Scott E.R.D. (1999) Origin of carbonate-magnetite-sulphide assemblages in Martian meteorite ALH84001. *Jour. Geophys. Res.*, **104**, 3803-3813.

Steele, A., Toporski, J., Avci, R., Guidry, S.A., and McKay, D.S. (2001): Time of Flight – Secondary Ion Mass Spectrometry (ToF-SIMS) of a number of bacterial hopanoids. *Chemical Geology* (In press).

Steele A., Goddard D.T., Stapleton D., Toporski J.K.W., Peters V., Bassinger, V., Sharples G., Wynn-Williams D.D. & McKay D.S. (2000a) Imaging of an unknown organism on the Martian Meteorite ALH84001. *Meteorit. Plan. Sci.*, **35**, 237-241.

Steele A., Toporski J., Westall F., Thomas-Keprta K., Gibson E., Avci R., Whitby C., Griffin C., & McKay D.S. (2000b) The microbiological contamination of meteorites: a null hypothesis. *Lunar Planet. Sci. XXXI*, LPI, Houston (abstr.).

Steele A., Goddard D.T., Beech I.B., Tapper R.C., Stapleton D. and Smith. J.R. (1998) Atomic force microscopy imaging of fragments from the Martian meteorite ALH84001. *J. Microsc.* **189**: 2-7

Stephan T., Heiss C.H., Rost D., and Jessberger E.K. (1999) Polycyclic aromatic hydrocarbons in meteorites: Allan Hills 84001, Murchison, and Orgueil. *Lunar Planet. Sci. XXX*, LPI, Houston, #1569 (Abstr.).

Summons R.E and Walter M.R. (1990) Molecular fossils and microfossils of prokaryotes and protists from Proterozoic sediments *Am. J. Sci.*, **290-A**, 212-244.

Thomas-Keprta K.L., Clemett S.J., Bazylinski D.A., Kirschvink J.L., McKay D.S., Wentworth S.J., Vali H., Gibson E.K., McKay M.F. & Romanek C.S. (2001) Truncated hexa-octahedral magnetite crystals in ALH84001: Presumptive biosignatures, *Proc. Natl. Acad. Sci.*, **98**, 2164-2169

Thomas-Keprta K.L., Bazylinski D.A., Kirschvink J.L., Clemett S.J., McKay D.S., Wentworth S.J., Vali H., Gibson E.K. Jr. and Romanek C.S. (2000) Elongated prismatic magnetite crystals in ALH84001 carbonate globules: Potential Martian magnetofossils. *Geochim. Cosmochim. Acta*, **64**, 4049-4081.

Timofejev B.W. (1963) *Grana Palynol.*, **4**, 92 (in Russian).

Toporski J., Steele A., Westall F., Avci R., Martill M., and McKay D.S. (2001) Morphological and spectral investigation of exceptionally well preserved bacterial biofilms from the Oligocene Enspel formation, Germany. *Geochim. Cosmochim. Acta*, accepted.

Toporski J., Steele A., Westall F., Griffin C., Whitby C., Avci R. & McKay D.S. (2000a) Electron microscopy studies, surface analysis and microbial culturing experiments on a depth profile through Martian meteorite Nakhla. *Lunar Planet. Sci. XXXI*, LPI, Houston, #1636 (abstr.).

Toporski J., Steele A., Westall F., Avci R. & McKay D.S. (2000b) A multi-disciplinary exemplifying case study for in-situ biomarker detection on a 25 ma old bacterial biofilm using in-situ surface analysis and imaging techniques, *Abstr. GSA 2000*, 12-16 Nov. 2000, Reno, Nevada, USA.

Treiman A.H. (2001) A hypothesis for the abiotic & non-Martian origins of putative signs of ancient Martian life in ALH84001. *Lunar. Planet. Sci XXXII*, LPI, Houston, Texas, #1304 (abstr.).

Treiman A.H. and Romanek C.S. (1998) Bulk and stable isotopic compositions of carbonate minerals in Allan Hills 84001: No proof of high formation temperature. *Meteorit. Planet. Sci.* **33**, 737-742.

Westall F., Steele A., Toporski J., Walsh M., Allen C., Guidry S., McKay D.S., Gibson E. & Chafetz H. (2000) Polymeric substances and biofilms as biomarkers in terrestrial materials: implications for extraterrestrial samples. *J. Geophys. Res. – Planets*, **105**, 24,511-24,527.

Whitby C., Griffin C., Saunders J.R., Steele A., Toporski J., Allen C. & McKay D.S. (2000) The direct extraction of DNA from Allende. *LPSC XXXI*, LPI, 13-17 March 2000, Houston, Texas, USA, #1732 (abstr.).

Zhang C., Vali H., Romaneck C.S., Roh Y., Sears S.K. and Phelps T.J (1999) Chemical and morphological characterisation of siderite formed by iron reducing bacteria. *Lunar Planet. Sci. XXX*, Lunar and Planetary Institute, Houston, #1855 (Abstr.).

Zhmur S. & Gerasimenko L. (1999) Biomorph forms in carbonaceous meteorite Allende and possible ecological system-producer of organic matter chondrites. *Instruments, Methods and Missions for Astrobiology II,* (ed.) RB Hoover, *Proc. SPIE*, **3755**, 48-58.

Zolotov M.Y. and Shock E.L. (2000) An abiotic origin for hydrocarbons in the Allan Hills 84001 Martian meteorite through cooling of magmatic and impact-generated gases. *Meteoritics and Planetary Science,* **35**, 629-638.

Zolotov M. and Shock E. (1999) Abiotic synthesis of polycyclic aromatic hydrocarbons on Mars. *J. Geophys. Res.*, **104** E6, 14.033-14,050.

Chapter 32

FOSSIL BIOFILMS AND THE SEARCH FOR LIFE ON MARS

Frances WESTALL
Lunar and Planetary Institute
3600 Bay Area Boulevard
Houston, TX 77058, USA
westall@cnrs-orleans.fr

Maud M. WALSH
Department of Agronomy
104 Sturgis Hall
Louisiana State University
Baton Rouge, LA 70803-5705, USA

Jan TOPORSKI and Andrew STEELE
School of Earth and Environmental Sciences
University of Portsmouth
Porstmouth P01 3QL, UK

1. INTRODUCTION

Microbial biofilms and mats are documented as fossils in rocks throughout the 3.5 b.y.-old morphological fossil record of life on Earth (Westall et al., 2000). The polymer-rich biofilms are, *per se*, highly robust structures capable of great resistance and durability. Moreover, the abundance of active groups in the polymers which can chelate mineral ions in solution, assures their ready preservation in the rock record. These active groups include the carboxylate, hydroxyl, amine and phosphate groups (Geesey and Jang, 1989). Precipitation of minerals within a microbial biofilm can be influenced by microbial metabolic control of the microenvironment. Furthermore, the presence of an organic template with active nucleation sites also contributes towards bio-catalysed precipitation of minerals. Some of the best-known examples of mineralised biofilms in the geological record are calcified and silicified stromatolites (Krumbein, 1983). Mineralisation of organic templates can occur very rapidly (within a day, Toporski et al., 2001a). Experiments to silicify microorganisms also document the potential faithfulness of reproduction of the original organism, and the fineness of detail obtainable, by silica impregnation (Westall et al., 1995; Westall, 1999; Toporski et al., 2001a).

One of the constant problems faced in bacterial palaeontology is the potential for confusion between truly biogenic microfossils and abiogenic bacteriomorphs (Westall, 1999). Identification of fossil bacteria is based on careful analysis of the following features: (a) morphological characteristics including size, shape, evidence for cell division, and cell surface texture; and (b) colonial characteristics such as the fact that bacteria always occur in communities of a few to thousands of individuals, often in consortia of more than one type of species which may have different morphologies, and the fact that they occur in biofilms or complexly structured microbial mats consisting of large amounts of extracellular polymeric substances (EPS). Certainly, additional information such as characteristic C, S, or Fe isotope ratios indicative of microbial metabolic fractionation, or the presence of biomarker molecules associated with the microfossils (Schidlowski, 1988; Canfield et al., 2000; Toporski et al., 2001b, and this volume) greatly improves the probability of correct interpretation of potentially biogenic structures. The carbon isotope and carbon biomarker analyses necessarily rely on the preservation of sufficient carbonaceous material in the fossils. However, the amount of remnant carbon associated with a microfossil is variable and microorganisms may also be almost completely replaced by a mineral (Westall et al., 1995; Westall et al., 2001). Thus, sometimes there remains only morphological evidence for biogenicity. Another aspect of abiotic mimics, which is particularly relevant to extraterrestrial studies, concerns the distinction between potential films formed by prebiotic molecules and microbial polymer films. Experimental work in this field (Westall et al., 2000) documented the similarities between simple films and globular structures, formed by an organic extract of the Murchison meteorite (consisting mostly of PAHs,) and simple microbial polymer films. The presence of an organic film of simple structure is not sufficient evidence on its own for biogenicity. It is necessary to be able to place a film in a context in which there is additional evidence for life, such as the association of fossil microorganisms, or a thick, structured film, typical of microbial mats.

2. EARLY EARTH

The oldest morphological evidence for life on Earth consists of microbial biofilms and mats in 3.3-3.5 b.y.-old cherts from the Barberton and Pilbara greenstone belts of South Africa and Australia, respectively (Awramik et al., 1983; Walsh and Lowe, 1985; Byerly et al., 1986; Walsh, 1992; Westall and Gerneke, 1998; Hofmann et al., 1999; Westall et al., 2001). Some of the mats from the Barberton greenstone belt are described in detail below. However, in order to underline their usefulness as analogues for potential microbial mats, it is first necessary to summarise the geological context in which they occur. This information will then be extrapolated to the martian situation.

The Barberton greenstone belt is one of the oldest terrains with extant sedimentary horizons to have survived the destructive effects of global plate tectonics. The 3.3-3.5 b.y.-old Onverwacht Group succession forms part of the Swaziland Supergroup and is coeval with parts of the Pilbara greenstone belt in NW Australia. Only the sediments of the Isua/Akilia Greenstone Belts in Greenland are older (>3.75 b.y.-old), but their highly metamorphosed state precludes the finding of morphological fossils (Westall, 2000) although there is some equivocal carbon isotope indication for the presence of life (Schidlowski, 1988; Mojzsis et al., 1996; Rosing, 1999). Recent mapping in the Pilbara and Barberton suggests that the Early Archaean sediments were deposited in broad, collapsed basinal structures on protocontinental material (Nijman et al., 1998a,b; Lowe and Byerly, 1999). The sedimentary horizons are intercalated with thick successions of basic to felsic volcanic rocks and volcaniclastic deposits. The almost ubiquitous presence of tourmaline and silica (many of the Early Archaean strata, including volcanic lithologies are silicified), as well as the presence of fossil hydrothermal vents testify to the importance of low temperature hydrothermal activity in the area (de Wit et al., 1982; Byerly and Palmer, 1991; de Ronde et al., 1994; Nijman et al., 1998a,b). Geochemical evidence for the recycling of evaporite-rich fluids through hydrothermal systems documents common evaporitic conditions (Byerly and Palmer, 1991; de Ronde et al., 1997).

The lack of material from this earliest period in the history of the Earth makes interpretation of more general environmental parameters difficult. However, modelling suggests that the early CO_2 rich atmosphere had to have been tempered by another greenhouse gas such as CH_4 to prevent early total glaciation of the Earth (Pavlov et al., 2000; Sleep and Zahnle, 2001). There is much debate concerning the concentration of O_2 in the early atmosphere with estimates ranging from <1% present atmospheric levels (PAL) to somewhat higher concentrations (Holland, 1984; Ohmoto et al., 1993; Canfield et al., 2000). The concensus of opinion seems to point to low oxygen levels. The oxidation state of the early Earth's atmosphere would have had a profound effect on UV penetration to the surface. Low oxygen levels would have meant no protective ozone layer around the Earth (Cockell et al., 2000). Thus, early life would have had to be able to cope with a radiation environment considered extreme by modern standards. On the other hand, the environmental conditions in which early life flourished were "extremely normal" for at least the first 1.5 b.y. of Earth's history. As in modern, extreme environments which resemble to a certain extent those of the Early Archaean (e.g. the sabkha), the survival of microbial mats depends very much on the protective attributes of the composition and structure of the mat in which the microbial colonies are embedded: the outer layers form a tough, resistant, protective "skin", under which the water-rich polymer layers provide a moist environment for the microorganisms (Gorbuschina and Krumbein, 2000).

Early Archaean seawater appears to have been more saline than modern seawater (de Ronde et al., 1997). Estimates of the pH range from highly alkaline (Kempe and Degens, 1985) to normal (8.1) or slightly acidic (Grotzinger and Kasting, 1993).

In terms of a global context, the discovery of 4.4. b.y-old, inherited zircons in younger rocks in W. Australia (Wilde et al., 2001) demonstrates that there was not only water on the surface of the Earth by that date, but also that there had been recycling of hydrated crustal rocks within the deep crust/upper mantle to produce granitoid rocks (protocontinents) containing zircons. Conditions on the Earth by 4.4 b.y. ago may have been very similar to those of the Early Archaean Earth between 3.3-3.5 b.y. ago. In theory, then, life could have already been present on the Earth by 4.4. b.y. One important extraterrestrial influence on the presence and distribution of life on early Earth concerns the possibility of extinction due to impacts by bodies large enough to cause the wholesale sterilisation of the planet (Maher and Stevenson, 1988). Comparison with the lunar cratering record shows that the period of heavy bombardment occurred between about 4.0-3.85 b.y. ago (Ryder et al., 2000). If such a planet sterilising event occurred, life must have survived in some protected niche, perhaps within (not too hot) hydrothermal vents, to flourish on the surface once again as soon as conditions were amenable (Sleep and Zahnle, 1998). This must have happened very rapidly because by 3.47 b.y. ago, if not earlier, microbial mats were widespread in any suitable, shallow water environment. Alternatively, the bolide spike, although severe, may not have been devastating (Ryder et al., 2000), thus life could have had almost one billion years to develop before producing the Early Archaean microbial mats of the Barberton and Pilbara greenstone belts which are so similar to modern microbial mats.

3. EARLY MARS

Conditions on early Mars were "warm and wet" compared to modern Mars and, as on Earth, the early atmosphere was mostly CO_2, perhaps with another greenhouse gas component (Carr, 1999). However, the term "warm and wet" refers to the fact that atmospheric pressures and temperatures were high enough to sustain liquid water at the surface of the planet (Carr, 1996). It has been suggested that the early thicker atmosphere was maintained by the volcanism that produced the Tharsis bulge and that the cessation of the major period of volcanism at the end of the Noachian was one of the prime causes for a decrease in the atmospheric pressure (Phillips et al, 2001). Although there is evidence for significant water at the surface of Mars during its early history (Noachian period), there is no evidence for large-scale standing bodies of water: the so-called northern ocean in the Northern Lowland Plains was probably at most the size of the Caspian Sea (Jakosky and Phillips, 2001). However, there is plentiful evidence for flowing water and for lakes in impact basins (Cabrol and Grin, 1999). Although Mars lost its surface volatiles and cooled down at the end of the Noachian, there are geomorphological indications of a substantial cryosphere below the surface (Clifford, 1993). In fact, evidence for continued, intermittent presence of water at the surface has been spectacularly revealed through the recent Mars Orbiter Camera (MOC) images (Malin and Edgett, 2000).

Thus, one of the major requirements for the appearance of life (liquid water) was met on early Mars. The other requirements, namely organics and an energy source were also present, as they were on primordial Earth: the organics originated from within the material of the planetesimals which consolidated to form the planets, and a large portion was delivered though cometary, meteorite and interstellar dust particles (IDPs) (Chyba et al., 1990). Potential energy sources include heat (hydrothermal activity), sunlight and chemical reactions (Jakosky and Shock, 1998).

However, there were important differences in conditions at the surfaces of the early Mars and early Earth. As we have seen above, the young Earth was apparently covered by water. It had an unknown quantity of mostly submerged protocontinents providing shallow water to partly subaerial habitats, adjacent to and in the vicinity of hydrothermal vent openings. In these environments microbial mats thrived. Early Mars was not covered with water and it did not have protocontinents providing shallow water habitats. However, it is possible that there were even more shallow water locations with hydrothermal activity on early Mars in the form of numerous water-filled impact craters (Cabrol and Grin, 1999), than on the early Earth. Early martian biota would have had to contend with harmful UV radiation, as did the terrestrial biofilms (Cockell, 2000). In a similar fashion, the structure and composition of potential, well-developed, martian biofilms and microbial mats would have helped them to survive these relatively extreme conditions.

4. MICROBIAL BIOFILMS AND MATS IN THE 3.3-3.5 B.Y.-OLD ONVERWACHT GROUP, BARBERTON GREENSTONE BELT, SOUTH AFRICA

4.1 Methods

Selected sedimentological samples showing evidence of fine layering were thin sectioned for petrological study. They were then delicately etched in the fumes of HF, along with fresh rock chips (Westall and Gerneke, 1998; Westall et al., 2001). The etched rock chips and thin sections, as well as unetched rock chips, were examined with scanning electron microscopes (Westall and Gerneke, 1998; Westall, 1999; Westall et al., 2001). The microscopes used included a Philips 515 with low-resolution EDX attachment (Univ. Bologna), a Leo Steroscan 440 with EDX attachment at the Univ. Cape Town, a Philips FEGSEM XL40 with light element detector at the Johnson Space Center, Houston, and a JEOL JSM-840 with EDX at Louisiana State Univ.

4.2 Microbial mats

Microbial mats from the sedimentary cherts of the Onverwacht Group display exceptional preservation and contain additional information pertaining to the immediate environment of formation. We will describe two examples of superbly preserved microbial mats, both formed on the bedding planes of finely laminated, silicified, volcaniclastic sediments (Figs. 1, 2).

4.2.1 Sample SA1 (Kromberg Formation, 3.3-3.42 b.y. old)

The sediment surface of this sample is coated with a texturally and compositionally complex, cracked film, ranging in thickness from 1-5 μm (Fig. 1d). The texture of the film ranges from fine to coarsely granular (0.1-20 μm), or smooth (Figs. 1e, h). A number of precipitated minerals are embedded in its surface. They include isolated and aggregated lozenge to fan-shaped crystals, ranging from 1-10 μm in size (Fig. 1c), and aggregates of rectangular crystals (Fig. 1e). From their morphology, these crystals resemble gypsum and calcite, respectively (G. Vai, pers. comm. 2001). Other crystals with a rectangular habit sometimes display a stepped depression on their surfaces (Fig. 1f). This morphology, known as "hopper", is typical of evaporite minerals, such as halite or sylvite. Rare, acicular, euhedral crystals with the morphology of aragonite are embedded in, and coated by, smooth surfaced film (Fig. 1g). In one, smooth-surfaced area of the film there are numerous oval to slightly irregularly shaped “bubble”-like structures, 3-5 μm in diameter, consisting of raised rims around hollow centers (Fig. 1h).

The association of both calcite and aragonite in the same sedimentary layer is somewhat unusual since aragonite is metastable in normal oceanic conditions and converts to calcite. It also forms in modern hot spring environments where it can co-exist with calcite (Folk, 1993). The presence of aragonite in the early Archaean sediments is not surprising; according to a recent re-evaluation of the oxygen isotopes from early Archaean cherts, the ocean waters of that period were considerably warmer (55-85°C) than modern oceans.

No microbial fossils have yet been identified in this film. Could it have formed abiotically? Although films of abiotic organic molecules can form on mineral surfaces, they are simple layers, which do not exhibit the complexity of structure or composition found in a microbial mat (Westall et al., 2000). In terms of complexity of texture, structure and composition, and based on its similarities to modern, microbial films, this film coating on a surface of the silicified volcaniclastic sediments from the Kromberg Formation is interpreted as representing a fossilised microbial mat. The characteristics linking this film to microbial mats include: (1) the spatial location of the film parallel to the bedding laminae of the chert sample (Figs. 1a, b, d), which suggests that it formed on a bedding plane, (2) its thick, robust structure (Fig. 1d), (3) its mixture of smooth to granular texture (Figs. 1e, h) similar to patchy microcrystalline precipitations in modern polymeric microbial films (Westall et al., 2000), (4) the embedding of precipitated minerals within the film and on its surface (Figs. 1c, e-g), and (5) the plastic texture of the "bubble" area of the mat which imitates the comportment of organic material (Fig. 1h). Furthermore, thin section study of this chert sample documents fine, undulating laminae having a brown/black colour (Fig. 1b), typical of carbonaceous biofilms (Walsh, 1992, Walsh and Lowe, 1998; Walsh and Westall, this volume). Moreover, other, similar films in this chert sample and in additional cherts from the Onverwacht Group, analysed using a light element EDX detector attached to a FESEM, contain carbon (Fig. 1i) (this particular sample preparation was not available for analysis with a light element EDX detector). Although microfossils were not found directly in association with the film, microfossils do occur in other parts of this sample. Filamentous structures about 1 μm in width and some tens of micrometers in

Figure 1. Microbial biofilms and mats in Sample SA1 (Kromberg Formation, Onverwacht Group, Barberton greenstone belt, South Africa, 3.3-3.42 b.y.-old). (a) Hand specimen view of finely laminated, black chert (silicified, fine volcaniclastic sediment). Arrow is parallel to the bedding planes. (b) Thin section view of finely undulating, carbonaceous laminate. (c) Gypsum (?) crystals associated with microbial mat in (d). (d) General view of the thick, cracked microbial mat on the bedding surface. (e) Aggregate of rhombic crystals (calcite?) embedded in the microbial mat. (f) Evaporite (halite?) hopper crystal embedded in the mat. (g) Euhedral aragonite crystal. (h) Smooth-surfaced area of the mat containing "bubble" structures of equivocal origin. (i) Cross section across carbonaceous mat layer. Three layers consisting of granular carbonaceous matter can be seen running from lower left to upper right. (j) Granular carbonaceous filament which possibly represents the remnants of a filamentous (carbonaceous filaments were seen in thin section). (k) Silicified chains of probable coccoidal bacteria in an HF-etched thin section (the spikes represent the C-axis of quartz). In the center, three coccoidal forms are encased in a common, rugged-surfaced envelope. In the upper left and upper right are coccoidal structures (S) which have been cut in half by the surface of the thin section (thus demonstrating that these structures are indigenous to the chert). A smaller ball attached by twisted material (arrow) to the group of sectioned coccoids in the upper right may represent lysed cytoplasm. (l) Small, oval coccoid structures embedded in a smooth film: possible silicified coccoidal bacteria?

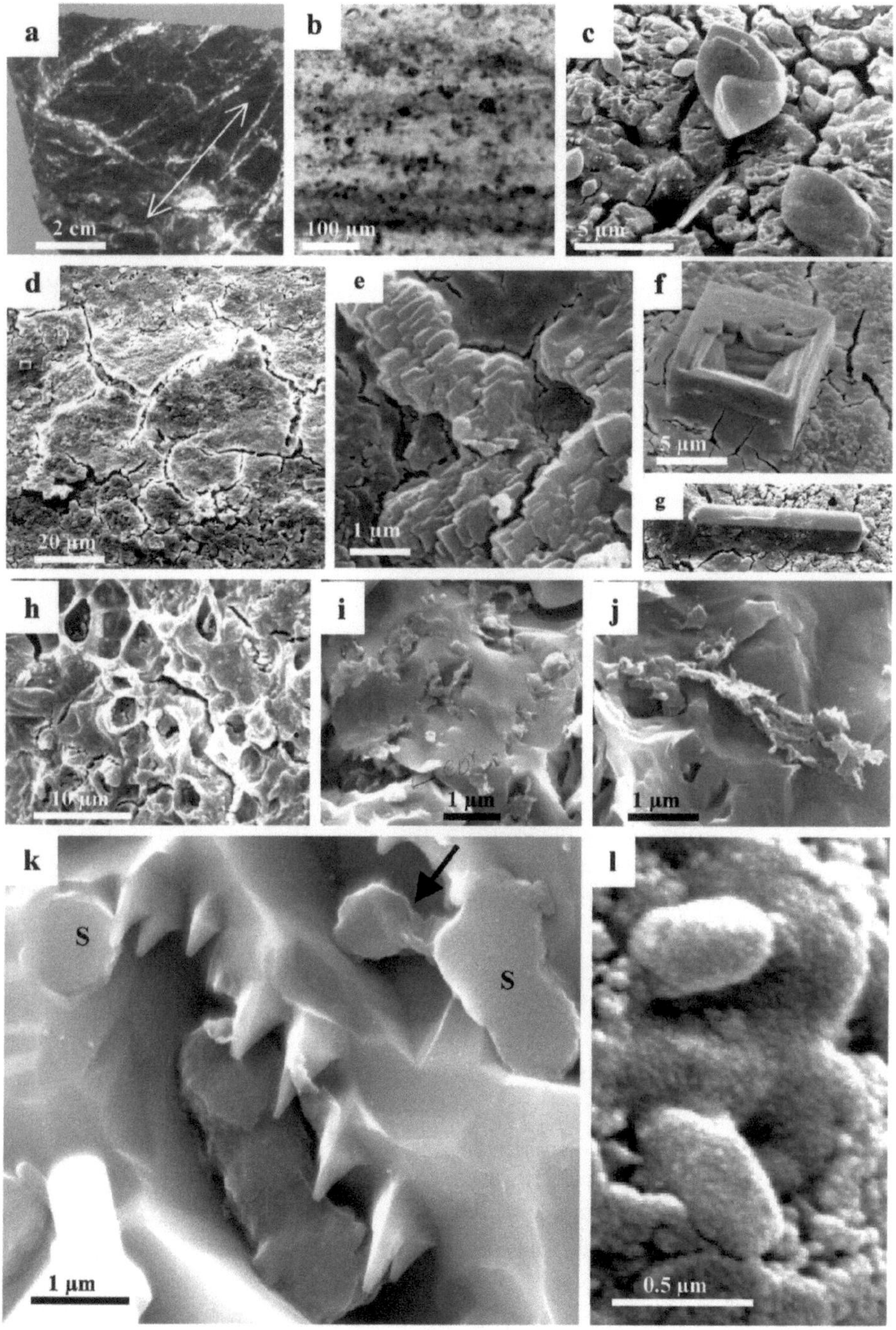
a
2 cm
b
100 µm
c
5 µm
d
20 µm
e
1 µm
f
5 µm
g
h
10 µm
i
1 µm
j
1 µm
k
S
S
1 µm
l
0.5 µm

length, resembling the carbonaceous filaments described by Walsh (1992) and Walsh and Westall (this volume), were observed in thin section. Moreover, strands of granulated carbonaceous matter observed with the FESEM (and light element EDX detector) may represent the degraded remnants of such filaments (Fig. 1j). Other, probable microfossils include small (0.6-1 μm in size), oval to spherical structures, interpreted as representing silicified, coccoidal bacteria (Figs. 1k, l) (Westall et al., 2001).

The "bubble" structures (Fig. 1h) are intriguing. They were originally interpreted as possibly representing fossilised gas bubbles (bacterial colonies in a biofilm produce large amounts of gas as a by-product which can form pockets in the polymer film, or escape through "volcano"-like openings in the film, Westall et al., 2000). There are two other possible explanations for these structures. They may represent the moulds of minerals embedded in the film, which were subsequently removed by dissolution. It is curious, though, that other embedded minerals have resisted dissolution. Moreover, the oval to slightly irregular (but rounded) shapes of the structures argues against their being crystal moulds. Another explanation is that the "bubbles" represent the moulds of microorganisms or small microcolonies of microorganisms, which had lysed and disappeared before the silicification of the film. Krumbein and Stal (1991) noted that the microorganisms that create microbial mats are often not preserved.

One important aspect of this film concerns its cracked appearance, which indicates dehydration of an originally hydrated structure. This is an additional reason for hypothesising that the original structure was a polymer film, since polymers consist of about 90% water (Decho, 1994). Dehydration could have occurred either during the mineralisation process and/or during subaerial exposure (*n.b.* these are very shallow water deposits). Evidence for subaerial exposure is provided by the assemblage of minerals precipitated onto, and within, the film. In particular, the presence of evaporite hoppers is indicative of an evaporitic environment. Furthermore, the edges of the cracked portions of the film curl slightly upwards, a phenomenon suggestive of subaerial dessication rather than mineralogical dehydration.

A biogenic origin for this film appears to be the most likely explanation of all the characteristics listed above. However, is it possible that this film could have been introduced into the cherts at a later date, *i.e.*, that it is not syngenous with the formation of the chert? Endolithic biofilms are known to form in fissures in rocks (Golubic and Schneider, this volume; Westall and Folk, 2003). A number of lines of evidence, however, indicate that this

Folk, 2003). A number of lines of evidence, however, indicate that this biofilm is most likely to be indigenous: (1) the film formed on a sedimentary bedding plane surface that is coherent with adjacent bedding planes in the rock; (2) the film was formed in an evaporite environment whichdoes not exist in the Barberton greenstone belt, neither at present, nor in the recent past; (3) aragonite could not have formed under present environmental conditions; and (4) the outer surface of the exposed rock was minutely examined in order to detect the presence of a film with similar characteristics. Although the rock sported a weathering rind, it contained no halite, calcite, gypsum and aragonite mineral assemblage and had a very different texture and structure.

Thus, we conclude that this film most likely represents a microbial mat formed at the surface of Early Archaean volcaniclastic sediments. These sediments had been subaerially exposed in evaporitic conditions, provoking the precipitation of a calcite-aragonite-gypsum-evaporite hopper (halite?) suite from hot fluids (either from a hot ocean or from hydrothermal fluids). We have no information about the original microorganisms which created the microbial mat. They appeared to have lysed and disappeared before silicification of the mat, although one explanation for the "bubble" structures is that they could represent the moulds of coccoidal microorganisms of small colonies of microorganisms. Microfossils found in other parts of the sample include 1 μm wide filaments (tens of micrometers in length) and 0.6-1 μm diameter oval to spherical coccoids.

4.2.2 Sample SA5 (Hooggenoeg Formation, 3.45-3.72 b.y. old)

Thin sections of this finely laminated chert sample (Figs. 2a, b) show brown/black, wavy laminations, similar to the carbonaceous films documented by Walsh (1992), Walsh and Lowe (1998) and Walsh and Westall (this volume). Part of the surface of this chert sample is coated with a torn, ropy textured film (Fig. 2c). The film is thick (up to 10 μm) and consists of a ropy, mostly smooth surfaced material, which is finely granular in places (Figs. 2d-f). A number of minerals are embedded in the film including: (1) acicular crystals1-2 μm in length), similar to the aragonite crystals in the previous example; (2) larger, irregularly shaped, 3-5 μm sized crystals of an Mg-aluminosilicate (Fig. 2d, e); and (3) a few isolated, rectangular crystals similar to the evaporite hopper crystals in the previous

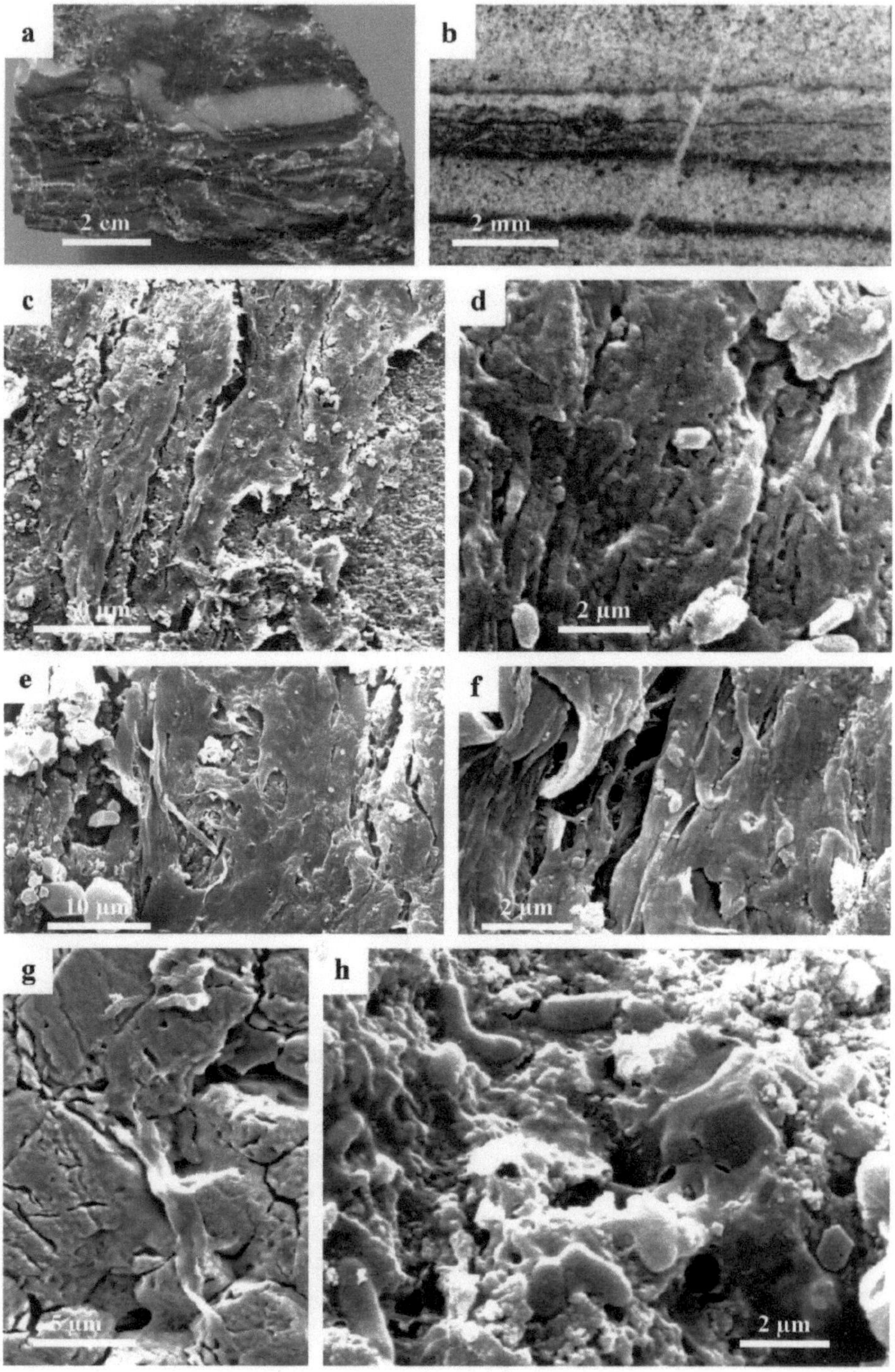
a
2 cm
b
2 mm
c
10 µm
d
2 µm
e
10 µm
f
2 µm
g
5 µm
h
2 µm

example. In many places tiny acicular crystals of <0.1 μm (probably aragonite) coat the surface of the film or are embedded in it. EDS analyses suggest that the film was probably carbonaceous in composition although a pervasive silica background indicates that the whole film has been silicified.

The most interesting characteristic of this film is its ropy texture. Although the fibers of the film generally exhibit a pronounced uniformity of orientation (Fig. 2d), many of them interweave with or cross each other (Fig. 2f). The individual fibers or making up the film actually consist of a filamentous core, 0.2 μm in width, coated with a thick film (Fig. 2d). It is impossible to know the total length of the filaments because the thick film coating masks them, but in some places they can be followed for about 20 μm before merging completely with the overlying film.

The mainly smooth surface of this film, with its partly torn, twisted, and partly interweaving ropy texture, strongly resembles a film of organic material. The occult filaments embedded in the film are similar to filamentous bacteria in morphology. Together, the presence of the filamentous bacteriomorph fibers, the carbonaceous composition of the film, and the complex texture of the film, suggests that this is a microbially produced biofilm.

Figure 2. Microbial biofilms and mats in Sample SA5 (Hooggenoeg Formation, Onverwacht Group, Barberton greenstone belt, South Africa, 3.45-3.72 b.y.-old). (a) Hand specimen view of black and green/white laminated chert (silicified volcaniclastic sediments). (b) Thin section view of finely undulating carbonaceous layers. (c) General view of a ragged-looking, thick film on the surface of bedding plane. (d) Detail of the film showing its "ropey" texture. Note the exposed filament (center right). Other occult polymer-coated filaments occur in the film. Note also the parallel orientation of the filaments. (e) Torn, smooth-surfaced area of the film with embedded crystals (calcite?). (f) Interwoven fibrous texture. (g) HF-etched thin section of carbonaceous filament, possibly similar to those making up the microbial mat. (h) Another area of bedding plane coated by a granular to smooth film. In the film are embedded 2-3.8 μm long, 1 μm wide, curved rod-shaped structures, interpreted as fossil vibroid-shaped bacteria. Some of the rods appear to be undergoing cell division (center right).

Two facts concerning the state of preservation of this microbial mat are important. In the first place the textural details of the microbial mat are superbly preserved by silicification. Second, the microbial mat shows evidence of rough treatment before being silicified: the mat as a whole has a patchy, torn aspect (Fig. 2, e). This textural state, coupled with the fact that the whole mat and the filaments forming it exhibit a marked, linear orientation, suggests that the mat formed under conditions of relatively strong current flow. In modern environments where there is relatively strong current flow, the microorganisms develop thick polymer layers as protection (Westall and Rincé, 1994). The dessicated aspect of this film indicates that it has been (subaerially?) dehydrated. The presence of some rectangular crystals similar to the evaporite hoppers in Sample SA1 points again to the possibility of subaerial, evaporitic conditions.

Another film on the surface of this bedded chert is represented by a finely granular (0.1-0-2 μm) deposit. One part of this film consists of a smooth-surface in which are embedded curved rod-shaped structures ranging in length from 2-3.8 μm and about 1 μm in width (Fig. 2h). The rods have rounded cross sections, and rounded terminations. Some of the curved rods are joined to each other at their terminations. All of these characteristics suggest that they probably represent vibroid-shaped bacteria, some obviously in the act of cell division. EDX analysis indicates that the organisms and film are silicified, although the finely granular deposit was probably calcium carbonate. Again, in this case, the association of probable microbial fossils with the bedding plane film, as well as the smooth to granular texture of the film, strongly argues for an interpretation as a microbial biofilm.

The arguments that these films represent indigenous microbial mats and biofilms on an Early Archaean bedding plane, as opposed to endolithic films, are similar to those put forward for the film on Sample SA1. In addition, neither the pronounced orientation of the ropey-textured film, indicating the influence of relatively strong current flow, nor the presence of dessication cracks and evaporite hoppers, could not be explained by the growth of an endolithic biofilm in a crack in the chert.

5. CONCLUSIONS

The cherts of the Early Archaean Onverwacht Group of the Barberton Greenstone Belt, South Africa, document the common formation of microbial mats in shallow water to intermittently subaerial habitats. The record of evaporitic conditions demonstrates that microbial life was already capable of surviving exposed conditions in the Early Archaean atmosphere (little or no O_2, UV radiation, impact events). In addition, the microbial biofilms and mats unexpectedly show themselves to be recorders of current flow direction.

Given the apparent robustness of life in Early Archaean conditions, how does this compare with the possibility of life on early Mars? Although the early geological evolution of Mars and Earth differed significantly, there were probably more shallow water habitats on Mars that were conducive to life (if life managed to evolve from prebiotic molecules), at least until the loss of surface volatiles and the freezing of the planet between about 3.8-3.5 b.y. ago. Numerous lakes in impact and crater basins and along the margins of other standing bodies of water (the Northern Sea, for instance) would have provided suitable habitats for potential martian life. As on Earth, thick, robust biofilms or mats would have protected the microbial colonies from dehydration and UV radiation.

In terms of the search for life on Mars, shallow water localities close to ancient hydrothermal activity would have provided habitats and the possibility of silicification and preservation of potential mats. The Early Archaean biofilms therefore represent ideal analogues for potential martian biofilms. It is not the aim of this paper to discuss extant martian life, but the possibility of reanimation of cryofrozen cells or spores in the frozen subsurface aquifers through intermittent hydrothermal warming, due either to impacts or volcanism, emphasises the fact that potential martian microfossils and biofilms could be any age and are not only restricted to the early period of Mars' history (Noachian period) (c.f. Gibson et al., 2001).

6. ACKNOWLEDGEMENTS

F.W. acknowledges kind support from M. de Wit, J. Högbom, the NRC and the NAI (through D. McKay) for part of this research. MW gratefully acknowledges support from the Louisiana State University Council for Research for a summer stipend and Louisiana Space Consortium and the Board of Regents Support fund, and the NASA Space Grant Program under NGT5-40035.

REFERENCES

Awramik, S.M., Schopf, J.W., and Walter, M.R. (1983) Filamentous fossil bacteria from the Archaean of Western Australia. *Precambrian Res.,* 20, 357-374.

Byerly, G., Lowe, D.R., and Walsh, M.M. (1986) Stromatolites from the 3300-3500 Myr Swaziland Supergroup, Barberton Mountain Land, South Africa. *Nature*, 319, 489-491.

Byerly, G.R. and Palmer, M.R. (1991) Tourmaline mineralisation in the Barberton greenstone belt, South Africa: Early Archaean metasomatism by evaporite-derived boron. *Contrib. Minerol. Petrol.,* 107, 387-402.

Cabrol, N.A. and Grin, E.A. (1999) Distribution, classification, and ages of martian impact crater lakes. *Icarus*, 142, 160-172.

Canfield, D., Habicht, K.S., and Thamdrup, B. (2000) The Archaean sulfur cycle and the early history of atmospheric oxygen. *Science*, 288, 658-661.

Carr, M.H. (1996) *Water on Mars*. Oxford, N.Y.

Carr, M.H. (1999) Retention of an atmosphere on early Mars. *J. Geophys. Res.*, 104, 21,897-21,909.

Chyba, C.F., Thomas, P.J., Brookshaw, L., and Sagan, C. (1990) Cometary delivery of organic molecules to early Earth. *Science*, 249, 366-373.

Clifford, S.M. (1993) A model for the hydrologic and climatic behaviour of water on Mars. *J. Geophys. Res.*, 98, 10,973-11,016.

Cockell, C.S., 2000. The ultraviolet history of the terrestrial planets - implications for biological evolution. *Planetary Space Sci.,* 48: 203-214.

Decho, A. (1994) Molecular-scale events influencing the macro-scale cohesiveness of exoplymers. In *Biostabilisation of Sediments* (ed. W.E. Krumbein, Patterson, D., and Stal, L.-J.), BIS, Oldenburg, pp.135-149.

De Ronde, C.E.J., de Wit, M.J., and Spooner, E.T.C. (1994) Early Archaean (>3.2 Ga) iron-oxide-rich, hydrothermal discharge vents in the Barberton greenstone belt, South Africa. *Geol. Soc. Amer. Bull.*, 106, 86-104.

De Ronde, C.E.J., DeR. Channer, D.M., Faure, K., Bray, C.J., and Spooner, E.T.C. (1997) Fluid chemistry of Archaean seafloor hydrothermal vents : Implications for the composition of circa 3.2 Ga seawater. *Geochim. Cosmochim. Acta*, 61, 4025-4042.

De Wit, M.J., Hart, R.A., Martin, A., and Abbott, P. (1982) Archaean abiogenic and probable biogenic structures associated with mineralised hydrothermal vent systems and regional metasomatism with implications for greenstone belt studies. *Econ. Geol.*, 77, 1783-1802.

Folk, R.L. (1993) SEM imaging of bacteria and nannobacteria in carbonate sediments and rocks. *J. Sedimentary Petrol.,* 63, 990-999.

Gibson, E.K., McKay, D.S., Thomas-Keprta, K., Wentworth, S.J., Westall., F., Steele, A., Romanek, C.S., Bell, M.S., and Toporski, J .(2001) Life on Mars: evaluation of the evidence within Martian meteorites ALH84001, Nakhla, and Shergotty. *Precambrian Res.* 106, 13-32.

Geesey, G.G. and Jang, L. (1989) Interactions between metal ions and capsular polymers. In *Metal Ions and Bacteria* (ed. T.J. Beveridge and D.J. Doyle) John Wiley, N.Y., 325-358.

Golubic, S. and Schneider, J. (2001) Microbial endoliths as internal biofilms. In *Fossil and Recent Biofilms* (ed. W.E. Krumbein, T. Dornieden, and M. Volkmann), Kluwer, Amsterdam, this volume.

Gorbuschina, A.A. and Krumbein, W.E. (2000) Subaerial microbial mats and their effects on soil and rock. In *Microbial Sediments* (ed. R.E. Riding and S.M. Awramik), Springer, Berlin, pp. 161-170.

Grotzinger, J.P. and Kasting, J.F. (1993) New constraints on Precambrian ocean composition. *J. Geology*, 101, 235-243.

Hofmann, H.J., Grey, K., Hickman, A.H., Thorpe, R.I. (1999) Origin of 3.45 Ga coniform stromatolites in the Warrawoona Group, Western Australia. *Geol. Soc. Amer. Bull.*, 111: 1256-1262.

Holland, H. D. (1984) *The chemical evolution of the atmosphere and the oceans.* Princeton, Princeton.

Jakosky, B.M. and Shock, E.L. (1998) The biological potential of Mars, the early Earth, and Europa. *J. Geophys. Res.*, 103, 19,359-19,364.

Jakosky, B.M. and Phillips, R.J. (2001) Mars volatile and climate history : Water the observational constraints? *Nature*, in press.

Kasting, J.F. (1993). Earth's early atmosphere. *Science*, 259: 920-926.

Kempe. S., and Degens, E.T. (1985) An early soda ocean ?. *Chem. Geol.*, 53, 95-108.

Knauth, L.P. and Lowe, D.R. (2003) High Archaean climatic temperature inferred from oxygen isotope geochemistry of cherts in the 3.5 Ga Swaziland Supergroup, South Africa. *GSA Bulletin*, 115, 566-580.

Krumbein, W.E. (1983) Stromatolites – the challenge of a term in space and time. *Precambrian Res.*, 20, 493-531.

Krumbein, W.E. and Stahl, L.-J. (1991) The geophysiology of marine cyanobacterial mats and biofilms. *Kieler Meeresforsch.*, 8, 158-163.

Maher, K.A. and Stevenson, D.J. (1988) Impact frustration of the origin of life. *Nature*, 331, 612-614.

Malin, M.C. and Edgett, K.S. (2000) Evidence for recent ground water seepage and surface runoff on Mars. *Science*, 288, 2330-2335.

Mojzsis, S.J., Arrhenius, G., Keegan, K.D., Harrison, T.H., Nutman, A.P. and Friend, C.L.R., 1996. Evidence for life on earth 3,800 million years ago. *Nature*, 384, 55-58.

Nijman, W., K.H. de Bruijne, M. Valkering, (1998) Growth fault control of Early Archaean cherts, barite mounds and chert-barite veins, North Pole Dome, Eastern Pilbara, Western Australia, *Precambrian Res.*, 88, 25-52.

Nijman, W., Willigers, B.J.A., and Krikke, A. (1998) Tensile compressive growth structures: relationships between sedimentation, deformation, and granite intrusion in the

Archaean Coppin Gap greenstone belt, Eastern Pilbara, Western Australia. *Precambrian Res.*, 88, 83-108.

Ohmoto, H., Kakegawa, T. and Lowe, D.R. (1993) 3.4-billion-year-old biogenic pyrites from Barberton, South Africa *Science*, 262, 555-558.

Pavlov, A.A., Kasting, J.F., Brown, L.L., Rages, K.A., and Freedman, R. (2000) Greenhouse warming by CH4 in the atmosphere of the early Earth. *J. Geophys. Res.*, 105, 11,981-11,990.

Phillips, R.J., Zuber, M.T., Solomon, S.C., Golombek, M.P., Jakosky, B.M., Banerdt, W.B., Smith, D.E., Williams, R.M.E., Hynek, B.M., Aharonson, O., and Hauck, S.A. (2001) Ancient geodynamics and global-scale hydrology on Mars. *Science*, in press.

Rosing, M.T (1999) 13C depleted carbon microparticles in >3700-Ma seafloor sedimentary rocks from Westa Greenland. *Science*, 283, 674-676.

Ryder, G., Koeberl, C., and Mojzsis, S.J. (2000) Heavy bombradment of the Earth at ~3.85 Ga: The search for petrographic and geochemical evidence. In *Origins of the Earth and Moon*

Schidlowski, M. (1988) A 3,800-million-year isotopic record of life from carbon in sedimentary rocks. *Nature*, 333, 313-335.

Sleep, N.H. and Zahnle, K.I (1998) Refugia from asteroid impacts on early Mars and early Earth. *J. Geophys. Res.*, 103, 28,529-28,544.

Sleep, N.H. and Zahnle, K. (2001) Carbon dioxide cycling and implications for climate on ancient Earth. *J. Geophys. Res.*, 106, 1373-1399.

Toporski J., Westall F., Steele A., Thomas-Keprta, K., and McKay D.S. (2001a). The simulated silicification of bacteria – new clues to the modes and timing of bacterial preservation and implications for the search for extraterrestrial microfossils. *Astrobiology*, in press.

Toporski J., Steele A., Westall F., Avci R., Martill M., and McKay D.S., (2001b). Morphological and spectral investigation of exceptionally well preserved bacterial biofilms from the Oligocene Enspel formation, Germany. *Geochim. Cosmochim. Acta*, in press.,

Toporski J., Steele A., McKay D.S., and Westall F. (2001c) Bacterial biofilms in Astrobiology: the importance of life detection. In *Fossil and Recent Biofilms* (ed. W.E. Krumbein, T. Dornieden, and M. Volkmann), Kluwer, Amsterdam, this volume.

Walsh, M.M., 1992. Microfossils and possible microfossils from the Early Archean Onverwacht Group, Barberton Mountain Land, South Africa. *Precambrian Res.*, 54, 271-293.

Walsh, M.M. and Lowe, D.R. (1985) Filamentous microfossils from the 3,500 Myr-old Onverwacht Group, Barberton Mountain Land, South Africa. *Nature*, 314: 530-532.

Walsh, M.M. and Lowe, D.R. (1998) Modes of accumulation of carbonaceous matter in the early Archaean: A petrographic and geochemical study of the carbonaceous cherts of the Swaziland Supergroup, in *Geologic evolution of the Barberton Greenstone Belt, South Africa*, edited by D.R. Lowe and G.R. Byerly, Geol. Soc. Am. Sp. Paper 329, pp. 115-132.

Walsh, M.M. and Westall, F. (2001) Archean biofilms preserved in the 3.2-3.6 Ga Swaziland Supergroup, South Africa. In *Fossil and Recent Biofilms* (ed. W.E. Krumbein, T. Dornieden, and M. Volkmann), Kluwer, Amsterdam, this volume.

Westall, F. (1999). The nature of fossil bacteria. *J. Geophys. Res.*, 104: 16,437-16,451.

Westall, F. (2000) . A high resolution SEM investigation for fossil life in cherts and BIFs from the >3.7 b.y.-old Isua supracrustals. *Amer. Geophys. Union, San Francisco, Dec. 2000*, (Abst.).

Westall, F. and Rincé, Y. (1994). Biofilms, microbial mats and microbe-particle interactions: electron microscope observations from diatomaceous sediments. *Sedimentology*, 41, 147-162.

Westall, F., Boni. L., and Guerzoni, M.E. (1995) The experimental silicification of microorganisms. *Palaeontol.*, 38, 495-528.

Westall, F. and Gerneke, D. (1998) Electron microscope methods in the search for the earliest forms of life on earth (in 3.5-3.3 Ga cherts from the Barberton greenstone belt, South Africa): applications for extraterrestrial life studies. *Proc. SPIE, Intl. Soc. Opt. Eng.*, 3114, 158-169.

Westall, F., Steele, A., Toporski, J., Walsh, M., Allen, C., Guidry, S., McKay, D., Gibson, E., and Chafetz, H. (2000). Polymeric substances and biofilms as biomarkers in terrestrial materials: Implications for extraterrestrial materials. *J. Geophys. Res.*, 105, 24,511-24,527.

Westall, F., de Wit, M.J., Dann, J., de Ronde, C.E.J., Gaast, S.J. and Gerneke, D. (2001) Early Archaean fossil bacteria and biofilms in in hydrothermally-influenced, shallow water sediments, Barberton greenstone belt, South Africa. *Precambrian Res.*, 106, 93-116.

Westall, F. and Folk, R.L. (2003) Exogenous carbonaceous microstructures in Early Archaean cherts and BIFs from the Isua greenstone belt: Implications for the search for life in ancient rocks. *Precambrian Res.*, in press.

Wilde, S.A., Valley, J.W., Peck, W.H., and Graham, C.M. (2001) Evidence from detrital zircons for the existence of continental crust and oceans on the Earth 4.4. Gyr ago. *Nature*, 409, 175-178.

Some aspects of studying of fossil mats and biofilms

A. Yu. Rozanov, E. A. Zhegallo, G.T.Ushatinskaya
Paleontological Institute
Russian Academy of Sciences
123, Profsoyuznaya str.
117 647 Moscow
Russia

1. Many years of studying of fossil mats and biofilms have resulted us to conclusion that almost always, when these formations were present in basins, they could be well preserved also in rocks. Thus necessary to mean, that preservation of fossilized bacteria rather different depending on a composition of rocks. The best preservation of microbes is watched in cherts and phosphorites, the preservation in carbonates is much worse.

2. The study of the mineralization of recent cyanobacteria and their morphology was made in laboratory conditions. A galophilic cyanobacterium *Microcoleus chthonoplastes* from the hypersaline lagoons of lake Sivash was chosen as a test object in studying phosphatization.

Preliminary experiments showed that although in natural waters phosphate in low concentrations, *Microcoleus* is able to scavenge traces of usable phosphate sources from the surrounding medium. The study of $\mathbf{P}_i$ (inorganic phosphorus) assimilation by cyanobacteria in laboratory showed that $\mathbf{P}_i$ concentration in cells varied and changed during cell growth (fig. 1). On sufficient phosphate concentration the highest amount of phosphorus assimilated by *Microcoleus* achieved 2% on dry weigh. $\mathbf{P}_i$ uptake rates were much higher at first hours of the growth. With increasing phosphorus content in the medium higher then needed for normal growth of cyanobacteria (36 mg **P**/l), mineralized sheaths were formed around each trichome, through the culture remains alive and the control over the phosphorus movements in the system "cell-medium" is kept constant. Mineralized sheaths are the result of two steps: 1. Formation of mucilaginous sheaths; 2 – trapping and precipitation of phosphate particles on the mucilaginous sheaths. When the concentration increases over 70 mg **P**/l the culture dies and spontaneous phosphatization of the trichomes themselves begins.

The mineral received is a poorly crystallized apatite resembling francolite.

This pattern is very similar to the phosphorus nodules from the Khusugul phosphorite Deposit.

Large rates of phosphorus absorption in a first few minutes explain the good preservation of morphological structures of cyanobacteria in ancient phosphorites. This indicate an exceedingly rapid phosphatization that occurs faster than the cells decay.

3. One of the best model objects for studying of fossilized mats and biofilms are the ancient phosphorites. Studying of them allows not only to see microorganisms perfectly preserved, but also to watch several types of preservation.

Long time Lower Cambrian (Tommotian) phosphorites from Northern Mongolia were considered as a classic example of ancient chemogenic phosphatic accumulation. Our researches (Rozanov, Zhegallo, 1989; Zhegallo, Rozanov, Ushatinskaya et al, 2000) have shown, that all phosphorites from this largest deposit constructed by nodules, consisting of cyanobacterial assemblages (fig. 2,3). The morphological features and sizes allow to consider, that the filaments are most similar to modern cyanobacteria *Microcoleus*. Present coccoid and dumb-bell shaped forms, depending on the sizes, represents either coccoid of cyanobacteria (up to 10 mkm in diameter) or with smaller sizes (n100 nm) – purple bacteria. In many cases it is possible to observe the perfectly preserved as construction of fossil cyanobacterial mat (fig 4).

Magnificent biofilms were watched inside the shells of Cambrian phosphate inarticulate brachiopods, redistribution of phosphorus in which one is connected with activity of any bacteria (Ushatinskaya, 1995) (fig 5).

4. One of the key problem is an identification of abiogenous globuls and of the fossilized coccoid forms of bacteria. From this point of view the huge concern represents studying of schungites, high-carbonaceous rocks from the Lower Proterozoic of Kareliya. Many authors studied on this example patterns of globulization of carbon materials (Sokolov et. al., 1984; Philippov et. al., 1994; Yushkin, 1994). Specially great interest schungites had evoked after detection in them fullerens. The additional studying has enabled to suppose, that from definite dimensional class, the globuls of carbon matter are pseudomorphes on bacteria (fig 6).

It was confirmed also at analysing of phosphorites and other rocks. The difference only in what dimensional class abiogenous globulization is replaced by bacteria.

5. A study of highly carbonaceous rocks (combustible and black shales) from the Early Proterozoic to Kz showed that these rocks were mainly formed in an environment with cyanobacterial benthic communities.

Good examples are schungites from Karelia and black shales of Kursk Magnet Anomaly, shales from Vendian of East European Platform, Cambrian of South Australia and Siberia Platform and many others.

6. Studying of fossil cyanobacterial mats and other microbial assemblages, as well as biofilms (in narrow sense), results us in the very relevant conclusion: probably, in epicontinental basins in past all processes

of sedimentations were connected with activity of bacteria (or more precisely, microorganisms), and no confident that without their participation were formed anything of sedimentary rocks (except for conglomerates and gritstones) (Rozanov, 1999).

This study was supported by the Russian Foundation for Basic Research (projects 02-04-48094, 03-05-64499), N.T. 974.2003.5.

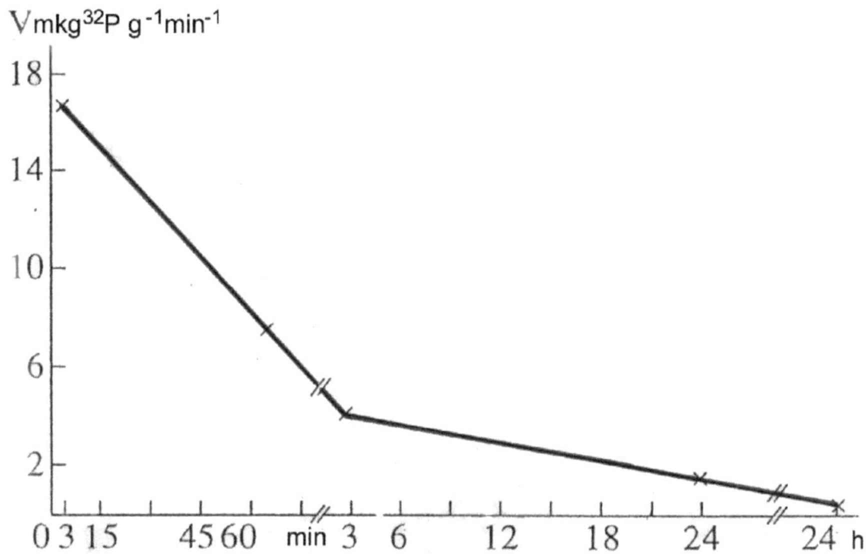

Figure1: The velocity of the accumulation of ^{32}**P** in the *Microcoleus chthonoplastes* cells.

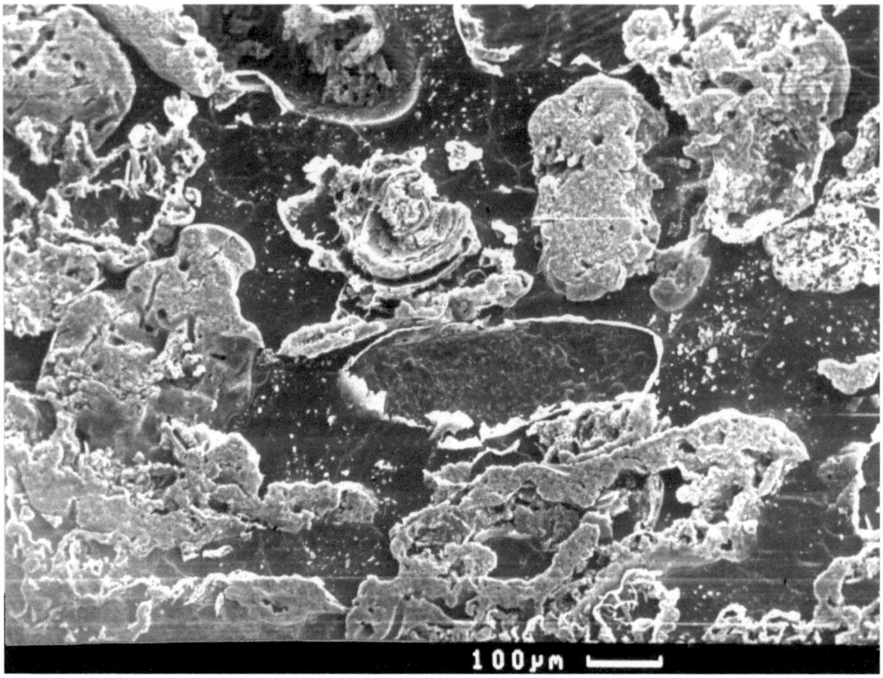

Figure 2: Fragment of Khubsugul phosphorite consisting of several micronodules.

Figure 3: Fragment of cyanobacterial mat from phosphorite.

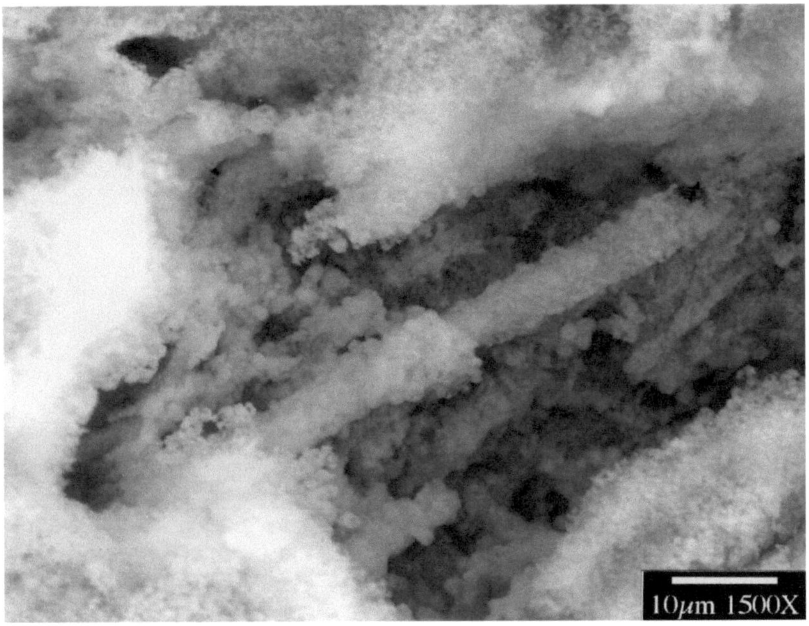

Figure 4: Structure of cyanobacterial mat is seen on 10-15 microns of depth. Concentration of pseudomorphs of purple bactjeria and cyanobacteria filaments, replaced by purple bacteria as well.

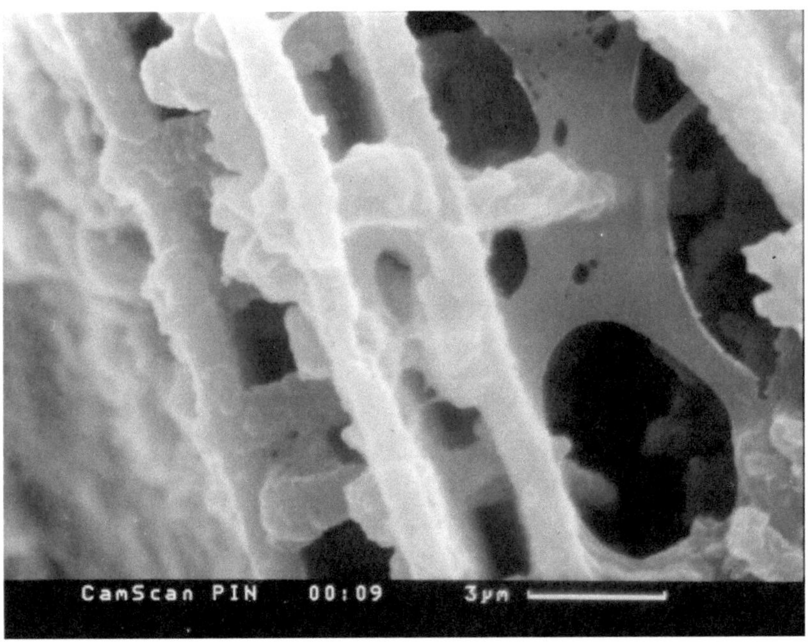

Figure 5: Phosphatized biofilms in the shell of Cambrian brachiopod.

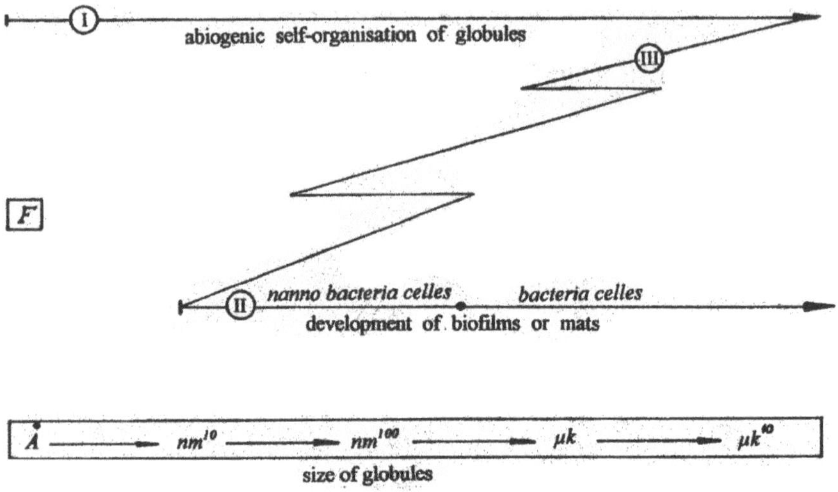

Figure 6: The theoretical curve of transition from abiogenetic to biogenetic globulization depending on composition of studied material.

REFERENCES

Philippov M.M., Golubev A.I., Medvedev P.V. et al. Organic matter from schungite rocks from Kareliya (genesis, evolution, methods of studying)/ Petrozavodsk: Kareliya Scientific Center of RAS, 1994. 208 p.

Rozanov A.Yu. Fossil bacteria and new view on the sedimentation. Soros Educational Journal. 1999, No. 10 (47), p.63-67.

Rozanov A.Yu., Zhegallo E.A. Problem of Origin of the Ancient Phosphorites of Asia. Lithology and Mineral Resources. 1989, N 3, pp.62-82 (in Russian).

Sokolov V.A., Kalinin Yu.K., Dukkiev E.F. (eds.) Shungite is new carbonic raw material. Petrozavodsk: Kareliya, 1984. 182 p. (in Russian)

Ushatinskaya G.T. The earliest lingulates. M.: Nauka. 1995. 91 p.

Yushkin N.P. Globular supermolecular structure of schungite: the data by scanning tunneling microscopy. Doklady Akademii Nauk, 1994, v.337, No.6, pp.800-803 (in Russian).

Zhegallo E.A., Rozanov A.Yu., Ushatinskaya G.T., Hoover R.B., Gerasimenko L.M., Ragozina A.L. Atlas of Microorganisms from Ancient Phosphorites of Khubsugul (Mongolia). Huntsvill, Alabama, USA, NASA/TP, 2000, 167 p.

Index

abiogenic, abiotic 433, 453
acetate, 365
acidophiles, 405
actinomycetes, 244
adhesion, 31, 33
aerobic, 14
Africa, 307
Alcyonidium gelatinosum, 190
algae, 35, 161, 173
algal mat, 4, 8
algobacterial films, 353
alkaline lakes, 147
alkaliphiles, 405
Allende meteorite, 431, 435
alumosilicates, 146
ammonification, 131-133
amoebae, xvii, 162
amphipods, 199, 207
Anabaena, 324, 325
anaerobes, 14-20, 144
anaerobic, 14
anoxia, 144
anoxygenic photosynthesis, 16-21, 85
anoxyphototrophic bacteria, 16-20, 69, 71, 142-145, 147
Antarctica, 407, 439
Anthophysa, 165
antibodies, 438-439
apatite, 286
Aphanizomenon flos-aquae, 156
aquifer, 373
Aquificales, xv
aragonite, 221
Archaea, xiv, 2, 132, 144, 403
Archaean biofilms, 307-313
Archaean, 307f, 449, 450, 457, 459-461
Arenicola marina, 210
artificial microbial mats, 77f, 151f
Aspergillus spp., 435
asteroid impacts, 309
Asterostomatid*ae*, 222
astrobiology, 407-412, 429-440
attachment, 29-33, 166-169
Australia, 10, 448
Azotobacter sp., 145
Bacillus alcaldophilus, 409
Bacillus firmus, 409
Bacillus licheniformis
Bacillus marismortui, 404
Bacillus megaterium, 375
Bacillus pumilis, 435
Bacillus spp., 435
Bacillus subtilis, 435
bacteria, 5, 30, 336, 344, 346-350, 365
bacteria, extraterrestrial, 410
bacteria, field guide, 344-350
bacterial biomass, 81
bacterial fossils, 437
bacterial mat, 222
bacterial paleontology, 448
bacteria-like, 432-434
bacteriophage, xvii
Bacteroides, 243
baffling, 271
Bahamas, 70
Baja California, 65f
banded iron formation (BIF), 10, 17
Barberton greenstone belt, 308, 449, 452, 456
Bathybius haeckelii, (urschleim), 3
bedding planes, 273
Beggiatoa, 12, 242, 283
benthic fauna, 79f, 180, 198f
Big Soda Lake, Mongolia, 146
bioaggregates, 318
bioastronomy, 406
biocorrosion, 254
biodictyon, 1, 8-10

bioerosion, 254
biofilm catena, 270
biofilm, benthic, 47f, 104f
biofilm, definition, 7, 48, 335
biofilm, evaporitic, 13
biofilm, evolution, 317f
biofilm, fungal, 363f
biofilm, genetic structure, 317f
biofilm, v, xiv-xx, 1, 3, 7-10, 29-41, 47- 61, 249, 281-286, 320, 343, 353, 357-359, 371f, 394, 433-440, 450-461,467
biofilm, marine, 65f
biofilm, structure, 2-20, 59, 65f, 172
biofilm, sub-aerial, 7-8
biofilm, sub-aquatic, 7, 229
biofilm, transient, 47-61
biogenic, biogenicity 5-20, 273, 434, 448, 456
biogeochemical cycles, 10-20, 141-149
biogeomorphogenesis, 20
bioherm, 17
biolaminated deposit, 1, 212
biological reactors, 167
biomarker, 436-439, 448
biomass, 53-61, 81, 142
biomat, 1f
biomineralization, 281f
biopatina, 8, 241
biosedimentary structures, 273f
biosignature, 272, 429-432
biosphere, 8, 21, 103, 141-149, 416, 422-427
biostabilisation, 270, 273
bioturbation, 198-209
bits, 321, 324
black crusts, 381-385
Black forest, 9
black fungi, 382
Black Sea, 358
Bledius angustus, 205
Bledius capra, 204
Bledius spectabilis, 204
Bledius spp., 79, 203f
Bledius subniger, 204
Bodo designis, 164
Bodo saltans, 164, 165
Bodo sp., 164
boreholes, 373
boring, 250f
boron isotopes, 312, 456
Brevibacteria, 349
bryostromatolite, 190f
bryozoa, 181f
butyroyl- homoserinlacton, 155
byte, 324
Caecitellus parvulus, 169
Cafeteria sp., 165
calcification, 77f, 227-232, 448
calcite, 88, 92, 365, 367, 459
calcium carbonate, 84-89, 90-95, 109f, 146-147, 228f,
calcium oxalate, 298, 365
calcium phosphate, 286
calcrete, 291, 293
Calothrix thermalis, 79
Calothrix, 79, 82, 88, 145
CAMP, 155
Canada, 10
capture, 282
carbohydrates, 3
carbon biomarker, 448
carbon cycle, 87f
carbonaceous matter, 454
carbonate buildups, 108
carbonate globules, 432
carbonate rocks, 364
carbonate spheres, 5, 156
Carboniferous, 10, 293, 300
Catena, 267
CCD-camera, 439
Cerithia californica, 79
Cerithium, 198f

Challenger expedition, 3
Chaperiopsis cervicornis, 188
chelation, 229
Chelicerata, 199
chemoautotroph, xvii
chemolithoautotroph, 371
chemotaxis, 151, 156
chemotroph, 12, 21, 435
chert, 307, 459
Chlorella pyrenoidosa, 407
Chlorella saccharophyla, 407
Chlorella vulgaris, 407
Chloroflexus sp., 145
chlorophyll fluorescence, 49, 51, 86
chlorophytes, 258
choanoflagellates, 164-171
Chondrites, 205
chondrites, 3, 431
Chromatium sp., 145
Chroococcidiopsis, 251, 402, 410
chrysomonads, 171
ciliates, 165-166
Cladophora sp., 143
Cladosporium, 382
Cletocampus dietersi, 79
cloning, 243
CLSM, 33
Cohn, Ferdinand, 11
coleopterans, 79
collagen, 438
Collwellia, 243
Cologne cathedral biofilms, 164
community structure, 327
confocal laser scanning microscopy, 30f
conidia, 302
conidiophore, 302
Coniosporium sp., 396
consortia, 48
contamination, of meteorites, 434
continous culture, 320, 328
copepods, 79
Corophium arenarium, 199, 205-206
Corophium volutator, 199, 209
Corynebacterium minutissimum, 435
Corynebacterium spp. 435
crenophilic, 145
Cretaceous, 285, 294, 295
Crimea, 364
crustaceans, 79, 199f
cryophilic, 411
Cyanidium caldarium, 405
cyanobacteria, 12, 17, 55-58, 70-73, 77f, 82f, 129f, 335, 403, 410, 431, 468
cyanobacterial bloom, 210
cyanobacterial mat, 13,77-95, 141-148, 198, 209, 425
cyanobacterial microbial mat (CBM), 425
Cystoseira, 358
Cytophaga, 243
D'Arcy Thompson, 20, 391
Daisy world, 317
Darwin, 11, 21, 142, 317, 323, 325, 403, 409
Darwinian evolution 21, 142f
death mask, 287
decay inhibition, 282
definitions, descriptive, 337
definitions, prescriptive, 337
Delos, 364
denitrification, 132f
desert crust, 252
desert, cold and hot, 424
dessication, 268
Desulfovibrio sp., xvi, 145
Devonian fungi, 293, 298
DGGE, 243
diatoms, 17, 50-58, 69, 78, 130
Dickinsonia, 287, 288
digital data, 388f
digital imaging, 388f
dipterans, 79

diversity, 141f
DNA repair, 402
DNA, 326, 344, 346, 349, 375, 402, 435, 438
DNA/RNA hybridisation 438
DOM (dissolved organic matter), 35
Dormancy, 404
dumbbell calcite, 90
Dunaliella acidophila, 405
dyssymmetry, 19
early atmosphere, 449
early Earth, 429, 449
echinoderms, 198, 217, 241, 254
echinoid hindgut, 217f
ecosystem function, 47f
Ediacara, 287
Efremova meteorite, 431
Ehrenberg, 11
ELISA, 217f
endolithophilic, 145
endoliths, 249f, 292, 456
endosymbiosis, 409
Eohyella, 258
epibenthic biofilms, 266
epilithophilic, 145
epiliths, 249f
EPS (extracellular polymeric substances), 7, 29-31, 35-41 48-50, 144, 197, 227f, 281, 448, 456
Erastophrya sp., 165
Ernietta, 287
ESA (European Space Agency), 439
Escherichia coli, 172, 346
Euglena mutabilis, 407
Euglena proxima, 60
euglenids, 56
eukaryogenesis, 409
Euplotes, 169, 171
Europa, 406
evaporitic environment, 456
evolution, 318, 329
exobiology, 1, 440
exopolymer, 105
extracellular material, 2
extracellular polymeric substances, see EPS
extraterrestrial bacteria, 410
extraterrestrial life, 410-412, 431
extraterrestrial materials, 434
extraterrestrial studies, 6, 448
extremophile, 8, 401-412
Farbstreifen-Sandwatt, 11-14, 269
felt mat, 143
ferric iron, 17, 242
field guide to bacteria, 343-50
Fig Tree Group, 308
filamentous bacteria, 336
Fimbriae, 30
FISH, 244
fish, 282
fish, browsing, 198,
flagellates, 162, 164-170
Flexibacter maritimus, 243
floating grains, 16-20, 271
Flora Danica, 5-6
fluorescence measurement, 51
fluorescent probes, 34
food web, 161f
foraminiferans, 162, 294
fossil bacteria, 17, 402, 437, 448, 457
fossil biofilm, 5, 9, 312, 376, 435-438, 447- 461
fossil filaments, 310
fossil microorganisms, 9-10, 430, 453
fossilisation of biofilms, 281, 313, 447-461
fractal, 356
fumaroles, 426
fungal biofilm, 363f
fungal calcification, 366
fungal communities, 17, 295
fungal metabolism, 365
fungal networks, 8, 189, 339, 363-367

fungal sediments, 17, 297
fungal stromatolite, 339
fungi, Carboniferous, 300
fungi, xvi, 5, 7, 20, 30, 71, 184, 270, 291f, 363f, 381f, 389, 392, 396, 397
Gaia theory, 1-21, 141, 317
Galileo Galilei, 330
Gallionella ferruginea, 377-378
gastropods, 79, 199, 254
GC-MS, 437
gene regulation, xvii, 346
gene transfer, xviii
genetic structure, 327
geobiology, 333
geochemical cycles, 17-21
geomicrobiology, 1-20, 103f
geophysiology, 18-21
gliding motility, 143
gliding trichomes, 147, 152
global mechanisms, 19
global tectonics, 20-21
gloeocapsin, 252
Gloeothece sp., 144
glycocalyx, 2, 6
glycogen, 130
grain separation, 271
gram negative bacteria, 70
grazing, 79, 169
green algae, 35, 71, 335
green cover (of Earth), v, 21, 419f
green house gas, 449
Greenland, 448
greigite, 432
groundwater, 371f
Gulf of Aqaba, 191, 203
Gunflint chert, 10
Gyrosigma, 55-58
Haeckel, 3
halite hoppers, 453, 456
halite, 457
halophiles, 404
harpactoids, 80
heliozoans, 162f
helminthoids, 205
hemoglobine, 438
Herpyzonema intermedium, 257
Heterocerus flexuosus, 204
heterocysts, 130
heterotrophic bacteria, 16-20, 70f., 92
honey pump, v
Hooke, 2, 6
hopane, 437
hopanoids, 437
Hormathonema spp, 257
Humboldt, 11
Hutton, 18
Huxley, 3
hybridisation, 438
Hydrobia ulvae, 199, 201, 207
hydroids, 358
hydrophilic, 2
hydrophobic, 2
hydrothemal vents, xv, 13, 449
Hyella spp, 256, 257
hypersaline environments, 66f, 79f
hypersaline ponds, 130
hyphae, 299, 302, 365, 396
ichnodiversity, 198f
ichnology of mats, 210f
ideonomy, 347
image analysis, 387f
image processing, 388f
image quantification, 387
imprinting, 271
insects, 199
internal biofilms, 249
interstellar dust, 451
iron deposition, 246, 339
iron incrustation, 5, 241f
iron oxidizing bacteria, 241f, 377
iron stromatolite, 5, 339
isotope ratios, 94, 448
isotope signature, 94, 367, 432
Isua greenstone belt, 448

Kalkowsky, 13
keratin, 438
kerogen, 141, 307, 309
killermat 209
Klebsiella sp., 145
korarchaeota, xv
Kromberg formation, 308, 453
Kyrthuthrix, 257
Laguna figueroa, 65f
Lake Khylganta, 147
Lake Magadi, 147
laminated mat 11-17, 66-67, 110, 115, 182
laminated sedimentary structures, 266
Laomedia exigna, 358
Leather mat, 143
lectin, 32
lichen, 8, 251-253, 364
life detection, 429f
life in meteorites, 431
Lineus ruber, 205
Lithonema adriaticum, 257
littoral, 357
living matter, constancy, 21, 417-422
lotic biofilms, 29-39
Lovelock, J., 3, 416
Loveniid*ae*, 222
Lumbrificoides, 205
Lunar cratering, 450
Lyngbya lagerheimeii, 145
Lyngbya sp., 145, 151f
macrofossil, 284
macroorganisms 197-212
macroscopic approach,343-348
magnetite, 432
Marinilabilia, 244
Mars, 407-412, 430-440, 451, 461
Martian meteorites, 431
Martian microorganisms, 432-437
mathematical approaches, 387-398
meiofauna, 79
melanin, 381-385
Mellum Island, 135, 199-205, 266-274
Menzenschwand uranium deposit, 9
Merismopedia punctata, 267
Merismopedia, 15, 267
mesocosm, 78
Messel shale, 283
meteor paper, 3
meteorite, 407, 410, 431-435, 448
methane, xvii, 449
methanogens, 142f
Mezozoic, 439
Mg-calcite, 90, 92
micritic crust, 109
microarray assay, 438-440
microautoradiography, 374
microbe-mineral, 87, 103-108
microbial biofilm, 334-338, 447, 452, 454, 457-458, 460-461
microbial carpet, 11-14
microbial mat 1, 2, 4, 11-18, 79f, 189, 219, 282, 334-336, 448, 450, 452, 456, 460
microbial mat farming, 217f
microbial spheres, 5, 151f
microbialite, 1, 17-19
microbialites, 21
microboring, 109, 235, 250-260
Micrococcus luteus, 435
Micrococcus spp., 435
Micrococcus, 244
Microcodium, 292
Microcoleus chthonoplastes, 66-69, 82, 143-148, 209, 267,467
microcolony, xvii, 33
microfossil, 284, 307, 454
micromycetes, 292f
microphythobenthos, 358
microprobes, 85
microreef, 186f, 252
microsensor, 48, 85f
mineralisation, 17, 284

MISS (microbially induced sedimentary structures), 266
Montacuta ferruginosa, 241
Montagne Noire, 267
Moraxella, 244
morphogenesis, 20, 391
morphometric index, 354f
morphoprocess, 190-192
mucilago, mucus, 2, 92, 425
mucopolysaccharide, 425
mullets, 201
multilayered microbial communities, 335
Murchison meteorite, 431, 435, 448
mutualism, 191-193
mycosporine, 252
mykorrhiza, 6
Myxosarcina, 144
nanobacteria, 410, 412, 434
NASA (National Aeronautic and Space Agency), 439
Navicula salinarum, 56
nematods, 79
Neoproterozoic, 141
Nereis diversicolor, 205-209
nested proliferation, 326
networks, 8, 20
Neurospora spp., 435
Nitella sp., 143
Nitocra sp., 80
nitrate, 134
nitrification, 142f
nitrogen cycling 129-146
nitrogen fixation, 143f
nitrogenase, 133-136
Noachian, 451
Oceanospirillum, 243
onkolites, 1
Onvervacht Group, 308, 452, 456
Oogonium, 301, 302
ooid, 4, 5, 6
oolite, 1, 4, 5
oolite, definition, 2
Origin of life, 408-412
Oscillatoria limosa, 267
Oscillatoria, 12, 70, 72, 79, 134-136, 145, 267
osmotic pressure, 268
ostracods, 79, 199
Ostreobium quekettii, 258
oxalates, 299, 365f
oxohexanoyl-homoserinlacton, 155
oxygen penetration, 16-20, 85f
oxygenic photosynthesis, 16-21, 85f
ozone, 408
Palaeomicrocodium, 292
paleosols, 291f
Paracelsus, 2, 13, 18
Paracoenia turbida, 79
parahistology, 1, 18-19, 143
Paraphysomonas sp., 165
Parasmittina tropica, 191
patina, 8, 21, 381
PCR, 438
Pelecanimimus, 286
Penicillium corylophilum, 382
Penicillium spp, 301, 302
Petalomonas minuta, 164
Phaeococcomyces sp., 396
Phaeophila dendroides, 258
Phaeophila tenuior, 258
Phanerozoic, 258, 282, 427
phase portrait, 420
philosophy, 334f
Phoma glomerata, 382
Phormidium, 67, 69, 82, 90, 151f
phosphorus, 467
phosphorus, inorganic, 467
phosphatisation, xviii, 285f
Photobacterium, 244
photoinhibition, 58-60
photon laser scanning microscopy (2-PLSM), 33
photophysiology, 49

photosynthesis, 49
phototroph, 12
Phragmites, 358
phycobiliprotein, 86
phycobiont, 252
Phycochromaceae, 12
Phyllozoon, 287
phytoplankton, 358-360
pigments, 382-384, 396
Pilbara greenstone belt, 448
pili, 30
Pirenella conica, 202
pixel, 390, 395, 396
planetary biology, 6, 20-22, 410-412
planetary climate, 415-427
planetary cover (skin), v, 21, 418, 421
Planktomycetales, 132
plankton, 4, 165-172, 358,
Plectonema, 151f
Pleurocapsa minor, 79
Pleurocapsa, 79
Pleurosigma, 55-58
Plinius, 147
poikilophile, 8, 364
poikilotroph, 8, 364
polychaetes, 199, 205-210
polyglucose, 15-16
polymer film, 456
POM (particulate organic matter), 35
Potamocypris sp., 79
prebiotic molecules, 461
Precambrian, 10, 292
predation, 161
primordial soup, 3
proliferation theory, 326
proteinaceous materials, 438
proteins, 230
proteobacteria, 243
proteomic techniques, xvi
Proterozoic Belt Supergroup, 284
Proterozoic, 141f, 198, 212, 288f
Proteus, xviii
protists, 71, 164
protoplasm, 3
protozoa, 30, 161f, 218f
Pseudanabaena galatea, 156
Pseudanabaena, 81-83, 88-90
Pseudoalteromonas, 243
Pseudomonas aeruginosa, xv, 394
pseudomorphing, 283
pseudoscorpions, 199
psychrophiles, 404
Puellina spp, 188
purple bacteria, 16, 77, 90
Pygospio, 211
pyrhotite, 432
pyrite, 432
Pyrolobus fumarii, 403
quantification of images, 396
quantification of spatial relationships, 387-398
rhodophytes, 258
ripple marks, 271
RNA, 71-72, 243, 438
Rocky Mountains, 292
rotating annular reactors (RAR), 37-39
Rynchomonas nasuta, 169
salt lakes, 406
Sarcynomyces sp., 396
Saturn, 407
Scenedesmus quadricauda, 407
Schizasteridae, 222
Schizomavella auriculata, 186, 192
Schizothrix gebeleinii, 70f
Schizothrix lacustris, 253
Schizothrix, 70f, 105, 228f, 235, 253
Schnoz (slime), 3
Scytonema endolithicum, 257
scytonemin, 252
Severn Estuary, 53
siliciclastic biofilms, 265
silicification, 307, 448, 456-458

silicified bacteria, 454
silicified biofilms, 448-453
similarity, 20, 387, 391
SIMS, 436-437
slime, 2-4, 144, 350
soda continent, 146f
soda lake, 147
soda ocean, 146
soft tissue, 284
solar energy, 21, 417
solar pump, v, 21, 417
Solar system, 430, 435
Solentia paulocellulare, 257
Solentia, 70f, 106f, 234, 257
Solnhofen (Jura), 210, 285
South Africa, 148, 307, 452
spatangoids, 222
spatial relations, 20, 347, 391
species, 344, 345
spherical biofilms, 6, 105, 151
spherulites, 6, 105, 151f
Spirulina platensis, 405
Spirulina, 12, 142, 143, 147
Spongomonas, 166
sporangium, 302
Spumella, 165
Staphylococcus aureus, xvi
Staphylococcus capitis, 435
Staphylococcus spp., 435
Staurophora, 55, 57
Streptococcus mutans, xv
Streptococcus spp., 435
stromatolites 1, 13, 70-75, 80-87, 103-115, 232f, 266f, 308, 312, 448
struvite, xviii
subaerial exposure, 7, 143, 291, 396, 456
subaerial mats, 7, 143, 291
subaquatic, 7
Suess, E., 416
Sulfolobus, 405
Sulphate reduction, 109, 142f, 219f
sulphur cycle, 144, 221f
sulphur oxidizing bacteria, 377
supralittoral, 357
Swaziland Soupergroup, 308
symbiotic community, 6, 217-224, 347
symbiotic relation, 157, 222
Symploca sp., 144
Synechococcus, 79, 144
Synechocystis, 12
TEM (transmission electron microscopy of microbial mats), 66f
template, 448
temporal relation, 347, 391
terminology, 333-340
terraforming, 410
Tertiary, 10
theory, 337
thermodynamics, v
thermophiles, 403
thermophilic mats, 145
Thermoplasma, 405
Thermus aquaticus, 403
Thiocapsa, 13, 81, 144
Thiopedia, 12, 16
Thiothrix, 217-220, 242
Thiovolum, 12
thrombolites, 199
tidal flats, 266
Tintenstrich, 252
tissue, 17
Titan, 407
ToF-SIMS, 436-437
tourmaline, 312, 459
toxic cyanobacterial blooms, 210
Trichocorix sp., 79
Trichodina sp., 166
trichomes, 147,467
trophic levels, 170
tryciclazole, 384
unicellular algae, 355
uranium deposit, 9

Ur-schleim, 3
UV-radiation, xvii, 130, 408, 411, 425, 450, 460-461
Uzon caldera, 145
Vanella septemtrionalis, 169
Vaucheria sp., 143
vegetation, 423
Vendian, 197
Ventrogyrus, 287
Venus, 410
Vernadsky, 416-417
vertebrate traces, 198f
viscoelasticity, xvi
volcanism, 145
Vostok station, Antarctica, 408, 411
voxels, 390
Wadden Sea, 13, 144-146, 156
weathering, 20, 146, 419, 422
weddellite, 365
whewellite, 365
wrinkle structures, 266
Xenococcus, 144
X-ray diffraction, 245
Yellowstone hot springs, 10, 145
ZoBell, 2
Zoogloea, 324-326
zooid, 184f
Zostera, 358

MIX
Papier aus verantwortungsvollen Quellen
Paper from responsible sources
FSC® C105338

If you have any concerns about our products,
you can contact us on
ProductSafety@springernature.com

In case Publisher is established outside the EU,
the EU authorized representative is:
Springer Nature Customer Service Center GmbH
Europaplatz 3, 69115 Heidelberg, Germany

Printed by Libri Plureos GmbH
in Hamburg, Germany